An Introduction to Mathematical Logic and Type Theory:
To Truth Through Proof

APPLIED LOGIC SERIES

VOLUME 27

Managing Editor

Dov M. Gabbay, *Department of Computer Science, King's College, London, U.K.*

Co-Editor

Jon Barwise†

Editorial Assistant

Jane Spurr, *Department of Computer Science, King's College, London, U.K.*

SCOPE OF THE SERIES

Logic is applied in an increasingly wide variety of disciplines, from the traditional subjects of philosophy and mathematics to the more recent disciplines of cognitive science, computer science, artificial intelligence, and linguistics, leading to new vigor in this ancient subject. Kluwer, through its Applied Logic Series, seeks to provide a home for outstanding books and research monographs in applied logic, and in doing so demonstrates the underlying unity and applicability of logic.

An Introduction to Mathematical Logic and Type Theory: To Truth Through Proof

Second Edition

by

PETER B. ANDREWS

Department of Mathematical Sciences,
Carnegie Mellon University,
Pittsburgh, Pennsylvania, U.S.A.

KLUWER ACADEMIC PUBLISHERS
DORDRECHT / BOSTON / LONDON

A C.I.P. Catalogue record for this book is available from the Library of Congress.

ISBN 978-90-481-6079-2

Published by Kluwer Academic Publishers,
P.O. Box 17, 3300 AA Dordrecht, The Netherlands.

Sold and distributed in North, Central and South America
by Kluwer Academic Publishers,
101 Philip Drive, Norwell, MA 02061, U.S.A.

In all other countries, sold and distributed
by Kluwer Academic Publishers,
P.O. Box 322, 3300 AH Dordrecht, The Netherlands.

Printed on acid-free paper

Printed in the Netherlands.

To Cate, Lyle, and Bruce

Contents

Preface to the Second Edition

For the benefit of readers who are already familiar with the first (1986) edition of this book, we briefly explain here how this edition differs from that one.

The basic structure of the book is unchanged, but improvements and enhancements may be found throughout the text. There is more introductory material and motivation. Historical notes and bibliographic references have been added. The discussion of vertical paths and normal forms in §14 is now much more comprehensive. A number of consequences of Gödel's Second Incompleteness Theorem are now discussed in §71.

Many exercises have been added, and at the end of the book there is a new collection of Supplementary Exercises which are not explicitly tied to any particular section. Information about obtaining the ETPS program, which facilitates work on exercises which consist of proving theorems in the object language, will be found near the end of the Preface.

Preface to the Second Edition

Preface

This book is an introductory text on mathematical logic and type theory. It is aimed primarily at providing an introduction to logic for students of mathematics, computer science, or philosophy who are at the college junior, senior, or introductory graduate level. It can also be used as an introduction to type theory by researchers at more advanced levels.

The first part of the book (Chapters 1 and 2, supplemented by parts of Chapters 3 and 4) is suitable for use as a text in a one-semester introduction to mathematical logic, and the second part (Chapters 5 – 7) for a second semester course which introduces type theory (more specifically, typed λ-calculus) as a language for formalizing mathematics and proves the classical incompleteness and undecidability theorems in this context. Persons who wish to learn about type theory and have had a prior introduction to logic will have no difficulty starting with Chapter 5.

The book is oriented toward persons who wish to study logic as a vehicle for formal reasoning. It may be particularly useful to those who are interested in issues related to the problem of constructing formal proofs as models of informal reasoning or for use in computerized systems which involve automated deduction. Proofs, which are often the chief end products and principal manifestations of mathematical reasoning, constitute highly significant pathways to truth and are a central concern of this book. Our choice of the title *To Truth Through Proof* is motivated by the consideration that while in most realms one needs more than logic to achieve an understanding of what is true, in mathematics the primary and ultimate tool for establishing truth is logic.

Of course, the study of logic involves reasoning about reasoning, and it is not surprising that complex questions arise. To achieve deep understanding and proper perspective, one must study a variety of logical systems as mathematical objects and look at them from a variety of points of view. We are thus led to study the interplay between syntax and semantics, questions of consistency and independence, formal rules of reasoning, various formats for proofs and refutations, ways of representing basic mathematical concepts in a formal system, the notion of computability, and the completeness, incompleteness, and undecidability theorems which illuminate both the power and the limitations of logic.

One of the basic tasks of mathematical logic is the formalization of mathematical reasoning. Both type theory (otherwise known as higher-order logic) and axiomatic set theory can be used as formal languages for this purpose, and it is really an accident of intellectual history that at present most logicians and mathematicians are more familiar with axiomatic set theory than with type theory. It is hoped that this book will help to remedy this situation.

In logic as in other realms, there is a natural tendency for people to prefer that with which they are most familiar. However, those familiar with both type theory and axiomatic set theory recognize that in some ways the former provides a more natural vehicle than the latter for formalizing what mathematicians actually do. Both logical systems are necessarily more restrictive than naive axiomatic set theory with the unrestricted Comprehension Axiom, since the latter theory is inconsistent. Axiomatic set theory achieves consistency by restricting the Comprehension Axiom and introducing the distinction between sets and classes. Mathematicians often find this distinction unnatural, ignore the technicalities about existence axioms, and leave it to the specialists to show that their reasoning can be justified. Type theory achieves consistency by distinguishing between different types of objects (such as numbers, sets of numbers, collections of sets of numbers, functions from numbers to numbers, sets of such functions, etc.). Mathematicians make such distinctions too, and even use different letters or alphabets to help them distinguish between different types of objects, so the restrictions which type theory introduces are already implicit in mathematical practice. While some formulations of type theory may seem cumbersome, the formulation $\mathcal{Q}_0$ introduced in Chapter 5 is a very rich and expressive language, with functions (which need not be regarded as sets of ordered pairs) of all types as primitive objects, so most of what mathematicians write can be translated into $\mathcal{Q}_0$ very directly. $\mathcal{Q}_0$ is a version of typed λ-calculus, and the availability of λ-notation in $\mathcal{Q}_0$ enables definitions to be handled very conveniently and eliminates the need for axioms asserting the existence of sets and functions.

One's choice of a formal language will generally depend on what one wishes to do with it. If one is choosing a language for expressing mathematics in computerized systems which deal with mathematics on a theoretical level (such as mathematically sophisticated information retrieval systems, proof checkers, or deductive aids), there are several reasons why type theory may be preferable to set theory. First, the primitive notation of set theory is so economical that this language is only practical to actually use if one expands it by introducing various abbreviations. This is usually done in an informal way (i.e., in the meta-language), but in an automated system it is important that the formal language be the one actually used to express mathematical statements, since this is the language which must be studied rigorously and manipulated systematically. Thus, the formal language should not only be adequate in principle for expressing mathematical ideas, but it should also be convenient in practice. Indeed, the translation from familiar mathematical notations to the formal language should be as simple and direct as possible. Second, in set theory one can write all sorts of expressions which are not legal in type theory, and which a mathematician would reject as nonsense almost instinctively. A computer program for manipulating mathematical expressions would certainly be very awkward and inefficient if it did not have some way of distinguishing between different types of mathematical objects and rejecting nonsensical expressions. Third, one of the basic operations in automated theorem proving is that of finding appropriate substitutions for variables, and an understanding of the types of various entities allows one to avoid much useless search. Finally, it has been found (see [Andrews *et al.*, 1984] [Andrews *et al.*, 1996]) that unification algorithms for type theory (such as [Huet, 1975] and [Jensen and Pietrzykowski, 1976]) are powerful tools in automating the proofs of simple mathematical theorems.

As noted in [Hanna and Daeche, 1985], [Hanna and Daeche, 1986], and [Gordon, 1986], typed λ-calculus is particularly well suited for specifying and verifying hardware and software. Familiarity with typed λ-calculus also provides fundamental background for the study of denotational semantics for functional programming languages.

While Gödel's Completeness Theorem guarantees that all valid wffs of first-order logic have proofs in first-order logic, extraordinary reductions in the lengths of such proofs can sometimes be achieved by using higher-order logic. (See §54.)

It is easy to discuss the semantics of type theory, and to explain the crucial distinction between standard and nonstandard models which sheds so much light on the mysteries associated with the incompleteness theorems, Skolem's paradox, etc. As will be seen in Chapter 7, in the context of the language $\mathcal{Q}_0$ it is easy to present the incompleteness theorems very elegantly and to show that their significance extends far beyond the realm of formalizing arithmetic.

Obviously, serious students of mathematical logic ought to know about both type theory and axiomatic set theory. Pedagogically, it makes good sense to introduce them to elementary logic (propositional calculus and first-order logic), then type theory, then set theory. In spite of the superficial appearance of simplicity possessed by set theory, its models are more complicated, and from a logical point of view it is a more powerful and complex language. In [Marshall and Chuaqui, 1991] it is argued that the set-theoretical sentences which are preserved under isomorphisms are of fundamental importance, and it is shown that these are the sentences which are equivalent to sentences of type theory.

A brief explanation of the labeling system used in this book may be helpful. The sections in Chapter 1 (for example) are labeled §10, §11, ..., §16. Of course, an alternative labeling would be §1.0, §1.1, ..., §1.6, but the periods are quite redundant once one understands the labeling system. These tags should be regarded as labels, not numbers, although they do have a natural ordering. Similarly, the theorems in §11 (for example) are labeled 1100, 1101, 1102, ..., rather than 1.1.00, 1.1.01, 1.1.02, ... Exercises are labeled in a similar way, but their labels start with an X. Thus, the exercises for section 11 are labeled X1100, X1101, X1102, ... Labeling starts with 0 rather than 1 because 0 is the initial ordinal. (Chapter 0 is the Introduction.)

The reader should examine the table of contents in parallel with the discussion of the individual chapters below. Much that is in this book is simply what one would expect in an introductory text on mathematical logic, so the discussion will concentrate on features one might not expect.

Chapter 1 introduces the student to an axiomatic system $\mathcal{P}$ of propositional calculus and later to more general considerations about propositional calculus. In §14 a convenient two-dimensional notation is introduced to represent wffs in negation normal form. This enables one to readily comprehend the structure of such wffs, find disjunctive and conjunctive normal forms by inspection, and check whether the wffs are contradictory by examining the "vertical paths" through them.

Chapter 2 contains what everyone ought to know about first-order logic. Students can learn a great deal about how to construct proofs by doing the exercises at the end of §21. The discussion of prenex normal form in §22 shows students how to pull out the quantifiers of a wff all at once as well as how to bring them out past each connective one step at a time. This discussion is facilitated by the use of the notation $\exists\!\!\!/\,$ for the ambiguous quantifier.

In §25 Gödel's Completeness Theorem is proved by an elegant generalization of Henkin's method due to Smullyan. The notion of consistency in Henkin's proof is replaced by abstract consistency, and with very little extra effort a proof is obtained for Smullyan's Unifying Principle, from which readily follow Gödel's Completeness Theorem in §25, Gentzen's Cut-Elimination Theorem in §31, the completeness of the method of semantic tableaux in §32, the completeness of a method for refuting universal sentences in §34, and the Craig–Lyndon Interpolation Theorem in §41. In their first encounter with the proof of Gödel's Completeness Theorem, students may need a less abstract approach than that of Smullyan's Unifying Principle, so as an alternative in §25A an elegant and direct simplified proof of the Completeness Theorem is given. It actually parallels the abstract approach very closely and provides a good introduction to the abstract approach. The Löwenheim–Skolem Theorem and the Compactness Theorem are derived in the usual way from the Completeness Theorem. It is shown how the Compactness Theorem can be applied to extend the Four Color Theorem of graph theory from finite to infinite graphs and to prove the existence of nonstandard models for number theory.

Chapter 3 is concerned with the logical principles underlying various techniques for proving or refuting sentences of first-order logic, and provides an introduction to the fundamental ideas used in various approaches to automated theorem proving.

The chapter starts in §30 with a system of natural deduction which is essentially a summary of the most useful derived rules of inference that were obtained in §21. Methods of proving existential formulas are discussed. A cut-free system of logic and semantic tableaux are then introduced in §31 and §32 as systems in which one can search more systematically for a proof (or refutation). When one constructs a semantic tableau, one often notices that it would be nice to postpone deciding how to instantiate universal quantifiers by just instantiating them with variables for which substitutions could be made later. This is facilitated when one eliminates existential quantifiers to obtain a universal sentence by Skolemization, which is the topic of §33. It is noted that several methods of Skolemization can be used.

§34 introduces a method for refuting universal sentences which avoids the branching on disjunctions of semantic tableaux and uses the cut rule to reduce formulas to the contradictory empty disjunction. This method has many elements in common with the resolution method [Robinson, 1965] of automated theorem-proving. Students are often successful at establishing theorems by this method even though they were unable to find proofs for them in natural deduction format.

The proof methods just discussed all yield highly redundant proof structures in which various subformulas may occur many times. §35 introduces a version of Herbrand's Theorem which enables one to establish that a wff is contradictory simply by displaying a suitable compound instance of it.

One of the basic processes of theorem-proving in first- or higher-order logic is that of substituting terms for variables (or instantiating quantified variables) in such a way that certain expressions become the same. §36 provides a description of Robinson's Unification Algorithm for first-order logic and a proof of its correctness. Since substitutions have been defined §10) to be functions of a certain sort which map formulas to formulas, certain facts about substitutions (such as the associativity of composition) can be used without special justification.

Chapter 4 discusses a few additional topics in first-order logic. In §40 the topic of Duality is introduced with a parable about two scholars who argue about an ancient text on logic because of a fundamental ambiguity in the notations for truth and falsehood. Of course, much of this material could appropriately be discussed immediately after §14.

Chapter 5 introduces a system $\mathcal{Q}_0$ of typed λ-calculus which is essentially the system introduced by Alonzo Church in [Church, 1940], except that (following [Henkin, 1963b]) equality relations rather than quantifiers and propositional connectives are primitive. Type theory is first introduced in §50 with a discussion of a rather traditional formulation $\mathcal{F}^{\omega}$ of type theory in which propositional connectives and quantifiers are primitive. It is shown that equality can be defined in such a system in two quite natural ways. It is then shown that this system can be simplified to obtain naive axiomatic set theory, which is inconsistent because Russell's paradox can be derived in it. Thus, one needs some device such as type distinctions or restrictions on the Comprehension Axiom to obtain a consistent formalization of naive set theory. The discussion turns to finding as simple, natural, and expressive a formulation of type theory as possible. (The choice of equality as the primitive logical notion is influenced not only by the desire for simplicity but also by the desire for a natural semantics, as discussed at the end of [Andrews, 1972a].) This leads to an exposition of the basic ideas underlying $\mathcal{Q}_0$. While these ideas usually come to be regarded as very natural, they often seem novel at this stage, and need to be absorbed thoroughly before one proceeds to the next section.

By the end of §52 the student should be ready to prove theorems in $\mathcal{Q}_0$, so notations for some simple but basic mathematical ideas are introduced, and in the exercises the student is asked to give formal proofs of some simple mathematical theorems about sets and functions. An exercise in §53 asks for a proof of a concise type-theoretic formulation of Cantor's Theorem.

Henkin's Completeness Theorem is proved for $\mathcal{Q}_0$ in §55. Skolem's paradox (which was first discussed in §25) is discussed again in the context of type theory and resolved with the aid of the important distinction between standard and nonstandard models. It is shown that theories which have infinite models must have nonstandard models.

Chapter 6 is intended to make it clear that mathematics really can be formalized within the system $\mathcal{Q}_0^{\infty}$ which is the result of adding an axiom of infinity to $\mathcal{Q}_0$. It is shown how cardinal numbers can be defined, Peano's Postulates can be derived, and recursive functions can be represented very elegantly in $\mathcal{Q}_0^\infty$. Proofs of theorems of $\mathcal{Q}_0^\infty$ are presented in sufficient detail so that students should have no difficulty providing formal justifications for each step. Naturally, each teacher will decide how (or whether) to treat these proofs in class. It is good experience for students to present some proofs, or parts of proofs, in class. By the end of this chapter students should have a firm grasp of the connection between abbreviated formal proofs and the informal proofs seen in other mathematics courses, and should therefore be much more confident in dealing with such proofs.

Chapter 7 presents the classical incompleteness, undecidability and undefinability results for $\mathcal{Q}_0^\infty$. In §70 it is shown that the numerical functions representable in any consistent recursively axiomatized extension of $\mathcal{Q}_0^\infty$ are precisely the recursive functions, and this leads to an argument for Church's Thesis. In §71 Gödel's First and Second Incompleteness Theorems are presented for $\mathcal{Q}_0^\infty$, along with Löb's Theorem about the sentence which says "I am provable". A number of consequences of Gödel's Second Incompleteness Theorem are discussed. In §72 the Gödel–Rosser Incompleteness Theorem is presented for $\mathcal{Q}_0^\infty$, and it is shown how this implies that there is no recursively axiomatized extension of $\mathcal{Q}_0$ whose theorems are precisely the wffs valid in all standard models. §73 is concerned with the unsolvability of the decision problems for $\mathcal{Q}_0$ and $\mathcal{Q}_0^\infty$, and the undefinability of truth. §74 is a brief epilogue reflecting on the elusiveness of truth.

The exercises at the end of each section generally provide opportunities for using material from that section. However, students need to learn how to decide for themselves what techniques to use to solve problems. Therefore, at the end of the book there is a collection of Supplementary Exercises which are not explicitly tied to any particular section.

Some exercises (see §21 and §52, for example) involve applying rules of inference to prove theorems of the formal system being discussed. A computer program called ETPS (which is reviewed in [Goldson and Reeves, 1993] and [Goldson *et al.*, 1993]) has been developed to facilitate work on such exercises. The student using ETPS issues commands to apply rules of inference in specified ways, and the computer handles the details of writing the appropriate lines of the proof and checking that the rules can be used in this way. Proofs, and the active lines of the proof, are displayed using the special symbols of logic in proof windows which are automatically updated as the proof is constructed. The program thus allows students to concentrate on the essential logical problems underlying the proofs, and it gives them immediate feedback for both correct and incorrect actions. Experience shows that ETPS enables students to construct rigorous proofs of more difficult theorems than they would otherwise find tractable.

ETPS permits students to work forwards, backwards, or in a combination of these modes, and provides facilities for rearranging proofs, deleting parts of proofs, displaying only those parts of proofs under active consideration, saving incomplete proofs, and printing proofs on paper. A convenient formula editor permits the student to extract needed formulas which occur anywhere in the proof, and build new formulas from them. Teachers who set up ETPS for use by a class can take advantage of its facilities for keeping records of completed exercises and processing information for grades.

Information about obtaining ETPS without cost can be obtained from http://gtps.math.cmu.edu/tps.html; alternatively, connect to the web page http://www.cmu.edu/ for Carnegie Mellon University or http://www.cs.cmu.edu/afs/cs/project/pal/www/pal.html for CMU's Pure and Applied Logic Program and from there find the author's web page.

I have used versions of this book for a course in mathematical logic at Carnegie Mellon University for about thirty years. Experience has shown that students are best prepared for the course if they have had at least one rigorous mathematics course or a course in philosophy or computer science that has prepared them to appreciate the importance of understanding the nature of deductive reasoning and the art of proving theorems. I have found that if one covers the material rather thoroughly in class, it is difficult to cover all of Chapters 1 – 4 in the first semester, so I normally cover Chapter 1, most of Chapter 2 (using §25A and omitting §24 and most of §26), and also §30, §33, and §34. Naturally, in a course primarily composed of graduate students, or students with more previous exposure to logic, one could move faster, ask students to read some material outside class, and cover all of Chapters 1 – 4 in one semester. In the second semester I normally cover Chapters 5 – 7 fairly completely.

A few comments about notation may be helpful. "U is a subset of V" is written as "$U \subseteq V$", and "U is a proper subset of V" as "$U \subset V$". The composition of functions f and g is written "$f \circ g$". "iff" is an abbreviation for "if and only if". Ends of proofs are marked with the sign ■.

I wish to thank the many people who contributed directly or indirectly to the development of this book. Alonzo Church, John Kemeny, Raymond Smullyan, and Abraham Robinson taught me logic. It will be obvious to all who have read them that this book has been deeply influenced by [Church, 1940] and [Church, 1956]. Numerous students at Carnegie Mellon helped shape the book with their questions, comments, and interests. Special thanks go to Ferna Hartman for her heroic work typing the difficult manuscript.

Chapter 0

Introduction

The ability to reason is one of the marvels of human nature. In many situations it is not easy to think logically, but the benefits of doing so are a familiar fact of human experience. Since at least the time of Aristotle (384-322 B.C.), many questions have been raised by inquisitive and reflective minds about the nature of reasoning, and about ways of improving our use of reason. From such questions and attempts to answer them has arisen the intellectual discipline known as logic.

Let us put the study of logic into perspective. Humans are born with hands, but the development of tools has vastly expanded what humans can do. Humans are born with eyes, but the study of optics and other relevant disciplines has made it possible to develop sophisticated telescopes and microscopes which vastly expand what we can see, leading to amazing advances in fields such as astronomy and biology. We are endowed with the ability to reason, and the study of logic can serve the following purposes:

(1) enhance our ability to reason correctly, deeply, and with well founded confidence;
(2) prepare us to use, and perhaps help develop, the automated reasoning tools which are being developed and which will have revolutionary impact on many intellectual activities;
(3) prepare us to study the many deep and fascinating questions in mathematics, philosophy, and computer science which have their roots in logic.

Note that as the development of automated reasoning tools makes it increasingly practical to use very rigorous reasoning in various disciplines, there will be increasing demand that such standards of rigor be adhered to. For example, the day will come when computer hardware and software which has simply been tested on many cases will be regarded as unacceptably unreliable. Only hardware and software which has been proved to meet its specifications will be regarded as acceptable.

Intellectual disciplines, such as scientific or philosophical theories, usually evolve over many years, often many centuries. Starting with isolated statements expressing various facts, they gradually become more organized and systematic as people struggle to find order in the phenomena under consideration. Gradually, general principles emerge which enable one to understand many particular facts as manifestations of these general principles. If the process progresses sufficiently far, one may obtain a systematic presentation of these general principles in which all of them can be derived from a few very basic and well specified principles called the *postulates* of the theory that has developed. If one goes a little farther, one specifies not only the postulates, but also the exact language in which the theory is to be expressed and the rules of logical reasoning which are to be used. One thus obtains what logicians call a *formal theory*, or *logistic system*. It can be seen that formalization of a theory is the ultimate act of systematization in a long chain of developments. It removes some of the last potential sources of confusion and vagueness. When one formalizes a theory, one may clarify what assumptions are tacitly made by those who habitually use the theory. (The Axiom of Choice is a striking example of an assumption which was used tacitly in mathematics for a number of years before it was explicitly recognized.[1])

The development of automated reasoning tools makes it likely that formalization of theories will eventually be regarded as a practical necessity, since such formalization will facilitate the development of comprehensive and sophisticated information systems which incorporate scientific and technical theories, and which can derive relevant consequences of these theories. The ability to derive logical consequences from fundamental principles provides an enormously economical way of storing information. Indeed, it permits one to obtain information which was never explicitly put into the system, and obtain it in a variety of forms.

Of course, attempts to understand the nature of reasoning and to build sophisticated information systems which can draw logical conclusions may be regarded as part of an endeavor to fashion more powerful intellectual tools for coping with the increasingly complex problems which confront mankind.

It is sometimes felt that in order to make certain disciplines (such as social sciences) more scientific, one should make them more "mathematical".

[1]See [Moore, 1982, Chapter 1].

To the unsophisticated observer of mathematics, it may appear that the use of numbers is the critical characteristic of disciplines which use mathematics, and that one can thus make a discipline more scientific by involving numbers in it more deeply. However, the really fundamental feature of mathematics is not numbers but logical rigor. While arguments involving numbers are often rigorous, it will be seen that a standard of rigor can be defined and maintained whether numbers are involved or not.

Our study of logic will be concerned with only certain aspects of the reasoning process. Our focus will not be on the psychology or physiology of human reasoning, but on reasoning as an abstract process which can be used to derive new information from known information.

A striking feature of the reasoning process is the "magic leap" which a reasoner makes from what is known to conclusions which "follow logically" from what is known. As we shall see, this magic leap can be analyzed and studied in a mathematically precise way.

How can something as abstract and apparently nebulous as thoughts be analyzed with mathematical precision? It turns out that there is a correspondence between our thoughts and the words we use to express them, and between our logical processes and manipulations of strings of words. Thus, we can analyze logical inferences in terms of operations on symbolic expressions, and these can be defined precisely and treated as objects of mathematical study.

Indeed, it has been found that valid arguments can be justified by showing that they have correct logical forms, even if one does not understand the meanings of many of the words in the arguments. It is important to learn to focus on the form rather than the content of an argument when one is analyzing its correctness, because in complicated situations one can easily be led into error by relying on one's intuition about what is true instead of using careful logical deduction.

Chapter 1

Propositional Calculus

§10. The Language of $\mathcal{P}$

As discussed in the Introduction, we can study the nature of reasoning by studying logistic systems in which reasoning is represented in a precise way. We shall commence our study with a rather simple logistic system called $\mathcal{P}$, which is one formulation of propositional calculus. Many of the definitions and concepts which we shall introduce while studying $\mathcal{P}$ will also be applicable to richer logical systems. In addition, $\mathcal{P}$ will occur as a subsystem of some of these richer systems. Once we have discussed $\mathcal{P}$, which is an example of a logistic system, we will be in a good position to give a general explanation of what a logistic system is.

In order to make clear what aspects of reasoning will be represented in $\mathcal{P}$, we start by considering three examples of logical inferences.

EXAMPLE 1: Jack is asked whether he will go to the picnic on Saturday. He doesn't like to go to picnics in the rain, and he remembers that if it does not rain, he will be playing in a tennis match which conflicts with the picnic. Hence he replies,

(A_1) "If it rains, I will not go to the picnic."

(B_1) "If it does not rain, I will not go to the picnic."

(C_1) "Therefore, I will not go to the picnic."

EXAMPLE 2: We are given that w, z, and y are sets such that $z - y = \emptyset$ and $w \cap \tilde{z} \subseteq y$ (where $\tilde{z}$ is the complement of the set z), and we wish to show that $w \subseteq y$. So we assume that $z - y = \emptyset$ and $w \cap \widetilde{z} \subseteq y$ and $x \in w$, and try to prove $x \in y$. Using the information that is given, it is not hard to show that

(A_2) If $x \in z$, then $x \in y$.

(B_2) If $x \notin z$, then $x \in y$.

(C_2) Therefore, $x \in y$.

EXAMPLE 3: Let us use ⋆⋆ as a name for the sentence "The first sentence in this chapter which contains an occurrence of the symbol @ is not true." One may argue that:

(A_3) If ⋆⋆ is true, then ⋆⋆ is not true.

(B_3) If ⋆⋆ is not true, then ⋆⋆ is not true.

(C_3) Therefore, ⋆⋆ is not true.

The reader need not be concerned with all the details of the arguments above, or even with whether the statements (A_i) and (B_i) were correctly inferred in each argument. The important point is that (C_i) is indeed a logical consequence of (A_i) and (B_i) in each argument.

Note that the main inference in each argument has the following form:

(A) If p, then q.

(B) If not p, then q.

(C) Therefore, q.

Obviously, the key words involved in this inference are "if", "then", and "not". These are connectives, words which can be used to construct more complex statements from simpler statements. Statements are sometimes called propositions (though some prefer to say that statements express propositions), and certain connectives which are important for logical purposes are represented in logistic systems as *propositional connectives.* A propositional calculus is a logistic system which formalizes the logical use of propositional connectives.

It is customary to use special symbols instead of English words for propositional connectives. This is very convenient, since the connectives are used very often and the symbols are quicker to write than the words (once one becomes accustomed to them). Also, the symbols allow one to exhibit very clearly the logical structure of a proposition which might be expressed in a variety of ways in English. Logistic systems which use symbols in this way are also referred to as systems of symbolic logic.

We shall express "If p, then q" symbolically as "$p \supset q$", and "not p" as "$\sim p$". In this notation, the inference above can be expressed as follows:

(A') $p \supset q$

(B') $\sim p \supset q$

(C') q

(The word "therefore" indicates that a conclusion is being reached, but has no logical content itself, so we have omitted it from (C').)

Before discussing the details of the logistic system $\mathcal{P}$, let us make a brief survey of the most important propositional connectives.

Negation ($\sim$)

We write the negation of the statement p as $\sim p$. In English this is generally expressed by inserting the word "not" into the sentence at an appropriate point. If w is the statement "everyone gets wet", $\sim w$ can be expressed by the statement "not everyone gets wet", but if r is the statement "it is raining", $\sim r$ is expressed by "it is not raining".

If p is true, then the statement $\sim p$ is false, while if p is a false statement, then $\sim p$ is a true statement. This is summarized in Figure 1.1. which we call the truth table for $\sim$. We write T for True, and F for False.

p	$\sim p$
T	F
F	T

Figure 1.1: Truth table for $\sim$

Conjunction ($\wedge$)

The conjunction of statements p and q is written as $[p \wedge q]$, and is generally expressed in English as "p and q" or "p, but q". Note that the statements "It is raining, and Jack is going to the picnic" and "It is raining, but Jack is going to the picnic" are *logically* equivalent, though they differ in emphasis. They both mean that the statement "It is raining" is true and

p	q	$p \wedge q$	$p \vee q$	$p \supset q$	$p \equiv q$	$p \not\equiv q$
T	T	T	T	T	T	F
T	F	F	T	F	F	T
F	T	F	T	T	F	T
F	F	F	F	T	T	F

Figure 1.2: Truth tables for $\wedge$, $\vee$, $\supset$, $\equiv$, $\not\equiv$

the statement "Jack is going to the picnic" is true. The statement $p \wedge q$ is true precisely when p and q are both true.

In Figure 1.2 we display the truth table for $\wedge$ and the other connectives which we will discuss below. Each horizontal line of the truth table corresponds to one of the logical possibilities with regard to the truth or falsity of the statements p and q. For example, in line 1 (below the bar) of the truth table, p is true and q is true and $[p \wedge q]$ is true, while in line 2, p is true and q is false and $[p \wedge q]$ is false.

Disjunction ($\vee$)

The disjunction of p and q is written as $[p \vee q]$, and generally expressed in English as "p or q". As can be seen from Figure 1.2, we regard $[p \vee q]$ as true if either or both of the statements p and q are true. Thus, we use $\vee$ for *inclusive* disjunction; we include the case where p and q are both true as one of the cases in which $[p \vee q]$ is true. For mathematical purposes, this is the generally accepted usage of the word "or". For example, 24 is regarded as a member of the set of integers which are multiples of 2 or multiples of 3. However, in ordinary usage the word "or" is sometimes used in the *exclusive sense*, in which "p or q" means p is true or q is true, but not both. For example, if Jack says, "I will go to the picnic or I will go to the tennis match", he may mean that he will go to one or the other, but not both. We will discuss exclusive disjunction below.

Implication ($\supset$)

$[p \supset q]$ means that the statement p implies the statement q, i.e., q is true whenever p is true. This can be translated into English in a variety of ways, including "if p, then q"; "q, if p"; "p only if q"; "p is sufficient for q"; "q is necessary for p" (i.e., "it is necessary for q to be true in order for p to be true").

We intend $\supset$ to represent *material implication*, so that the truth of $[p \supset q]$ depends only on the truth or falsity of the statements p and q.

Other notions of implication, in which the truth of "p implies q" may depend on such factors as the relevance of p to q, may be useful for certain purposes, but are much more complex than material implication, and seem not to be necessary for the formalization of mathematics, so we shall not consider them further here.

Note from Figure 1.2 that $[p \supset q]$ is regarded as true when p is false, without regard for q. The appropriateness of this way of defining material implication may be illuminated by supposing that p means "it rains", and q means "the street gets wet", so $[p \supset q]$ means "if it rains, then the street gets wet". If the street is protected from the rain in some way, so that on a certain day it rains, but the street stays dry, then on that day the statement "if it rains, then the street gets wet" is clearly false. However, if the street is not so protected, we would expect the statement "if it rains, then the street gets wet" to be true every day, whether or not it rains that day. In particular, if on a certain day it does not rain but the street gets wet because a pail of water is spilled on it, the statement "if it rains, then the street gets wet" is still true. Similarly, the statement is true on a day when it does not rain and the street remains dry. Thus the only case in which $[p \supset q]$ is false is that in which p is true and q is false.

Equivalence ($\equiv$)

$[p \equiv q]$ means that the statements p and q are equivalent with respect to truth; that is, they are both true or both false.

Figure 1.3 displays truth tables for $[[p \supset q] \wedge [q \supset p]]$ and $[p \equiv q]$. In each line of the truth table, the truth of each component statement is displayed under the main connective of that statement. It is easy to see that in every possible situation, $[[p \supset q] \wedge [q \supset p]]$ and $[p \equiv q]$ are both true or both false; that is, they are equivalent. Thus, p is equivalent to q means that p implies q and that q implies p. This leads to several alternative ways of expressing the assertion that p is equivalent to q which are encountered very frequently in mathematics.

p	q	$[[p \supset q]$	$\wedge$	$[q \supset p]]$	$[p \equiv q]$
T	T	T	T	T	T
T	F	F	F	T	F
F	T	T	F	F	F
F	F	T	T	T	T

Figure 1.3: Truth tables for $[[p \supset q] \wedge [q \supset p]]$ and $[p \equiv q]$

p	q	$\sim [p \equiv q]$		$p \not\equiv q$
T	T	F	T	F
T	F	T	F	T
F	T	T	F	T
F	F	F	T	F

Figure 1.4: Truth table for $\not\equiv$

If p is equivalent to q, then q implies p, so we may say "p, if q"; also p implies q, so we may say "p only if q"; combining these, we say "p, if q, and p only if q", which is shortened to "p if and only if q". (For example, in discussing Figure 1.3 we could have said that in each line of the table, $[[p \supset q] \wedge [q \supset p]]$ is true if and only if $[p \equiv q]$ is true.) In mathematical writing it is common to shorten "p if and only if q" even further to "p iff q".

Similarly, if p is equivalent to q, then q implies p, so we may say "p is necessary for q"; also p implies q, so we may say "p is sufficient for q"; combining these, we say "p is necessary and sufficient for q".

Non-equivalence, exclusive disjunction ($\not\equiv$)

As can be seen from Figures 1.4 and 1.2, $\sim [p \equiv q]$ expresses the *exclusive* disjunction of p and q; it is true precisely when one of the statements p and q is true, while the other is false, and excludes the case where they are both true. $[p \not\equiv q]$ is often expressed in English by saying "p unless q". For example, if Jack says "I will go to the picnic unless it rains", he probably means that he will go to the picnic and it will not rain, or it will rain and he will not go to the picnic.

Now that we have some familiarity with propositional connectives, we turn our attention to the details of defining the logistic system $\mathcal{P}$ in which those aspects of reasoning which involve propositional connectives can be formally represented, clarified, and studied rigorously. We shall make $\mathcal{P}$ as simple as possible while still being adequate for our purposes.

Certain propositional connectives can be defined in terms of others. Indeed, we saw in Figure 1.3 how $\equiv$ could be defined in terms of $\wedge$ and $\supset$. We shall see that all propositional connectives can be defined in terms of $\sim$ and $\vee$, and for the sake of simplicity these will be the only connectives in $\mathcal{P}$.

Since the exact nature of the statements to which we will apply the propositional connectives need not concern us for the present, we shall introduce variables, called ***propositional variables***, to play the role of statements.

This is analogous to using variables like x and y to play the role of arbitrary numbers when discussing laws of arithmetic such as $x + y = y + x$.

The first step in defining the system $\mathcal{P}$ is to define precisely what sequences of symbols may be used as the language of $\mathcal{P}$. We start by describing the symbols.

The *primitive symbols* of $\mathcal{P}$ are the following:

(a) Improper symbols: [ ] $\sim$ $\vee$
(b) Proper symbols: denumerably many propositional variables:
$p\ q\ r\ s\ p_1\ q_1\ r_1\ s_1\ p_2\ q_2 \ldots$

A *formula* is a finite sequence of primitive symbols. If $\mathbf{X}$ and $\mathbf{Y}$ are formulas, we shall denote by $\mathbf{XY}$ the formula which can be constructed by concatenating the sequence $\mathbf{Y}$ to the end of the sequence $\mathbf{X}$. Thus, if $\mathbf{X}$ is $\sim [\ \sim p$ and $\mathbf{Y}$ is q (the sequence of length one whose sole term is q), $\mathbf{XY}$ is $\sim [\ \sim pq$.

It will sometimes be important to distinguish carefully between the language $\mathcal{P}$, which is the object of our study, and the language (English supplemented by mathematical terminology and symbolism) in which we discuss $\mathcal{P}$. The former we call the *object language*, and the latter the *meta-language*. (For example, if a class which is conducted in French is studying Chinese, then the object language is Chinese, and the meta-language is French.) We shall prove theorems in both the object language and the meta-language. Indeed, in the meta-language we shall prove theorems about the theorems of the object language. For emphasis, we may call a theorem of the meta-language a *metatheorem*.

We shall use formulas of the object language as names for themselves in the meta-language. (Indeed, this has already been done above.) We shall sometimes find it convenient to have symbols which stand for arbitrary propositional variables, just as in mathematics one sometimes uses a letter such as x to refer to an arbitrary real number. We shall henceforth use $\mathbf{p}$, $\mathbf{q}$, $\mathbf{r}$ and $\mathbf{s}$, with or without subscripts, to stand for arbitrary propositional variables. Technically, they are variables of our meta-language. That portion of the meta-language which is used to describe the syntax of $\mathcal{P}$ is called the *syntax language* for $\mathcal{P}$, and $\mathbf{p}$, $\mathbf{q}$, etc., are called *syntactical variables* for the propositional variables of $\mathcal{P}$.

From among the formulas we shall single out for special attention the *well-formed formulas* (*wffs*), which will be the "meaningful expressions" of $\mathcal{P}$. We shall define the set of wffs inductively, so that a formula is a wff iff its being so is a consequence of the following *formation rules*:

(1) A propositional variable standing alone is a wff.
(2) If **A** is a wff, then $\sim$**A** is a wff.
(3) If **A** and **B** are wffs, then [**A** $\vee$ **B**] is a wff.

DEFINITION. The *set of wffs* is the intersection of all sets $\mathcal{S}$ of formulas such that:

(i) **p** $\in \mathcal{S}$ for each propositional variable **p**.
(ii) For each formula **A**, if **A** $\in \mathcal{S}$, then $\sim$**A** $\in \mathcal{S}$.
(iii) For all formulas **A** and **B**, if **A** $\in \mathcal{S}$ and **B** $\in \mathcal{S}$, then [**A** $\vee$ **B**] $\in \mathcal{S}$.

A *wff* is a member of the set of wffs.

This formal definition of wffs is mathematically precise, and provides a firm foundation for rigorous proofs of metatheorems about wffs. The definition says that a formula is a wff iff it is in *every* set $\mathcal{S}$ which satisfies conditions (i), (ii), and (iii). Clearly, the set of wffs is the "smallest" set $\mathcal{S}$ satisfying conditions (i), (ii), and (iii) if we use "smaller than" in the sense that $\mathcal{S}_1$ is smaller than $\mathcal{S}_2$ iff $\mathcal{S}_1 \subseteq \mathcal{S}_2$.

Using this definition, it is easy to prove (Exercise X1000) the formation rules which were stated above.

EXAMPLE: Let $\mathcal{S}$ be any set satisfying conditions (i), (ii), and (iii). Then $p \in \mathcal{S}$ by (i), and $\sim p \in \mathcal{S}$ by (ii), and $[p \vee \sim p] \in \mathcal{S}$ by (iii). Thus $[p \vee \sim p]$ is in every set satisfying the three conditions, so it is a wff.

EXAMPLE: Let $\mathcal{S}_1$ be the set of all formulas in which propositional variables occur, and let $\mathcal{S}_2$ be the set of all formulas in which [and] occur the same number of times. It is easy to see that $\mathcal{S}_1$ and $\mathcal{S}_2$ both satisfy conditions (i), (ii), and (iii), so they both contain all wffs. The formula [[$\vee$] $\vee$] is not in $\mathcal{S}_1$ (though it is in $\mathcal{S}_2$), so it is not a wff. The formula $\sim p\,[\sim q]$ is also not in *all* sets $\mathcal{S}$ satisfying (i), (ii), and (iii) (as seen in Exercise X1004), although it is in both $\mathcal{S}_1$ and $\mathcal{S}_2$, so it is not a wff either.

We shall use **A, B, C, D, E**, etc., as syntactical variables for wffs.

As an immediate consequence of the definition of the set of wffs, we have the following useful induction rule:

1000 Principle of Induction on the Construction of a Wff. Let $\mathcal{R}$ be a property of formulas, and let $\mathcal{R}$(**A**) mean that **A** has property $\mathcal{R}$. Suppose

(1) $\mathcal{R}(\mathbf{p})$ for each propositional variable $\mathbf{p}$.
(2) Whenever $\mathcal{R}(\mathbf{A})$, then $\mathcal{R}(\sim\mathbf{A})$.
(3) Whenever $\mathcal{R}(\mathbf{A})$ and $\mathcal{R}(\mathbf{B})$, then $\mathcal{R}([\mathbf{A} \vee \mathbf{B}])$.

Then every wff has property $\mathcal{R}$.

Proof: Let $\mathcal{S}$ be the set of wffs which have property $\mathcal{R}$. It is easily seen that $\mathcal{S}$ has properties (i), (ii), and (iii) in the definition of the set of wffs, so every wff is in $\mathcal{S}$, and therefore has property $\mathcal{R}$. ■

Induction is used very frequently in logic, mathematics, and computer science, and our inductive definition of the set of wffs, with the associated Principle of Induction embodied in Theorem 1000, provides a typical example which is worth reflecting on. We have a certain set $\mathcal{W}$ (the set of wffs) which we wish to define. $\mathcal{W}$ must contain certain *initial objects* (the propositional variables) and be closed[1] under certain *operations* (negation and disjunction). Moreover, $\mathcal{W}$ is to contain only those objects that can be obtained from the initial objects by applying the operations an arbitrary number of times. We therefore define $\mathcal{W}$ to be the intersection of all sets which contain the initial objects and are closed under the operations. When $\mathcal{W}$ is defined in this way, it contains the initial objects and is closed under the operations. Moreover, there is always a Principle of Induction on the Construction of the Objects in $\mathcal{W}$ analogous to Theorem 1000: any property which is possessed by all the initial objects and preserved by the operations is possessed by all the members of $\mathcal{W}$.

We will see many examples of this method in this book. We will use it to define the wffs of a variety of logistic systems, and to define the primitive recursive functions and the recursive functions. In Chapter 6, where we show how to define various mathematical concepts in our object language, we will use it to define the set of natural numbers (with 0 as the sole initial object, the successor function as the sole operation, and the familiar Principle of Mathematical Induction as the induction principle), the order relation on the set of natural numbers, and a recursion operator.

Once one knows some basic facts about wffs, the Principle of Induction on the Construction of a Wff can be given an alternative justification in terms of numerical induction, and applications of it can be replaced by arguments involving numerical induction. This is a somewhat clumsier approach, but

[1] We say that a set $\mathcal{W}$ is *closed* under an operation if the result of applying the operation to members of $\mathcal{W}$ always produces members of $\mathcal{W}$. For example, the disjunction of wffs is a wff, so the set of wffs is closed under disjunction.

p	q	$\sim p$	$\vee$	q	$p \supset q$
T	T	F	T		T
T	F	F	F		F
F	T	T	T		T
F	F	T	T		T

Figure 1.5: $[\sim p \vee q]$ and $[p \supset q]$

p	q	$\sim$	$[\sim p$	$\vee$	$\sim q]$	$p \wedge q$
T	T	T	F	F	F	T
T	F	F	F	T	T	F
F	T	F	T	T	F	F
F	F	F	T	T	T	F

Figure 1.6: $\sim [\sim p \vee \sim q]$ and $[p \wedge q]$

it is widely used and may be comforting to those who prefer to do induction on numbers rather than wffs, so we show how it can be done in Supplement §10A.

A wff **B** which occurs as a (consecutive) part of a wff **A** will be known as a *well-formed* (*wf*) *part* of **A**. As a special case, **A** is a wf part of itself.

We next wish to introduce several abbreviations which will allow us to readily express in $\mathcal{P}$ ideas which are most naturally expressed with the aid of propositional connectives which are not primitive connectives of $\mathcal{P}$. Note from Figure 1.5 that $[\sim p \vee q]$ and $[p \supset q]$ have the same truth value in every possible situation. Similarly, Figure 1.6 shows that $\sim [\sim p \vee \sim q]$ and $[p \wedge q]$ have the same truth value in every possible situation. Also recalling Figure 1.3, we are thus led to make the following definitions:

$[\mathbf{A} \supset \mathbf{B}]$ stands for $[\sim \mathbf{A} \vee \mathbf{B}]$.
$[\mathbf{A} \wedge \mathbf{B}]$ stands for $\sim [\sim \mathbf{A} \vee \sim \mathbf{B}]$.
$[\mathbf{A} \equiv \mathbf{B}]$ stands for $[[\mathbf{A} \supset \mathbf{B}] \wedge [\mathbf{B} \supset \mathbf{A}]]$.

We say that **A** is the *scope* of the exhibited occurrence of $\sim$ in the wff $\sim$**A**, and that **A** and **B** are the *left* and *right scopes*, respectively, of the exhibited occurrence of $\vee$ in the wff $[\mathbf{A} \vee \mathbf{B}]$. The scopes of the other connectives are defined similarly. $\sim$ always has the smallest possible scope, so the scope of the exhibited occurrence of $\sim$ in a wff of the form $[\sim \mathbf{p} \vee \mathbf{B}]$ is **p**.

For the sake of convenience, we introduce the following conventions which permit us to omit brackets when writing wffs and abbreviations for wffs:

(a) The outermost brackets of a wff may be omitted.

(b) A square dot (▪) stands for a left bracket, whose mate is as far to the right as is possible without altering the pairing of left and right brackets already present.[2]

(c) In restoring brackets associated with several occurrences of the same connective, use the convention of association to the left, so that the connective on the left has the smaller scope. Thus,

$\mathbf{A} \supset \mathbf{B} \supset \mathbf{C}$	stands for	$[\mathbf{A} \supset \mathbf{B}] \supset \mathbf{C}$ and
$\mathbf{A} \supset \mathbf{B} \supset \mathbf{C} \supset \mathbf{D}$	stands for	$[[\mathbf{A} \supset \mathbf{B}] \supset \mathbf{C}] \supset \mathbf{D}$.

(d) When the relative scopes of several connectives of different kinds must be determined, $\sim$ is to be given the smallest possible scope, then $\wedge$ the next smallest possible scope except for $\sim$, then $\vee$, then $\supset$, then $\equiv$.

In summary, one may proceed as follows to restore brackets to an abbreviated wff, always taking care not to alter the pairing of left and right brackets already present:

(i) First restore brackets associated with the square dots (▪) in a wff; place a left bracket in place of the dot, and its mate as far to the right as possible.

(ii) Secondly, starting from the left, restore the brackets associated with each occurrence of $\wedge$ so as to give these connectives the smallest possible scope without enlarging the scope of any occurrence of $\sim$. (Thus $\sim p \wedge q$ stands for $[\sim p \wedge q]$ rather than $\sim [p \wedge q]$. Note that by proceeding from left to right one automatically implements the convention of association to the left.)

(iii)-(v) Similarly restore all brackets associated with occurrences of $\vee$, then $\supset$, then $\equiv$.

Note that the scope of $\sim$ in the wff $[\sim p \vee q \vee r]$ is p, while the scope of $\sim$ in the wff $[\sim [p \vee q] \vee r]$ is $[p \vee q]$, and the scope of $\sim$ in the wff $[\sim ▪p \vee q \vee r]$ is $[p \vee q \vee r]$.

EXAMPLES:

(1) $p \vee q \supset \sim r \wedge s \supset r$ stands for $[[[p \vee q] \supset [\sim r \wedge s]] \supset r]$.

(2) $p \wedge [\sim r \vee ▪q \supset ▪r \vee s] \vee p$ stands for $[[p \wedge [\sim r \vee [q \supset [r \vee s]]]] \vee p]$.

[2]The use of dots to replace brackets was introduced by Peano and was used by Whitehead and Russell in [Whitehead and Russell, 1913]. The different dot convention we use here was introduced by Alonzo Church and used in [Church, 1940] and [Church, 1956].

(3) $\sim r \wedge \sim p \vee \boldsymbol{\cdot} q \wedge r \vee [\sim s \supset \boldsymbol{\cdot} p_1 \wedge q \supset p] \equiv s \vee p \supset s$ stands for $[[\sim r \wedge \sim p] \vee [[[q \wedge r] \vee [\sim s \supset [[p_1 \wedge q] \supset p]]] \equiv [[s \vee p] \supset s]]]$.

Sometimes it is convenient to write certain brackets as parentheses to make it easier to see how brackets are paired. When this is done, brackets which are written as parentheses should always have their mates treated in the same way. For example, we may regard

$$[(p \equiv q) \equiv r] \wedge \sim [p \equiv (q \equiv r)]$$

as an abbreviation for

$$[[p \equiv q] \equiv r] \wedge \sim [p \equiv [q \equiv r]].$$

DEFINITION. A function θ from the set of formulas into the set of formulas is a *substitution* iff

(1) $\theta\mathbf{X}$ is the empty formula iff $\mathbf{X}$ is the empty formula.
(2) for all formulas $\mathbf{X}$ and $\mathbf{Y}$, $\theta(\mathbf{XY}) = (\theta\mathbf{X})(\theta\mathbf{Y})$; i.e., θ applied to a formula $\mathbf{XY}$ which is the concatenation of $\mathbf{X}$ and $\mathbf{Y}$ is the result of concatenating $\theta\mathbf{X}$ and $\theta\mathbf{Y}$.

Clearly a substitution is determined by its behavior on unit formulas; that is, if θ_1 and θ_2 are substitutions such that $\theta_1\mathbf{x} = \theta_2\mathbf{x}$ for every primitive symbol $\mathbf{x}$, then $\theta_1 = \theta_2$. Note that $\theta\mathbf{x}$ is the formula which we substitute for $\mathbf{x}$. A substitution θ is *finite* iff the set of primitive symbols $\mathbf{x}$ such that $\theta\mathbf{x} \neq \mathbf{x}$ is finite. For the most part we shall be concerned with *substitutions for variables*, i.e., substitutions θ for which the only primitive symbols $\mathbf{x}$ such that $\theta\mathbf{x} \neq \mathbf{x}$ are variables.

NOTATION. Let $\mathbf{x}_1, \ldots, \mathbf{x}_n$ be distinct primitive symbols and let $\mathbf{Y}_1, \ldots, \mathbf{Y}_n$ be formulas. $\mathsf{S}^{\mathbf{x}_1 \ldots \mathbf{x}_n}_{\mathbf{Y}_1 \ldots \mathbf{Y}_n}$ is that (finite) substitution θ such that $\theta\mathbf{x}_i = \mathbf{Y}_i$ for $1 \leq i \leq n$ and $\theta\mathbf{y} = \mathbf{y}$ for any primitive symbol $\mathbf{y}$ distinct from $\mathbf{x}_1, \ldots,$ and $\mathbf{x}_n$. If $\mathbf{Z}$ is a formula, we say that $(\mathsf{S}^{\mathbf{x}_1 \ldots \mathbf{x}_n}_{\mathbf{Y}_1 \ldots \mathbf{Y}_n}\mathbf{Z})$ is the result of simultaneously substituting $\mathbf{Y}_1$ for $\mathbf{x}_1, \ldots,$ and $\mathbf{Y}_n$ for $\mathbf{x}_n$ in $\mathbf{Z}$.

Now that we have discussed the language of $\mathcal{P}$, let us make a few remarks on what we are about to do next, and illustrate them with Figure 1.7 In §11 we will explain the way this language is to be used by specifying certain axioms (initial principles) and rules of inference (rules of reasoning). We will thus have specified the *syntax* of the language. The syntactical aspects of a formal language are those aspects of it which can be discussed purely in

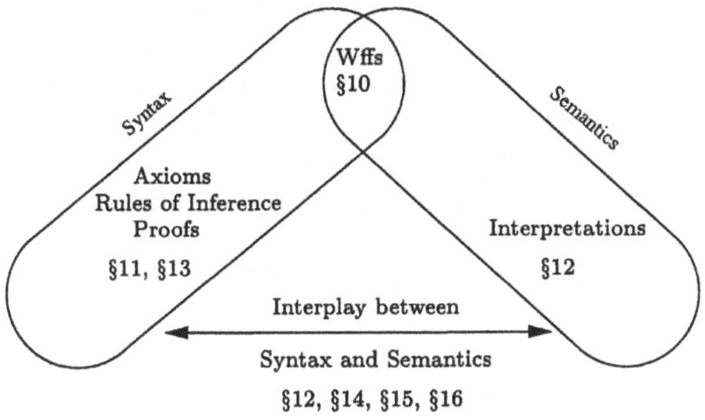

Figure 1.7: Plan for Chapter 1

terms of the symbols and sequences of symbols of the language, without any regard for what these might mean. In §12 we shall give a mathematically precise method of assigning meanings to the well-formed formulas, and thus specify the *semantics* of the language. After having specified the syntax and the semantics in ways which are quite independent of each other, except that they concern the same set of well-formed formulas, we will investigate how the syntax and semantics are related to each other. It will be seen that the study of the interplay between syntax and semantics is fundamental to the study of logical systems.

§10A. Supplement on Induction

In this supplement we shall discuss induction in a very informal way, and this discussion can be skipped by readers already familiar with induction and complete induction. In Chapter 6 we shall give a formal definition of the natural numbers, and prove the Principle of Mathematical Induction (Theorem 6102).

We may informally describe the set of natural numbers as the set which contains 0, 1, 2, 3, 4, 5, 6, etc. (The definition of $\mathbb{N}$ in §60 shows that the meaning of the word "etc." here can be made very explicit, but we shall not do that now.) One of the fundamental facts about the natural numbers is the Principle of Mathematical Induction (PMI), which says that for any property P of numbers, if 0 has property P, and $n + 1$ has property P

whenever n is a natural number which has property P, then every natural number has property P. If we abbreviate "n has property P" by "Pn", "**A** and **B**" by "[**A** $\wedge$ **B**]", "if **A**, then **B**" by "[**A** $\supset$ **B**]", "for all natural numbers n" by "$\forall n$", and "for any property P" by "$\forall P$", then we can use the following symbolic form of the Principle of Mathematical Induction:

$$\forall P[[P0 \ \wedge \ \forall n[Pn \ \supset \ P(n+1)]] \ \supset \ \forall nPn]$$

A related principle, called the Principle of Complete Induction (PCI), can be symbolized as follows:

$$\forall R[\forall n[(\forall j < n)Rj \ \supset \ Rn] \ \supset \ \forall nRn]$$

This can be paraphrased as saying that for any property R, if every natural number n has property R whenever all the natural numbers less than n have property R, then every natural number has property R.

By using some elementary facts about natural numbers we can derive each of these principles from the other. We shall indicate how this can be done, using symbolic notation for the sake of brevity.

First, assume that we have accepted PCI, and wish to justify PMI. Thus, we assume $P0 \ \wedge \ \forall n[Pn \ \supset \ P(n+1)]$, and try to prove $\forall nPn$. Let us try to show that $\forall n[(\forall j < n)Pj \ \supset \ Pn]$, so that we can use PCI. Let n be any natural number such that $(\forall j < n)Pj$. If $n = 0$, then we already know Pn, since we have assumed $P0$. Otherwise, $n = m + 1$ for some natural number m. We know Pm since $(\forall j < m+1)Pj$, so $P(m+1)$ follows from our assumption $\forall n[Pn \ \supset \ P(n+1)]$. Thus in each case we have established Pn, so since n was arbitrary, we have shown that $\forall n[(\forall j < n)Pj \ \supset \ Pn]$. Hence $\forall nPn$ by PCI, which is the desired conclusion.

Next we show how PCI can be derived from PMI. Assume R is a property such that $\forall n[(\forall j < n)Rj \ \supset \ Rn]$, and try to prove $\forall nRn$. In order to apply PMI, we shall let P be the property such that Pn means $(\forall j \leq n)Rj$, i.e., every natural number less than or equal to n has property R. We must prove $P0$ and $\forall n[Pn \ \supset \ P(n+1)]$. We are given that $(\forall j < 0)Rj \ \supset \ R0$, but there are no natural numbers less than 0, so $(\forall j < 0)Rj$ is vacuously true, so $R0$, and hence $P0$. Next assume Pn and try to derive $P(n+1)$. Pn means that $(\forall j \leq n)Rj$, and hence that $(\forall j < n+1)Rj$, and hence $R(n+1)$ by our assumption about R. From $(\forall j \leq n)Rj$ and $R(n+1)$ we infer $(\forall j \leq n+1)Rj$, which is $P(n+1)$. Thus from Pn we derive $P(n+1)$, so $\forall n[Pn \ \supset \ P(n+1)]$. Hence by PMI we obtain $\forall nPn$, which implies $\forall nRn$, and the proof of PCI is complete.

Finally, we show how metatheorem 1000 can be proved by complete induction and some basic facts about wffs. Assume that $\mathcal{R}$ is a property of wffs with properties (1), (2), and (3) of metatheorem 1000. For any wff **D**, let #**D** (the length of **D**) be the number of occurrences of primitive symbols in **D**. Let P be the property of natural numbers such that Pn means $\forall \mathbf{D}[\#\mathbf{D} = n \supset \mathcal{R}(D)]$, i.e., every wff of length n has property $\mathcal{R}$. We must show that $\forall n[(\forall j < n)Pj \supset Pn]$, so let n be any natural number, and assume that $(\forall j < n)Pj$. To show Pn, let **D** be any wff such that $\#\mathbf{D} = n$, and show $\mathcal{R}(\mathbf{D})$.

Now we use the facts that **D** has one of the forms **p**, $\sim$**A**, or [**A** $\vee$ **B**] (Exercise X1002), and that in the latter cases $\#\mathbf{A} < \#\ \mathbf{D}$ and $\#\ \mathbf{B} < \#\mathbf{D}$. If **D** is **p**, then $\mathcal{R}(\mathbf{D})$ by (1). If **D** is $\sim$**A**, then from $(\forall j < n)Pj$ we infer Pj, where $j = \#\mathbf{A}$, so $\mathcal{R}(\mathbf{A})$ (by the definition of P), so $\mathcal{R}(\mathbf{D})$ by (2). Similarly, if **D** is [**A** $\vee$ **B**], we infer $P(\#\mathbf{A})$ and $P(\#\mathbf{B})$, and obtain $\mathcal{R}(\mathbf{D})$ by (3). Thus in all cases we have shown $\mathcal{R}(\mathbf{D})$.

In summary, we have shown that $\forall\ \mathbf{D}[\ \#\ \mathbf{D} = n \supset \mathcal{R}(\mathbf{D})]$, which is Pn. Since n was arbitrary, we have shown that $\forall n[(\forall j < n)Pj \supset Pn]$, so $\forall n\ Pn$ follows by PCI. Now let **D** be any wff, and let $n = \#\mathbf{D}$. We have established that Pn, so $\mathcal{R}(\mathbf{D})$ follows. This completes the proof of metatheorem 1000 using PCI.

EXERCISES

X1000. Using the definition of the set of wffs, show that:

(a) Every formula consisting of a propositional variable standing alone is a wff.
(b) If **A** is a wff, then $\sim$**A** is a wff.
(c) If **A** and **B** are wffs, then [**A** $\vee$ **B**] is a wff.

X1001. A *formation sequence for a formula* **Y** is a finite sequence $\mathbf{Y}_1, \ldots,$ $\mathbf{Y}_m$ of formulas such that $\mathbf{Y}_m$ is **Y** and for each i $(1 \leq i \leq m)$, one of the following conditions holds:

(a) $\mathbf{Y}_i$ is a propositional variable;
(b) there is $j < i$ such that $\mathbf{Y}_i$ is $\sim\mathbf{Y}_j$;
(c) there are indices $j < i$ and $k < i$ such that $\mathbf{Y}_i$ is $[\mathbf{Y}_j \vee \mathbf{Y}_k]$.

Prove that a formula **Y** is well-formed iff it has a formation sequence. (*Hint*: Metatheorem 1000 may be useful.)

X1002. Prove that every wff is of the form **p** for some propositional variable **p**, or of the form $\sim\mathbf{A}$ or $[\mathbf{A} \lor \mathbf{B}]$ for some wffs **A** and **B**.

X1003. Using the results above, describe an effective procedure which can be used to determine of an arbitrary formula **Y** whether or not it is wf, and prove that this procedure is correct.

X1004. Decide whether the following formulas are wffs, and provide proofs to justify your answers:

(a) $\sim p[\sim q]$
(b) $[\ ]p$
(c) $[\sim p \lor]q$
(d) $\sim [\sim\sim q \lor r]$

X1005. The following procedure is useful in checking whether complex expressions are wffs, or abbreviations of wffs. Given a formula **A**, consider the occurrences of symbols in **A** one-by-one in left-to-right order, and, starting with a count of zero, increase the count by one every time a [is encountered, and decrease the count by one every time a] is encountered.

EXAMPLE:

$$\underset{1}{[}\underset{2}{[} \sim r \lor \sim p \underset{1}{]} \lor \sim \underset{2}{[}\underset{3}{[}\underset{4}{[} q \lor r \underset{3}{]} \lor \underset{4}{[} p \lor \sim s \underset{3}{]}\underset{2}{]} \lor q \underset{1}{]}\underset{0}{]}$$

Prove that if the expression is a wff, then the final count is zero, and the count never becomes negative. *Hint:* Use metatheorem 1000.

X1006. Show that if θ is any substitution for variables and **A** and **B** are wffs,

(a) $(\theta \sim \mathbf{A}) = \sim (\theta\mathbf{A})$
(b) $(\theta[\mathbf{A} \lor \mathbf{B}]) = [(\theta\mathbf{A}) \lor (\theta\mathbf{B})]$

X1007. Show that if $\mathbf{p}_1, \ldots, \mathbf{p}_n$ are distinct variables and $\mathbf{B}_1, \ldots, \mathbf{B}_n$, **A** are wffs, then $\left(\mathsf{S}^{\mathbf{p}_1 \ldots \mathbf{p}_n}_{\mathbf{B}_1 \ldots \mathbf{B}_n} \mathbf{A}\right)$ is a wff. *Hint*: Use metatheorem 1000.

X1008. Evaluate $\mathsf{S}^{p}_{[q \lor r]}{}^{q}_{\sim p}[\sim [\sim p \lor q] \lor \sim\sim p]$.

X1009. Show that if σ and τ are substitutions, then the composition $\sigma \circ \tau$ of σ with τ is also a substitution.

X1010. Express the statements below in symbolic notation. Let
j mean "John can do the job",

a mean "Alice can do the job",
w mean "Alice is willing to do the job",
g mean "Alice will get the job",
b mean "Bill will help John",
v mean "John will be willing to do the job",
d mean "John will do the job".

(a) If John can't do the job, then Alice can't either.
(b) If Alice is willing and able to do the job, then she will get the job.
(c) John can do the job only if Bill will help him.
(d) If Bill will help John, then John will be willing to do the job.
(e) John won't do the job unless Bill will help him.
(f) John will do the job if and only if Bill will help him.

X1011. Prove that if **X**[**Y**]**Z** is a wff, where **X**, **Y**, and **Z** are formulas of length ≥ 0, then some propositional variable occurs in **Y**. (*Hint:* Use metatheorem 1000 to prove some assertion which implies this.)

§11. The Axiomatic Structure of $\mathcal{P}$

Before discussing the details of the axiomatic structure of $\mathcal{P}$, let us make some general remarks about the axiomatic method and the role of syntax. When one uses the axiomatic method, one chooses as *axioms* certain statements whose consequences one wishes to investigate, and then derives their consequences (which one calls *theorems*) by the methods of logic. The set of theorems which can be derived from the axioms is called a *theory*, though the theory is sometimes identified with its set of axioms. It should be noted that people sometimes set up theories in order to explore ideas which are still in the process of being clarified. Thus, while the expressions used in the theory are not likely to be completely meaningless, the semantics of the theory may be vague or subject to revision. It is important to have a way of exploring the consequences of the axioms which depends only on syntax, and is not dependent on perhaps vague or shifting semantics.

While words may mean different things to different people, and it is sometimes difficult to know whether different people really attach the same meanings to various words and phrases, one way of understanding what people mean by words and phrases is to observe (over a long period) how they use these words and phrases in various situations. When one is concerned with the communication of mathematical ideas, which may be very abstract, people may have different inner mental "images" of mathematical entities (including relations, functions, sets, etc.), but effective communication about these entities is still possible if there is agreement about how to use the language which describes these entities.

Thus, the syntactical aspect of a language plays an extremely important role even if people start out with an understanding of the language which is guided by some semantical perspective. The axiomatic method provides a way of specifying how a language describing some logically coherent discipline is to be *used* by giving rules for asserting certain statements.

Note the role of the axiomatic method in dealing with complexity. Many of the phenomena (including mathematical phenomena) which we would like to understand are extremely complicated, and even if one has a clear understanding of the meanings of the words in a sentence, one may not know whether or not the sentence is true. One needs to have a method of reasoning from simple statements whose truth is evident to complex statements whose truth may not be evident. The axiomatic method provides a way of doing this.

We shall illustrate how the axiomatic method works by introducing axioms and a rule of inference for the system $\mathcal{P}$, and deriving theorems which express general truths involving propositional connectives. The reader can probably think of a variety wffs which might be taken as axioms, such as :

- $p \lor \sim p$
- $p \supset p$
- $p \supset {\scriptstyle\blacksquare} p \lor q$
- $p \supset {\scriptstyle\blacksquare} q \lor p$
- $p \land q \supset p$
- $p \lor p \supset p$
- $q \lor p \supset p \lor q$
- $p \supset {\scriptstyle\blacksquare} \sim p \supset q$
- $[p \supset q] \land [q \supset r] \supset {\scriptstyle\blacksquare} p \supset r$

Actually, there are many different sets of axioms and rules of inference which one might reasonably choose, and each choice yields a different logistic system. Once one has chosen some axioms and specified one or more rules of inference which can be used to derive theorems from the axioms, one may find that some of the axioms are derivable from others, and decide to modify the axiom set. Axiom systems tend to evolve as one learns more about them. The axioms and rule of inference for $\mathcal{P}$ were chosen to be quite simple, independent (a matter which will be discussed in §13), and still adequate for quickly deriving the theorems that will be needed later.[3]

The axioms of $\mathcal{P}$ consist of all wffs having one of the following three forms:

(1) $\sim [\mathbf{A} \lor \mathbf{A}] \lor \mathbf{A}$

(2) $\sim \mathbf{A} \lor {\bullet} \mathbf{B} \lor \mathbf{A}$

(3) $\sim [\sim \mathbf{A} \lor \mathbf{B}] \lor {\bullet} \sim [\mathbf{C} \lor \mathbf{A}] \lor {\bullet} \mathbf{B} \lor \mathbf{C}$

Note that $\mathcal{P}$ has infinitely many axioms. We leave it to the reader to verify that there is an effective test for determining of any wff whether or not it is an axiom. We refer to the three forms of the axioms specified above as the axiom schemata for $\mathcal{P}$. Using the definition of $\supset$ we may write them as follows:

Axiom Schema 1. $\mathbf{A} \lor \mathbf{A} \supset \mathbf{A}$

Axiom Schema 2. $\mathbf{A} \supset {\bullet} \mathbf{B} \lor \mathbf{A}$

Axiom Schema 3. $\mathbf{A} \supset \mathbf{B} \supset {\bullet} \mathbf{C} \lor \mathbf{A} \supset {\bullet} \mathbf{B} \lor \mathbf{C}$

$\mathcal{P}$ has just one rule of inference, which is:

Modus Ponens (MP). From $\mathbf{A}$ and $\sim\mathbf{A} \lor \mathbf{B}$, infer $\mathbf{B}$.

Using the definition of $\supset$, we can also state Modus Ponens as follows: From $\mathbf{A}$ and $\mathbf{A} \supset \mathbf{B}$, infer $\mathbf{B}$.

The rationale for using Modus Ponens as a rule of inference is provided by the meanings of the propositional connectives. In every situation where the statements $\mathbf{A}$ and $\mathbf{A} \supset \mathbf{B}$ are true, $\mathbf{B}$ must also be true.

We shall use $\mathcal{H}$ (with or without subscripts) as a syntactical variable for a set of wffs. An especially important case is that in which $\mathcal{H}$ denotes the empty set of wffs.

---

[3] $\mathcal{P}$ is very closely related to the system discussed in [Rasiowa, 1949].

DEFINITIONS.

A proof of a wff $\mathbf{B}$ *from the set* $\mathcal{H}$ *of hypotheses* is a finite sequence $\mathbf{B}_1, \ldots, \mathbf{B}_m$ of wffs such that $\mathbf{B}_m$ is $\mathbf{B}$ and for each j $(1 \leq j \leq m)$ at least one of the following conditions is satisfied:

(1) $\mathbf{B}_j$ is an axiom.
(2) $\mathbf{B}_j$ is a member of $\mathcal{H}$.
(3) $\mathbf{B}_j$ is inferred by Modus Ponens from wffs $\mathbf{B}_i$ and $\mathbf{B}_k$, where $i < j$ and $k < j$.

An alternative way of expressing condition (3) is to say that there exist $i < j$ and $k < j$ such that $\mathbf{B}_k$ is $[\mathbf{B}_i \supset \mathbf{B}_j]$.

A *proof* of a wff $\mathbf{B}$ is a proof of $\mathbf{B}$ from the empty set of hypotheses. A *theorem* is a wff which has a proof.

We write $\mathcal{H} \vdash \mathbf{B}$ to indicate that there is a proof of $\mathbf{B}$ from $\mathcal{H}$ and say that $\mathbf{B}$ is *derivable from* $\mathcal{H}$. In place of $\{\mathbf{A}_1, \ldots, \mathbf{A}_n\} \vdash \mathbf{B}$ we write $\mathbf{A}_1, \ldots, \mathbf{A}_n \vdash \mathbf{B}$, and in place of $\mathcal{H} \cup \{\mathbf{A}\} \vdash \mathbf{B}$ we write $\mathcal{H}, \mathbf{A} \vdash \mathbf{B}$. We write $\vdash \mathbf{B}$ to indicate that $\mathbf{B}$ is derivable from the empty set, i.e., that $\mathbf{B}$ is a theorem. In order to avoid confusion when several logistic systems are under consideration, we may write $\vdash_{\mathcal{P}} \mathbf{B}$ (instead of $\vdash \mathbf{B}$) to indicate that $\mathbf{B}$ is a theorem of $\mathcal{P}$.

We have now completed the specification of $\mathcal{P}$ as a *logistic system.* To define a logistic system, one must specify its primitive symbols, formation rules, axioms, and rules of inference. A logistic system must be defined in such a way that there is an effective test (algorithm) for determining of any formula whether or not it is in fact a wff.[4] A *proof* in a logistic system $\mathcal{L}$ of a wff $\mathbf{B}$ is a sequence of wffs, each of which is an axiom of $\mathcal{L}$ or is inferred from preceding wffs in the sequence by a rule of inference of $\mathcal{L}$, with $\mathbf{B}$ as the last wff in the sequence. A *theorem* of $\mathcal{L}$ is a wff which has a proof in $\mathcal{L}$. It is part of the definition of a logistic system that there must be an effective test for determining of a sequence of formulas whether or not it is a proof.[5] (The reader is asked to show this for the system $\mathcal{P}$ in Exercise X1100.) The rationale for this is clear: a proof should carry conviction to one who accepts the axioms and the methods of reasoning embodied in the rules of inference. However, the basic purpose of a proof is defeated if one cannot always determine whether an alleged proof is actually a proof. As a matter of fact, mathematicians sometimes do offer proofs which their colleagues do not find

[4]The notion of effectiveness will be discussed in detail in §65.

[5]For more discussion of these matters the reader is referred to the Introduction of [Church 1956].

convincing. Occasionally such proofs are actually incorrect, but more often they are just so abbreviated that they are hard to follow. The idea of a logistic system is to provide a standard of rigor which will guarantee that when a proof is presented within a logistic system, one can verify that it is in fact a proof, and (if so) be confident that the theorem is really proved. While mathematical proofs are rarely presented with all the details that strict conformity to the definition of a proof in a logistic system demands, once one has an appropriate logistic system in mind, one can regard a mathematical proof as an abbreviated proof in the logistic system, and provide more details (thus making the proof less abbreviated) whenever questions about it arise. In principle, sufficient repetition of this process would eventually produce an unabbreviated proof in the logistic system. We can anticipate the development of automated theorem proving systems which will provide these additional details automatically, thus verifying the correctness of proofs and permitting readers to examine the proofs at whatever level of detail they find most congenial. (For convenience, we have referred to "mathematical proofs" in the discussion above, but the same considerations apply to proofs in other disciplines which use logic.)

While $\mathcal{P}$ is only one example of a logistic system, the methods and concepts we shall use while investigating $\mathcal{P}$ will be relevant to a variety of logistic systems.

One of the surprising things about many logistic systems is how much can be derived from simple axioms. This is also one of the reasons such systems are so useful; they provide an extremely economical way of storing a great wealth of information. We next explore the power of $\mathcal{P}$ by proving certain theorems of $\mathcal{P}$ and metatheorems about $\mathcal{P}$.

1100. If $\mathcal{H}_1 \vdash \mathbf{A}$ and $\mathcal{H}_1 \subseteq \mathcal{H}_2$, then $\mathcal{H}_2 \vdash \mathbf{A}$.

Proof: Any proof of $\mathbf{A}$ from $\mathcal{H}_1$ must also be a proof of $\mathbf{A}$ from $\mathcal{H}_2$. ∎

Corollary. If $\mathcal{H}_1 \vdash \mathbf{A}$ and $\mathcal{H}_2 \vdash \mathbf{A} \supset \mathbf{B}$ and $\mathcal{H}_1 \subseteq \mathcal{H}$ and $\mathcal{H}_2 \subseteq \mathcal{H}$, then $\mathcal{H} \vdash \mathbf{B}$.

REMARK. Metatheorem 1100 tells us that when using the system $\mathcal{P}$ we can enlarge the set of hypotheses (which we can think of as expressing the information we consider) without losing any consequences we have derived from these hypotheses. Logistic systems which have this property are called *monotonic*. If one wishes to formalize reasoning in a context where additional information may lead one to change one's conclusions, one needs to use a *nonmonotonic* logistic system. We shall not study nonmonotonic logics in this book.

1101 Rule of Substitution (Sub). If $\mathcal{H} \vdash \mathbf{A}$, and if $\mathbf{p}_1, \ldots, \mathbf{p}_n$ are distinct variables which do not occur in any wff in $\mathcal{H}$, then $\mathcal{H} \vdash \mathrm{S}^{\mathbf{p}_1 \ldots \mathbf{p}_n}_{\mathbf{B}_1 \ldots \mathbf{B}_n} \mathbf{A}$.

Proof: Let θ be the substitution $\mathrm{S}^{\mathbf{p}_1 \ldots \mathbf{p}_n}_{\mathbf{B}_1 \ldots \mathbf{B}_n}$. It is easy to see that if $\mathbf{C}_1, \ldots, \mathbf{C}_m$ is a proof of $\mathbf{A}$ from $\mathcal{H}$, then $\theta\mathbf{C}_1, \ldots, \theta\mathbf{C}_m$ is a proof of $\theta\mathbf{A}$ from $\mathcal{H}$. Note how the condition on the variables comes into play when $\mathbf{C}_i$ is a member of $\mathcal{H}$. ∎

Of course, the sentence above is really a rather brief sketch of a proof of this metatheorem. For a more complete proof, show by complete induction on i that $\mathcal{H} \vdash \theta\mathbf{C}_i$ for each i with $1 \leq i \leq m$. Break the proof into cases according to how $\mathbf{C}_i$ was justified in the original proof.

In proving theorems of $\mathcal{P}$, we shall indicate at the right-hand margin how each step of the proof is obtained. However, we shall often omit explicit mention of the rule of substitution and metatheorem 1100. We shall also sometimes combine several steps into one.

1102. If $\mathcal{H} \vdash \mathbf{A} \supset \mathbf{B}$ and $\mathcal{H} \vdash \mathbf{C} \lor \mathbf{A}$, then $\mathcal{H} \vdash \mathbf{B} \lor \mathbf{C}$.

Proof: Suppose we are given proofs $\mathbf{D}_1, \ldots, \mathbf{D}_i$ of $\mathbf{A} \supset \mathbf{B}$ and $\mathbf{E}_1, \ldots, \mathbf{E}_j$ of $\mathbf{C} \lor \mathbf{A}$ from the set $\mathcal{H}$ of hypotheses. Thus $\mathbf{D}_i$ is $[\mathbf{A} \supset \mathbf{B}]$ and $\mathbf{E}_j$ is $[\mathbf{C} \lor \mathbf{A}]$. It is easy to see that the following is a proof from $\mathcal{H}$ of $\mathbf{B} \lor \mathbf{C}$:

(1)	$\mathbf{D}_1$	
	$\vdots$	
(i)	$\mathbf{D}_i$	
$(i+1)$	$\mathbf{A} \supset \mathbf{B} \supset . \mathbf{C} \lor \mathbf{A} \supset . \mathbf{B} \lor \mathbf{C}$	Axiom 3
$(i+2)$	$\mathbf{C} \lor \mathbf{A} \supset . \mathbf{B} \lor \mathbf{C}$	MP: $i, i+1$
$(i+3)$	$\mathbf{E}_1$	
	$\vdots$	
$(i+j+2)$	$\mathbf{E}_j$	
$(i+j+3)$	$\mathbf{B} \lor \mathbf{C}$	MP: $i+2, i+j+2$

∎

Henceforth, we shall abbreviate proofs such as that above, and simply say something like "by Axiom 3 and MP" to indicate the main ideas used in the proof.

Note that metatheorem 1102 can be used as a rule of inference, since it shows that one can infer **B** $\vee$ **C** from **A** $\supset$ **B** and **C** $\vee$ **A**. Moreover, the proof of 1102 shows explicitly how to construct a proof of **B** $\vee$ **C** if one is given proofs of **A** $\supset$ **B** and **C** $\vee$ **A**. Metatheorems such as 1102 which can be used as rules of inference are called *derived rules of inference.* For contrast, we refer to Modus Ponens as the *primitive* rule of inference of $\mathcal{P}$.

1103. $\vdash p \vee \sim p$

Proof:

(.1) $\vdash p \vee p \supset p \supset \centerdot \sim p \vee [p \vee p] \supset \centerdot p \vee \sim p$ Axiom 3

(.2) $\vdash p \vee \sim p$ MP: Axiom 1, .1, Axiom 2

■

1104. $\vdash p \supset \sim\sim p$ Sub: 1103

■

1105. $\vdash \sim\sim p \supset p$

Proof:

(.1) $\vdash \sim p \supset \sim\sim\sim p \supset \centerdot p \vee \sim p \supset \centerdot \sim\sim\sim p \vee p$ Axiom 3

(.2) $\vdash \sim\sim p \supset p$ Sub: 1104; MP: .1; MP: 1103

■

1106. $\vdash p \supset p$

REMARK. The proof below is quite easy to understand if one has a good intuitive undertanding of 1102. When one has proved **A** $\supset$ **B**, one can use 1102 to

replace **C** $\vee$ **A**
by **B** $\vee$ **C**.

In the proof below, (.1) and (.2) are preliminary results. The essential part of the proof consists of systematically transforming Axiom 2 in line (.3) into line (.8), from which (.11) is easily obtained.

Proof of 1106:

(.1) $\vdash$	$p \vee p \supset \centerdot \sim\sim p \vee p$	MP: 1104, Axiom 3
(.2) $\vdash$	$\sim\sim p \vee p \supset \centerdot \sim\sim p \vee \sim\sim p$	MP: 1104, Axiom 3
(.3) $\vdash$	$\sim p \vee [p \vee p]$	Axiom 2
(.4) $\vdash$	$[\sim\sim p \vee p] \vee \sim p$	1102: .1, .3
(.5) $\vdash$	$\sim\sim\sim p \vee [\sim\sim p \vee p]$	Sub: 1104; 1102: .4
(.6) $\vdash$	$[\sim\sim p \vee \sim\sim p] \vee \sim\sim\sim p$	1102: .2, .5
(.7) $\vdash$	$\sim p \vee [\sim\sim p \vee \sim\sim p]$	Sub: 1105; 1102: .6
(.8) $\vdash$	$\sim\sim p \vee \sim p$	1102: Axiom 1, .7
(.9)	$\sim\sim\sim p \vee \sim\sim p$	Sub: .8
(.10)	$p \vee \sim\sim\sim p$	1102 : 1105, .9
(.11) $\vdash$	$\sim p \vee p$	Sub: 1105; 1102: .10

■

In order to see the usefulness of derived rules of inference, the reader is invited to write out a complete proof of 1106 using only the primitive rule of inference, and compare it with the abbreviated proof using derived rules of inference which is given above.

1107. $\vdash q \vee p \supset p \vee q$ MP: 1106, Axiom 3

■

1108. If $\mathcal{H} \vdash \mathbf{A} \vee \mathbf{B}$, then $\mathcal{H} \vdash \mathbf{B} \vee \mathbf{A}$.

Proof: By 1107. ■

1109 Transitive Law of Implication (Trans). If $\mathcal{H} \vdash \mathbf{A}_1 \supset \mathbf{A}_2$, $\mathcal{H} \vdash \mathbf{A}_2 \supset \mathbf{A}_3, \ldots$, and $\mathcal{H} \vdash \mathbf{A}_{n-1} \supset \mathbf{A}_n$, then $\mathcal{H} \vdash \mathbf{A}_1 \supset \mathbf{A}_n$.

Proof: It suffices to consider the case in which $n = 3$, since the general case follows easily from this one by mathematical induction.

(.1)	$\mathcal{H} \vdash \sim\mathbf{A}_1 \vee \mathbf{A}_2$	given
(.2)	$\mathcal{H} \vdash \mathbf{A}_2 \supset \mathbf{A}_3$	given
(.3)	$\mathcal{H} \vdash \mathbf{A}_3 \vee \sim\mathbf{A}_1$	1102: .2, .1
(.4)	$\mathcal{H} \vdash \mathbf{A}_1 \supset \mathbf{A}_3$	1108: .3

■

1110. If $\mathcal{H} \vdash \mathbf{A} \supset \mathbf{C}$ and $\mathcal{H} \vdash \mathbf{B} \supset \mathbf{C}$, then $\mathcal{H} \vdash \mathbf{A} \vee \mathbf{B} \supset \mathbf{C}$.

Proof:

(.1) $\mathcal{H} \vdash \mathbf{B} \supset \mathbf{C}$	given
(.2) $\mathcal{H} \vdash \mathbf{A} \vee \mathbf{B} \supset \mathbf{C} \vee \mathbf{A}$	MP: .1, Axiom 3
(.3) $\mathcal{H} \vdash \mathbf{A} \supset \mathbf{C}$	given
(.4) $\mathcal{H} \vdash \mathbf{C} \vee \mathbf{A} \supset \mathbf{C} \vee \mathbf{C}$	MP: .3, Axiom 3
(.5) $\mathcal{H} \vdash \mathbf{C} \vee \mathbf{C} \supset \mathbf{C}$	Axiom 1
(.6) $\mathcal{H} \vdash \mathbf{A} \vee \mathbf{B} \supset \mathbf{C}$	Trans: .2, .4, .5

■

1111. If $\mathcal{H} \vdash \mathbf{A} \supset \mathbf{C}$ and $\mathcal{H} \vdash {\sim}\mathbf{A} \supset \mathbf{C}$, then $\mathcal{H} \vdash \mathbf{C}$.

Proof: By 1110 and 1103. ■

1112. If $\mathcal{H} \vdash \mathbf{A} \supset \mathbf{B}$, then $\mathcal{H} \vdash \mathbf{A} \supset \centerdot\, \mathbf{C} \vee \mathbf{B}$ and $\mathcal{H} \vdash \mathbf{A} \supset \centerdot\, \mathbf{B} \vee \mathbf{C}$.

Proof:

(.1) $\mathcal{H} \vdash \mathbf{A} \supset \mathbf{B}$	given
(.2) $\mathcal{H} \vdash \mathbf{A} \supset \centerdot\, \mathbf{C} \vee \mathbf{B}$	Trans: .1, Axiom 2
(.3) $\mathcal{H} \vdash \mathbf{A} \supset \centerdot\, \mathbf{B} \vee \mathbf{C}$	Sub: 1107; Trans: .2

■

1113. $\vdash [p \vee q] \vee r \supset \centerdot\, p \vee \centerdot\, q \vee r$

Proof:

(.1) $\vdash p \supset \centerdot\, p \vee \centerdot\, q \vee r$	1112: 1106
(.2) $\vdash q \supset \centerdot\, q \vee r$	1112: 1106
(.3) $\vdash q \supset \centerdot\, p \vee \centerdot\, q \vee r$	1112: .2
(.4) $\vdash p \vee q \supset \centerdot\, p \vee \centerdot\, q \vee r$	1110: .1, .3
(.5) $\vdash r \supset \centerdot\, p \vee \centerdot\, q \vee r$	1112: 1106
(.6) $\vdash [p \vee q] \vee r \supset \centerdot\, p \vee \centerdot\, q \vee r$	1110: .4, .5

■

1114. If $\mathcal{H} \vdash [\mathbf{A} \vee \mathbf{B}\,] \vee \mathbf{C}$, then $\mathcal{H} \vdash \mathbf{A} \vee \centerdot\, \mathbf{B} \vee \mathbf{C}$.

Proof: By 1113. ■

REMARK. The converse of 1114 is established in exercise X1101.

1115. If $\mathcal{H} \vdash \mathbf{A} \supset \mathbf{B}$ and $\mathcal{H} \vdash \mathbf{A} \supset \centerdot \mathbf{B} \supset \mathbf{C}$, then $\mathcal{H} \vdash \mathbf{A} \supset \mathbf{C}$.

Proof:

(.1) $\mathcal{H} \vdash \mathbf{A} \supset \mathbf{B}$	given
(.2) $\mathcal{H} \vdash \sim\mathbf{A} \vee \centerdot \sim\mathbf{B} \vee \mathbf{C}$	given
(.3) $\mathcal{H} \vdash [\sim\mathbf{B} \vee \mathbf{C}] \vee \sim\mathbf{A}$	1108: .2
(.4) $\mathcal{H} \vdash \mathbf{B} \supset \centerdot \mathbf{C} \vee \sim\mathbf{A}$	1114: .3
(.5) $\mathcal{H} \vdash \mathbf{A} \supset \centerdot \mathbf{C} \vee \sim\mathbf{A}$	Trans: .1, .4
(.6) $\mathcal{H} \vdash \mathbf{C} \vee \sim\mathbf{A} \vee \sim\mathbf{A}$	1108: .5
(.7) $\mathcal{H} \vdash \mathbf{C} \vee \centerdot \sim\mathbf{A} \vee \sim\mathbf{A}$	1114: .6
(.8) $\mathcal{H} \vdash \sim\mathbf{A} \vee \mathbf{C}$	1102: Axiom 1, .7

■

1116. Deduction Theorem. If $\mathcal{H}$, $\mathbf{A} \vdash \mathbf{B}$, then $\mathcal{H} \vdash \mathbf{A} \supset \mathbf{B}$.

Proof: Let $\mathbf{C}_1, \ldots, \mathbf{C}_m$ be a proof of $\mathbf{B}$ from the hypotheses $\mathcal{H}$, $\mathbf{A}$. We prove $\mathcal{H} \vdash \mathbf{A} \supset \mathbf{C}_i$ for all i $(1 \leq i \leq m)$ by complete induction on i.

Case 1. $\mathbf{C}_i$ is an axiom or a member of $\mathcal{H}$.

(.11) $\mathcal{H} \vdash \mathbf{C}_i$	$\mathbf{C}_i$ is an axiom or hypothesis
(.12) $\mathcal{H} \vdash \mathbf{C}_i \supset \centerdot \mathbf{A} \supset \mathbf{C}_i$	Axiom 2
(.13) $\mathcal{H} \vdash \mathbf{A} \supset \mathbf{C}_i$	MP: .11, .12

Case 2. $\mathbf{C}_i$ is $\mathbf{A}$.

(.21) $\vdash \mathbf{A} \supset \mathbf{C}_i$	Sub: 1106
(.22) $\mathcal{H} \vdash \mathbf{A} \supset \mathbf{C}_i$	1100: .21

Case 3. In the proof from $\mathcal{H} \cup \{ \mathbf{A}\}$, $\mathbf{C}_i$ is inferred by Modus Ponens from $\mathbf{C}_j$ and $\mathbf{C}_k$, where $j < i$, $k < i$, and $\mathbf{C}_k$ is $[\mathbf{C}_j \supset \mathbf{C}_i]$.

(.31) $\mathcal{H} \vdash \mathbf{A} \supset \mathbf{C}_j$	by inductive hypothesis
(.32) $\mathcal{H} \vdash \mathbf{A} \supset \centerdot \mathbf{C}_j \supset \mathbf{C}_i$	by inductive hypothesis
(.33) $\mathcal{H} \vdash \mathbf{A} \supset \mathbf{C}_i$	1115: .31,.32

Since by the definition of a proof from hypotheses $\mathbf{C}_i$ must fall under one of these cases, the inductive proof is complete. Thus $\mathcal{H} \vdash \mathbf{A} \supset \mathbf{C}_m$, and $\mathbf{C}_m$ is $\mathbf{B}$, so $\mathcal{H} \vdash \mathbf{A} \supset \mathbf{B}$. ■

1117. $\vdash \sim p \supset \centerdot \sim q \supset \sim \centerdot p \vee q$

Proof:

(.1) $\vdash p \supset \centerdot \sim\sim p \vee \sim\sim q$	1112: 1104
(.2) $\vdash q \supset \centerdot \sim\sim p \vee \sim\sim q$	1112: 1104
(.3) $\vdash p \vee q \supset \centerdot \sim\sim p \vee \sim\sim q$	1110: .1, .2
(.4) $\sim\sim p \vee \sim\sim q \vee \sim \centerdot p \vee q$	1108: .3
(.5) $\sim\sim p \vee \centerdot \sim\sim q \vee \sim \centerdot p \vee q$	1114: .4

■

1118. $\vdash p \supset \centerdot \sim p \supset q$ 1112: 1104

■

DEFINITION. Let $\mathcal{R}$ be any property of wffs. We say that a rule of inference *preserves* property $\mathcal{R}$ iff a wff inferred by the rule from wffs which have the property must also have the property.

1119. Principle of Induction on Proofs. Let $\mathcal{L}$ be any logistic system, and let $\mathcal{R}$ be any property of wffs of $\mathcal{L}$. If every axiom of $\mathcal{L}$ has property $\mathcal{R}$, and each rule of inference of $\mathcal{L}$ preserves property $\mathcal{R}$, then every theorem of $\mathcal{L}$ has property $\mathcal{R}$.

Proof: Let $\mathbf{C}_1, \ldots, \mathbf{C}_m$ be a proof of $\mathbf{B}$ in $\mathcal{L}$. It is easy to establish that $\mathbf{C}_i$ has property $\mathcal{R}$ for all i $(1 \leq i \leq m)$ by complete induction on i. Either $\mathbf{C}_i$ is an axiom, and therefore has property $\mathcal{R}$, or $\mathbf{C}_i$ is inferred by a rule of inference from preceding wffs in the proof. In the latter case the preceding wffs have property $\mathcal{R}$ by inductive hypothesis, so $\mathbf{C}_i$ has the property since the rules of inference preserve the property. In particular, $\mathbf{C}_m$, which is $\mathbf{B}$, must have property $\mathcal{R}$. ■

EXERCISES

X1100. Show that there is an effective test for determining whether an arbitrary finite sequence of formulas is a proof.

X1101. Show that in any system of logic in which 1108 and 1114 are primitive or derived rules of inference, from $\mathbf{A} \vee \centerdot \mathbf{B} \vee \mathbf{C}$ one may infer $[\mathbf{A} \vee \mathbf{B}] \vee \mathbf{C}$.

X1102. Let $\mathcal{L}$ be a formulation of propositional calculus in which the sole connectives are negation and disjunction, the sole rule of inference is Modus

Ponens, and with the following axiom schema:

$$\sim [\mathbf{A} \vee \mathbf{B}] \vee [\mathbf{B} \vee \mathbf{C}]$$

Show that every wff of $\mathcal{L}$ is a theorem of $\mathcal{L}$. (Such a logistic system is said to be *absolutely inconsistent.*) (*Hint:* You should think of this as a purely syntactic problem involving symbol manipulation, since $\mathcal{L}$ obviously does not make sense when the connectives are interpreted in the usual way. It may be helpful to prove some derived rules of inference for $\mathcal{L}$.)

X1103. Let $\mathcal{M}$ be the system which has the same wffs and rule of inference as $\mathcal{P}$, and the single axiom schema

$$[\mathbf{A} \supset \mathbf{B} \supset \centerdot \mathbf{C} \vee \centerdot \mathbf{D} \vee \mathbf{E}] \supset \centerdot \mathbf{D} \supset \mathbf{A} \supset \centerdot \mathbf{C} \vee \centerdot \mathbf{E} \vee \mathbf{A}.$$

Show that each theorem of $\mathcal{P}$ is a theorem of $\mathcal{M}$. (For help see [Meredith, 1953].)

§12. Semantics, Consistency, and Completeness of $\mathcal{P}$

In §§10 – 11 we have been concerned with the syntax of $\mathcal{P}$. In this section we shall define the semantics of $\mathcal{P}$ in a mathematically precise way, and discuss some important aspects of the relation between the syntax and the semantics of $\mathcal{P}$.

We said earlier that the propositional variables of $\mathcal{P}$ would play the role of statements. However, in any particular context, a statement is either true or false, so in defining the formal semantics of $\mathcal{P}$ we shall simply assign truth values to these variables.

DEFINITION. An *assignment* of truth values to propositional variables is a function from the set of variables to the set {T,F} of truth values. (We use T as an abbreviation for Truth, and F for Falsehood.) Occasionally we shall also consider assignments which have as domain some subset of the set of all variables.

Once one has specified the values of the variables in a wff, the wff itself becomes either true or false, and we say that its *value* is truth or falsehood. This is made precise in the following definition.

DEFINITION. The *value* $\mathcal{V}_\varphi\mathbf{A}$ of a wff $\mathbf{A}$ with respect to the assignment φ is defined as follows by induction on the construction of $\mathbf{A}$:

(1) $\mathcal{V}_\varphi \mathbf{p} = \varphi \mathbf{p}$

$$\begin{aligned}(2)\quad \mathcal{V}_\varphi \sim\mathbf{B} &= \mathsf{T} \text{ if } \mathcal{V}_\varphi\mathbf{B} = \mathsf{F} \\ &= \mathsf{F} \text{ if } \mathcal{V}_\varphi\mathbf{B} = \mathsf{T}\\ (3)\quad \mathcal{V}_\varphi [\mathbf{B} \lor \mathbf{C}] &= \mathsf{T} \text{ if } \mathcal{V}_\varphi\mathbf{B} = \mathsf{T} \text{ or } \mathcal{V}_\varphi\mathbf{C} = \mathsf{T}\\ &= \mathsf{F} \text{ if } \mathcal{V}_\varphi\mathbf{B} = \mathsf{F} \text{ and } \mathcal{V}_\varphi\mathbf{C} = \mathsf{F}.\end{aligned}$$

We say that $\mathbf{A}$ is *true* with respect to φ if $\mathcal{V}_\varphi\mathbf{A} = \mathsf{T}$, and *false* with respect to φ if $\mathcal{V}_\varphi\mathbf{A} = \mathsf{F}$.

Since we shall have many occasions to use the definition of $\mathcal{V}_\varphi$ in proofs, it is convenient to extend our meta-language so that the definition can be expressed more concisely. To this end, let us introduce $\sim$ and $\lor$ into our meta-language as operators on truth values defined as follows: $\sim \mathsf{T} = \mathsf{F}$; $\sim \mathsf{F} = \mathsf{T}$; $\mathsf{T} \lor \mathsf{T} = \mathsf{T}$; $\mathsf{T} \lor \mathsf{F} = \mathsf{T}$; $\mathsf{F} \lor \mathsf{T} = \mathsf{T}$; $\mathsf{F} \lor \mathsf{F} = \mathsf{F}$. Thus, as symbols of the meta-language $\lor$ and $\sim$ now have the same meaning as we have just given them in the object language.

Of course, we thus introduce some ambiguity into the meta-language, since $\sim$ and $\lor$ also serve in the meta-language as names for themselves as symbols of the object language. This ambiguity could be obviated by using different symbols in the meta-language, or by coloring symbols a different color when they are used as symbols of the meta-language, and we leave it to the reader to mentally color the symbols of this text in an appropriate way.

Using the notation just introduced, we can express the inductive definition of $\mathcal{V}_\varphi$ as follows:

(1) $\mathcal{V}_\varphi \mathbf{p} = \varphi \mathbf{p}$
(2) $\mathcal{V}_\varphi \sim \mathbf{B} = \sim \mathcal{V}_\varphi\mathbf{B}$
(3) $\mathcal{V}_\varphi [\mathbf{B} \lor \mathbf{C}] = (\mathcal{V}_\varphi\mathbf{B}) \lor (\mathcal{V}_\varphi\mathbf{C})$.

The reader should take care to note how each occurrence of $\sim$ or $\lor$ is being used here, and provide appropriate mental colorings of these statements.

DEFINITIONS.

(1) $\mathbf{A}$ is a *tautology* iff $\mathcal{V}_\varphi\mathbf{A} = \mathsf{T}$ for all assignments φ. We write $\models \mathbf{A}$ to indicate that $\mathbf{A}$ is a tautology.
(2) $\mathbf{A}$ is a *contradiction* iff $\mathcal{V}_\varphi\mathbf{A} = \mathsf{F}$ for all assignments φ.
(3) $\mathbf{A}$ *implies* $\mathbf{B}$ (by propositional calculus) iff $[\mathbf{A} \supset \mathbf{B}]$ is a tautology.
(4) $\mathbf{A}$ *is equivalent to* $\mathbf{B}$ (by propositional calculus) iff $[\mathbf{A} \equiv \mathbf{B}]$ is a tautology. Thus, $\mathbf{A}$ is equivalent to $\mathbf{B}$ iff $\mathcal{V}_\varphi\mathbf{A} = \mathcal{V}_\varphi\mathbf{B}$ for all assignments φ.

p	q	$p \wedge q$	$\supset$	$p \vee q$
T	T	T	T	T
T	F	F	T	T
F	T	F	T	T
F	F	F	T	F

Figure 1.8: A simple tautology

(5) An assignment φ *falsifies* **A** iff $\mathcal{V}_\varphi\mathbf{A} = \mathsf{F}$.

(6) An assignment φ *satisfies* **A** iff $\mathcal{V}_\varphi\mathbf{A} = \mathsf{T}$. We write $\models_\varphi \mathbf{A}$ to indicate that φ satisfies **A**. An alternative notation for this is $\models \mathbf{A}[\varphi]$.

(7) An assignment φ *satisfies* a set $\mathcal{S}$ of wffs iff φ satisfies every wff of $\mathcal{S}$. (Every assignment satisfies the empty set of wffs.)

(8) A set $\mathcal{S}$ of wffs is *satisfiable* iff there is an assignment which satisfies it. The set is *unsatisfiable* or *contradictory* iff it is not satisfiable.

In a truth table, each horizontal line corresponds to some assignment. Thus a wff is a tautology iff the column under its main connective contains only T'S (as in Figure 1.8, for example).

It is easy to see that if n propositional variables occur in a wff A, then there are 2^n possible assignments of values to the variables of **A**, since each variable can be assigned either of two values. Wffs with many variables have very large truth tables, so it is nice to be able to show that a wff is a tautology without writing out the entire truth table. A method which enables one to do this for some wffs is illustrated in Figure 1.9. One wishes to investigate whether $p \wedge [p \supset q] \supset q \vee r$ is a tautology, so one supposes (1) that there is an assignment with respect to which it is false. One then considers what the values of various wf parts of the given wff must be with respect to this assignment. Since $[\mathbf{A} \supset \mathbf{B}]$ can be false only when **A** is true and **B** is false, (2) $p \wedge [p \supset q]$ must be true, while (3) $q \vee r$ must be false. Since $[\mathbf{A} \wedge \mathbf{B}]$ is true only if **A** and **B** are both true, (4) p must be true, and (5) $p \supset q$ must be true, so (6) q must be true. This makes $q \vee r$ true (regardless of the value of r), which contradicts (3). This contradiction shows that there is no assignment which falsifies the wff, so it must be a tautology.

Of course, when one applies the method above to certain examples, one may eventually reach a point where one must consider several possible cases, but one still avoids writing out the entire truth table. If one applies the method to a wff which is actually not a tautology, one will obtain an assignment which falsifies the wff, and thus see that it is not a tautology.

DEFINITION. We say that a wff of a logistic system is *valid* with respect to an interpretation iff the value of that wff is truth (under that interpretation) for all assignments of values to its (free) [6] variables. We shall write $\models \mathbf{A}$ to indicate that $\mathbf{A}$ is valid. An interpretation of a logistic system is *sound* iff, under that interpretation, the axioms are all valid and the rules of inference preserve validity. Clearly, in this case the theorems must all be valid.

Note that a wff of $\mathcal{P}$ is a tautology iff it is valid under the interpretation which is implicit in the definition of the valuation function $\mathcal{V}_\varphi$.

1200 Soundness Theorem. Every theorem of $\mathcal{P}$ is a tautology.

Proof: It is easy to check that each axiom is a tautology. Also, if $\mathcal{V}_\varphi\mathbf{A} = \mathsf{T}$ and $\mathcal{V}_\varphi[\sim \mathbf{A} \lor \mathbf{B}] = \mathsf{T}$ then $\mathcal{V}_\varphi\mathbf{B} = \mathsf{T}$. Hence if $\mathbf{A}$ and $[\sim\mathbf{A} \lor \mathbf{B}]$ are tautologies, $\mathbf{B}$ must also be a tautology. Therefore we see by 1119 that every theorem is a tautology. ∎

If every wff is a theorem of a particular logistic system, then a wff gains no significance by being a theorem of that system. Such systems are called *inconsistent*. Actually, this notion can be defined in several ways.

DEFINITION. A logistic system is *absolutely inconsistent* iff every wff is a theorem. A logistic system is *inconsistent with respect to negation* iff there is a wff $\mathbf{A}$ such that both $\mathbf{A}$ and the negation of $\mathbf{A}$ are theorems. A logistic system is *absolutely consistent* iff it is not absolutely inconsistent, and it is *consistent with repect to negation* iff it is not inconsistent with respect to negation. A set $\mathcal{H}$ of wffs is *consistent* iff there is a wff $\mathbf{A}$ such that not $\mathcal{H} \vdash \mathbf{A}$.

[6] In propositional calculus, all occurrences of variables are free. We insert the qualifying adjective "free" here so that this definition will be applicable to systems which we shall consider later.

p	$\land$	$[p$	$\supset$	$q]$	$\supset$	q	$\lor$	r
T	T	T	T	T	F	T	F	
4	2	4	5	6	1	6	3	
							x	

Figure 1.9: A short method of verifying a tautology

By using 1118 and 1101 it is easy to see that every wff of $\mathcal{P}$ of the form $[\mathbf{A} \supset \centerdot \sim \mathbf{A} \supset \mathbf{B}]$ is a theorem of $\mathcal{P}$. The expression $[\mathbf{A} \supset \centerdot \sim \mathbf{A} \supset \mathbf{B}]$, which contains syntactical variables and belongs to the meta-language, is called a *theorem schema* of $\mathcal{P}$.

1201 Theorem. Let $\mathcal{S}$ be any logistic system in which $\mathbf{A} \supset \centerdot \sim \mathbf{A} \supset \mathbf{B}$ is a theorem schema and in which Modus Ponens is a primitive or derived rule of inference. Then $\mathcal{S}$ is absolutely consistent iff $\mathcal{S}$ is consistent with respect to negation.

Proof: If $\mathcal{S}$ is inconsistent with respect to negation there is a wff $\mathbf{A}$ such that $\vdash \mathbf{A}$ and $\vdash \sim \mathbf{A}$. Hence by two applications of Modus Ponens to $[\mathbf{A} \supset \centerdot \sim \mathbf{A} \supset \mathbf{B}]$ we obtain $\vdash \mathbf{B}$ for any wff $\mathbf{B}$, so $\mathcal{S}$ is absolutely inconsistent. Conversely, if $\mathcal{S}$ is absolutely inconsistent, then $\vdash \mathbf{A}$ and $\vdash \sim \mathbf{A}$ for every wff, so $\mathcal{S}$ is inconsistent with respect to negation.

1202 Consistency Theorem. $\mathcal{P}$ is absolutely consistent and consistent with respect to negation.

Proof: If $\mathbf{A}$ is any theorem of $\mathcal{P}$, then $\mathbf{A}$ is a tautology, so $\sim\mathbf{A}$ is not a tautology, so $\sim\mathbf{A}$ is not a theorem. Thus $\mathcal{P}$ is consistent with respect to negation, and hence, by 1118, 1101, and 1201, absolutely consistent. ∎

Next we prepare to prove the converse of 1200.

1203 Lemma. Let $\mathbf{A}$ be a wff and let all the variables in $\mathbf{A}$ be among $\mathbf{p}_1, \ldots, \mathbf{p}_n$. Let φ be any assignment. For any wff $\mathbf{B}$, let

$$\begin{aligned} \mathbf{B}^{\varphi} &= \ \mathbf{B} \text{ if } \mathcal{V}_{\varphi}\mathbf{B} = \mathsf{T} \\ &= \sim\mathbf{B} \text{ if } \mathcal{V}_{\varphi}\mathbf{B} = \mathsf{F}. \end{aligned}$$

Then $\mathbf{p}_1^{\varphi}, \ldots, \mathbf{p}_n^{\varphi} \vdash \mathbf{A}^{\varphi}$.

EXAMPLE: $p, \sim q, r \vdash \sim \centerdot p \supset q \wedge r$

Proof: The proof is by induction on the construction of $\mathbf{A}$.

Case 1. $\mathbf{A}$ is $\mathbf{p}_i$. Then $\mathbf{p}_1^{\varphi}, \ldots, \mathbf{p}_n^{\varphi} \vdash \mathbf{p}_i^{\varphi}$.

Case 2a. $\mathbf{A}$ is $\sim \mathbf{B}$ and $\mathcal{V}_{\varphi}\mathbf{B} = \mathsf{T}$.

$\mathbf{p}_1^{\varphi}, \ldots, \mathbf{p}_n^{\varphi} \vdash \mathbf{B}$ by inductive hypothesis
$\mathbf{p}_1^{\varphi}, \ldots, \mathbf{p}_n^{\varphi} \vdash \mathbf{B} \supset \sim\sim \mathbf{B}$ Sub: 1104

$\mathbf{p}_1^\varphi, \ldots, \mathbf{p}_n^\varphi \vdash \sim\sim \mathbf{B}$ **MP**

Since $\mathbf{A}^\varphi$ is $\sim\sim\mathbf{B}$, this is the desired conclusion.

Case 2b. $\mathbf{A}$ is $\sim \mathbf{B}$ and $\mathcal{V}_\varphi \mathbf{B} = \mathsf{F}$.

$\mathbf{p}_1^\varphi, \ldots, \mathbf{p}_n^\varphi \vdash \sim \mathbf{B}$ by inductive hypothesis

But $\mathbf{A}^\varphi$ is $\sim \mathbf{B}$, so this is the desired conclusion.

Case 3a. $\mathbf{A}$ is $[\mathbf{B} \vee \mathbf{C}]$ and $\mathcal{V}_\varphi\ \mathbf{B} = \mathsf{T}$.

$\mathbf{p}_1^\varphi, \ldots, \mathbf{p}_n^\varphi \vdash \mathbf{B}$ by inductive hypothesis

$\mathbf{p}_1^\varphi, \ldots, \mathbf{p}_n^\varphi \vdash \mathbf{B} \supset \blacksquare\, \mathbf{B} \vee \mathbf{C}$ 1112: 1106

$\mathbf{p}_1^\varphi, \ldots, \mathbf{p}_n^\varphi \vdash \mathbf{B} \vee \mathbf{C}$ MP

Case 3b. $\mathbf{A}$ is $[\mathbf{B} \vee \mathbf{C}]$ and $\mathcal{V}_\varphi\ \mathbf{C} = \mathsf{T}$.

The proof proceeds as in Case 3a.

Case 3c. $\mathbf{A}$ is $[\mathbf{B} \vee \mathbf{C}]$ and $\mathcal{V}_\varphi\ \mathbf{B} = \mathsf{F}$ and $\mathcal{V}_\varphi \mathbf{C} = \mathsf{F}$.

$\mathbf{p}_1^\varphi, \ldots, \mathbf{p}_n^\varphi \vdash \sim \mathbf{B}$ by inductive hypothesis

$\mathbf{p}_1^\varphi, \ldots, \mathbf{p}_n^\varphi \vdash \sim \mathbf{C}$ by inductive hypothesis

$\mathbf{p}_1^\varphi, \ldots, \mathbf{p}_n^\varphi \vdash \sim \mathbf{B} \supset \blacksquare \sim \mathbf{C} \supset \sim \blacksquare\, \mathbf{B} \vee \mathbf{C}$ Sub: 1117

$\mathbf{p}_1^\varphi, \ldots, \mathbf{p}_n^\varphi \vdash \sim \blacksquare\, \mathbf{B} \vee \mathbf{C}$ MP

■

1204 Completeness Theorem. Every tautology is a theorem of $\mathcal{P}$.

Proof: Let $\mathbf{A}$ be a tautology and let n be the number of variables which occur in $\mathbf{A}$. Let $\{\mathbf{p}_1, \ldots, \mathbf{p}_j\}$ (where $0 \leq j \leq n$) be a set of distinct variables which occur in $\mathbf{A}$. We prove that $\mathbf{p}_1^\varphi, \ldots, \mathbf{p}_j^\varphi \vdash \mathbf{A}$ for any assignment φ by induction on $n - j$.

For $n - j = 0$ this follows from 1203.

For the induction step, we consider assignments φ' and φ'' such that $\varphi'\mathbf{p}_i = \varphi\mathbf{p}_i = \varphi''\mathbf{p}_i$ for $i \leq j$ and $\varphi'\mathbf{p}_{j+1} = \mathsf{T}$ and $\varphi''\mathbf{p}_{j+1} = \mathsf{F}$ to see that:

(.1) $\mathbf{p}_1^\varphi, \ldots, \mathbf{p}_j^\varphi, \mathbf{p}_{j+1} \vdash \mathbf{A}$ by inductive hypothesis (using φ')

(.2) $\mathbf{p}_1^\varphi, \ldots, \mathbf{p}_j^\varphi, \sim \mathbf{p}_{j+1} \vdash \mathbf{A}$ by inductive hypothesis (using φ'')

(.3) $\mathbf{p}_1^\varphi, \ldots, \mathbf{p}_j^\varphi \vdash \mathbf{p}_{j+1} \supset \mathbf{A}$ Deduction Theorem (1116): .1

(.4) $\mathbf{p}_1^\varphi, \ldots, \mathbf{p}_j^\varphi \vdash \sim \mathbf{p}_{j+1} \supset \mathbf{A}$ Deduction Theorem (1116): .2

(.5) $\mathbf{p}_1^\varphi, \ldots, \mathbf{p}_j^\varphi \vdash \mathbf{A}$ 1111: .3, .4

■

1205 Theorem. A wff is a theorem of $\mathcal{P}$ if and only if it is a tautology.

Proof: By 1200 and 1204. ∎

The *decision problem* for a logistic system is the problem of finding an effective procedure for determining of any wff **A** whether or not **A** is a theorem, and, if so, constructing a proof of **A**. Theorem 1205 provides a solution of the decision problem for $\mathcal{P}$, since one can effectively determine whether or not a wff is a tautology. Moreover, implicit in the proof of 1204 is a method for actually constructing a proof of any wff which is a tautology.

In showing that the theorems of $\mathcal{P}$ are precisely the tautologies, we have established an important connection between the syntax and the semantics of $\mathcal{P}$. The set of theorems is characterized in two radically different ways, one syntactic (as the set of wffs which have proofs) and one semantic (as the set of tautologies). Note that the decision problem is stated in purely syntactic terms, but that a solution for it which relied solely on syntactic arguments about proofs might be very difficult. The introduction of semantic concepts greatly facilitates a solution to this syntactic problem.

EXERCISES

In exercises X1200–X1208, decide whether the wff is a tautology, a contradiction, or neither. If it is neither, give a falsifying assignment and a satisfying assignment. Use efficient methods.

X1200. $[\sim [p \wedge \sim q] \wedge [\sim r \vee s]] \supset \centerdot [[s \supset q] \wedge \sim p] \supset \sim r$

X1201. $[r \vee \centerdot q \wedge \sim \centerdot p \supset s] \supset \centerdot \sim [s \supset r] \supset \centerdot q \supset p$

X1202. $p \equiv [p \vee p]$

X1203. $[p \supset q] \equiv \centerdot [\sim p \vee q] \wedge \sim \centerdot p \wedge \sim q$

X1204. $[p \supset \centerdot q \vee r] \supset \sim \centerdot [q \wedge \centerdot p \vee r] \supset \centerdot \sim r \supset \sim p$

X1205. $\sim [p \vee q] \equiv \centerdot \sim p \wedge \sim q$

X1206. $p \supset q \supset \sim \centerdot \sim q \supset \sim p$

X1207. $p \supset q \supset \centerdot q \supset [p \vee r] \supset \centerdot [p \vee r] \equiv q$

X1208. $p \wedge [q \vee r] \equiv \centerdot [p \wedge q] \vee [p \wedge r]$

X1209. If **C** and **D** are wffs, we say that **D** is obtained from **C** *by identifying* *certain propositional variables* if $\mathbf{D}$ is of the form $\mathsf{S}^{\mathbf{p}_1 \ldots \mathbf{p}_n}_{\mathbf{q}_1 \ldots \mathbf{q}_n}\mathbf{C}$, where $\mathbf{p}_1, \ldots, \mathbf{p}_n$ are distinct propositional variables of $\mathbf{C}$ and $\mathbf{q}_1, \ldots, \mathbf{q}_n$ are propositional variables of $\mathbf{C}$ which are distinct from $\mathbf{p}_1, \ldots,$ and $\mathbf{p}_n$. Prove that if $\mathbf{C}$ is a tautology, and $\mathbf{D}$ is obtained from $\mathbf{C}$ by identifying certain propositional variables, then $\mathbf{D}$ is a tautology.

Justify your answers to the next two exercises with very rigorous proofs, using induction whenever the opportunity arises.

X1210. Does $\mathcal{P}$ have any theorems in which there are no occurrences of disjunction?

X1211. Does $\mathcal{P}$ have any theorems in which there are no occurrences of negation?

X1212. Let $\mathcal{P}'$ be the system obtained from $\mathcal{P}$ by adding the single wff $[p \supset q]$ to the axioms of $\mathcal{P}$. Show that $\mathcal{P}'$ is consistent.

X1213. Suppose we add to $\mathcal{P}$ the axiom schema $[\mathbf{A} \supset \mathbf{B}]$. Is the resulting system consistent?

X1214. Let $\mathcal{P}_0$ be the logistic system which has the same wffs and rule of inference as $\mathcal{P}$, and the following axiom schemas:

(1) $[\mathbf{A} \wedge \sim \mathbf{A}\,] \supset \mathbf{B}$
(2) $[\mathbf{A} \supset \mathbf{B}\,] \supset \mathbf{.B} \supset \mathbf{A}$

Is $\mathcal{P}_0$ consistent?

X1215. Discuss the consistency and completeness of the following logistic system:

The wffs are those of $\mathcal{P}$. The axioms are all wffs of the form $[\mathbf{A} \supset \mathsf{S}^{\mathbf{p}}_{\mathbf{B}}\mathbf{A}]$, where $\mathbf{A}$ and $\mathbf{B}$ are wffs and $\mathbf{p}$ is any propositional variable. The sole rule of inference is Modus Ponens.

X1216. Let $\mathcal{R}$ be a system of propositional calculus with a single binary connective $*$. Thus the wffs of $\mathcal{R}$ are defined inductively as follows:

(a) Every propositional variable is a wff.
(b) If $\mathbf{A}$ and $\mathbf{B}$ are wffs, so is $[\mathbf{A} * \mathbf{B}]$.

The axioms of $\mathcal{R}$ are all wffs of the form $[\mathbf{A} * \mathbf{A}]$. The sole rule of inference is: from $\mathbf{A}$ and $[\mathbf{B} * \mathbf{A}]$ to infer $\mathbf{B}$. Solve the decision problem for $\mathcal{R}$, and prove that your solution is correct.

X1217. Let $\mathcal{Q}$ be the system of propositional calculus whose sole primitive connectives are $\sim$ (negation) and $\wedge$ (conjunction), and which has the following rule of inference and axiom schemata:
Rule of Inference. From **A** and $\sim$ **A** $\wedge$ **B** to infer **B**.
Axiom Schemata.

(1) $\sim [\mathbf{A} \wedge \mathbf{A}] \wedge \mathbf{A}$
(2) $\sim \mathbf{A} \wedge \ {}_{\blacksquare}\mathbf{B} \wedge \mathbf{A}$
(3) $\sim [\sim \mathbf{A} \wedge \mathbf{B}] \wedge {}_{\blacksquare}{\sim} [\mathbf{C} \wedge \mathbf{A}] \wedge {}_{\blacksquare}\mathbf{B} \wedge \mathbf{C}$

Prove that a wff of $\mathcal{Q}$ is a theorem of $\mathcal{Q}$ if and only if it is a contradiction.

X1218. If $\mathcal{Q}$ is any absolutely consistent system of propositional calculus, are the theorems of $\mathcal{Q}$ necessarily all tautologies? Give a complete justification for your answer.

X1219. Let $\mathcal{P}_1$ be a logistic system with the same primitive symbols, wffs, abbreviations and rule of inference as $\mathcal{P}$. The two axiom schemata for $\mathcal{P}_1$ are:
$[[\mathbf{A} \supset \mathbf{B}] \supset \mathbf{A}] \supset \mathbf{A}$
$[\mathbf{A} \vee [\mathbf{B} \wedge \mathbf{C}]] \supset {}_{\blacksquare}\mathbf{C} \vee \mathbf{A}$
Is every theorem of $\mathcal{P}_1$ a theorem of $\mathcal{P}$?

X1220. Let $\mathcal{P}_2$ be obtained from $\mathcal{P}$ by adding to $\mathcal{P}$ as axioms all wffs of the form $\sim\mathbf{p}$ (where **p** is a propositional variable).

(a) Is $\mathcal{P}_2$ consistent?
(b) Solve the decision problem for $\mathcal{P}_2$.

X1221. Prove that if **A** is any wff of $\mathcal{P}$ and $\varphi\mathbf{p} = \psi\mathbf{p}$ for each variable **p** which occurs in **A**, then $\mathcal{V}_\varphi\mathbf{A} = \mathcal{V}_\psi\mathbf{A}$.

X1222. For any wff **A** (written in primitive notation without abbreviations) of $\mathcal{P}$, let $\mathbf{A}'$ (which is called the *dual* of **A**) be the wff whose abbreviation is obtained when the symbol $\vee$ is replaced by $\wedge$ everywhere in **A**. (Thus if **A** is $\sim p \vee q$, then $\mathbf{A}'$ is $\sim p \wedge q$, i.e., $\sim [\sim \sim p \vee \sim q]$.) You may use abbreviations freely in your work on this problem.

(a) Give a definition of $\mathbf{A}'$ by induction on the construction of **A**.

If φ is any assignment of truth values to propositional variables, let φ' be the "opposite" assignment, i.e., $\varphi'\mathbf{p} = \mathsf{F}$ if $\varphi\mathbf{p} = \mathsf{T}$, and $\varphi'\mathbf{p} = \mathsf{T}$ if $\varphi\mathbf{p} = \mathsf{F}$.

By using $\sim$ in the meta-language to denote the negation operation on truth values, we may concisely define φ' thus: $\varphi'\mathbf{p} = \sim \varphi\mathbf{p}$ for every propositional variable $\mathbf{p}$.

(b) Compute $\mathcal{V}_{\varphi'} \sim \mathbf{A}$ and $\mathcal{V}_{\varphi}\ \mathbf{A}$ when $\mathbf{A}$ is $[\sim [p \vee \sim q] \vee p]$ and $\varphi p = \mathsf{F}$ and $\varphi q = \mathsf{T}$.

(c) Give a rigorous proof that if $\mathbf{A}$ is any wff of $\mathcal{P}$ and φ is any assignment, then $\mathcal{V}_{\varphi'} \sim \mathbf{A}' = \mathcal{V}_{\varphi}\ \mathbf{A}$.

X1223. Let $\mathcal{K}$ be the system which has the same wffs and rule of inference as $\mathcal{P}$, and the following axiom schemata:

1. $\mathbf{A} \supset \mathbf{A}$.
2. $\mathbf{A} \supset \mathbf{B} \supset . \sim \mathbf{B} \supset \sim \mathbf{A}$

Is the wff $[p \supset q] \supset . \sim p \supset \sim q$ a theorem of $\mathcal{K}$?

X1224. Let $\mathcal{P}^*$ be the system obtained from $\mathcal{P}$ by adding the single wff $[p \supset \sim p]$ (which we shall call Axiom 4) to the axioms of $\mathcal{P}$. (In this problem $\vdash A$ means A is provable in $\mathcal{P}^*$.) Consider the following arguments:

Argument A: $\mathcal{P}^*$ is consistent with respect to negation. For it is easy to see by induction on proofs that every theorem of $\mathcal{P}^*$ is satisfied by the particular assignment φ which gives the value falsehood to every propositional variable. If $\vdash \mathbf{A}$ and $\vdash \sim \mathbf{A}$, then $V_{\varphi}\mathbf{A} = \mathsf{T}$ and $V_{\varphi} \sim \mathbf{A} = \mathsf{T}$, so $V_{\varphi}\mathbf{A} = \mathsf{F}$, which is a contradiction. Thus, there can be no such wff $\mathbf{A}$, and $\mathcal{P}^*$ is consistent with respect to negation.

Argument B: $\mathcal{P}^*$ is inconsistent with respect to negation.

(1) $\vdash p \supset \sim p$ — Axiom 4

(2) $\vdash [p \supset \sim p] \supset \sim [p \supset \sim p]$ — Sub: 1

(3) $\vdash \sim [p \supset \sim p]$ — MP: 1,2

Thus, there is a wff $\mathbf{A}$ such that (1) $\vdash \mathbf{A}$ and (3) $\vdash \sim \mathbf{A}$, so $\mathcal{P}^*$ is inconsistent with respect to negation.

Find a flaw in one of these arguments, and explain exactly what the problem is.

§13. Independence

An axiom, axiom schema, or rule of inference of a logistic system is said to be *independent* iff the set of theorems is diminished when it is deleted from the system. More precisely, an axiom of a logistic system is *independent* iff it cannot be derived from the other axioms by the rules of inference. An axiom schema is *independent* iff not every instance of it is a theorem of the system obtained by deleting from the set of axioms all instances of the schema in question. A rule of inference is *independent* if there is a theorem which cannot be derived without using that rule of inference.

We remark that in order to show the independence of an axiom, it suffices to find some property which that axiom does not have but which all the other axioms of the system do have, and which is preserved by the rules of inference. For in this case all theorems derived without using the axiom in question must have the property, and the axiom in question cannot be one of these theorems. It often happens that an appropriate property for this purpose is that of being valid under some particular (unsound) interpretation of the system.

1300 Theorem. The axiom schemata and the rule of inference of $\mathcal{P}$ are all independent.

Proof: It is trivial to see the Modus Ponens is independent, since it is the sole rule of inference. Simply note that $p \lor \sim p$ is a theorem (1103), but not an axiom.

To prove the idependence of the axiom schemata, we use three truth values, 0, 1, and 2. In each case we *designate* certain truth values, and call a wff *special* (under the given interpretation) iff its value (under that interpretation) is designated for all assignments of values from the set $\{0, 1, 2\}$ to its variables.

For Axiom Schema 1, we use the interpretation of $\lor$ and $\sim$ expressed in Figure 1.10. (Note that here $p \lor q = p \times q$ mod 4.) Here 0 is designated. It is easy to check that all instances of Axiom Schemata 2 and 3 are special. (To simplify the analysis, use the fact that if **A** or **B** is 0, then $[\mathbf{A} \lor \mathbf{B}]$ is 0 also.) Also, if **A** is 0 and $\sim \mathbf{A} \lor \mathbf{B}$ is 0, then **B** must be 0, so Modus Ponens preserves specialness. However $\sim [p \lor p] \lor p$ is 2 when p is 2, so Axiom Schema 1 is independent.

To prove the independence of Axiom Schema 2, we let 0 and 1 be designated and use the interpretation of $\lor$ and $\sim$ in Figure 1.11. It is easily

$\lor$	0	1	2	$\sim$
0	0	0	0	1
1	0	1	2	0
2	0	2	0	2

Figure 1.10:
Table for Axiom 1

$\lor$	0	1	2	$\sim$
0	0	0	0	2
1	0	1	2	1
2	0	2	2	0

Figure 1.11:
Table for Axiom 2

$\lor$	0	1	2	$\sim$
0	0	0	0	1
1	0	1	1	2
2	0	0	2	0

Figure 1.12:
Table for Axiom 3

checked that all instances of Axiom Schemata 1 and 3 are special. If **A** is 0 or 1 then $\sim$ **A** is 2 or 1, so in order that [$\sim$ **A** $\lor$ **B**] be 0 or 1 **B** must be 0 or 1, so Modus Ponens preserves specialness. However, $\sim p \lor \centerdot q \lor p$ is 2 when p is 1 and q is 2, so Axiom Schema 2 is independent.

To prove the independence of Axiom Schema 3, we use Figure 1.12[7], with 0 designated. All instances of Axiom Schemata 1 and 2 are special, and if **A** is 0 and $\sim$ **A** $\lor$ **B** is 0, then **B** must be 0, so Modus Ponens preserves specialness. However $\mathcal{V}_\varphi[\sim [\sim p \lor q] \lor \centerdot \sim [r \lor p] \lor \centerdot q \lor r] = 1$ when $\varphi p = 1 = \varphi q$ and $\varphi r = 2$, so this instance of Axiom Schema 3 is not derivable from Axioms 1 and 2. ■

EXERCISES

X1300. Let $\mathcal{P}'$ be the logistic system with the same wffs as $\mathcal{P}$, with Modus Ponens and Substitution (i.e., 1101 for the case where $\mathcal{H}$ is empty) as primitive rules of inference, and with the following three axioms:

(1) $p \lor p \supset p$
(2) $p \supset \centerdot q \lor p$
(3) $p \supset q \supset \centerdot r \lor p \supset \centerdot q \lor r$

(a) Prove that $\mathcal{P}'$ has exactly the same theorems as $\mathcal{P}$.
(b) Prove the independence of the rules of inference of $\mathcal{P}'$.

X1301. Using the methods of the proofs above, show that if any of the three axiom schemata of $\mathcal{P}$ were replaced by the schema **A** $\lor \sim$ **A** or by $\sim$ **A** $\lor$ **A**, the resulting system would not be complete (i.e., would not have all tautologies as theorems).

[7]This table was found by Yuh-yun Marjorie Hsu.

X1302. Let $\mathcal{P}^*$ be the system of propositional calculus with primitive connectives $\sim$ and $\supset$, with Modus Ponens (from $\mathbf{A}$ and $[\mathbf{A} \supset \mathbf{B}]$ to infer $\mathbf{B}$) as sole role of inference, and the following axiom schemata:

$$\mathbf{A} \supset \centerdot \mathbf{B} \supset \mathbf{A}$$
$$\mathbf{A} \supset [\mathbf{B} \supset \mathbf{C}] \supset \centerdot \mathbf{A} \supset \mathbf{B} \supset \centerdot \mathbf{A} \supset \mathbf{C}$$
$$\sim\sim \mathbf{A} \supset \mathbf{A}.$$

Prove that $p \supset \sim\sim p$ is not derivable in $\mathcal{P}^*$. (*Hint*: consider interpretations with two truth values in which $\supset$ has the usual meaning, but $\sim$ denotes a truth function other than negation.)

§14. Propositional Connectives

We now leave the system $\mathcal{P}$ and turn to considerations which are relevant to a variety of formulations of propositional calculus. The reader who wishes to see some other formulations of propositional calculus may look in [Church, 1956], §§10, 25, 26, 29, and references cited therein.

An n-ary *truth-function* is a function from n-tuples of truth values to truth values. As a special case we regard truth values as 0-ary truth functions. A *propositional connective* is a symbol of a formalized language which, in the intended interpretation, denotes a truth function. Thus $\sim$, the negation connective, denotes the negation function, which maps truth to falsehood and falsehood to truth. Figure 1.13 lists all the truth functions of two arguments, together with connectives which denote them.

Various other names and notations are sometimes used for these connectives. Peirce's dagger is sometimes called "nor", and Sheffer's stroke is sometimes called "nand". It is common to use $\Rightarrow$ for implication and $\Leftrightarrow$ for equivalence, but we prefer to reserve these for informal use in our meta-language.

By a *wff of propositional calculus* we mean a wff of a formulation of propositional calculus which has among its primitive symbols all propositional connectives. Semantic notions such as tautology, $\mathcal{V}_\varphi\mathbf{A}$, etc., are defined for arbitrary wffs of propositional calculus in a manner analogous to that in §12. Some basic tautologies are summarized at the end of this section.

An operator (connective or function) is called *unary* or *monadic* if it takes one argument (like $\sim$ or cosine), *binary* if it takes two arguments (like $\supset$ or $+$), and *n-ary* if it takes n arguments. It is customary to write

	p	T	T	F	F
	q	T	F	T	F
Falsehood	f	F	F	F	F
Non-disjunction, Peirce's dagger	$p \downarrow q$	F	F	F	T
Converse non-implication	$p \not\subset q$	F	F	T	F
Negation	$\sim p$	F	F	T	T
Non-implication	$p \not\supset q$	F	T	F	F
Negation	$\sim q$	F	T	F	T
(Material) non-equivalence, exclusive disjunction	$p \not\equiv q$	F	T	T	F
Non-conjunction, Sheffer's stroke	$p \mid q$	F	T	T	T
Conjunction	$p \wedge q$	T	F	F	F
(Material) equivalence	$p \equiv q$	T	F	F	T
	q	T	F	T	F
(Material) implication	$p \supset q$	T	F	T	T
	p	T	T	F	F
Converse implication	$p \subset q$	T	T	F	T
(Inclusive) disjunction, alternation	$p \vee q$	T	T	T	F
Truth	t	T	T	T	T

Figure 1.13: Binary truth functions.

n-ary operators in *prefix* position *in front of* their arguments (as in $\sim p$ and cos π and $f(x_1, \ldots, x_n)$) except when $n = 2$. Binary operators are customarily written in *infix* position *between* their arguments (as in $[p \supset q]$ and $3 + 5$). However, in theoretical discussions it is sometimes convenient to pretend that all operators are written in prefix position, and write $* \, \mathbf{B}_1 \, \mathbf{B}_2$ as a shorthand for $[\mathbf{B}_1 * \mathbf{B}_2]$. Thus, every wff of propositional calculus may be regarded as having the form $\mathbf{p}$ or $*\mathbf{B}_1 \ldots \mathbf{B}_n$, where $*$ is an n-ary propositional connective. Notice that when one uses prefix notation and knows the number of arguments of each operator, one does not really need brackets. In prefix notation $\supset\supset pqr$ means $[[p \supset q] \supset r]$, while $\supset p \supset qr$ means $[p \supset [q \supset r]]$.[8]

Theorems 1401 and 1402 below present facts involving substitutions and tautologies which seem so obvious that they are sometimes used without justification. Theorem 1400 serves as a lemma for 1401.

[8]Prefix notation for binary propositional connectives is sometimes referred to as *Polish* notation, since it was actually used by certain Polish logicians.

DEFINITION. A *proper substitution* is a substitution of the form $\mathsf{S}^{\mathbf{p}_1 \dots \mathbf{p}_n}_{\mathbf{C}_1 \dots \mathbf{C}_n}$.

DEFINITION. If θ is a proper substitution, and φ is any assignment, the assignment $\varphi \circ \theta$ is defined as follows:

$$(\varphi \circ \theta)\, \mathbf{p} = \mathcal{V}_\varphi \theta \mathbf{p} \text{ for every propositional variable } \mathbf{p}.$$

EXAMPLE: If $\varphi q = \mathsf{T}$ and $\varphi r = \mathsf{F}$ and $\theta = \mathsf{S}^{\,p}_{[q \supset r]}$, then $(\varphi \circ \theta)p = \mathcal{V}_\varphi(\theta p) = \mathcal{V}_\varphi[q \supset r] = \mathsf{F}$.

1400 Substitution-Value Theorem. Let φ be any assignment and θ be any proper substitution. If $\mathbf{A}$ is any wff of propositional calculus, then $\mathcal{V}_\varphi \theta \mathbf{A} = \mathcal{V}_{\varphi\circ\theta} \mathbf{A}$.

EXAMPLE: Let φ and θ be as in the previous example. Note that $\theta q = q$ and $\theta r = r$ so $(\varphi \circ \theta)q = \mathsf{T}$ and $(\varphi \circ \theta)r = \mathsf{F}$. Let W1400 be $[[p \supset r] \wedge \blacksquare q \vee p]$.
Then $\mathcal{V}_\varphi \theta$ W1400 $= \mathcal{V}_\varphi$ $\begin{array}{cccccccccccccccccc} [[[& q & \supset & r &] & \supset & r &] & \wedge & \blacksquare q & \vee & \blacksquare q & \supset & r &] = \mathsf{T} \\ & \mathsf{T} & \mathsf{F} & \mathsf{F} & & \mathsf{T} & \mathsf{F} & & \mathsf{T} & \mathsf{T} & \mathsf{T} & \mathsf{T} & \mathsf{F} & \mathsf{F} & \end{array}$
and $\mathcal{V}_{\varphi\circ\theta}$ W1400 $= \mathcal{V}_{\varphi\circ\theta}$ $\begin{array}{ccccccccc} [[& p & \supset & r &] & \wedge & \blacksquare q & \vee & p &] = \mathsf{T} \\ & \mathsf{F} & \mathsf{T} & \mathsf{F} & & \mathsf{T} & \mathsf{T} & \mathsf{T} & \mathsf{F} & \end{array}$

Proof of 1400 *by induction on the construction of* $\mathbf{A}$:

Case: $\mathbf{A}$ is a propositional variable $\mathbf{p}$. $\mathcal{V}_\varphi \theta \mathbf{A} = \mathcal{V}_\varphi \theta \mathbf{p} = (\varphi \circ \theta)\mathbf{p}$ (by the definition of $\varphi \circ \theta$) $= \mathcal{V}_{\varphi\circ\theta}\mathbf{A}$

Case: $\mathbf{A}$ has the form $*\mathbf{B}_1 \dots \mathbf{B}_n$, where $*$ is an n-ary propositional connective denoting the truth function $\circledast$, with $n \geq 1$.

$$\begin{aligned}
\mathcal{V}_\varphi \theta \mathbf{A} &= \mathcal{V}_\varphi * (\theta \mathbf{B}_1) \dots (\theta \mathbf{B}_n) && \text{(by the definition of a substitution)} \\
&= \circledast\, (\mathcal{V}_\varphi \theta \mathbf{B}_1) \dots (\mathcal{V}_\varphi \theta \mathbf{B}_n) && \text{(by the definition of } \mathcal{V}_\varphi\text{)} \\
&= \circledast\, (\mathcal{V}_{\varphi\circ\theta} \mathbf{B}_1) \dots (\mathcal{V}_{\varphi\circ\theta} \mathbf{B}_n) && \text{(by inductive hypothesis)} \\
&= \mathcal{V}_{\varphi\circ\theta} * \mathbf{B}_1 \dots \mathbf{B}_n && \text{(by the definition of } \mathcal{V}_{\varphi\circ\theta}\text{)} \\
&= \mathcal{V}_{\varphi\circ\theta} \mathbf{A} && \blacksquare
\end{aligned}$$

1401 Corollary If $\models \mathbf{A}$, and θ is any proper substitution, then $\models \theta \mathbf{A}$.

Proof: Let φ be any assignment. $\mathcal{V}_\varphi \theta \mathbf{A} = \mathcal{V}_{\varphi\circ\theta} \mathbf{A}$ (by 1400) $= \mathsf{T}$ (since $\models \mathbf{A}$). Thus every assignment satisfies $\theta \mathbf{A}$, so $\models \theta \mathbf{A}$. ■

p	q	p	$\supset$	q	$\supset$	q
T	T		T		T	
F	F		F		T	
F	T		T		T	
F	F		T		F	

Figure 1.14: $\vee$ expressed in terms of $\supset$

1402 Substitutivity of Equivalence. If **A**, **B**, **C**, and **D** are wffs of propositional calculus, and **D** is obtained from **C** by replacing zero or more occurrences of **A** in **C** by occurrences of **B**, and $\models [\mathbf{A} \equiv \mathbf{B}]$, then $\models [\mathbf{C} \equiv \mathbf{D}]$.

Proof: By induction on the number of occurrences of propositional connectives in **C**. Consider the possible forms that **C** can have, and also the special cases where **D** is **C**, and where **C** is **A** and **D** is **B**.

NOTATION. If $\mathbf{p}_1, \ldots, \mathbf{p}_n$ are distinct propositional variables including all variables in **A**, we let $[\lambda \mathbf{p}_1 \ldots \lambda \mathbf{p}_n \mathbf{A}]$ denote that truth function whose value on any n-tuple $(x_1, \ldots, x_n)$ of truth values is $\mathcal{V}_\varphi \mathbf{A}$ when $\varphi \mathbf{p}_i = x_i$ for $1 \leq i \leq n$.

EXAMPLE: It can be seen with the aid of Figure 1.14 that $[\lambda p \lambda q \,\blacksquare\, p \supset q \supset q]$ is the truth function h such that $h(\mathsf{T}, \mathsf{T}) = \mathsf{T}$, $h(\mathsf{T}, \mathsf{F}) = \mathsf{T}$, $h(\mathsf{F}, \mathsf{T}) = \mathsf{T}$, and $h(\mathsf{F}, \mathsf{F}) = \mathsf{F}$. This is commonly known as the disjunction function, and denoted by $\vee$.

Clearly, $[\lambda \mathbf{p}_1 \ldots \lambda \mathbf{p}_n \mathbf{A}]$ and $[\lambda \mathbf{p}_1 \ldots \lambda \mathbf{p}_n\, \mathbf{B}]$ denote the same truth function iff $[\mathbf{A} \equiv \mathbf{B}]$ is a tautology. (Exercise X1418.) For example, $\models [p \vee q] \equiv [p \supset q \supset q]$, so $[\lambda p \lambda q \,\blacksquare\, p \vee q] = [\lambda p \lambda q \,\blacksquare\, p \supset q \supset q]$.

NOTATION. We shall use $\bigwedge_{i=1}^{m} \mathbf{A}_i$ as a notation for $[\mathbf{A}_1 \wedge \ldots \wedge \mathbf{A}_m]$, and $\bigvee_{i=1}^{m} \mathbf{A}_i$ as a notation for $[\mathbf{A}_1 \vee \ldots \vee \mathbf{A}_m]$. For $m \geq 1$ these notations can be defined inductively: $\bigwedge_{i=1}^{1} \mathbf{A}_i = \mathbf{A}_1$ and $\bigwedge_{i=1}^{m+1} \mathbf{A}_i = \bigwedge_{i=1}^{m} \mathbf{A}_i \wedge \mathbf{A}_{m+1}$; similarly, $\bigvee_{i=1}^{1} \mathbf{A}_i = \mathbf{A}_1$ and $\bigvee_{i=1}^{m+1} \mathbf{A}_i = \bigvee_{i=1}^{m} \mathbf{A}_i \vee \mathbf{A}_{m+1}$.

Sometimes we need more general notations for a disjunction or conjunction of a finite set of wffs. We discuss general notations for disjunctions; similar notations will be used for conjunctions. If P is a finite index set $\{i_1, \ldots, i_n\}$ or a finite sequence $< i_1, \ldots, i_n >$, and $\mathbf{A}_k$ is a wff for each $k \in P$, then $\bigvee_{k \in P} \mathbf{A}_k$ stands for $[\mathbf{A}_{i_1} \vee \ldots \vee \mathbf{A}_{i_n}]$. If P is a finite set or sequence of wffs (so that $\mathbf{A}_k = k$ in the notation above), instead of

$\bigvee_{k \in P} k$ we may simply write $\bigvee P$ for the disjunction of the wffs in P. Of course, in case P is a finite *set*, these notations introduce mild ambiguity, since (for example) $\{1,3,2,1\} = \{3,1,2\}$, but disjunction and conjunction are associative and commutative, so $\bigvee P$ with one representation of P will be equivalent to $\bigvee P$ with any other representation of P.

In certain contexts it is convenient to form conjunctions and disjunctions when the index set P is empty or $m = 0$, in the notation above. An empty conjunction is always true, and may be symbolized by t, while an empty disjunction is always false, and may be symbolized by f or $\Box$. These conventions are justified by the fact that adding on an empty conjunction or disjunction to a conjunction or disjunction does not alter its truth, since $\mathbf{A} \wedge \mathbf{B} \wedge \mathsf{t} \equiv \mathbf{A} \wedge \mathbf{B}$ and $\mathbf{A} \vee \mathbf{B} \vee \mathsf{f} \equiv \mathbf{A} \vee \mathbf{B}$.

The following simple but very useful tautologies are known as DeMorgan's laws: $[\sim [\mathbf{A}_1 \wedge \mathbf{A}_2] \equiv {}_\blacksquare \sim \mathbf{A}_1 \vee \sim \mathbf{A}_2]$ and $[\sim [\mathbf{A}_1 \vee \mathbf{A}_2] \equiv {}_\blacksquare \sim \mathbf{A}_1 \wedge \sim \mathbf{A}_2]$. Using the notation above, they can be stated in more general form as $[\sim \bigwedge_{i=1}^{m} \mathbf{A}_i \equiv \bigvee_{i=1}^{m} \sim \mathbf{A}_i]$ and $[\sim \bigvee_{i=1}^{m} \mathbf{A}_i \equiv \bigwedge_{i=1}^{m} \sim \mathbf{A}_i]$. Note that the generalized DeMorgan laws are correct even when $m = 0$ or $m = 1$.

DEFINITIONS.

(1) A *literal* is a wff of the form $\mathbf{p}$ or of the form $\sim \mathbf{p}$.

(2) A wff $\mathbf{D}$ is in *disjunctive normal form (dnf)* iff it is a disjunction of conjunctions of literals. Thus it is of the form $\bigvee_{i=1}^{m} \bigwedge_{j=1}^{n_i} \mathbf{P}_{ij}$, where the $\mathbf{P}_{ij}$ are literals. The conjunctions of literals $\bigwedge_{j=1}^{n_i} \mathbf{P}_{ij}$ which are the scopes of the disjunctions of $\mathbf{D}$ are called the *disjuncts* of $\mathbf{D}$. As a special case we admit the possibility that m or any n_i may be 1. Similarly, a wff is in *conjunctive normal form (cnf)* iff it is a conjunction of disjunctions of literals, and the disjunctions of literals which are the scopes of its conjunctions are called its *conjuncts*.

(3) A wff is in *full* disjunctive normal form iff it has the form $\bigvee_{i=1}^{m} \bigwedge_{j=1}^{n} \mathbf{P}_{ij}$, where for each i, $\mathbf{P}_{ij}$ is $\mathbf{p}_j$ or $\sim \mathbf{p}_j$, and $\mathbf{p}_1 \ldots, \mathbf{p}_n$ are distinct propositional variables. Full conjunctive normal form is defined similarly.

(4) A wff $\mathbf{A}$ is a *disjunctive normal form for* a wff $\mathbf{B}$ iff $\mathbf{A}$ is in disjunctive normal form and is equivalent to $\mathbf{B}$. Similarly, $\mathbf{A}$ is a *conjunctive normal form for* a wff $\mathbf{B}$ iff $\mathbf{A}$ is in conjunctive normal form and is equivalent to $\mathbf{B}$.

EXAMPLES: $p \vee [q \wedge \sim p] \vee [\sim q \wedge r \wedge \sim r]$ is in disjunctive normal form, and $\sim p \wedge [\sim q \vee p] \wedge [q \vee \sim r \vee r]$ is in conjunctive normal form.

q	r	p	$\sim\centerdot [q \equiv r] \supset p$	Conjunctions of literals for disjunctive normal form
T	T	T	F	
T	T	F	T	$q \wedge r \wedge \sim p$
T	F	T	F	
T	F	F	F	
F	T	T	F	
F	T	F	F	
F	F	T	F	
F	F	F	T	$\sim q \wedge \sim r \wedge \sim p$

Figure 1.15: Conjunctions of literals for dnf of $\sim \centerdot [q \equiv r] \supset p$

$[p \wedge q \wedge \sim r] \vee [p \wedge q \wedge r] \vee [\sim p \wedge \sim q \wedge r]$ is in full disjunctive normal form.

EXAMPLE: $p \equiv q$ has disjunctive normal form $[p \wedge q] \vee [\sim p \wedge \sim q]$ and conjunctive normal form $[\sim p \vee q] \wedge [p \vee \sim q]$.

Next we shall show that every wff of propositional calculus has disjunctive and conjunctive normal forms. The basic idea in the proof below can be illustrated by using the truth table in Figure 1.15 to find a disjunctive normal form for $\sim \centerdot [q \equiv r] \supset p$. We find the lines of the truth table (which correspond to assignments of values to variables) which make the wff true, and represent these as conjunctions of literals. The disjunction of these conjunctions of literals is a dnf for the wff. In our example, examination reveals that the wff is true iff q and r are true, and p is false, or q, r, and p are all false, so the disjunctive normal form is

$$[q \wedge r \wedge \sim p] \vee [\sim q \wedge \sim r \wedge \sim p].$$

1403 Theorem. Let h be any truth function with n arguments, where $n \geq 1$, and let $\mathbf{p}_1, \ldots, \mathbf{p}_n$ be distinct propositional variables. Then there is a wff $\mathbf{A}$ in disjunctive normal form such that $h = [\lambda \mathbf{p}_1, \ldots, \lambda \mathbf{p}_n \mathbf{A}]$.

Proof: If the value of h is F for all arguments, let $\mathbf{A}$ be $\mathbf{p}_1 \wedge \sim \mathbf{p}_1$.

Otherwise, for each assignment φ let $\mathbf{A}_\varphi = \mathbf{p}_1^\varphi \wedge \ldots \wedge \mathbf{p}_n^\varphi$, where

$$\begin{aligned} \mathbf{p}_i^\varphi &= \mathbf{p}_i \text{ if } \varphi \mathbf{p}_i = \mathsf{T} \\ &= \sim \mathbf{p}_i \text{ if } \varphi \mathbf{p}_i = \mathsf{F}. \end{aligned}$$

Thus $\mathcal{V}_\varphi \mathbf{A}_\varphi = \mathsf{T}$, and if ψ is any assignment such that $\mathcal{V}_\psi \mathbf{A}_\varphi = \mathsf{T}$, then $\psi \mathbf{p}_i = \varphi \mathbf{p}_i$ for each $i \leq n$.

Let $\mathbf{A} = \bigvee_{\varphi \in \{\varphi | h(\varphi \mathbf{p}_1, \ldots, \varphi \mathbf{p}_n) = \mathsf{T}\}} \mathbf{A}_\varphi$; thus $\mathbf{A}$ is the disjunction of those wffs $\mathbf{A}_\varphi$ such that $h(\varphi \mathbf{p}_1, \ldots, \varphi \mathbf{p}_n) = \mathsf{T}$. Note that for any assignment ψ, $\mathcal{V}_\psi \mathbf{A} = h(\psi \mathbf{p}_1, \ldots, \psi \mathbf{p}_n)$, since $\mathcal{V}_\psi \mathbf{A} = \mathsf{T}$ iff $h(\psi \mathbf{p}_1, \ldots, \psi \mathbf{p}_n) = \mathsf{T}$, and there are only two truth values.

To show that $[\lambda \mathbf{p}_1 \ldots \lambda \mathbf{p}_n \mathbf{A}] = h$, we let $(x_1, \ldots, x_n)$ be any n-tuple of truth values, and show that $[\lambda \mathbf{p}_1 \ldots \lambda \mathbf{p}_n \mathbf{A}](x_1 \ldots x_n) = h(x_1, \ldots, x_n)$. Define ψ so that $\psi \mathbf{p}_i = x_i$ for $1 \leq i \leq n$. Then $[\lambda \mathbf{p}_1 \ldots \lambda \mathbf{p}_n \mathbf{A}](x_1 \ldots x_n) = \mathcal{V}_\psi \mathbf{A} = h(\psi \mathbf{p}_1, \ldots, \psi \mathbf{p}_n) = h(x_1, \ldots, x_n)$. ■

EXAMPLE: Let $h(x_1, x_2) = \mathsf{T}$ iff $x_1 = x_2$. Let $\varphi_1 \mathbf{p}_1 = \varphi_1 \mathbf{p}_2 = \mathsf{T}$ and $\varphi_2 \mathbf{p}_1 = \varphi_2 \mathbf{p}_2 = \mathsf{F}$. Then $\mathbf{A} = \mathbf{A}_{\varphi_1} \vee \mathbf{A}_{\varphi_2} = [\mathbf{p}_1 \wedge \mathbf{p}_2] \vee [\sim \mathbf{p}_1 \wedge \sim \mathbf{p}_2]$.

1404 Corollary. Every wff of propositional calculus has a disjunctive normal form.

Proof: Let $\mathbf{B}$ be an arbitrary wff, let $\mathbf{p}_1, \ldots, \mathbf{p}_n$ be the propositional variables in $\mathbf{B}$, and let $h = [\lambda \mathbf{p}_1 \ldots \lambda \mathbf{p}_n \mathbf{B}]$. By 1403 there is a wff $\mathbf{A}$ in disjunctive normal form such that $h = [\lambda \mathbf{p}_1 \ldots \lambda \mathbf{p}_n \mathbf{A}]$, so $\models \mathbf{B} \equiv \mathbf{A}$.

REMARK. The following tautologies, known as distributive laws, are often useful for putting wffs into disjunctive or conjunctive normal form:

$$[\mathbf{A} \vee \mathbf{B}] \wedge \mathbf{C} \quad \equiv \quad [\mathbf{A} \wedge \mathbf{C}] \vee [\mathbf{B} \wedge \mathbf{C}]$$

$$[\mathbf{A} \wedge \mathbf{B}] \vee \mathbf{C} \quad \equiv \quad [\mathbf{A} \vee \mathbf{C}] \wedge [\mathbf{B} \vee \mathbf{C}].$$

EXAMPLE: Find a disjunctive normal form for $\sim \blacksquare\, [q \equiv r] \supset p$. We have already shown how to do this using the method implicit in the proofs of theorems 1403 and 1404. Another method is to transform the given wff into disjunctive normal form step by step, using simple tautologies and the Substitutivity of Equivalence, as follows:

$$\sim \blacksquare\, [q \equiv r] \supset p$$

$$[q \equiv r] \wedge \sim p$$

$$[[q \wedge r] \vee \blacksquare \sim q \wedge \sim r] \wedge \sim p$$

$$[q \wedge r \wedge \sim p] \vee \blacksquare \sim q \wedge \sim r \wedge \sim p.$$

Informally, we say that a set of propositional connectives is complete iff every truth function can be expressed in terms of those connectives. This can be stated more formally as follows:

DEFINITION. A set $\mathcal{S}$ of propositional connectives is *complete* iff for each integer $n \geq 1$ and for each truth function h of n arguments, there is a wff $\mathbf{A}$ in which only connectives of $\mathcal{S}$ occur such that $h = [\lambda p_1 \ldots \lambda p_n \mathbf{A}]$.

1405 Theorem. The following sets of propositional connectives are complete:

(a) $\{\vee, \sim\}$
(b) $\{\wedge, \sim\}$
(c) $\{\supset, \sim\}$
(d) $\{\supset, \mathsf{f}\}$
(e) $\{|\}$
(f) $\{\downarrow\}$
(g) $\{\wedge, \not\equiv, \mathsf{t}\}$

Proof: It follows directly from theorem 1403 that $\{\wedge, \vee, \sim\}$ is complete. However, since for any wffs $\mathbf{C}$ and $\mathbf{D}$, $[\mathbf{C} \wedge \mathbf{D}]$ is equivalent to $\sim [\sim \mathbf{C} \vee \sim \mathbf{D}]$, any wff in which $\sim$, $\vee$, and $\wedge$ are the only connectives is equivalent to some wff in which $\sim$ and $\vee$ are the only connectives. This proves (a). The remaining assertions follow by similar arguments using the following equivalences:

(b) $[\mathbf{A} \vee \mathbf{B}] \equiv \sim\!\centerdot\! \sim \mathbf{A} \wedge \sim \mathbf{B}$
(c) $[\mathbf{A} \vee \mathbf{B}] \equiv \centerdot\! \sim \mathbf{A} \supset \mathbf{B}$
(d) $\sim \mathbf{A} \equiv \centerdot\, \mathbf{A} \supset \mathsf{f}$
(e) $\sim \mathbf{A} \equiv [\mathbf{A} \mid \mathbf{A}]$ and $[\mathbf{A} \wedge \mathbf{B}] \equiv \sim [\mathbf{A} \mid \mathbf{B}]$
(f) $\sim \mathbf{A} \equiv [\mathbf{A} \downarrow \mathbf{A}]$ and $[\mathbf{A} \vee \mathbf{B}] \equiv \sim [\mathbf{A} \downarrow \mathbf{B}]$.
(g) $\sim \mathbf{A} \equiv \centerdot\, \mathbf{A} \not\equiv \mathsf{t}$ ■

DEFINITION. A wff of propositional calculus is in *negation normal form* (*nnf*), and is called a *negation normal formula* (which we also abbreviate as *nnf*), iff it contains no propositional connectives other than $\wedge$, $\vee$, and $\sim$, and the scope of each occurrence of $\sim$ is a propositional variable.

It follows from 1404 that every wff of propositional calculus is equivalent to a wff in nnf. However, a negation normal form of a wff may be much shorter than any disjunctive normal form of that wff. A wff can be put into nnf very easily by expressing all its connectives in terms of $\wedge$, $\vee$, and $\sim$, and then applying DeMorgan's Laws and eliminating double negations until the scopes of all negations are propositional variables.

$$\left[\begin{bmatrix} p \\ q \end{bmatrix} \vee \begin{bmatrix} \sim p \\ \sim q \end{bmatrix}\right]$$

Figure 1.16: dnf of $p \equiv q$

$$\begin{bmatrix} \sim p \vee q \\ p \vee \sim q \end{bmatrix}$$

Figure 1.17: cnf of $p \equiv q$

$$\left[\begin{bmatrix} p \vee q \\ \sim r \\ \sim q \vee s \end{bmatrix} \vee \begin{bmatrix} \sim p \\ r \vee \sim s \end{bmatrix}\right]$$

Figure 1.18: Vpform for W1401: $[[p \vee q] \wedge \sim r \wedge {\scriptstyle\blacksquare} \sim q \vee s] \vee {\scriptstyle\blacksquare} \sim p \wedge {\scriptstyle\blacksquare} r \vee \sim s$

It is sometimes illuminating to display nnfs in a two-dimensional format called a *vpform* (vertical path form), in which disjunctions are written horizontally in the usual way, but conjunctions are written vertically, so that $[\mathbf{A} \wedge \mathbf{B}]$ is written as $\begin{bmatrix} \mathbf{A} \\ \mathbf{B} \end{bmatrix}$.

Vpforms for the dnf and cnf of $p \equiv q$ are displayed in Figures 1.16 and 1.17. A vpform for $[[p \vee q] \wedge \sim r \wedge {\scriptstyle\blacksquare} \sim q \vee s] \vee {\scriptstyle\blacksquare} \sim p \wedge {\scriptstyle\blacksquare} r \vee \sim s$, which we call W1401, is displayed in Figure 1.18.

It is useful to define the *vertical* and *horizontal paths* through a nnf **A** as certain sequences of occurrences of literals in that nnf. We shall let $\mathcal{VP}(\mathbf{A})$ be the set of vertical paths through **A**, and $\mathcal{HP}(\mathbf{A})$ be the set of horizontal paths through **A**. Essentially, if one erases the left scope or the right scope of each occurrence of $\vee$ (perhaps making different choices for different occurrences of $\vee$) in a nnf, the remaining literal occurrences form a vertical path. Similarly, if one one erases the left scope or the right scope of each occurrence of $\wedge$ in a nnf, the remaining literal occurrences form a horizontal path.

EXAMPLE: $\mathcal{VP}(\text{W1401}) = \{<p, \sim r, \sim q>, <p, \sim r, s>, <q, \sim r, \sim q>,$
$<q, \sim r, s>, <\sim p, r>, <\sim p, \sim s>\}$.
$\mathcal{HP}(\text{W1401}) = \{<p, q, \sim p>, <p, q, r, \sim s>, <\sim r, \sim p>,$
$<\sim r, r, \sim s>, <\sim q, s, \sim p>, <\sim q, s, r, \sim s>\}$.
The reader who traces these paths on Figure 1.18 will see that the ideas underlying vertical and horizontal paths are very simple and natural. We shall see that they are also very useful.

In order to prove theorems about vertical and horizontal paths, we need careful definitions of these concepts. Instead of directly defining what a vertical (or horizontal) path is, we define the *set* of vertical (or horizontal) paths through a nnf. (This is a technique which is very often useful in mathematics.) If $P =< P_1, \ldots, P_m >$ and $Q =< Q_1, \ldots, Q_n >$ are paths, we shall use the notation PQ for the concatenation $< P1, \ldots, P_m, Q1, \ldots, Q_n >$ of these paths.

DEFINITIONS. Let **A** be a nnf. We define $\mathcal{VP}(\mathbf{A})$ and $\mathcal{HP}(\mathbf{A})$ by induction on the construction of **A** as a negation normal formula:

(1) If **A** is a literal, $\mathcal{VP}(\mathbf{A}) = \{< \mathbf{A} >\}$, i.e., the set containing the sequence of length 1 whose first and only member is the occurrence of **A** in **A**.
(2) If **A** has the form $[\mathbf{B} \vee \mathbf{C}]$, then $\mathcal{VP}(\mathbf{A}) = \mathcal{VP}(\mathbf{B}) \cup \mathcal{VP}(\mathbf{C})$.
(3) If **A** has the form $[\mathbf{B} \wedge \mathbf{C}]$, then $\mathcal{VP}(\mathbf{A}) = \{\mathbf{PQ} \mid \mathbf{P} \in \mathcal{VP}(\mathbf{B})$ and $\mathbf{Q} \in \mathcal{VP}(\mathbf{C})\}$.
(1′) If **A** is a literal, $\mathcal{HP}(\mathbf{A}) = \{< \mathbf{A} >\}$.
(2′) If **A** has the form $[\mathbf{B} \vee \mathbf{C}]$, then then $\mathcal{HP}(\mathbf{A}) = \{\mathbf{PQ} \mid \mathbf{P} \in \mathcal{HP}(\mathbf{B})$ and $\mathbf{Q} \in \mathcal{HP}(\mathbf{C})\}$.
(3′) If **A** has the form $[\mathbf{B} \wedge \mathbf{C}]$, $\mathcal{HP}(\mathbf{A}) = \mathcal{HP}(\mathbf{B}) \cup \mathcal{HP}(\mathbf{C})$.

Associated with each sequence P of *occurrences* of literals in a wff is a sequence P' of *literals*. If a term of P is an *occurrence* of the literal L, then the corresponding term of P' is the literal L itself. We shall generally not distinguish carefully between sequences of literal occurrences and sequences of literals, use the same notation for P and P', and rely on the reader to determine from the context which is intended.[9] Thus, if P is a path through a nnf and φ is an assignment, we say that φ satisfies [falsifies] every literal in P iff φ satisfies [falsifies] every literal which has an occurrence in P.

1406 Theorem. Let **A** be a wff in negation normal form, and let φ be an assignment. Then

(a) φ satisfies **A** iff there is a path $P \in \mathcal{VP}(\mathbf{A})$ such that φ satisfies every literal in P.
(b) φ falsifies **A** iff there is a path $P \in \mathcal{HP}(\mathbf{A})$ such that φ falsifies every literal in P.

Proof of 1406(a) *by induction on the construction of* **A**:

[9] It might seem simpler to define vertical and horizontal paths as sequences of literals or as sets of literals, but this would not permit adequate descriptions of certain algorithms involving paths.

Case 1. **A** is a literal. This case is trivial.

Case 2. **A** is [**B** $\lor$ **C**]. The following are equivalent:

(i) $\mathcal{V}_\varphi$ **A** $=$ T.
(ii) $\mathcal{V}_\varphi$ [**B** $\lor$ **C**] $=$ T.
(iii) $\mathcal{V}_\varphi$ **B** $=$ T or $\mathcal{V}_\varphi$ **C** $=$ T (by the definition of $\mathcal{V}_\varphi$)
(iv) there is a path $P \in \mathcal{VP}(\mathbf{B})$ such that φ satisfies every literal in P, or there is a path $P \in \mathcal{VP}(\mathbf{C})$ such that φ satisfies every literal in P (by the inductive hypotheses, which we can apply since **B** and **C** are in nnf).
(v) there is a path $P \in \mathcal{VP}(\mathbf{B}) \cup \mathcal{VP}(\mathbf{C})$ such that φ satisfies every literal in P (by the definition of $\cup$).
(vi) there is a path $P \in \mathcal{VP}(\mathbf{A})$ such that φ satisfies every literal in P (by the definition of $\mathcal{VP}($ [**B** $\lor$ **C**])).

Case 3. **A** is [**B** $\land$ **C**]. The following are equivalent:

(i) $\mathcal{V}_\varphi$ **A** $=$ T.
(ii) $\mathcal{V}_\varphi$ [**B** $\land$ **C**] $=$ T.
(iii) $\mathcal{V}_\varphi$ **B** $=$ T and $\mathcal{V}_\varphi$ **C** $=$ T (by the definition of $\mathcal{V}_\varphi$)
(iv) there is a path $P \in \mathcal{VP}(\mathbf{B})$ such that φ satisfies every literal in P, and there is a path $Q \in \mathcal{VP}(\mathbf{C})$ such that φ satisfies every literal in Q (by the inductive hypotheses).
(v) there is a path $R \in \mathcal{VP}(\mathbf{A})$ such that φ satisfies every literal in R.

To prove that (*iv*) implies (*v*), note that if we have paths $P \in \mathcal{VP}(\mathbf{B})$ such that φ satisfies every literal in P and $Q \in \mathcal{VP}(\mathbf{C})$ such that φ satisfies every literal in Q, then their concatenation PQ is a path in $\mathcal{VP}([\mathbf{B} \land \mathbf{C}]$ such that φ satisfies every literal in PQ. The proof that (*v*) implies (*iv*) is also a direct consequence of the definition of $\mathcal{VP}([\mathbf{B} \land \mathbf{C}]$.

This completes the proof of (a). As might be expected, the proof of (b) mirrors the proof of (a).

Proof of 1406(b) *by induction on the construction of* **A**:

Case 1. **A** is a literal. This case is trivial.

Case 2. **A** is [**B** $\lor$ **C**]. The following are equivalent:

(i) $\mathcal{V}_\varphi$ **A** $=$ F.

(ii) $\mathcal{V}_\varphi\ [\mathbf{B}\ \lor \mathbf{C}\ ] = \mathsf{F}$.
(iii) $\mathcal{V}_\varphi\ \mathbf{B} = \mathsf{F}$ and $\mathcal{V}_\varphi\ \mathbf{C} = \mathsf{F}$ (by the definition of $\mathcal{V}_\varphi$).
(iv) there is a path $P \in \mathcal{HP}(\mathbf{B})$ such that φ falsifies every literal in P, and there is a path $Q \in \mathcal{HP}(\mathbf{C})$ such that φ falsifies every literal in Q (by the inductive hypotheses).
(v) there is a path $R \in \mathcal{HP}(\mathbf{A})$ such that φ falsifies every literal in R.

To prove that *(iv)* implies *(v)*, note that if we have paths $P \in \mathcal{HP}(\mathbf{B})$ such that φ falsifies every literal in P and $Q \in \mathcal{HP}(\mathbf{C})$ such that φ falsifies every literal in Q, then their concatenation PQ is a path in $\mathcal{HP}([\mathbf{B}\ \lor\ \mathbf{C}]$ such that φ falsifies every literal in PQ. The proof that *(v)* implies *(iv)* is also a direct consequence of the definition of $\mathcal{HP}([\mathbf{B}\ \lor\ \mathbf{C}]$.

Case 3. $\mathbf{A}$ is $[\mathbf{B}\ \land\ \mathbf{C}]$. The following are equivalent:

(i) $\mathcal{V}_\varphi\ \mathbf{A} = \mathsf{F}$.
(ii) $\mathcal{V}_\varphi\ [\mathbf{B}\ \land \mathbf{C}\ ] = \mathsf{F}$.
(iii) $\mathcal{V}_\varphi\ \mathbf{B} = \mathsf{F}$ or $\mathcal{V}_\varphi\ \mathbf{C} = \mathsf{F}$ (by the definition of $\mathcal{V}_\varphi$).
(iv) there is a path $P \in \mathcal{HP}(\mathbf{B})$ such that φ falsifies every literal in P, or there is a path $P \in \mathcal{HP}(\mathbf{C})$ such that φ falsifies every literal in P (by the inductive hypotheses).
(v) there is a path $P \in \mathcal{HP}(\mathbf{B}) \cup \mathcal{HP}(\mathbf{C})$ such that φ falsifies every literal in P.
(vi) there is a path $P \in \mathcal{HP}(\mathbf{A})$ such that φ falsifies every literal in P (by the definition of $\mathcal{HP}(\ [\mathbf{B} \land \mathbf{C}])$.) ■

Next we shall see that we can construct a disjunctive normal form for a nnf simply by looking at its vpform, forming the conjunctions of the literals in each vertical path, and forming the disjunction of these conjunctions. A conjunctive normal form can be obtained by a similar procedure.

1407 Theorem. Let $\mathbf{A}$ be a wff in negation normal form. Then

(a) $\bigvee_{P \in \mathcal{VP}(\mathbf{A})} \bigwedge P$ is a disjunctive normal form of $\mathbf{A}$.
(b) $\bigwedge_{P \in \mathcal{HP}(\mathbf{A})} \bigvee P$ is a conjunctive normal form of $\mathbf{A}$.

EXAMPLE:
A dnf of W1401 is $[[p \land \sim r \land \sim q] \lor [p \land \sim r \land s] \lor [q \land \sim r \land \sim q]$ $\lor [q \land \sim r \land s] \lor [\sim p \land r] \lor [\sim p \land \sim s]]$.
A cnf of W1401 is $[[p \lor q \lor \sim p] \land [p \lor q \lor r \lor \sim s] \land [\sim r \lor \sim p]$ $\land [\sim r \lor r \lor \sim s] \land [\sim q \lor s \lor \sim p] \land [\sim q \lor s \lor r \lor \sim s]]$.
The reader should note how the paths in Figure 1.18 occur in these wffs.

Proof of 1407(a): Clearly $\bigvee_{P \in \mathcal{VP}(\mathbf{A})} \bigwedge P$ is in disjunctive normal form. To show that it is equivalent to $\mathbf{A}$, let φ be any assignment. The following are equivalent:

(i) φ satisfies $\mathbf{A}$.
(ii) There is a path $P \in \mathcal{VP}(\mathbf{A})$ such that φ satisfies every literal in P (by 1406(a)).
(iii) There is a path $P \in \mathcal{VP}(\mathbf{A})$ such that φ satisfies $\bigwedge P$.
(iv) φ satisfies $\bigvee_{P \in \mathcal{VP}(\mathbf{A})} \bigwedge P$.

Proof of 1407(b): Clearly $\bigwedge_{P \in \mathcal{HP}(\mathbf{A})} \bigvee P$ is in conjunctive normal form. To show that it is equivalent to $\mathbf{A}$, let φ be any assignment. The following are equivalent:

(i) φ falsifies $\mathbf{A}$.
(ii) There is a path $P \in \mathcal{HP}(\mathbf{A})$ such that φ falsifies every literal in P (by 1406(b)).
(iii) There is a path $P \in \mathcal{HP}(\mathbf{A})$ such that φ falsifies $\bigvee P$.
(iv) φ falsifies $\bigwedge_{P \in \mathcal{HP}(\mathbf{A})} \bigvee P$. ■

EXAMPLE: Find a disjunctive normal form for $[[p \equiv q] \equiv \sim r]$.

The following are equivalent:

$$[[p \equiv q] \equiv \sim r]$$

$$\left[\begin{bmatrix} p \equiv q \\ \sim r \end{bmatrix} \vee \begin{bmatrix} p \not\equiv q \\ r \end{bmatrix}\right]$$

$$\left[\begin{bmatrix} p \equiv q \\ \sim r \end{bmatrix} \vee \begin{bmatrix} p \equiv \sim q \\ r \end{bmatrix}\right]$$

$$\begin{bmatrix} \begin{bmatrix} p \\ q \end{bmatrix} \vee \begin{bmatrix} \sim p \\ \sim q \end{bmatrix} \\ \sim r \end{bmatrix} \vee \begin{bmatrix} \begin{bmatrix} p \\ \sim q \end{bmatrix} \vee \begin{bmatrix} \sim p \\ q \end{bmatrix} \\ r \end{bmatrix}$$

$$[[p \wedge q \wedge \sim r] \vee [\sim p \wedge \sim q \wedge \sim r] \vee [p \wedge \sim q \wedge r] \vee [\sim p \wedge q \wedge r]].$$

Note that we now have three methods for putting a wff into disjunctive normal form:

(1) Use the method in the proof of 1403 and 1404.

(2) Put it into nnf and use distributive laws.
(3) Put it into nnf and use 1407.

Of course, there are close relationships between these methods.

Next we turn our attention to the "method of vertical paths" for seeing that a wff is a contradiction (and hence that it is the negation of a tautology) and the "method of horizontal paths" for seeing that a wff is a tautology. In each case, all one has to do is to look at a vpform for the wff.

DEFINITION. Two literals are *complementary* or *mated* iff one is the negation of the other.

1408 Theorem. Let **A** be a wff in negation normal form. Then

(a) **A** is a contradiction iff every path in $\mathcal{VP}(\mathbf{A})$ contains complementary literals.
(b) **A** is a tautology iff every path in $\mathcal{HP}(\mathbf{A})$ contains complementary literals.

Proof of (a): If every path in $\mathcal{VP}(\mathbf{A})$ contains complementary literals, then no assignment can satisfy every literal on any path in $\mathcal{VP}(\mathbf{A})$, so by 1406(a) **A** is not satisfiable, i.e., it is a contradiction. If some path in $\mathcal{VP}(\mathbf{A})$ does not contain complementary literals, then by 1406(a) **A** is satisfied by the assignment which makes every literal in that vertical path true, so **A** is not a contradiction.

Proof of (b): If every path in $\mathcal{HP}(\mathbf{A})$ contains complementary literals, then no assignment can falsify every literal on any path in $\mathcal{HP}(\mathbf{A})$, so by 1406(b) **A** is not falsified by any assignment, so it is a tautology. If some path in $\mathcal{HP}(\mathbf{A})$ does not contain complementary literals, then by 1406(b) **A** is falsified by the assignment which makes every literal in that horizontal path false, so **A** is not a tautology. ∎

EXAMPLE: One can easily see from Figure 1.18 that the vpform for W1401 has vertical paths which do not contain complementary literals, and it has horizontal paths which do not contain complementary literals. Therefore, it is neither a contradiction nor a tautology.

EXAMPLE: Let W1402 be the wff

$$[p \lor q] \land \sim [p \lor r] \supset \sim r \land {\scriptstyle\blacksquare} s \supset q$$

$$\begin{bmatrix} p \lor q \\ \sim p \\ \sim r \\ r \lor \begin{bmatrix} s \\ \sim q \end{bmatrix} \end{bmatrix}$$

Figure 1.19: Vpform for $\sim$ W1402.

We use the method of vertical paths to show that $\models$ W1402.

The following are equivalent:

$$\sim W1402$$

$$\sim \bullet [p \lor q] \land \sim [p \lor r] \supset \sim r \land \bullet s \supset q$$

$$[p \lor q] \land [\sim p \land \sim r] \land \bullet r \lor s \land \sim q$$

The vpform for this wff is displayed in in Figure 1.19. One can see that all its vertical paths contain complementary literals, so by 1408, $\sim$ W1402 is a contradiction, so $\models$ W1402.

EXAMPLE: Let W1403 be the wff $[(p \equiv q) \equiv r] \supset \bullet p \equiv (q \equiv r)$. We use the method of vertical paths to show that $\models$ W1403.

The following are equivalent:

$$\sim W1403$$

$$[(p \equiv q) \equiv r] \land \sim [p \equiv (q \equiv r)]$$

$$[(p \equiv q) \equiv r] \land [\sim p \equiv (q \equiv r)]$$

$$[((p \equiv q) \land r) \lor ((p \not\equiv q) \land \sim r)] \land \bullet (\sim p \land \bullet q \equiv r) \lor (p \land \bullet q \not\equiv r)$$

$$[([(p \land q) \lor (\sim p \land \sim q)] \land r) \lor ([(p \land \sim q) \lor (\sim p \land q)] \land \sim r)] \land$$

$$\bullet (\sim p \land \bullet (q \land r) \lor (\sim q \land \sim r)) \lor (p \land \bullet (q \land \sim r) \lor (\sim q \land r))$$

The vpform for this wff is displayed in in Figure 1.20. One can see that all its vertical paths contain complementary literals, so by 1408 $\sim$ W1403 is a contradiction, so $\models$ W1403.

In summary, to determine whether a wff **B** is a tautology by the method of vertical paths, one may put $\sim$ **B** into negation normal form, display it in a

$$
\left[\begin{array}{c}
\left[\begin{array}{c} \left[\begin{array}{c} p \\ q \end{array}\right] \vee \left[\begin{array}{c} \sim p \\ \sim q \end{array}\right] \\ r \end{array}\right] \vee \left[\begin{array}{c} \left[\begin{array}{c} p \\ \sim q \end{array}\right] \vee \left[\begin{array}{c} \sim p \\ q \end{array}\right] \\ \sim r \end{array}\right] \\
\left[\begin{array}{c} \sim p \\ \left[\begin{array}{c} q \\ r \end{array}\right] \vee \left[\begin{array}{c} \sim q \\ \sim r \end{array}\right] \end{array}\right] \vee \left[\begin{array}{c} p \\ \left[\begin{array}{c} q \\ \sim r \end{array}\right] \vee \left[\begin{array}{c} \sim q \\ r \end{array}\right] \end{array}\right]
\end{array}\right]
$$

Figure 1.20: Vpform for $\sim$ W1403.

vpform, and check whether each vertical path through this vpform contains complementary literals.

We conclude this section with a list of basic tautologies which the reader should verify and become familiar with.

BASIC TAUTOLOGIES

LAWS OF NEGATION

$$
\begin{array}{rcll}
\sim\sim p & \equiv & p & \text{(Law of double negation)} \\
\sim [p \wedge q] & \equiv & \bullet \sim p \vee \sim q & \text{(De Morgan's Law)} \\
\sim [p \vee q] & \equiv & \bullet \sim p \wedge \sim q & \text{(De Morgan's Law)} \\
\sim [p \supset q] & \equiv & \bullet p \wedge \sim q & \\
\sim [p \equiv q] & \equiv & \bullet \sim p \equiv q & \\
\sim [p \equiv q] & \equiv & \bullet p \equiv \sim q & \\
\sim p & \equiv & \bullet p \mid p & \\
\sim p & \equiv & \bullet p \downarrow p &
\end{array}
$$

LAWS DEFINING CONNECTIVES

$$
\begin{array}{rcl}
p \vee q & \equiv & \bullet \sim p \supset q \\
p \vee q & \equiv & \bullet p \supset q \supset q \\
p \wedge q & \equiv & \bullet p \equiv \bullet p \supset q \\
p \supset q & \equiv & \bullet \sim p \vee q \\
p \supset q & \equiv & \sim \bullet p \wedge \sim q \\
p \supset q & \equiv & \bullet p \equiv \bullet p \wedge q \\
p \equiv q & \equiv & \bullet [p \supset q] \wedge [q \supset p] \\
p \equiv q & \equiv & \bullet [\sim p \vee q] \wedge [p \vee \sim q] \\
p \equiv q & \equiv & \bullet [p \wedge q] \vee [\sim p \wedge \sim q] \\
p \mid q & \equiv & \sim \bullet p \wedge q \\
p \downarrow q & \equiv & \sim \bullet p \vee q
\end{array}
$$

DISTRIBUTIVE LAWS

$$p \wedge [q \vee r] \equiv \centerdot [p \wedge q] \vee [p \wedge r]$$
$$p \vee [q \wedge r] \equiv \centerdot [p \vee q] \wedge [p \vee r]$$

COMMUTATIVE LAWS

$$p \wedge q \equiv \centerdot q \wedge p$$
$$p \vee q \equiv \centerdot q \vee p$$
$$p \equiv q \equiv \centerdot q \equiv p$$
$$p \not\equiv q \equiv \centerdot q \not\equiv p$$
$$p \mid q \equiv \centerdot q \mid p$$
$$p \downarrow q \equiv \centerdot q \downarrow p$$

ASSOCIATIVE LAWS

$$[p \wedge q] \wedge r \equiv \centerdot p \wedge \centerdot q \wedge r$$
$$[p \vee q] \vee r \equiv \centerdot p \vee \centerdot q \vee r$$
$$[p \equiv q] \equiv r \equiv \centerdot p \equiv \centerdot q \equiv r$$
$$[p \not\equiv q] \not\equiv r \equiv \centerdot p \not\equiv \centerdot q \not\equiv r$$

TRANSITIVE LAWS

$$[p \supset q] \wedge [q \supset r] \supset \centerdot p \supset r$$
$$[p \equiv q] \wedge [q \equiv r] \supset \centerdot p \equiv r$$

LAWS OF t AND f

$$\sim \mathsf{t} \equiv \mathsf{f}$$
$$\sim \mathsf{f} \equiv \mathsf{t}$$
$$p \wedge \mathsf{t} \equiv p$$
$$p \wedge \mathsf{f} \equiv \mathsf{f}$$
$$p \vee \mathsf{t} \equiv \mathsf{t}$$
$$p \vee \mathsf{f} \equiv p$$

$$p \supset \mathsf{t} \equiv \mathsf{t}$$
$$p \supset \mathsf{f} \equiv \sim p$$
$$\mathsf{t} \supset p \equiv p$$
$$\mathsf{f} \supset p \equiv \mathsf{t}$$
$$p \equiv \mathsf{t} \equiv p$$
$$p \equiv \mathsf{f} \equiv \sim p$$

OTHER LAWS

$\sim p \lor p$	(Law of excluded middle)
$p \supset p$	
$p \equiv p$	
$p \equiv \bullet p \land p$	(IdempotentLaw)
$p \equiv \bullet p \lor p$	(IdempotentLaw)
$p \equiv \bullet p \land [q \lor p]$	(AbsorptionLaw)
$p \equiv \bullet p \lor [q \land p]$	(AbsorptionLaw)
$p \land q \equiv \bullet p \land [q \lor \sim p]$	
$p \lor q \equiv \bullet p \lor [q \land \sim p]$	
$p \equiv q \equiv \bullet \sim p \equiv \sim q$	
$p \supset q \equiv \bullet \sim q \supset \sim p$	(Contrapositive law)
$p \land q \supset r \equiv \bullet p \supset \bullet q \supset r$	
$[p \supset q] \land [\sim p \supset r] \equiv \bullet [p \land q] \lor [\sim p \land r]$	
$p \supset q \supset p \supset p$	(Peirce's law)
$p \supset \bullet p \lor q$	
$p \land q \supset p$	

EXERCISES

X1400. Prove that the only binary connectives which, taken alone, constitute complete sets of connectives are $\mid$ and $\downarrow$.

X1401. Prove that $\{\sim, \equiv\}$ is not a complete set of connectives . (*Hint:* can you find a wff constructed from these connectives and containing exactly two propositional variables which is true in an odd number of lines of the truth table?)

X1402. Let $\mathcal{P}'$ be a system obtained from $\mathcal{P}$ by adding propositional constants t and f to the list of proper symbols of $\mathcal{P}$. The formation rules of $\mathcal{P}'$ are those of $\mathcal{P}$ plus the additional rule: (0) t and f are wffs. The axiom schemata $\mathcal{P}'$ are the same as those of $\mathcal{P}$, but of course there are certain instances of these schemata which are axioms of $\mathcal{P}'$ but not wffs (and hence not axioms) of $\mathcal{P}$. In addition, $\mathcal{P}'$ has the wffs t and $\sim \mathsf{f}$ as axioms. For any assignment φ, $\mathcal{V}_\varphi \mathsf{t} = \mathsf{T}$ and $\mathcal{V}_\varphi \mathsf{f} = \mathsf{F}$. Show that every tautology of $\mathcal{P}'$ is a theorem of $\mathcal{P}'$.

X1403. Show that if $\mathbf{A}$ is a wff in which no propositional connective other than $\equiv$ occurs, and φ is an assignment, then $\mathcal{V}_\varphi \mathbf{A} = \mathsf{F}$ iff the number of occurrences of variables $\mathbf{p}$ in $\mathbf{A}$ such that $\varphi \mathbf{p} = \mathsf{F}$ is odd. (*Hint:* Consider some examples.)

X1404. Prove that a wff of propositional calculus containing only the connective $\equiv$ is a tautology iff each propositional variable occurs an even number of times. (*Hint:* What other exercise is related to this one?)

X1405. In a wff containing only $\wedge$, $\not\equiv$, t, and f as connectives, one may write $\wedge$ as $\times$, $\not\equiv$ as $+$, t as 1, and f as 0 (since operations on truth values may be regarded as operations of arithmetic modulo 2). Note that in this notation the following wffs are tautologies:

(a)	$[[\mathbf{P}+\mathbf{Q}]+\mathbf{R}] \equiv \centerdot \mathbf{P}+[\mathbf{Q}+\mathbf{R}]$	(f)	$[\mathbf{P}+\mathbf{P}] \equiv 0$
(b)	$[[\mathbf{P}\times\mathbf{Q}]\times\mathbf{R}] \equiv \centerdot \mathbf{P}\times[\mathbf{Q}\times\mathbf{R}]$	(g)	$[\mathbf{P}\times\mathbf{P}] \equiv \mathbf{P}$
(c)	$[\mathbf{P}+\mathbf{Q}] \equiv \centerdot \mathbf{Q}+\mathbf{P}$	(h)	$[\mathbf{P}+0] \equiv \mathbf{P}$
(d)	$[\mathbf{P}\times\mathbf{Q}] \equiv \centerdot \mathbf{Q}\times\mathbf{P}$	(i)	$[\mathbf{P}\times 0] \equiv 0$
(e)	$[[\mathbf{P}+\mathbf{Q}]\times\mathbf{R}] \equiv \centerdot [\mathbf{P}\times\mathbf{R}]+[\mathbf{Q}\times\mathbf{R}]$	(j)	$[\mathbf{P}\times 1] \equiv \mathbf{P}$

Corresponding to each schema of the form $[\mathbf{D} \equiv \mathbf{E}]$ one has the rule of inference: to replace any occurrence of **D** in a wff by an occurrence of **E**. By using rules (a)-(e) one can transform any wff into a polynomial in its propositional variables, which can be further simplified with the aid of rules (f)-(j).

Show that a wff **A** (in which $\times$, $+$, 1, and 0 are the only propositional connectives) is a tautology iff there is a sequence of applications of rules (a)-(j) which transforms **A** into 1.

X1406. Using the notation of X1405, write wffs equivalent to the following wffs, and simplify them:

(a) $p \supset q$

(b) $[p \supset q] \wedge [q \supset p]$

(c) $p \vee q$

(d) $[p \equiv q] \wedge [p \vee r] \supset \centerdot q \vee r$

(e) $[p \equiv \centerdot q \wedge r] \wedge [q \not\equiv r] \supset \sim p$

(f) $p \wedge [p \supset q] \supset q$

Find conjunctive and disjunctive normal forms for the wffs in Exercises X1407–X1409.

X1407. $p \equiv \sim [q \wedge r]$

X1408. $[q \vee r] \supset p$

X1409. $[p \vee r] \equiv [q \wedge \sim p]$

X1410. A binary connective $*$ is *associative* iff $[[p * q] * r] \equiv \centerdot p * \centerdot q * r$ is a tautology. Is Sheffer's stroke associative?

X1411. Find a wff equivalent to $[p \supset q]$ in which the only connective is Sheffer's stroke.

X1412. Let $\mathcal{P}_L$ be the system of propositional calculus which has Sheffer's stroke as sole connective, and the following axiom and rules of inference:

Axiom. $p \mid [q \mid r] \mid \centerdot\, p \mid [r \mid p] \mid \centerdot\, s \mid q \mid \centerdot\, p \mid s \mid \centerdot\, p \mid s$

Rules of Inference.

(1) (Substitution) From $\mathbf{A}$ infer $\mathsf{S}^{\mathbf{p}}_{\mathbf{B}}\mathbf{A}$.
(2) From $\mathbf{A} \mid \centerdot\, \mathbf{B} \mid \mathbf{C}$ and $\mathbf{A}$ infer $\mathbf{C}$.

Show that $\mathcal{P}_L$ is consistent, in the sense that there is no wff $\mathbf{A}$ such that both $\mathbf{A}$ and $\mathbf{A} \mid \mathbf{A}$ are theorems.

X1413. Let $\mathcal{P}_2$ be a formulation of propositional calculus in which the primitive connectives are $\wedge$, $\vee$, and $\sim$. Without using 1404, give a direct inductive proof that if $\mathbf{A}$ is any wff of $\mathcal{P}_2$, there is a wff $\mathbf{A}'$ of $\mathcal{P}_2$ in negation normal form such that $\models \mathbf{A} \equiv \mathbf{A}'$.

X1414. Is $\{\equiv, \supset, \wedge, \vee\}$ a complete set of propositional connectives? Justify your answer.

X1415. For each wff below, find an equivalent wff which contains no connectives other than $\sim$, $\vee$, $\wedge$, $\supset$, and $\equiv$, and which contains as few occurrences of propositional variables as possible.

(a) $p \supset q \supset q$
(b) $[p \vee q] \wedge p$
(c) $\sim [\sim p \wedge \centerdot\, p \supset q]$
(d) $[p \wedge q] \vee [p \wedge r]$
(e) $p \equiv\, \sim [q \equiv p]$
(f) $\sim [q \supset \sim \centerdot\, q \supset q]$

In exercises X1416–X1417 a wff W is given. In each case, do the following:

(a) Find a wff in negation normal form which is equivalent to $\sim W$.
(b) Display your answer to part (a) in a vpform.
(c) Decide whether W is a tautology. If it is not, give an assignment which falsifies it.

X1416. $[[p \wedge r] \vee [q \wedge \sim r]] \supset \centerdot [q \supset \centerdot\, r \wedge s] \supset r$

X1417. $[[p \wedge q] \vee [\sim p \wedge r]] \equiv \centerdot [p \supset q] \wedge [\sim p \supset r]$

X1418. Prove that $[\lambda \mathbf{p}_1 \ldots \lambda \mathbf{p}_n \mathbf{A}]$ and $[\lambda \mathbf{p}_1 \ldots \lambda \mathbf{p}_n\, \mathbf{B}]$ denote the same truth function iff $[\mathbf{A} \equiv \mathbf{B}]$ is a tautology.

§15. Compactness

In §12 we explored one aspect of the relationship between the semantics and syntax of $\mathcal{P}$ by showing that the tautologies are the same as the theorems. In this section we show that satisfiability and consistency are coextensive.

1500 Proposition. Let $\mathcal{H}$ be a finite set of wffs of $\mathcal{P}$. $\mathcal{H}$ is satisfiable iff $\mathcal{H}$ is consistent in $\mathcal{P}$.

Proof: Suppose $\mathcal{H}$ is not satisfiable. Since the empty set of wffs is trivially satisfiable, $\mathcal{H}$ is nonempty, so we can write $\mathcal{H}$ as $\{\mathbf{A}_1, \dots \mathbf{A}_n\}$. Since $\mathcal{H}$ is not satisfiable, $\mathbf{A}_1 \land \dots \land \mathbf{A}_n$ is a contradiction, so for any wff $\mathbf{B}$, $\models \mathbf{A}_1 \supset {\scriptstyle\blacksquare}\, \mathbf{A}_2 \supset \dots \supset {\scriptstyle\blacksquare}\, \mathbf{A}_n \supset \mathbf{B}$. By 1204 this tautology is a theorem, so $\mathcal{H} \vdash \mathbf{B}$ by Modus Ponens. Since $\mathbf{B}$ is an arbitrary wff, $\mathcal{H}$ is inconsistent.

Suppose $\mathcal{H}$ is satisfiable, but inconsistent. Then $\mathcal{H} \vdash p \land \sim p$. Let φ be an assignment which satisfies $\mathcal{H}$. Modus Ponens preserves satisfaction by φ (see the proof of 1200), so $\models_\varphi p \land \sim p$. However, $p \land \sim p$ is not satisfiable. Therefore $\mathcal{H}$ must be consistent. ■

We shall show that 1500 is also true when $\mathcal{H}$ is infinite. The main part of this argument is supplied by the Compactness Theorem below, which is a purely semantic theorem, and does not depend on the details of any axiomatic treatment.

1501 Compactness Theorem. Let $\mathcal{S}$ be an infinite set of wffs of propositional calculus such that every finite subset of $\mathcal{S}$ is satisfiable. Then $\mathcal{S}$ is satisfiable.

Proof: Let $\mathbf{p}_1, \mathbf{p}_2, \mathbf{p}_3, \dots$ be the variables which occur in wffs of $\mathcal{S}$. We shall define an assignment θ by defining $\theta\mathbf{p}_i$ by induction on i. Simultaneously we shall establish by complete induction on i that for every finite subset $\mathcal{H}$ of $\mathcal{S}$, θ is extendable to $\mathcal{H}$, i.e., there is an assignment φ which satisfies $\mathcal{H}$ such that $\varphi\mathbf{p}_j = \theta\mathbf{p}_j$ for all $j \leq i$.

Suppose this has been done for all $k < i$. If for every finite subset $\mathcal{H}$ of $\mathcal{S}$ there is an assignment φ which satisfies $\mathcal{H}$ such that $\varphi\mathbf{p}_j = \theta\mathbf{p}_j$ for all $j < i$ and $\varphi\mathbf{p}_i = \mathsf{T}$, let $\theta\mathbf{p}_i = \mathsf{T}$ and we are done at stage i. Otherwise, let $\theta\mathbf{p}_i = \mathsf{F}$.

In the latter case there is a finite subset $\mathcal{K}$ of $\mathcal{S}$ such that for every assignment φ satisfying $\mathcal{K}$, if $\varphi\mathbf{p}_j = \theta\mathbf{p}_j$ for all $j < i$, then $\varphi\mathbf{p}_i = \mathsf{F}$. Let $\mathcal{H}$ be any finite subset of $\mathcal{S}$. Then $\mathcal{K} \cup \mathcal{H}$ is also a finite subset of $\mathcal{S}$, so by inductive hypothesis (or by the hypothesis of the theorem when $i = 1$) there is an assignment φ satisfying $\mathcal{K} \cup \mathcal{H}$ such that $\varphi\mathbf{p}_j = \theta\mathbf{p}_j$ for all $j < i$. φ satisfies $\mathcal{K}$ as well as $\mathcal{H}$, so $\varphi\mathbf{p}_i = \mathsf{F} = \theta\mathbf{p}_i$. Thus the required assignment φ exists.

Finally we prove that θ satisfies $\mathcal{S}$. Let $\mathbf{A}$ be any wff in $\mathcal{S}$, and let i be the largest index of a variable $\mathbf{p}_i$ which occurs in $\mathbf{A}$. There is an assignment φ which satisfies $\{\mathbf{A}\}$ such that $\varphi\mathbf{p}_j = \theta\mathbf{p}_j$ for all $j \leq i$. Thus φ and θ agree on all variables which occur in $\mathbf{A}$, so $\mathcal{V}_\theta\mathbf{A} = \mathcal{V}_\varphi\mathbf{A} = \mathsf{T}$. ■

REMARK. Sometimes one gains insight into an argument by considering a situation to which it does *not* apply. Suppose in some generalized system there are "wffs" $\mathbf{A}$ such that $\mathcal{V}_\varphi\mathbf{A}$ depends on the values that φ assigns to infinitely many variables. For example, suppose $\mathcal{S} = \{\mathbf{A}_i \mid i = 1, 2, 3, \ldots\}$ where $\mathcal{V}_\varphi\mathbf{A}_i = \mathsf{T}$ iff $\varphi p_i = \mathsf{T}$ and there is a $j > i$ such that $\varphi p_j = \mathsf{F}$. (Informally we may write $\mathbf{A}_i = [p_i \wedge (\exists j > i) \sim p_j]$. Of course, this is not a wff of propositional calculus.)

It is easy to see that every finite subset of $\mathcal{S}$ is satisfiable, but $\mathcal{S}$ is not satisfiable. Thus it appears that the Compactness Theorem depends in an essential way on the fact that each wff of propositional calculus contains only finitely many propositional variables.

1502 Corollary Let $\mathcal{S}$ be any set of wffs of $\mathcal{P}$. $\mathcal{S}$ is satisfiable iff $\mathcal{S}$ is consistent in $\mathcal{P}$.

Proof: (See Figure 1.21.) If $\mathcal{S}$ is inconsistent, then there is a finite subset $\mathcal{H}$ of $\mathcal{S}$ such that $\mathcal{H} \vdash p \wedge \sim p$, so $\mathcal{H}$ is unsatisfiable by 1500, so $\mathcal{S}$ is unsatisfiable too.

Suppose $\mathcal{S}$ is consistent. Then every finite subset $\mathcal{H}$ of $\mathcal{S}$ is consistent, and hence satisfiable by 1500. Therefore $\mathcal{S}$ is satisfiable by 1501. ■

$\mathcal{S}$ is consistent	$\overset{1502}{\Longleftrightarrow}$	$\mathcal{S}$ is satisfiable
trivial $\Updownarrow$		1501 $\Uparrow\Downarrow$ trivial
Every finite subset of $\mathcal{S}$ is consistent	$\overset{1500}{\Longleftrightarrow}$	Every finite subset of $\mathcal{S}$ is satisfiable

Figure 1.21: Proof of 1502

APPLICATION: We shall show how the Compactness Theorem can be used to establish that every partial ordering of an infinite (but countable) set can be extended to a total ordering. We shall use the fact that every partial ordering of a finite set can be extended to a total ordering; this can easily be proved by induction on the size of the set.

A *partial ordering* of a set S is a relation on S which is reflexive, transitive, and antisymmetric. A set which is endowed with a specified partial ordering is often called a *poset* (partially ordered set). A partial ordering $\preceq$ on a set S is a *total ordering* or *linear ordering* iff for all x and y in S, $x \preceq y$ or $y \preceq x$. To say that a partial ordering $\preceq_1$ on S can be extended to a linear ordering means that there is a linear ordering $\preceq_2$ on S such that $x \preceq_2 y$ whenever $x \preceq_1 y$.

Now let a partial ordering $\preceq_1$ on a countable infinite set S be given. For each ordered pair $< x, y >$ of elements from S, specify a propositional variable $\mathbf{p}_{xy}$ so that distinct pairs correspond to distinct propositional variables. (We shall think of $\mathbf{p}_{xy}$ as meaning that $x \preceq y$.) Let $\mathcal{W}$ be the set of all wffs with one of the following forms, where where x, y, and z are in S:

$\mathbf{p}_{xy}$, where $x \preceq_1 y$;

$\mathbf{p}_{xx}$;[10]

$[\mathbf{p}_{xy} \wedge \mathbf{p}_{yz} \supset \mathbf{p}_{xz}]$;

$\sim [\mathbf{p}_{xy} \wedge \mathbf{p}_{yx}]$, where $x \neq y$;

$[\mathbf{p}_{xy} \vee \mathbf{p}_{yx}]$.

To apply the Compactness Theorem, let any finite subset $\mathcal{W}'$ of $\mathcal{W}$ be given. Let S' be the finite subset of S consisting of those elements x such that a variable of the form $\mathbf{p}_{xy}$ or $\mathbf{p}_{yx}$ (or $\mathbf{p}_{xx}$) occurs in a wff of $\mathcal{W}'$. The resriction $\preceq_1'$ of $\preceq_1$ to S' is a partial ordering of S', which can be extended to a total ordering $\preceq_2'$ of S' since S' is finite. Define the assignment φ as follows: $\varphi\mathbf{p}_{xy} = \mathsf{T}$ if x and y are elements of S' such that $x \preceq_2' y$; if $\mathbf{q}$ is any propositional variable which is not of this form, $\varphi\ \mathbf{q} = \mathsf{F}$. It is easy to see that φ satisfies $\mathcal{W}'$. Thus every finite subset of $\mathcal{W}$ is satisfiable, so by the Compactness Theorem there is an assignment ψ which satisfies $\mathcal{W}$.

Now define the relation $\preceq_2$ on S as follows: $x \preceq_2 y$ iff $\psi\mathbf{p}_{xy} = \mathsf{T}$. Since ψ satisfies $\mathcal{W}$, it is easy to see that $\preceq_2$ is an extension of $\preceq_1$ which is reflexive, transitive, antisymmetric, and is a total ordering of S.

[10] There is admittedly redundancy here, since wffs of this form also have the form above, but it is important to note that wffs of this form are in $\mathcal{W}$.

EXERCISES

X1500. Extend the proof of the Compactness Theorem so that it will apply to a formulation of the propositional calculus in which there are uncountably many propositional variables. Assume that the propositional variables are well ordered.

X1501. A set $\mathcal{S}$ of wffs of propositional calculus is *disjunctively valid* iff for every assignment φ, there exists $\mathbf{A} \in \mathcal{S}$ such that $\models_\varphi \mathbf{A}$. Prove that a set of wffs is disjunctively valid iff some finite subset of it is disjunctively valid.

X1502. Let $\mathcal{T}$ be the set of all assignments of truth values to propositional variables of $\mathcal{P}$. For any nonempty set $\mathcal{S}$ of wffs, let $\mathrm{Sat}(\mathcal{S}) = \{\varphi \in \mathcal{T} \mid (\forall \mathbf{A} \in \mathcal{S} \models_\varphi \mathbf{A}\}$, i.e., $\mathrm{Sat}(\mathcal{S})$ is the set of assignments which satisfy every wff in $\mathcal{S}$. A subset $\mathcal{C}$ of $\mathcal{T}$ is said to be *closed* iff $\mathcal{C} = \mathrm{Sat}(\mathcal{S})$ for some nonempty set $\mathcal{S}$ of wffs.

(1) Prove that every finite union of closed sets is a closed set.
(2) Prove that an arbitrary intersection of closed sets is a closed set.
(3) Prove that $\mathcal{T}$ is closed.

Thus $\mathcal{T}$ may be regarded as a topological space . A family $\mathcal{F}$ of subsets of $\mathcal{T}$ is said to have the *finite intersection property* iff the intersection of each finite subfamily of $\mathcal{F}$ is nonempty, and $\mathcal{T}$ is said to be *compact* iff each family of closed sets which has the finite intersection property has a nonempty intersection.

(4) Use 1501 to show that $\mathcal{T}$ is compact.

§16. Ground Resolution

Although it is in principle a routine problem to determine whether a wff is a tautology of a contradiction by constructing its truth table, this can be an extremely lengthy task if the wff contains a large number of propositional variables. Therefore, additional approaches are needed. In §14 we discussed the method of vertical paths. In this section we shall discuss the surprisingly useful "ground resolution" method for determining whether a wff in conjunctive normal form is a contradiction. Both of these methods gain added significance from the fact that they can be extended to theorem proving procedures for first- and higher-order logic.

If $\mathbf{C}$ and $\mathbf{D}$ are disjunctions of literals, we say that $\mathbf{C}$ and $\mathbf{D}$ are *equivalent as clauses* iff one can be transformed into the other by using the substitutivity of equivalence and the associative, commutative, and idempotent laws of disjunction ($[\mathbf{A} \lor \mathbf{B}] \lor \mathbf{C} \equiv \mathbf{A} \lor [\mathbf{B} \lor \mathbf{C}]$, $\mathbf{A} \lor \mathbf{B} \equiv \mathbf{B} \lor \mathbf{A}$, and $\mathbf{A} \lor \mathbf{A} \equiv \mathbf{A}$). Thus $[[p \lor r] \lor \sim q]$ and $[[r \lor [\sim q \lor r]] \lor [p \lor p]]$ and $[r \lor [\sim q \lor [p \lor \sim q]]]$ are all equivalent to each other. Note that $\mathbf{C}$ and $\mathbf{D}$ are equivalent as clauses iff they contain the same literals, so if $\mathbf{C}$ and $\mathbf{D}$ are not tautologies, they are equivalent as clauses iff $[\mathbf{C} \equiv \mathbf{D}]$ is a tautology. A *clause* is an equivalence class of disjunctions of zero or more literals. There is a natural one-to-one correspondence between clauses and sets of literals; namely, a clause corresponds to the set of literals which occur in any wff of that clause. Thus one may informally think of a clause as a disjunction of literals or as a set of literals. We shall denote a clause by any of its wffs. Thus we may say that the clause $[r \lor [\sim q \lor [p \lor \sim q]]]$ is the same as the clause $p \lor r \lor \sim q$. We denote the empty clause, which is always false, by $\Box$. The disjunction of two clauses is the equivalence class of disjunctions of wffs from the given clauses. It corresponds to the union of the associated sets of literals.

In some contexts it is natural to think of a clause as a wff (rather than as a class of equivalent wffs), and to apply to clauses terminology which is appropriate to wffs. We shall say that a set of clauses is *contradictory, satisfiable*, etc., iff the corresponding set of wffs is. Note that a finite set of clauses is contradictory iff the conjunction of the clauses (regarded as wffs) is a contradiction. Thus a finite set of clauses may be regarded as representing the conjunction of these clauses. Hence every wff in conjunctive normal form is represented by a finite set of clauses.

We may write the disjunction $\mathbf{A} \lor \mathbf{B}$ of clauses $\mathbf{A}$ and $\mathbf{B}$ as $\mathbf{A} \dot{\lor} \mathbf{B}$ when $\mathbf{A}$ and $\mathbf{B}$ are disjoint clauses (i.e., clauses with no common literal).

DEFINITION. A clause $\mathbf{C}$ is a (*ground*) *resolvent* of clauses $\mathbf{A}$ and $\mathbf{B}$ iff there is some propositional variable $\mathbf{p}$ and there are clauses $\mathbf{D}$ and $\mathbf{E}$ such that $\mathbf{A} = p \dot{\lor} \mathbf{D}$, $\mathbf{B} = \sim p \dot{\lor} \mathbf{E}$, and $\mathbf{C} = \mathbf{D} \lor \mathbf{E}$. Note that $\mathbf{C}$ may be $\Box$ if $\mathbf{D}$ and $\mathbf{E}$ are both $\Box$.

REMARK. The ground resolvent rule is a special case of a rule known as the *Cut* rule, which is to infer $\mathbf{D} \lor \mathbf{E}$ from $\mathbf{M} \lor \mathbf{D}$ and $\sim \mathbf{M} \lor \mathbf{E}$. Of course, when $\mathbf{D}$ is $\Box$, the Cut rule reduces to Modus Ponens.

DEFINITION. Let $\mathcal{C}$ be a set of clauses, and let $\mathbf{B}$ be a clause. A *deduction by* (*ground*) *resolution of* $\mathbf{B}$ *from* $\mathcal{C}$ is a sequence $\mathbf{S}_1, \ldots, \mathbf{S}_n$ of clauses such that

$\mathbf{S}_n$ is $\mathbf{B}$ and for each i $(1 \leq i \leq n)$, $\mathbf{S}_i$ is in $\mathcal{C}$ or $\mathbf{S}_i$ is a resolvent of clauses $\mathbf{S}_j$ and $\mathbf{S}_k$ such that $j < i$ and $k < i$. A *refutation of* $\mathcal{C}$ *(by resolution)* is a deduction of $\Box$ from $\mathcal{C}$. We shall write $\mathcal{C} \vdash \mathbf{B}$ to mean there is a deduction of $\mathbf{B}$ from $\mathcal{C}$.

1600 Ground Resolution Theorem. [Robinson, 1965] Let $\mathcal{C}$ be any set of clauses. Then $\mathcal{C}$ is contradictory iff $\mathcal{C} \vdash \Box$.

Proof:[11] Note that $[\mathbf{p} \lor \mathbf{D}] \land [\sim p \lor \mathbf{E}] \supset \centerdot \mathbf{D} \lor \mathbf{E}$ is a tautology even if $\mathbf{D}= \Box$ or $\mathbf{E}= \Box$. Hence resolution preserves truth (with respect to an assignment), so if φ satisfies $\mathcal{C}$ and $\mathcal{C} \vdash S$ then $\mathcal{V}_\varphi S = \mathsf{T}$. Since $\mathcal{V}_\varphi \Box = \mathsf{F}$ for all assignments φ, if $\mathcal{C} \vdash \Box$, then $\mathcal{C}$ must be contradictory.

To prove the converse, by the Compactness Theorem (1501) it clearly suffices to suppose that $\mathcal{C}$ is a finite contradictory set of clauses. For any finite nonempty set $\mathcal{C} = \{\mathbf{S}_1, \ldots, \mathbf{S}_n\}$ of clauses, define *the number of excess literals in* $\mathcal{C}$ to be $e(\mathcal{C}) = \#(\mathbf{S}_1) + \ldots + \#(\mathbf{S}_n) - n$, where $\#(\mathbf{S})$ is the number of (distinct) literals in the clause $\mathbf{S}$. Thus $e(\mathcal{C})$ measures the extent to which $\mathcal{C}$ differs from a set of one-literal clauses. The proof is by complete induction on $e(\mathcal{C})$.

If $e(\mathcal{C}) \leq 0$, then $\Box \in \mathcal{C}$ or every clause in $\mathcal{C}$ is a one-literal clause. In the latter case since $\mathcal{C}$ is contradictory there must be a contradictory pair $\mathbf{p}$ and $\sim \mathbf{p}$ in $\mathcal{C}$, from which one can derive $\Box$ directly.

Next suppose that we are given a contradictory set $\mathcal{C}$ with $e(\mathcal{C}) > 0$, and by inductive hypothesis $\mathcal{D} \vdash \Box$ for every contradictory set $\mathcal{D}$ such that $e(\mathcal{D}) < e(\mathcal{C})$.

If $\mathcal{C}$ contains $\Box$ we are done. Otherwise $\mathcal{C}$ must contain some clause $\mathbf{A}$ with at least two literals, so we may write $\mathbf{A}= \mathbf{B} \dot{\lor} \mathbf{L}$ with $\mathbf{L}$ a literal and $\mathbf{B}$ a nonempty clause. Let $\mathcal{S} = \mathcal{C} - \{\mathbf{A}\}$ so $\mathcal{C} = \mathcal{S} \cup \{\mathbf{B} \lor \mathbf{L}\}$. Let $\mathcal{C}_1 = \mathcal{S} \cup \{\mathbf{B}\}$ and $\mathcal{C}_2 = \mathcal{S} \cup \{\mathbf{L}\}$. It is easy to see that $\mathcal{C}_1$ and $\mathcal{C}_2$ are contradictory, for if φ satisfies $\mathcal{C}_2$ (for example), then φ satisfies $\mathcal{S}$ and $\mathcal{V}_\varphi \mathbf{L} = \mathsf{T}$ so $\mathcal{V}_\varphi [\mathbf{B} \lor \mathbf{L}] = \mathsf{T}$ so φ satisfies $\mathcal{C}$. Also $e(\mathcal{C}_1)$ and $e(\mathcal{C}_2)$ are less than $e(\mathcal{C})$ (even if $\mathbf{B} \in \mathcal{S}$ or $\mathbf{L} \in \mathcal{S}$), since $e(\mathcal{C}) = e(\mathcal{S}) + \#(\mathbf{B})$. Therefore by inductive hypothesis $\mathcal{C}_1 \vdash \Box$ and $\mathcal{C}_2 \vdash \Box$.

We next show that $\mathcal{C} \vdash \Box$ or $\mathcal{C} \vdash \mathbf{L}$. Let $\mathbf{A}_1, \ldots, \mathbf{A}_m$ be a deduction of $\Box$ from $\mathcal{C}_1$. We prove by induction on i that $\mathcal{C} \vdash \mathbf{A}_i$ or $\mathcal{C} \vdash \mathbf{A}_i \lor \mathbf{L}$ for $1 \leq i \leq m$.

Case 1. $\mathbf{A}_i \in \mathcal{C}_1$. Then $\mathbf{A}_i \in \mathcal{S} \subseteq \mathcal{C}$, or $\mathbf{A}_i = \mathbf{B}$ so $\mathbf{A}_i \lor \mathbf{L} = \mathbf{B} \lor \mathbf{L} \in \mathcal{C}$.

[11] This proof is taken from [Anderson and Bledsoe, 1970].

Case 2 $\mathbf{A}_i$ is a resolvent of $\mathbf{A}_j$ and $\mathbf{A}_k$, where $j < i$ and $k < i$. Thus $\mathbf{A}_j = \mathbf{D}_j \dot{\vee}\ \mathbf{p}$, $\mathbf{A}_k = \mathbf{D}_k \dot{\vee} \sim \mathbf{p}$, and $\mathbf{A}_i = \mathbf{D}_j \vee \mathbf{D}_k$. $\mathbf{L}$ might be $\mathbf{p}$, or a literal in $\mathbf{D}_j$ or $\mathbf{D}_k$, but in all cases the result of resolving $\mathbf{A}_j$ or $\mathbf{A}_j \vee \mathbf{L}$ with $\mathbf{A}_k$ or $\mathbf{A}_k \vee \mathbf{L}$ is seen to be $\mathbf{A}_i$ or $\mathbf{A}_i \vee \mathbf{L}$. Hence using the inductive hypothesis for $\mathbf{A}_j$ and $\mathbf{A}_k$ we see that $\mathcal{C} \vdash \mathbf{A}_i$ or $\mathcal{C} \vdash \mathbf{A}_i \vee \mathbf{L}$.

If $\mathcal{C} \vdash \square$ we are done. Otherwise $\mathcal{C} \vdash \mathbf{L}$ and $\mathcal{S} \subseteq \mathcal{C}$ and $\mathcal{S} \cup \{\mathbf{L}\} \vdash \square$ so $\mathcal{C} \vdash \square$. ■

EXAMPLE: Use the resolution method to show that the following wff is a tautology: $[p \supset \centerdot q \wedge \centerdot r \vee s] \wedge [\sim q \vee \sim r] \supset \centerdot p \supset s$.

Solution. The negation of the given wff $\mathbf{A}$ must be put into conjunctive normal form. We do this in steps as follows:

(**B**) $[\sim p \vee \centerdot q \wedge \centerdot r \vee s] \wedge [\sim q \vee \sim r] \wedge \sim \centerdot p \supset s$.

(**C**) $[\sim p \vee q] \wedge [\sim p \vee r \vee s] \wedge [\sim q \vee \sim r] \wedge p \wedge \sim s$

The wff $\mathbf{C}$ is equivalent to $\sim \mathbf{A}$, so $\mathbf{A}$ is a tautology iff $\mathbf{C}$ is contradictory, which is the case iff $\square$ can be derived by resolution from the following five clauses:

(1) $\sim p \vee q$
(2) $\sim p \vee r \vee s$
(3) $\sim q \vee \sim r$
(4) p
(5) $\sim s$

This we proceed to do as follows:

(6)	q	1,4
(7)	$\sim r$	3,6
(8)	$r \vee s$	2,4
(9)	s	7,8
(10)	$\square$	5, 9

EXERCISES

X1600. Show that the following wff is contradictory:

$$\begin{aligned}
&[\sim d \vee \sim e] \wedge [s \vee q] \\
&\wedge [h \vee \blacksquare\, k \wedge [m \vee e \vee a] \wedge [p \vee j \vee b]] \\
&\wedge [[\sim s \wedge \sim q] \vee \blacksquare \sim h \wedge [\sim r \vee \sim a] \wedge [r \vee a]] \\
&\wedge [\sim r \vee \blacksquare \sim m \wedge \blacksquare \sim k \vee d \vee a] \\
&\wedge [r \vee m \vee \blacksquare \sim p \wedge g \wedge \sim c] \\
&\wedge [\sim a \vee \blacksquare [\sim k \vee \sim m] \wedge [\sim b \vee c] \wedge [\sim j \vee \sim g]].
\end{aligned}$$

X1601. *Definition:* The clause **A** *subsumes* the clause **B** iff every literal which occurs in **A** also occurs in **B**. (Thus when we think of clauses as sets of literals, **A** subsumes **B** iff $\mathbf{A} \subseteq \mathbf{B}$.)

Suppose $\mathcal{S}$ is a set of clauses which has a refutation of length n, and $\mathcal{S}'$ is a set of clauses such that every clause in $\mathcal{S}$ is subsumed by some clause in $\mathcal{S}'$. Show that $\mathcal{S}'$ has a refutation of length $\leq n$.

In X1602-X1607, decide whether $\Box$ can be derived by resolution from the given clauses. If so, derive it; number the clauses and indicate from which clauses each derived clause is obtained. If $\Box$ cannot be derived, prove that it cannot be derived.

X1602.
(1) $p \vee r$
(2) $q \vee s$
(3) $\sim p \vee \sim q$
(4) $\sim r \vee \sim s$
(5) $q \vee r$
(6) $p \vee s$

X1603.
(1) $p \vee q$
(2) $\sim p \vee r \vee \sim q \vee p$
(3) $p \vee \sim r$
(4) $q \vee \sim p \vee \sim q$
(5) $q \vee \sim r \vee p$
(6) $r \vee q \vee \sim p \vee \sim r$
(7) $\sim r \vee q$

X1604.
(1) $\sim q \vee r$
(2) $\sim r \vee p$
(3) $\sim p \vee q$
(4) $\sim p \vee \sim q \vee \sim r$
(5) $p \vee q \vee r$

X1605.
(1) $p \vee q \vee r$
(2) $\sim q \vee s$
(3) $\sim r \vee s$
(4) $p \vee \sim s$
(5) $\sim p$
(6) $q \vee \sim r$

X1606.

(1) $\sim p \vee q \vee \sim s$
(2) $p \vee r$
(3) $\sim q \vee s$
(4) $r \vee q \vee s$
(5) $\sim p \vee \sim r \vee \sim q \vee s$
(6) $\sim r \vee q \vee \sim s$
(7) $p \vee \sim r \vee q \vee s$
(8) $\sim q \vee \sim s$

X1607.

(1) $p_1 \vee q_1 \vee r_1$
(2) $r \vee p_1 \vee s$
(3) $\sim p_1$
(4) p
(5) $\sim p \vee q$
(6) $\sim p \vee \sim r \vee \sim q_1$
(7) $\sim q \vee \sim p \vee \sim q_1 \vee \sim s$
(8) $\sim p \vee \sim q \vee \sim r \vee \sim r_1$
(9) $\sim p \vee \sim r_1 \vee \sim s$

In X1608-X1610, convert the negation of the given wff to a set of clauses and try to derive □. If you fail to derive □, show that the given wff is not a tautology.

X1608. $[p \wedge {\scriptstyle\blacksquare}\, q \supset r] \supset {\scriptstyle\blacksquare}[s \vee p] \wedge {\scriptstyle\blacksquare} \sim r \supset \sim q$

X1609. $[[p \vee q] \wedge [\sim p \vee \sim {\scriptstyle\blacksquare}\, r \supset s] \wedge [r \supset s] \wedge [r \vee \sim q]] \supset {\scriptstyle\blacksquare}\, q \wedge s$

X1610. $[[p \vee q] \wedge [\sim s \supset {\scriptstyle\blacksquare}\, p \supset r] \wedge [[r \vee s] \supset q]] \supset q$

X1611. Suppose we modify the definition of *ground resolvent* to read as follows: a clause **C** is a ground resolvent of clauses **A** and **B** iff there is some propositional variable **p** such that $\mathbf{A} = \sim \mathbf{p} \vee \mathbf{C}$ and $\mathbf{B} = \mathbf{p}$. With this modified definition is Theorem 1600 still provable? Justify your answer.

Chapter 2

First-Order Logic

We next consider first-order logic, which is also called quantification theory and predicate calculus.[1] Initially we discuss first-order logic without equality, and then we discuss first-order logic with equality in §26.

§20. The Language of $\mathcal{F}$

Having discussed in Chapter 1 how to combine statements using propositional connectives, we next turn to the problem of formally representing the statements themselves in a way that shows more of their structure. We shall focus on those aspects of the structure of statements which are most important for representing reasoning involving these statements.

Statements make assertions about entities (such as people, animals, plants, colors, qualities, numbers, abstractions, ideas, feelings, etc.), which are designated by nouns and pronouns. For convenience we shall refer to these entities as *objects* or *individuals*, slightly extending the usual meanings of these words in English. In our formal language we shall introduce *terms* (which will be defined below) to denote (serve as names for) these individuals.

Statements can be regarded as asserting that individuals have certain properties, or that individuals are related to each other in certain ways. For example, "Jack will go to the picnic" expresses the assertion that the individual Jack has the property of being an individual who will go to the

[1]Frege [Frege, 1879] is generally credited with first developing those aspects of first-order logic which go beyond propositional calculus, though others independently developed similar ideas somewhat later, and Frege's notation never found general favor. See [Church, 1956, pp. 288–294] for more information about the history of first-order logic.

picnic. "112 is prime" expresses the assertion (which happens to be false) that the individual 112 has the property of being a prime number. "Jill votes for Alice" expresses the assertion that the individuals "Jill" and "Alice" are related by the relation "votes for". "$5 < 112$" expresses the assertion that the individuals 5 and 112 are related by the relation " $<$ ". Another way of saying this is to assert that the relation " $<$ " *holds* between 5 and 112.

Since it is sometimes complicated to analyze a sentence in a natural language such as English or Chinese, it is convenient to have a quite simple and uniform syntax to represent the assertion that the entity t has the property P, or that the relation R holds between the entities s and t. We shall use Pt to represent the former, and Rst to represent the latter. (Sometimes it will be convenient to introduce commas or parentheses and write these as $P(t)$ and $R(s,t)$, but the commas and parentheses are in principle unnecessary.) Thus, if we let "Prime" be the property of being a prime number, we can express "112 is prime" as "Prime(112)". Similarly, we can express "Jill votes for Alice" as "VotesFor(Jill, Alice)". We call names for properties and relations *predicates*, and they play such a fundamental role in first-order logic that it is sometimes called predicate calculus.

In order to be able to deal adequately with reasoning involving individuals, one must be able to represent statements containing words like "all", "every", and "some". To illustrate the relevant ideas, let us suppose that a class of girls holds an election for class president. Each girl is given a ballot containing the names of all the girls in the class, and is allowed to vote for as many members of the class as she pleases. Suppose we want to express in a formal language the assertion that everyone votes for Alice. One way one might try to formalize this is to introduce a term "everyone" into the formal language, and express the assertion as "VotesFor(everyone, Alice)". In a similar way, to express the assertion that Jill votes for someone, one might introduce a term "someone" into the formal language, and write "VotesFor(Jill, someone)". If such statements were permitted in the formal language, it would be very natural to also permit "VotesFor(everyone, someone)". What should this mean? One obvious possibility is "Everyone votes for someone". However, there is another possibility: "There is someone for whom everyone votes". Note that these express quite distinct assertions; the latter statement asserts that everyone votes for the same person, while the former statement does not.

Contemplation of this and other examples (particularly those dealing with more complicated statements) leads to the conclusion that a better way of formally representing assertions involving words like "all", "every", and "some" is needed. The solution that has been developed is to use *indi-*

VotesFor(Jill, Alice)	Jill votes for Alice.
$\forall x$ VotesFor$(x, Alice)$	Everyone votes for Alice.
$\forall y$ VotesFor$(Jill, y)$	Jill votes for everyone.
$\forall x$ VotesFor(x, x)	Everyone votes for herself.
$\forall x \forall y$ VotesFor(x, y)	Everyone votes for everyone.
$\forall x \exists y$ VotesFor(x, y)	Everyone votes for someone.
$\exists y \forall x$ VotesFor(x, y)	There is someone for whom everyone votes.
$\exists x \forall y$ VotesFor(x, y)	There is someone who votes for everyone.
$\forall y \exists x$ VotesFor(x, y)	Everyone gets someone's vote.
$\exists x \exists y$ VotesFor(x, y)	Someone votes for someone.
$\exists x$ VotesFor(x, x)	Someone votes for herself.
$\exists x$ VotesFor$(x, Alice)$	Someone votes for Alice.
$\exists y$ VotesFor$(Jill, y)$	Jill votes for someone.

Figure 2.1: Statements about voting in symbolic form and in English.

vidual variables to refer to arbitrary (unspecified) individuals, and express the assertion that "Everyone votes for Alice" as "For every x, x votes for Alice", and express the assertion that "Jill votes for someone" as "There is a y such that Jill votes for y." Just as we introduced symbols for propositional connectives, we introduce the symbol $\forall$ to abbreviate the phrase "for every" (or "for all") and the symbol $\exists$ and write "$\exists y$" to abbreviate the phrase "there is a y such that" (where y is any individual variable). Now we can express "Everyone votes for someone" as "$\forall$x$\exists y$ VotesFor(x, y)" and "There is someone for whom everyone votes" as "$\exists y \forall x$ VotesFor(x, y)". With this symbolic representation, the distinction between these assertions is made very clear. In Figure 2.1 we list a variety of statements about voting in symbolic form and in English. Note how the symbolic representations of these statements make it easy to focus on the essential logical differences between them.

Expressions of the form $\forall$x and $\exists$x are called *quantifiers*. They play such a fundamental role in first-order logic that it is sometimes called quantification theory. The reason we actually use the name "first-order logic" will become evident when we discuss higher-order logic at the beginning of section §50.

Just as $\vee$ can be defined in terms of $\sim$ and $\wedge$, $\exists$ can be defined in terms of $\sim$ and $\forall$. We can see this from the following example. Suppose our class of girls decides to vote on some issue, but each girl has the option of voting or abstaining (not voting). If we write "x votes" as "Votes(x)", then "$\sim$Votes(x)" means "x does not vote" or "x abstains". Therefore

"$\forall$x$\sim$Votes(x)" means "everyone abstains", and "$\sim \forall x \sim$Votes(x)" means "not everyone abstains", which is of course equivalent to the assertion that "someone votes". From this and other examples it can be seen that $\exists xPx$ always has the same logical meaning as $\sim \forall x \sim Px$, provided that there is at least one individual.[2] This provides a contextual definition of $\exists$ in terms of $\sim$ and $\forall$. Therefore, for the sake of economy, $\exists$ will not be a primitive symbol of the formal system $\mathcal{F}$ of first-order logic which we shall soon introduce.

Sometimes we wish to refer to individuals by more complex expressions than we have used so far, such as "Jill's father", "$x + 3$", or "the value of the diamond". We can do this in a uniform way by having in our language notations for functions which map individuals to individuals. If we let f(x) mean "the father of x", g(x, y) mean "$x + y$", and V(x) mean "the value of x", then we can represent "Jill's father" as f(Jill), "$x + 3$" as g(x, 3), and "the value of the diamond" as V(the diamond).

The symbols we will use to refer to individuals, functions, and predicates will be *variables* or *constants.* Variables are used to refer to arbitrary elements of the appropriate type, whereas constants denote particular elements. Thus, variables may occur in quantifiers, but constants may not.

Now we are ready to start describing the system $\mathcal{F}$ of first-order logic. The *primitive symbols* of $\mathcal{F}$ are the following:

(a) Improper symbols: [ ] $\sim \lor \forall$
(b) Individual variables: $u\ v\ w\ x\ y\ z\ u_1\ v_1\ w_1\ \ldots\ u_2\ \ldots$
(c) n-ary function variables: $f^n\ g^n\ h^n\ f_1^n\ g_1^n\ h_1^n\ \ldots$ for each natural number $n \geq 1$.
(d) Propositional variables: $p\ q\ r\ s\ p_1\ q_1\ r_1\ s_1\ \ldots$
(e) n-ary predicate variables: $P^n\ Q^n\ R^n\ S^n\ P_1^n\ Q_1^n\ R_1^n\ S_1^n \ldots$ for each natural number $n \geq 1$.

An n-ary symbol is called *unary* or *monadic* if $n = 1$, and it is called *binary* if $n = 2$. There are denumerably many variables of each type. Individual variables may be regarded as 0-ary function variables, and propositional variables may be regarded as 0-ary predicate variables. In addition, there may be finitely or infinitely many individual, function, propositional, or predicate constants. We also admit systems in which there are no variables of certain types; however, individual variables must always be present, and there must be predicate variables or at least one predicate constant. Each choice of constants and variables determines a different formulation of the

[2]There is little need for a logical system to discuss nothing, so, as is customary, our logical system will be designed with the tacit assumption that there is at least one individual under discussion.

system $\mathcal{F}$, of course; thus we use $\mathcal{F}$ ambiguously as the name of any one of a variety of quite similar systems.

The *terms* and *wffs* of $\mathcal{F}$ are defined inductively by the following *formation rules*:

(a) Each individual variable or constant is a term.
(b) If $\mathbf{t}_1, \ldots, \mathbf{t}_n$ are terms and $\mathbf{f}^n$ is an n-ary function variable or constant, then $\mathbf{f}^n\mathbf{t}_1 \ldots \mathbf{t}_n$ is a term.
(c) If $\mathbf{t}_1, \ldots, \mathbf{t}_n$ are terms and $\mathbf{P}^n$ is an n-ary predicate variable or constant, then $\mathbf{P}^n\mathbf{t}_1 \ldots \mathbf{t}_n$ is a wff. Also, each propositional variable standing alone is a wff. (Wffs of these forms are called *atomic* or *elementary* wffs.)
(d) If $\mathbf{A}$ is a wff, so is $\sim\mathbf{A}$.
(e) If $\mathbf{A}$ and $\mathbf{B}$ are wffs, so is $[\mathbf{A} \lor \mathbf{B}]$.
(f) If $\mathbf{A}$ is a wff and $\mathbf{x}$ is an individual variable, then $\forall\mathbf{x}\mathbf{A}$ is a wff.

Sometimes $\forall\mathbf{x}\mathbf{A}$ is written as $(\forall\mathbf{x})\mathbf{A}$ or simply as $(\mathbf{x})\mathbf{A}$. We shall often omit the numerical superscripts from function and predicate symbols in terms and wffs, since the context will show what they must be.

We leave it to the reader (exercises X2006 and X2007) to formulate definitions of the sets of terms and wffs in the style of the definition of the set of wffs in §10 and to formulate and prove analogues of Theorem 1000 (the Principle of Induction on the Construction of a Wff) for terms and wffs of $\mathcal{F}$.

As in the case of propositional calculus, we shall use $\mathbf{A}$, $\mathbf{B}$, $\mathbf{C}$, $\mathbf{D}$, $\mathbf{E}$, etc., as syntactical variables ranging over wffs. Similarly, we use $\mathbf{u}$, $\mathbf{v}$, $\mathbf{w}$, $\mathbf{x}$, $\mathbf{y}$, $\mathbf{z}$, etc., as syntactical variables for individual variables.

While in first-order logic only individual variables may be quantified, in second-order logic the other types of variables may be quantified too. Since variables designate arbitrary elements of the appropriate type, it is normally appropriate to substitute arbitrary expressions of the appropriate type for variables, under suitable conditions. Such substitution for constants is generally not appropriate since it would have the effect of implying that all elements have the special properties of the element denoted by the constant in question.

For the sake of economy, starting with §25 we shall restrict our attention to formulations of $\mathcal{F}$ in which there are no variables other than individual variables.

We adopt from §10 the conventions for regarding $[\mathbf{A} \land \mathbf{B}]$, $[\mathbf{A} \supset \mathbf{B}]$, and $[\mathbf{A} \equiv \mathbf{B}]$ as abbreviations for wffs (now wffs of $\mathcal{F}$). In addition,

$$\exists \mathbf{x} \mathbf{A} \text{ shall stand for } \sim \forall \mathbf{x} \sim \mathbf{A}.$$

In addition to statements of the sort we have discussed thus far, we need to be able to symbolically represent statements such as "every natural number is a real number" and "some real numbers are natural numbers". We shall use $\mathbb{N}$ as a notation for the set of natural numbers, and $\mathbb{R}$ as a notation for the set of real numbers. Associated with every property P is the set S of entities which have property P, and associated with every set S is the property P of being in the set S. Whether one starts with P or with S, one sees that an entity x has property P (i.e., Px) if and only if $x \in S$, so it is sometimes convenient to identify a property P with the set S of entities which have that property, and use the same notation for both. For example, an entity is a natural number if and only if it is in the set of natural numbers, and we may write either $\mathbb{N}x$ or $x \in \mathbb{N}$ to express the assertion that x is a natural number.

We shall express the assertion that "every natural number is a real number" by "for every x in $\mathbb{N}$, x is a real number", which we write in symbolic form as $(\forall x \in \mathbb{N})\mathbb{R}x$. Similarly, we express the assertion that "some real numbers are natural numbers" by the statement "there is an x in $\mathbb{R}$ such that x is a natural number", and write this symbolically as $(\exists x \in \mathbb{R})\mathbb{N}x$. More generally, if $\mathbf{A}$ is a statement about $\mathbf{x}$, and $\mathbf{P}$ is a property, $(\forall \mathbf{x} \in \mathbf{P})\mathbf{A}$ means "the statement $\mathbf{A}$ is true for every element $\mathbf{x}$ having the property $\mathbf{P}$", and $(\exists \mathbf{x} \in \mathbf{P})\mathbf{A}$ means "there exists an element $\mathbf{x}$ having the property $\mathbf{P}$ such that the statement $\mathbf{A}$ is true".

Let us now observe that these statements can be expressed in terms of the notations we have already introduced in the system $\mathcal{F}$. We first establish that in every situation, $(\forall \mathbf{x} {\in} \mathbf{P})\mathbf{A}$ is true if and only if $\forall \mathbf{x}[\mathbf{Px} \supset \mathbf{A}]$ is true.

Let us start by supposing that $(\forall \mathbf{x} {\in} \mathbf{P})\mathbf{A}$ is true, and let $\mathbf{x}$ be any entity. If $\mathbf{Px}$ is false, then $[\mathbf{Px} \supset \mathbf{A}]$ is true, but if $\mathbf{Px}$ is true, then $\mathbf{x} {\in} \mathbf{P}$ so our assumption guarantees that $\mathbf{A}$ is true too, so $[\mathbf{Px} \supset \mathbf{A}]$ is true in this case also. Since we can reach this conclusion for every entity $\mathbf{x}$, we see that $\forall \mathbf{x}[\mathbf{Px} \supset \mathbf{A}]$ is true.

Next we suppose that $\forall \mathbf{x}[\mathbf{Px} \supset \mathbf{A}]$ is true. Let $\mathbf{x}$ be any entity in $\mathbf{P}$. Then our assumption guarantees that $\mathbf{A}$ is true, so we have established that $(\forall \mathbf{x} {\in} \mathbf{P})\mathbf{A}$, completing the proof of our assertion.

In a similar way, we can see that in every situation, $(\exists \mathbf{x} {\in} \mathbf{P})\mathbf{A}$ is true if and only if $\exists \mathbf{x}[\mathbf{Px} \wedge \mathbf{A}]$ is true. We first suppose that $(\exists \mathbf{x} {\in} \mathbf{P})\mathbf{A}$ is true, and choose some entity $\mathbf{x}$ in $\mathbf{P}$ such that $\mathbf{A}$. Clearly $[\mathbf{Px} \wedge \mathbf{A}]$ is true, so there is an entity $\mathbf{x}$ such that $[\mathbf{Px} \wedge \mathbf{A}]$ is true, which means that $\exists \mathbf{x}[\mathbf{Px} \wedge \mathbf{A}]$

is true. Conversely, if we suppose that $\exists\mathbf{x}[\mathbf{Px} \wedge \mathbf{A}]$ is true, we can choose an entity $\mathbf{x}$ such that $[\mathbf{Px} \wedge \mathbf{A}]$ is true, so $\mathbf{x}\in\mathbf{P}$, so it is indeed true that $(\exists\mathbf{x}\in\mathbf{P})\mathbf{A}$.

Motivated by these considerations, we introduce the following additional abbreviations for the system $\mathcal{F}$. Whenever $\mathbf{P}$ is a monadic predicate variable or constant and $\mathbf{A}$ is a wff,

$$(\forall\mathbf{x} \in \mathbf{P})\mathbf{A} \text{ shall stand for } \forall\mathbf{x}\,[\mathbf{P}x \supset \mathbf{A}], \quad \text{and}$$

$$(\exists\mathbf{x} \in \mathbf{P})\mathbf{A} \text{ shall stand for } \exists\mathbf{x}\,[\mathbf{P}x \wedge \mathbf{A}].$$

$(\forall\mathbf{x} \in \mathbf{P})$ and $(\exists\mathbf{x} \in \mathbf{P})$ are called *relativized quantifiers.*

The wff $\mathbf{A}$ is the *scope* of the quantifier $\forall\mathbf{x}$ in $\forall\mathbf{x}\mathbf{A}$, of $\exists\mathbf{x}$ in $\exists\mathbf{x}\mathbf{A}$, of $(\forall\mathbf{x} \in \mathbf{P})$ in $(\forall\mathbf{x} \in \mathbf{P})\mathbf{A}$, and of $(\exists\mathbf{x} \in \mathbf{P})$ in $(\exists\mathbf{x} \in \mathbf{P})\mathbf{A}$. We shall continue to use the conventions concerning dots and omissions of brackets introduced in §10, supplemented by the convention that in restoring brackets to an abbreviation of a wff, the quantifiers $\forall\mathbf{x}$ and $\exists\mathbf{x}$, like the connective $\sim$, are to be given the smallest possible scope.

EXAMPLE: Suppose we interpret the predicate constants P, T, Z, B, and R so that Px means "x is a person", Ty means "y is a ticket", Zz means "z is a prize", Bxy means "x buys y", and Rxz means "x receives z". Then $\exists y\,[Ty \wedge Bxy]$ means "x buys a ticket", $\exists z\,[Zz \wedge Rxz]$ means "x receives a prize", and

$$\forall x\,[Px \supset \centerdot\, \exists y\,[Ty \wedge Bxy] \supset \exists z\,\centerdot\, Zz \wedge Rxz]$$

means "everyone who buys a ticket receives a prize". Of course, this sentence could also be written as

$$(\forall x \in P)\,\centerdot\,(\exists y \in T)Bxy \supset (\exists z \in Z)Rxz.$$

EXAMPLE: Interpret P, T, D, and S so that Px means "x is a person", Ty means "y is trouble", Kuv means "u knows v" and Sxy means "x has seen y". Also, let the individual constant i denote the person who is speaking. Then

$$\sim \exists x\,[Px \wedge \forall y\,\centerdot\, Ty \wedge Siy \supset Kxy]$$

may be translated into idiomatic English as "Nobody knows the trouble I've seen". Of course, a more literal translation would be "Nobody knows all the trouble I've seen", but in the first translation the word "all" is tacitly understood.

When one interprets wffs of first-order logic, one has in mind some set which is called the *universe* or the *domain of individuals*, such as the set of real numbers or the set of all objects on earth, and one interprets a wff of the form $\forall \mathbf{x} \mathbf{A}$ as meaning "$\mathbf{A}$ is true for every element $\mathbf{x}$ in the domain of individuals". When this domain is finite and one has names for all the individuals, one can dispense with the quantifier $\forall \mathbf{x}$ and express the same idea in terms of conjunctions. For example, suppose there are just three individuals in the domain, and the individual constants, a, b, and c are used as names for them.[3] Then $[Pa \wedge Pb \wedge Pc]$ expresses the same idea as $\forall x Px$. Thus we see that $\forall$ can be regarded as a generalized conjunction operator. If the domain has infinitely many individuals, with names $a_1, a_2, a_3, \ldots$, then $\forall x Px$ expresses the idea of the infinite conjunction $[Pa_1 \wedge Pa_2 \wedge \ Pa_3 \ \wedge \ \ldots \ ]$. Of course, the latter expression is not a wff of first order logic.

In a similar way, $\exists$ may be regarded as a generalized form of disjunction, and $\exists x \ Px$ can be written instead of $[Pa \vee Pb \vee Pc]$ in the domain $\{a, b, c\}$. The relationship between $\forall$ and $\exists$ mirrors the relationship between $\wedge$ and $\vee$. Our definition of $\exists x Px$ as $\sim \forall x \sim Px$ mirrors the generalized DeMorgan Law

$$[Pa \vee Pb \vee Pc] \equiv \sim [\sim Pa \wedge \ \sim Pb \wedge \sim Pc].$$

This relationship also holds for the relativized quantifiers $(\forall \mathbf{x} \in \mathbf{S})$ and $(\exists \mathbf{x} \in \mathbf{S})$, and all wffs of the forms $(\exists \mathbf{x} \in \mathbf{S})\mathbf{A} \equiv \sim (\forall \mathbf{x} \in \mathbf{S}) \sim \mathbf{A}$ will be seen (X2139) to be theorems of $\mathcal{F}$.

Note that in the domain $\{a, b, c\}$, $\forall x \exists y Rxy$ can also be expressed as

$$\forall x \, [Rxa \vee Rxb \vee Rxc] \text{, as}$$

$$[\exists y Ray \wedge \exists y Rby \wedge \exists y Rcy] \text{, and as}$$

$$[[Raa \vee Rab \vee Rac] \wedge [Rba \vee Rbb \vee Rbc] \wedge [Rca \vee Rcb \vee Rcc]] .$$

The well-formed (wf) parts of a wff $\mathbf{B}$ are the consecutive subformulas of $\mathbf{B}$ (including $\mathbf{B}$ itself) which are wffs.

[3]The individual constants a, b, and c are symbols of the object language $\mathcal{F}$, but we shall also use them in our meta-language as names for the individuals in the domain, and thus refer to the domain as $\{a, b, c\}$. Of course, these symbols are also used in the meta-language as names for themselves as symbols of the object language. The context should make clear which usage is intended.

An occurrence of a variable $\mathbf{x}$ in a wff $\mathbf{B}$ is *bound* iff it is in a wf part of $\mathbf{B}$ of the form $\forall\mathbf{x}\mathbf{C}$; otherwise, it is *free*. For example, in the wff below, the occurrences of variables with "f" above them are free, while those with "b" below them are bound:

$$[\sim[\forall\underset{b}{x}P\overset{f}{u}\underset{b}{x} \lor \forall\underset{b}{v}\sim\forall\underset{b}{u}[P\underset{b}{v}\underset{b}{u} \lor \forall\underset{b}{u}P\underset{b}{u}\overset{f}{x}]] \lor \forall\underset{b}{y}P\underset{b}{y}\overset{f}{z}]$$

The bound [free] variables of a wff are those which have bound [free] occurrences in the wff. The bound variables in this example are x, v, u, and y, while the free variables are u, x, and z. Note that a variable can be both bound and free in a wff (at different occurrences). A wff or term without free individual variables is said to be *closed*. A closed wff is sometimes called a *cwff*. A *sentence* is a wff without free variables of any type.

Note that a wff can be regarded as a statement about (the objects designated by) its free variables, but that it should not be regarded as asserting anything about its bound variables, which simply serve as pronouns. For example, if the predicate constants C and L are interpreted so that Cy means "y is a cat" and Lxy means "x likes y", then $\forall y\,[Cy \supset Lxy]$ means "x likes all cats".

If $\mathbf{x}_1, \ldots, \mathbf{x}_n$ are distinct individual variables and $\mathbf{t}_1, \ldots, \mathbf{t}_n$ are terms, the notations $\left(\mathsf{S}^{\mathbf{x}_1\ldots\mathbf{x}_n}_{\mathbf{t}_1\ldots\mathbf{t}_n}\mathbf{A}\right)$ and $\left(\underset{\cdot}{\mathsf{S}}{}^{\mathbf{x}_1\ldots\mathbf{x}_n}_{\mathbf{t}_1\ldots\mathbf{t}_n}\mathbf{A}\right)$ shall denote, respectively, the results of simultaneously substituting $\mathbf{t}_i$ for $\mathbf{x}_i$ (for each i such that $1 \le i \le n$) at all, and at all free, occurrences of $\mathbf{x}_i$ in $\mathbf{A}$. (The parentheses will generally be omitted when the context makes them unnecessary.) The former concept was discussed in §10, and the latter may be defined inductively as follows: Let θ stand for $\underset{\cdot}{\mathsf{S}}{}^{\mathbf{x}_1\ldots\mathbf{x}_n}_{\mathbf{t}_1\ldots\mathbf{t}_n}$.

(a) $\theta\mathbf{A} = \mathsf{S}^{\mathbf{x}_1\ldots\mathbf{x}_n}_{\mathbf{t}_1\ldots\mathbf{t}_n}\mathbf{A}$ if $\mathbf{A}$ is atomic.
(b) $(\theta \sim \mathbf{B}) = \sim(\theta\mathbf{B})$.
(c) $\theta\,[\mathbf{B} \lor \mathbf{C}] = [(\theta\mathbf{B}) \lor (\theta\mathbf{C})]$.

(d)
$$\begin{aligned} \theta\forall\mathbf{y}\mathbf{B} &= \forall\mathbf{y}\theta\mathbf{B} \text{ if } \mathbf{y} \notin \{\mathbf{x}_1, \ldots, \mathbf{x}_n\}; \\ &= \forall\mathbf{y}\underset{\cdot}{\mathsf{S}}{}^{\mathbf{x}_1\ldots\mathbf{x}_{i-1}\mathbf{x}_{i+1}\ldots\mathbf{x}_n}_{\mathbf{t}_1\ldots\,\mathbf{t}_{i-1}\,\mathbf{t}_{i+1}\ldots\,\mathbf{t}_n}\mathbf{B} \text{ if } \mathbf{y} = \mathbf{x}_i. \end{aligned}$$

$\mathsf{S}^{\mathbf{x}_1\ldots\mathbf{x}_n}_{\mathbf{t}_1\ldots\mathbf{t}_n}$ and $\underset{\cdot}{\mathsf{S}}{}^{\mathbf{x}_1\ldots\mathbf{x}_n}_{\mathbf{t}_1\ldots\mathbf{t}_n}$ will be called *proper* substitutions.

Sometimes the following substitution notation is convenient. A wff $\mathbf{A}$ is written as $\mathbf{A}(\mathbf{x}_1, \ldots, \mathbf{x}_n)$, and then $\mathbf{A}(\mathbf{t}_1, \ldots, \mathbf{t}_n)$ is understood to stand for $\underset{\cdot}{\mathsf{S}}{}^{\mathbf{x}_1\ldots\mathbf{x}_n}_{\mathbf{t}_1\ldots\mathbf{t}_n}\mathbf{A}$.

In general, it is appropriate to substitute a term [or wff] $\mathbf{t}$ for the free occurrences of an individual [or propositional] variable $\mathbf{m}$ in a wff $\mathbf{A}$ only if this can be done without the free variables of **t** becoming bound by quantifiers of **A**. For example, most wffs of the form $\forall \mathbf{x}\mathbf{A} \supset \mathsf{S}^{\mathbf{x}}_{\mathbf{t}}\mathbf{A}$ (where **t** is a term) are valid (true in all interpretations). In particular, consider the wff $\forall x \exists y\, [x < y] \supset \exists y\, [5 < y]$, where $x < y$ is an abbreviation for Pxy and 5 is an individual constant. This is valid, and in the domain of real numbers it can be interpreted as saying, "If for every number x there is a number y which is greater than x, then there is a number greater than 5". However, if we let the term **t** be the variable y instead of the constant 5, we obtain the non-valid wff $\forall x \exists y\, [x < y] \supset \exists y\, [y < y]$, which can be interpreted as saying, "If for every number x there is a number y greater than x, then there is a number greater than itself".

Thus, we are led to formulate the following definition of the phrase "**t** is *free for* **m** *in* **A**", which expresses in a precise way the assertion that the free variables of **t** will not become bound by the quantifiers of **A** when **t** is substituted for the free occurrences of **m** in **A**, and that this substitution is therefore legitimate.

DEFINITION. If **m** is an individual variable and **t** is a term, or if **m** is a propositional variable and t is a wff, we say that t is *free for* **m** *in* the wff **A** iff no free occurrence of **m** in **A** is in a wf part of **A** of the form $\forall\mathbf{y}\mathbf{B}$, where **y** is a free variable of **t**.

EXERCISES

In X2000-X2001, proceed as follows:

(a) Copy the wff, and underline the free occurrences of variables in it.
(b) List the free variables of the wff.
(c) List the bound variables of the wff.
(d) Decide whether gx is free for u in the wff.
(e) Decide whether gx is free for z in the wff.
(f) Decide whether hxy is free for u in the wff.
(g) Decide whether u is free for x in the wff.

X2000. $\sim [[\forall u\, Puz \wedge \sim \exists x \,\blacksquare\, Qux \vee \forall z \sim Rxz] \supset \exists z Quz]$

X2001. $\forall x\, [Pxu \vee \exists y \,\blacksquare\, Qxy \wedge \sim \forall u Ruz] \supset \forall z Qzu$

In exercises X2002-X2005, interpret the predicate, function, and individual constants as indicated, and translate the given English expression [wff of $\mathcal{F}$] into a wff of $\mathcal{F}$ [English expression].

X2002. Mx means "x is a man"; Wx means "x is a woman"; Bxy means "x is better than y".

(a) $\forall x\,[Mx \supset \exists y\,.\,Wy \wedge Byx]$
(b) $\sim \exists z\,[Wz \wedge \forall u\,.\,Mu \supset Bzu]$

X2003. Sxy means "x and y are sisters", Lxy means "x loves y", Px means "x is a person", Dx means "x is a dog", fx means "the wife of x", j means "Jack".

(a) Somebody loves Jack, but Jack loves only dogs.
(b) Jack's wife has a sister who doesn't love anyone.

X2004. Hx means "x is a horse", tx means "the tail of x", mx means "the mane of x", Bx *means* "x is in the barn", Wy means "y is white", Ky means "y is black", Lxy means "x likes y".

(a) Every horse in the barn which has a white tail has a black mane.
(b) White horses don't like horses with black manes.
(c) No horse in the barn has a white tail.
(d) $(\exists x \in H)\,[Bx \wedge (\forall y \in H)\,\blacksquare\, By \wedge Kty \;\supset\; Lxy]$
(e) $\sim (\exists x \in H)\,[Bx \wedge \sim Wtx]$

X2005. Bxy means "x is a brother of y", r means "Robert", Uxy means "x is an uncle of y", f^1x means "the father of x", Cxy means "x is a cousin of y", Yxy means "x is younger than y".

(a) Any brother of Robert's father is an uncle of Robert.
(b) $\forall x \forall y\,[Bf^1xf^1y \;\supset\; Cxy]$
(c) Robert has a cousin who is younger than one of Robert's brothers.

X2006. Formulate a definition of the set of terms of $\mathcal{F}$ in the style of the definition of the set of wffs of $\mathcal{P}$ in §10, and state and prove an analogue of Theorem 1000.

X2007. Formulate a definition of the set of wffs of $\mathcal{F}$ in the style of the definition of the set of wffs of $\mathcal{P}$ in §10, and state and prove an analogue of Theorem 1000.

X2008. Show that every wf part of a wff $\sim \mathbf{B}$ or $\forall \mathbf{x}\mathbf{B}$ is either the entire wff or a wf part of $\mathbf{B}$.

X2009. Show that every wf part of a wff $[\mathbf{B} \vee \mathbf{C}]$ is either the entire wff or a wf part of **B** or a wf part of **C**. Would this still be true if brackets were omitted from the list of improper symbols and from formation rule (e)?

X2010. Show that if $[\mathbf{B} \vee \mathbf{C}]$ and $[\mathbf{D} \vee \mathbf{E}]$ are the same wff, then **B** is **D** and **C** is **E**.

X2011. Prove by induction on the construction of **C** that if **C** is any term or wff and **x** and **y** are distinct individual variables such that **y** is not free in **C** and **y** is free for **x** in **C**, then $\mathsf{S}^{\mathbf{y}}_{\cdot\mathbf{x}}\mathsf{S}^{\mathbf{x}}_{\cdot\mathbf{y}}\mathbf{C} = \mathbf{C}$. Note explicitly where the conditions on the variables are used in your proof. (*Hint:* when **C** is $\forall\mathbf{zB}$, consider separately the various possibilities concerning **z**.)

X2012. Let **A** and **C** be wffs and let **m** be a propositional variable. Let $\mathbf{x}_1, \ldots, \mathbf{x}_n$ be the free individual variables of **C**, and let $\mathbf{u}_1, \ldots, \mathbf{u}_n$ be distinct individual variables which do not occur in **C** or **A**. Let **B** be the wff $\mathsf{S}^{\mathbf{x}_1 \ldots \mathbf{x}_n}_{\cdot\mathbf{u}_1 \ldots \mathbf{u}_n}\mathbf{C}$. Prove that $\mathsf{S}^{\mathbf{u}_1}_{\cdot\mathbf{x}_1} \ldots \mathsf{S}^{\mathbf{u}_n}_{\cdot\mathbf{x}_n}\mathsf{S}^{\mathbf{m}}_{\cdot\mathbf{B}}\mathbf{A} = \mathsf{S}^{\mathbf{m}}_{\cdot\mathbf{C}}\mathbf{A}$. Also give some examples to illustrate this metatheorem, and to show the necessity of the condition on $\mathbf{u}_1, \ldots, \mathbf{u}_n$.

§21. The Axiomatic Structure of $\mathcal{F}$

$\mathcal{F}$ has two rules of inference, which are:

Modus Ponens (MP). From **A** and $\mathbf{A} \supset \mathbf{B}$ to infer **B**.

Generalization (Gen). From **A** to infer $\forall\mathbf{xA}$, where **x** is any individual variable.

We say that we *generalize on* **x** when we infer $\forall\mathbf{xA}$ from **A**.

The rationale for the Rule of Generalization is that if we are able to establish a statement **A** containing a free individual variable **x**, and we have not made any special assumptions about **x**, then our reasoning about **x** should apply to every individual in our universe, so the assertion $\forall\mathbf{xA}$ should also be true. We shall see that if **A** does not contain a free occurrence of **x**, then it is pointless but harmless to generalize on **x**.

Axiom Schemata for $\mathcal{F}$.

(1) $\mathbf{A} \vee \mathbf{A} \supset \mathbf{A}$
(2) $\mathbf{A} \supset \centerdot \mathbf{B} \vee \mathbf{A}$
(3) $\mathbf{A} \supset \mathbf{B} \supset \centerdot \mathbf{C} \vee \mathbf{A} \supset \centerdot \mathbf{B} \vee \mathbf{C}$

(4) $\forall \mathbf{x}\mathbf{A} \supset \mathsf{S}^{\mathbf{x}}_{\mathbf{t}}\mathbf{A}$ where $\mathbf{t}$ is a term which is free for the individual variable $\mathbf{x}$ in $\mathbf{A}$.

(5) $\forall \mathbf{x}[\mathbf{A} \lor \mathbf{B}] \supset \centerdot \mathbf{A} \lor \forall \mathbf{x}\mathbf{B}$ provided that $\mathbf{x}$ is not free in $\mathbf{A}$.

A simple example of Axiom 4 is $[\forall x Px \supset Pa]$, and a simple example of Axiom 5 is $\forall x [p \lor Qx] \supset \centerdot p \lor \forall x Qx$. In the domain $\{a, b, c\}$ the former wff has the same meaning as $[Pa \land Pb \land Pc] \supset Pa$, and the latter has the same meaning as

$$[[p \lor Qa] \land [p \lor Qb] \land [p \lor Qc] \supset \centerdot p \lor \centerdot Qa \land Qb \land Qc].$$

As in the case of the system $\mathcal{P}$, a *proof* in $\mathcal{F}$ of a wff $\mathbf{A}$ is a sequence of wffs, each of which is an axiom of $\mathcal{F}$ or is inferred from preceding wffs in the sequence by a rule of inference of $\mathcal{F}$, with $\mathbf{A}$ as the last wff in the sequence. A *theorem* of $\mathcal{F}$ is a wff which has a proof in $\mathcal{F}$. We now write $\vdash \mathbf{A}$ to mean that $\mathbf{A}$ is a theorem of $\mathcal{F}$. It is obvious that every wff of $\mathcal{P}$ is a wff of $\mathcal{F}$, and every theorem of $\mathcal{P}$ is a theorem of $\mathcal{F}$.

A variety of disciplines can be formalized by adding appropriate additional axioms, often called *postulates*, to those above. These postulates will express the basic definitions and assumptions of the discipline being formalized. Logistic systems obtained in this manner will be called *first-order theories* or *elementary theories*. Normally if the language contains a constant to denote a particular entity of some type, there will also be postulates which express the special properties of that entity.

As an example of a first-order theory we present the elementary theory of additive abelian groups, which we shall call $\mathcal{G}$. In this theory there are no function or predicate variables, but there is an individual constant 0, a binary function constant $+$, and a binary predicate constant $=$. As abbreviations we ordinarily write $= \mathbf{ts}$ as $[\mathbf{t} = \mathbf{s}]$ and $+\mathbf{ts}$ as $(\mathbf{t} + \mathbf{s})$. The postulates for the theory are the following sentences:

(*G1*) $\forall x\, x = x$

(*G2*) $\forall x \forall y \forall z \centerdot x = y \supset \centerdot x = z \supset y = z$

(*G3*) $\forall x \forall y \forall z \centerdot x = y \supset x + z = y + z$

(*G4*) $\forall x \forall y \forall z \centerdot x + (y + z) = (x + y) + z$

(*G5*) $\forall x \centerdot x + 0 = x$

(*G6*) $\forall x \exists y \centerdot x + y = 0$

(*G7*) $\forall x \forall y \centerdot x + y = y + x$

DEFINITION. The wff $\mathbf{A}$ is a *substitution instance of a tautology*, or is *tautologous*, iff there is a tautology $\mathbf{B}$ (of $\mathcal{P}$) such that $\mathbf{A}$ has the form $\mathsf{S}^{\mathbf{p}_1 \ldots \mathbf{p}_n}_{\mathbf{C}_1 \ldots \mathbf{C}_n}\mathbf{B}$.

2100 Rule P. If $\mathbf{B}$ is tautologous, then $\vdash \mathbf{B}$. If $\vdash \mathbf{A}_1, \ldots, \vdash \mathbf{A}_n$, and if $[\mathbf{A}_1 \wedge \ldots \wedge \mathbf{A}_n \supset \mathbf{B}]$ is tautologous, then $\vdash \mathbf{B}$.

Proof: If $\mathbf{B}$ is tautologous, there is a tautology $\mathbf{D}$ such that $\mathbf{B}$ is $\mathsf{S}^{\mathbf{p}_1 \ldots \mathbf{p}_n}_{\mathbf{C}_1 \ldots \mathbf{C}_n}\mathbf{D}$. Since $\mathbf{D}$ is a theorem of $\mathcal{P}$, there is a proof $\mathbf{D}_1, \ldots, \mathbf{D}_m$ of $\mathbf{D}$ in $\mathcal{P}$. It is easily seen by induction on i that $\mathsf{S}^{\mathbf{p}_1 \ldots \mathbf{p}_n}_{\mathbf{C}_1 \ldots \mathbf{C}_n}\mathbf{D}_i$ is a theorem of $\mathcal{F}$ for $1 \le i \le m$. This proves the first part of the metatheorem.

To prove the second part, note that if $[\mathbf{A}_1 \wedge \ldots \wedge \mathbf{A}_n \supset \mathbf{B}]$ is tautologous, then $[\mathbf{A}_1 \supset . \mathbf{A}_2 \supset . \ldots \supset . \mathbf{A}_n \supset \mathbf{B}]$ must be tautologous also, so it is a theorem of $\mathcal{F}$, so by n applications of Modus Ponens $\vdash \mathbf{B}$. ■

REMARK. Rule P is a very powerful derived rule of inference which embodies all the power of reasoning involving propositional connectives. For example, the following are special cases of Rule P:

(1) **Modus Ponens.** If $\vdash \mathbf{A}$ and $\vdash \mathbf{A} \supset \mathbf{B}$, then $\vdash \mathbf{B}$ (since $[\mathbf{A} \wedge [\mathbf{A} \supset \mathbf{B}] \supset \mathbf{B}]$ is tautologous).
(2) **Modus Tollens.** If $\vdash \mathbf{A} \supset \mathbf{B}$ and $\vdash {\sim}\mathbf{B}$, then $\vdash {\sim}\mathbf{A}$ (since $[\mathbf{A} \supset \mathbf{B}] \wedge {\sim}\mathbf{B} \supset {\sim}\mathbf{A}$ is tautologous).
(3) **Transitive Law of Implication (Trans).** If $\vdash \mathbf{A}_1 \supset \mathbf{A}_2$, $\vdash \mathbf{A}_2 \supset \mathbf{A}_3$, $\ldots$, and $\vdash \mathbf{A}_{n-1} \supset \mathbf{A}_n$, then $\vdash \mathbf{A}_1 \supset \mathbf{A}_n$ (since $[\mathbf{A}_1 \supset \mathbf{A}_2] \wedge [\mathbf{A}_2 \supset \mathbf{A}_3] \wedge \ldots \wedge [\mathbf{A}_{n-1} \supset \mathbf{A}_n] \supset [\mathbf{A}_1 \supset \mathbf{A}_n]$ is tautologous).

2101 Consistency Theorem. $\mathcal{F}$ is absolutely consistent and consistent with respect to negation.

Proof: For each wff $\mathbf{A}$ of $\mathcal{F}$, let $\natural\mathbf{A}$ be the result of erasing all quantifiers in $\mathbf{A}$, and replacing each occurrence of an atomic wff in $\mathbf{A}$ by the propositional variable p. It is easily seen by induction on proofs that if $\vdash \mathbf{A}$, then $\natural\mathbf{A}$ is a tautology. (That is, it is easily seen by induction on i that if $\mathbf{A}_1, \ldots, \mathbf{A}_i, \ldots, \mathbf{A}_n$ is any proof in $\mathcal{F}$, then $\natural\mathbf{A}_i$ is a tautology for each i from 1 to n.) However, $\natural \sim \mathbf{A}$ is $\sim \natural\mathbf{A}$, so $\mathcal{F}$ is consistent with respect to negation, and hence, by 1201 and 2100, absolutely consistent. ■

We next prove certain theorem schemata and derived rules of inference of $\mathcal{F}$. We remark that, in contrast to the situation for $\mathcal{P}$, the decision problem for $\mathcal{F}$ is unsolvable. (This deep metatheorem is known as *Church's Theorem* [Church, 1936a].) Therefore, techniques for proving theorems of $\mathcal{F}$, and for showing that certain wffs are not theorems of $\mathcal{F}$, are of permanent importance. (By contrast, the theorems of $\mathcal{P}$ proved in §11 were simply needed to prove Theorem 1205, and they are all subsumed by Theorem 1205.) Note that the proof of 2101 provides a technique for showing that certain wffs are not theorems of $\mathcal{F}$.

2102. $\vdash \forall \mathbf{x}_n \ldots \forall \mathbf{x}_1 \mathbf{A} \supset \mathbf{A}$.

Proof: By induction on n. For $n = 1$, this is an instance of Axiom Schema 4, since **x** is always free for **x** in **A**. For the induction step, proceed as follows:

(.1)	$\vdash \forall \mathbf{x}_{n+1} \forall \mathbf{x}_n \ldots \forall \mathbf{x}_1 \mathbf{A} \supset \forall \mathbf{x}_n \ldots \forall \mathbf{x}_1 \mathbf{A}$	Axiom 4
(.2)	$\vdash \forall \mathbf{x}_n \ldots \forall \mathbf{x}_1 \mathbf{A} \supset \mathbf{A}$	by inductive hypothesis
(.3)	$\vdash \forall \mathbf{x}_{n+1} \forall \mathbf{x}_n \ldots \forall \mathbf{x}_1 \mathbf{A} \supset \mathbf{A}$	Rule P: .1, .2

■

2103 $\supset\forall$ Rule. If $\vdash \mathbf{A} \supset \mathbf{B}$ and **x** is not free in **A**, then $\vdash \mathbf{A} \supset \forall \mathbf{x}\mathbf{B}$.

Proof:

(.1)	$\vdash \sim \mathbf{A} \vee \mathbf{B}$	given
(.2)	$\vdash \forall \mathbf{x} \centerdot \sim \mathbf{A} \vee \mathbf{B}$	Gen: .1
(.3)	$\vdash \forall \mathbf{x} [\sim \mathbf{A} \vee \mathbf{B}] \supset \centerdot \sim \mathbf{A} \vee \forall \mathbf{x}\mathbf{B}$	Axiom 5
(.4)	$\vdash \mathbf{A} \supset \forall \mathbf{x}\mathbf{B}$	MP: .2, .3

■

2104. $\vdash \forall \mathbf{x} [\mathbf{A} \supset \mathbf{B}] \supset \centerdot \forall \mathbf{x}\mathbf{A} \supset \forall \mathbf{x}\mathbf{B}$

ILLUSTRATION: If the street gets wet every day that it rains, then if it rains every day, the street gets wet every day.

Proof:

(.1)	$\vdash \forall \mathbf{x} [\mathbf{A} \supset \mathbf{B}] \supset \centerdot \mathbf{A} \supset \mathbf{B}$	2102
(.2)	$\vdash \forall \mathbf{x}\mathbf{A} \supset \mathbf{A}$	2102
(.3)	$\vdash \forall \mathbf{x} [\mathbf{A} \supset \mathbf{B}] \wedge \forall \mathbf{x}\mathbf{A} \supset \mathbf{B}$	Rule P: .1, .2
(.4)	$\vdash \forall \mathbf{x} [\mathbf{A} \supset \mathbf{B}] \wedge \forall \mathbf{x}\mathbf{A} \supset \forall \mathbf{x}\mathbf{B}$	2103: .3
(.5)	$\vdash \forall \mathbf{x} [\mathbf{A} \supset \mathbf{B}] \supset \centerdot \forall \mathbf{x}\mathbf{A} \supset \forall \mathbf{x}\mathbf{B}$	Rule P: .4

■

DEFINITION. Let **M** be a wf part of **A**. An occurrence of **M** in **A** is *positive* [*negative*] in **A** iff that occurrence is in the scope of an even [odd] number of occurrences of $\sim$ in **A**. (0 is even).

2105 Substitutivity of Implication. Let **M**, **N**, and **A** be wffs, and let **A**′ be the result of replacing **M** by **N** at zero or more occurrences (henceforth called designated occurrences) of **M** in **A**. Let $\mathbf{y}_1, \ldots, \mathbf{y}_k$ be a list including all individual variables which occur free in **M** or **N**, but are bound in **A** by quantifiers whose scopes contain designated occurrences of **M**.

If the designated occurrences of **M** are all positive in **A**, then

(a) $\vdash \forall \mathbf{y}_1 \ldots \forall \mathbf{y}_k [\mathbf{M} \supset \mathbf{N}] \supset \centerdot \mathbf{A} \supset \mathbf{A}'$, and
(b) if $\vdash \mathbf{M} \supset \mathbf{N}$, then $\vdash \mathbf{A} \supset \mathbf{A}'$.

If the designated occurrences of **M** are all negative in **A**, then

(c) $\vdash \forall \mathbf{y}_1 \ldots \forall \mathbf{y}_k [\mathbf{M} \supset \mathbf{N}] \supset \centerdot \mathbf{A}' \supset \mathbf{A}$, and
(d) if $\vdash \mathbf{M} \supset \mathbf{N}$, then $\vdash \mathbf{A}' \supset \mathbf{A}$.

EXAMPLE:
$\vdash \forall y [\forall x Pxy \supset \exists z Qyz] \supset \centerdot \exists y [\exists z Qyz \supset Ry] \supset \exists y [\forall x Pxy \supset Ry]$ by 2105c.

Proof: Clearly (b) and (d) follow from (a) and (c), respectively, by Gen and MP. We prove (a) and (c) simultaneously by induction on the number of occurrences of $\sim$, $\vee$, and $\forall$ in **A**.

Case 1. There are no designated occurrences of **M** in **A**. Then **A**′ is **A**, so Rule P establishes the result.

Case 2. **A** is **M**, and the sole occurrence of **M** in **A** is designated. Then **M** occurs positively in **A**, and **A**′ is **N**, so the desired theorem is

$$\forall \mathbf{y}_1 \ldots \forall \mathbf{y}_k [\mathbf{M} \supset \mathbf{N}] \supset \centerdot \mathbf{M} \supset \mathbf{N},$$

which is an instance of 2102.

In the remaining cases we assume, as part of the definition of these cases, that neither Case 1 nor Case 2 applies.

Case 3. **A** has the form $\sim$**B**. Since Cases 1 and 2 do not apply, all designated occurrences in **M** in **A** are occurrences in **B**, and **A**′ is $\sim$**B**′. In part (a) of the metatheorem the designated occurrences of **M** are all positive in **A**, hence negative in **B**, so

(.31) $\vdash \forall \mathbf{y}_1 \ldots \forall \mathbf{y}_k [\mathbf{M} \supset \mathbf{N}] \supset \centerdot \mathbf{B}' \supset \mathbf{B}$ by inductive hypothesis

(.32) $\vdash \forall \mathbf{y}_1 \ldots \forall \mathbf{y}_k [\mathbf{M} \supset \mathbf{N}] \supset \centerdot \sim \mathbf{B} \supset \sim \mathbf{B}'$ by Rule P

Part (c) is established similarly.

Case 4. **A** has the form [**B** $\vee$ **C**]. All designated occurrences of **M** in **A** must be in **B** or in **C**, and **A**′ is [**B**′ $\vee$ **C**′]. Again we establish (a) and leave (c) to the reader. Since the designated occurrences of **M** are all positive in **A**, they are positive in **B** and in **C**.

(.41) $\vdash \forall \mathbf{y}_1 \ldots \forall \mathbf{y}_k [\mathbf{M} \supset \mathbf{N}] \supset \centerdot \mathbf{B} \supset \mathbf{B}'$ by inductive hypothesis
(.42) $\vdash \forall \mathbf{y}_1 \ldots \forall \mathbf{y}_k [\mathbf{M} \supset \mathbf{N}] \supset \centerdot \mathbf{C} \supset \mathbf{C}'$ by inductive hypothesis

(Note that the conditions on the variables $\mathbf{y}_1 \ldots \mathbf{y}_k$ have been stated in such a way that the inductive hypothesis applies to **B** and **C**.)

(.43) $\vdash \forall \mathbf{y}_1 \ldots \forall \mathbf{y}_k [\mathbf{M} \supset \mathbf{N}] \supset \centerdot [\mathbf{B} \vee \mathbf{C}] \supset [\mathbf{B}' \vee \mathbf{C}']$ Rule P: .41, .42

Case 5. **A** is $\forall \mathbf{xB}$. All designated occurrences of **M** in **A** must be in **B**, so **A**′ is $\forall \mathbf{xB}'$. Note that if **x** is free in **M** or in **N**, then one of the variables $\mathbf{y}_i$ must be **x**, so **x** is not free in $\forall \mathbf{y}_1 \ldots \forall \mathbf{y}_k [\mathbf{M} \supset \mathbf{N}]$. Again we treat only part (a).

(.51) $\vdash \forall \mathbf{y}_1 \ldots \forall \mathbf{y}_k [\mathbf{M} \supset \mathbf{N}] \supset \centerdot \mathbf{B} \supset \mathbf{B}'$ by inductive hypothesis
(.52) $\vdash \forall \mathbf{y}_1 \ldots \forall \mathbf{y}_k [\mathbf{M} \supset \mathbf{N}] \supset \forall \mathbf{x} \centerdot \mathbf{B} \supset \mathbf{B}'$ 2103: .51
(.53) $\vdash \forall \mathbf{x} [\mathbf{B} \supset \mathbf{B}'] \supset \centerdot \forall \mathbf{xB} \supset \forall \mathbf{xB}'$ 2104
(.54) $\vdash \forall \mathbf{y}_1 \ldots \forall \mathbf{y}_k [\mathbf{M} \supset \mathbf{N}] \supset \centerdot \forall \mathbf{xB} \supset \forall \mathbf{xB}'$ Rule P: .52, .53

Since **A** must fall under one of these five cases, the proof is complete. ∎

2106 Substitutivity of Equivalence. Let **M, N, A, A**′, and $\mathbf{y}_1, \ldots, \mathbf{y}_k$ be as in 2105.

(a) $\vdash \forall \mathbf{y}_1 \ldots \forall \mathbf{y}_k [\mathbf{M} \equiv \mathbf{N}] \supset \centerdot \mathbf{A} \equiv \mathbf{A}'$
(b) If $\vdash \mathbf{M} \equiv \mathbf{N}$ then $\vdash \mathbf{A} \equiv \mathbf{A}'$.
(c) If $\vdash \mathbf{M} \equiv \mathbf{N}$ and $\vdash \mathbf{A}$, then $\vdash \mathbf{A}'$.

Proof: Clearly (b) and (c) follow from (a) by Gen and Rule P. To prove (a), let **B** be the result of replacing all positive designated occurrences of **M** in **A** by occurrences of **N**.

(.1) $\vdash \forall \mathbf{y}_1 \ldots \forall \mathbf{y}_k [\mathbf{M} \supset \mathbf{N}] \supset \centerdot \mathbf{A} \supset \mathbf{B}$ 2105(a)
(.2) $\vdash \forall \mathbf{y}_1 \ldots \forall \mathbf{y}_k [\mathbf{N} \supset \mathbf{M}] \supset \centerdot \mathbf{B} \supset \mathbf{A}$ 2105(a)

Clearly $\mathbf{A}'$ is the result of replacing certain negative occurrences of $\mathbf{M}$ in $\mathbf{B}$ by occurrences of $\mathbf{N}$, so

(.3) $\vdash \forall \mathbf{y}_1 \ldots \forall \mathbf{y}_k [\mathbf{M} \supset \mathbf{N}] \supset \centerdot \mathbf{A}' \supset \mathbf{B}$ — 2105(c)
(.4) $\vdash \forall \mathbf{y}_1 \ldots \forall \mathbf{y}_k [\mathbf{N} \supset \mathbf{M}] \supset \centerdot \mathbf{B} \supset \mathbf{A}'$ — 2105(c)
(.5) $\vdash [\mathbf{M} \equiv \mathbf{N}] \supset \centerdot \mathbf{M} \supset \mathbf{N}$ — Rule P
(.6) $\vdash \forall \mathbf{y}_1 \ldots \forall \mathbf{y}_k [\mathbf{M} \equiv \mathbf{N}] \supset \forall \mathbf{y}_1 \ldots \forall \mathbf{y}_k [\mathbf{M} \supset \mathbf{N}]$ — 2105(b): .5
(.7) $\vdash [\mathbf{M} \equiv \mathbf{N}] \supset \centerdot \mathbf{N} \supset \mathbf{M}$ — Rule P
(.8) $\vdash \forall \mathbf{y}_1 \ldots \forall \mathbf{y}_k [\mathbf{M} \equiv \mathbf{N}] \supset \forall \mathbf{y}_1 \ldots \forall \mathbf{y}_k [\mathbf{N} \supset \mathbf{M}]$ — 2105(b): .7
(.9) $\vdash \forall \mathbf{y}_1 \ldots \forall \mathbf{y}_k [\mathbf{M} \equiv \mathbf{N}] \supset \centerdot \mathbf{A} \equiv \mathbf{A}'$ — Rule P: .1, .2, .3, .4, .6, .8

■

2107. $\vdash \forall \mathbf{x} [\mathbf{M} \equiv \mathbf{N}] \supset \centerdot \forall \mathbf{x}\mathbf{M} \equiv \forall \mathbf{x}\mathbf{N}$ — 2106(a)

■

2108. $\vdash \forall \mathbf{x} [\mathbf{M} \equiv \mathbf{N}] \supset \centerdot \exists \mathbf{x}\mathbf{M} \equiv \exists \mathbf{x}\mathbf{N}$ — 2106(a)

■

2109. $\vdash \forall \mathbf{x}\mathbf{C} \equiv \forall \mathbf{y} \mathsf{S}^{\mathbf{x}}_{\cdot \mathbf{y}} \mathbf{C}$, provided that $\mathbf{y}$ is not free in $\mathbf{C}$ and $\mathbf{y}$ is free for $\mathbf{x}$ in $\mathbf{C}$.

EXAMPLE: $\vdash \forall x \exists u [Rxug^1 x \lor \forall y \centerdot Puy \supset \forall x Pxy]$
$\equiv \forall y \exists u [Ryug^1 y \lor \forall y \centerdot Puy \supset \forall x Pxy]$

ILLUSTRATION: Let us illustrate the restrictions in Theorem 2109 with the following wffs:

(A) $\forall x [Fx \supset Lzx]$
(B) $\forall y [Fy \supset Lzy]$
(C) $\forall z [Fz \supset Lzz]$
(D) $\forall x [[Px \land \exists y \centerdot Fy \land Hxy] \supset Lzx]$
(E) $\forall y [[Py \land \exists y \centerdot Fy \land Hyy] \supset Lzy]$

We shall interpret the predicate constants in these wffs so that Fx means "x is a cat" (or "x is feline"), Lzx means "z likes x", Px means "x is a person", and Hxy means "x has y". Then A and B both mean "z likes all cats", but C means "every cat likes itself". $\vdash A \equiv B$ by 2109, but $A \equiv C$ is not an instance of 2109, since z is free in $[Fx \supset Lzx]$. D means "z likes everyone who has a cat", but E means "z likes every person such that some cat has itself". $D \equiv E$ is not an instance of 2109, since y is not free for x in $[[Px \land \exists y \centerdot Fy \land Hxy] \supset Lzx]$.

Proof of 2109: We may assume $\mathbf{x}$ and $\mathbf{y}$ are distinct, since the theorem follows by Rule P if $\mathbf{x}$ is $\mathbf{y}$.

(.1) $\vdash \forall \mathbf{x} \mathbf{C} \supset \mathsf{S}^{\mathbf{x}}_{\cdot \mathbf{y}} \mathbf{C}$ — Axiom 4

(.2) $\vdash \forall \mathbf{x} \mathbf{C} \supset \forall \mathbf{y} \mathsf{S}^{\mathbf{x}}_{\cdot \mathbf{y}} \mathbf{C}$ — $\supset \forall$ Rule (2103): .1

It is easily verified that $\mathbf{x}$ is not free in $\mathsf{S}^{\mathbf{x}}_{\cdot \mathbf{y}} \mathbf{C}$ and that $\mathbf{x}$ is free for $\mathbf{y}$ in $\mathsf{S}^{\mathbf{x}}_{\cdot \mathbf{y}} \mathbf{C}$.

(.3) $\vdash \forall \mathbf{y} \mathsf{S}^{\mathbf{x}}_{\cdot \mathbf{y}} \mathbf{C} \supset \mathsf{S}^{\mathbf{y}}_{\cdot \mathbf{x}} \mathsf{S}^{\mathbf{x}}_{\cdot \mathbf{y}} \mathbf{C}$ — Axiom 4

Our conditions assure that $\mathsf{S}^{\mathbf{y}}_{\cdot \mathbf{x}} \mathsf{S}^{\mathbf{x}}_{\cdot \mathbf{y}} \mathbf{C}$ is $\mathbf{C}$ (see Exercise X2011), so

(.4) $\vdash \forall \mathbf{y} \mathsf{S}^{\mathbf{x}}_{\cdot \mathbf{y}} \mathbf{C} \supset \forall \mathbf{x} \mathbf{C}$ — $\supset \forall$ Rule: .3

The desired result follows by Rule P from .2 and .4. ■

2110 Rule of Alphabetic Change of Bound Variables (α Rule). Suppose that $\mathbf{A}'$ is obtained from $\mathbf{A}$ upon replacing an occurrence of $\forall \mathbf{x} \mathbf{C}$ in $\mathbf{A}$ by an occurrence of $\forall \mathbf{y} \mathsf{S}^{\mathbf{x}}_{\cdot \mathbf{y}} \mathbf{C}$, where $\mathbf{y}$ is not free in $\mathbf{C}$ and $\mathbf{y}$ is free for $\mathbf{x}$ in $\mathbf{C}$. If $\vdash \mathbf{A}$, then $\vdash \mathbf{A}'$.

Proof: By 2109 and the Substitutivity of Equivalence (2106). ■

DEFINITION. A *proof of a wff* $\mathbf{B}$ *from the finite set of hypotheses* $\{\mathbf{A}_1, \ldots, \mathbf{A}_n\}$ is a finite sequence $\mathbf{B}_1, \ldots, \mathbf{B}_m$ of wffs such that $\mathbf{B}_m$ is $\mathbf{B}$, and for each i one of the following conditions is satisfied:

(a) (*Ax*) $\mathbf{B}_i$ is an axiom;
(b) (*Hyp*) $\mathbf{B}_i$ is a hypothesis $\mathbf{A}_j$;
(c) (*MP*) $\mathbf{B}_i$ follows by Modus Ponens from preceding members of the sequence;
(d) (*Gen*) there exists $j < i$ such that $\mathbf{B}_i$ is $\forall \mathbf{x} \mathbf{B}_j$, and $\mathbf{x}$ is not free in any hypothesis;
(e) (α) $\mathbf{B}_i$ follows from some preceding member of the sequence by the Rule of Alphabetic Change of Bound Variables.

We write $\mathbf{A}_1, \ldots, \mathbf{A}_n \vdash \mathbf{B}$ to indicate that there is a proof of $\mathbf{B}$ from the hypotheses $\mathbf{A}_1, \ldots, \mathbf{A}_n$. If $\mathcal{S}$ is a set of wffs, $\mathcal{S} \vdash \mathbf{B}$ means there is a proof of $\mathbf{B}$ from some finite subset of $\mathcal{S}$. It is easily seen that $\vdash \mathbf{B}$ iff $\emptyset \vdash \mathbf{B}$. As in Chapter 1 we use $\mathcal{H}$ as a syntactical variable for a set of wffs. When we say **x** is not free in $\mathcal{H}$, we shall mean that **x** is not free in any wff in $\mathcal{H}$.

The restriction on the Rule of Generalization (Gen) in proofs from hypotheses is a very important and natural one. In general, if we prove a wff **B** in which **x** occurs free (which can, of course, be regarded as a statement about **x**), and have made no special assumptions about **x**, then the argument should apply to an arbitrary object which could be named by **x**, so it is intuitively clear that $\forall$**xB** should also be true. However, if **x** occurs free in some hypothesis **H**, then the proof may depend on the assumption about **x** which **H** expresses, so there is no reason to suppose that $\forall$**xB** will also be true.

Note that if $\mathcal{S}$ is a set of closed wffs which are the postulates for a first-order theory $\mathcal{T}$, then $\vdash_{\mathcal{T}}$ **B** iff $\mathcal{S} \vdash$ **B** (where $\vdash_{\mathcal{T}}$ **B** means that **B** is a theorem of $\mathcal{T}$). Thus, derived rules of inference concerning proofs from hypotheses can also be applied to proofs in first-order theories.

2111 Universal Instantiation ($\forall$**I**) If $\mathcal{H} \vdash \forall \mathbf{xA}$, then $\mathcal{H} \vdash \mathsf{S}^{\mathbf{x}}_{\mathbf{t}}\mathbf{A}$, provided that **t** is a term free for **x** in **A**.

Proof: By Axiom 4 and MP. ∎

2112 Lemma. Suppose $\mathcal{H} \vdash \mathbf{A}$, and let $\mathcal{U}$ be a finite set of individual variables which are not free in **A** or in $\mathcal{H}$. Then there is a proof of **A** from $\mathcal{H}$ in which there is no application of the rule of Generalization involving generalization on a member of $\mathcal{U}$.

Proof: It suffices to prove the lemma for the case where $\mathcal{H}$ is finite. Let $\mathcal{R}$ be a proof of **A** from $\mathcal{H}$, and let $\mathcal{U} = \{\mathbf{x}_1, \ldots, \mathbf{x}_n\}$. Let $\mathbf{y}_1, \ldots, \mathbf{y}_n$ be distinct individual variables which do not occur in $\mathcal{U}$ or any wff of $\mathcal{R}$ or $\mathcal{H}$. Let θ be the substitution $\mathsf{S}^{\mathbf{x}_1 \ldots \mathbf{x}_n}_{\mathbf{y}_1 \ldots \mathbf{y}_n}$. Clearly $\theta\mathcal{R}$ is a proof of $\theta\mathbf{A}$ from $\theta\mathcal{H}$, and by inserting appropriate alphabetic changes of bound variables it can be expanded to a proof of **A** from $\mathcal{H}$. The new proof contains no generalization on any variable in $\mathcal{U}$. ∎

2113 Theorem. If $\mathcal{H}_1 \vdash \mathbf{A}$ and $\mathcal{H}_1 \subseteq \mathcal{H}_2$, then $\mathcal{H}_2 \vdash \mathbf{A}$.

Proof: It suffices to consider the case where $\mathcal{H}_1$ and $\mathcal{H}_2$ are both finite. Let $\mathbf{y}_1, \ldots, \mathbf{y}_k$ be the free individual variables of **A** which are not free in $\mathcal{H}_1$ and let $\overline{\mathbf{A}} = \forall \mathbf{y}_1 \ldots \forall \mathbf{y}_k \mathbf{A}$. By Generalization, $\mathcal{H}_1 \vdash \overline{\mathbf{A}}$. Let $\mathcal{U}$ be the set of individual variables which occur free in $\mathcal{H}_2$ but not in $\mathcal{H}_1$. Then no member of $\mathcal{U}$ is free in $\overline{\mathbf{A}}$ (since the free variables of $\overline{\mathbf{A}}$ are all free in wffs of $\mathcal{H}_1$), so by 2112 there is a proof of $\overline{\mathbf{A}}$ from $\mathcal{H}_1$ in which no member of $\mathcal{U}$ is generalized on. This is a proof from $\mathcal{H}_2$, so $\mathcal{H}_2 \vdash \overline{\mathbf{A}}$, and $\mathcal{H}_2 \vdash \mathbf{A}$ by Universal Instantiation. ■

Corollary. If $\vdash \mathbf{A}$, then $\mathcal{H} \vdash \mathbf{A}$.

REMARK. The reader may have wondered why the Rule of Alphabetic Change of Bound Variables occurs in the definition of a proof from hypotheses. It will be observed that this rule was used in the proof of 2112, which was used to prove 2113. Actually, it would not be possible to prove metatheorem 2113 if the α Rule were omitted from the definition of a proof from hypotheses. To see this, note that $\vdash \forall yPy \supset \forall xPx$ (by 2109.2), so $[Qx \supset Qx] \vdash \forall yPy \supset \forall xPx$ by 2113. Suppose there were a proof of $\forall yPy \supset \forall xPx$ from $[Qx \supset Qx]$ in which the α Rule was not used. Since $[Qx \supset Qx]$ can be proved using only Axioms 1–3 and Modus Ponens, it would not actually be necessary to use $[Qx \supset Qx]$ as a premise in this proof. Hence the given proof would simply be a proof of $\forall yPy \supset \forall xPx$ in which no generalization is made on the variable x. However, we shall see in §24 that any proof in $\mathcal{F}$ of $[\forall yPy \supset \forall xPx]$ must contain a generalization upon x.

2114 Extended Rule P. If $\mathcal{H} \vdash \mathbf{A}_1, \ldots, \mathcal{H} \vdash \mathbf{A}_n$, and if $[\mathbf{A}_1 \wedge \ldots \wedge \mathbf{A}_n \supset \mathbf{B}]$ is tautologous, then $\mathcal{H} \vdash \mathbf{B}$. Also, if $\mathbf{B}$ is tautologous, $\mathcal{H} \vdash \mathbf{B}$.

Proof: $\mathcal{H} \vdash \mathbf{A}_1 \supset \centerdot \ldots \supset \centerdot\, \mathbf{A}_n \supset \mathbf{B}$ by 2100 and 2113. Hence $\mathcal{H} \vdash \mathbf{B}$ by n applications of MP. ■

2115 Extended $\supset\forall$ Rule. If $\mathcal{H} \vdash \mathbf{A} \supset \mathbf{B}$ and if $\mathbf{x}$ is not free in $\mathcal{H}$ or in $\mathbf{A}$, then $\mathcal{H} \vdash \mathbf{A} \supset \forall\mathbf{x}\mathbf{B}$.

The proof is similar to the proof of 2103.

2116 Deduction Theorem (Ded). If $\mathcal{H}, \mathbf{A} \vdash \mathbf{B}$ then $\mathcal{H} \vdash \mathbf{A} \supset \mathbf{B}$.

Proof. Let $\mathbf{B}_1 \ldots, \mathbf{B}_n$ be a proof of $\mathbf{B}$ from $\mathcal{H}, \mathbf{A}$. We prove that $\mathcal{H} \vdash \mathbf{A} \supset \mathbf{B}_i$ for $1 \leq i \leq n$ by induction on i.

Case 1. $\mathbf{B}_i$ is an axiom or a member of $\mathcal{H}$.

(.11) $\mathcal{H} \vdash \mathbf{B}_i$ axiom or hypothesis

(.12) $\mathcal{H} \vdash \mathbf{A} \supset \mathbf{B}_i$ Rule P: .11

Case 2. $\mathbf{B}_i$ is $\mathbf{A}$.

(.21) $\mathcal{H} \vdash \mathbf{A} \supset \mathbf{B}_i$ Rule P

Case 3. $\mathbf{B}_i$ is inferred by MP from $\mathbf{B}_j$ and $\mathbf{B}_k$, which is $[\mathbf{B}_j \supset \mathbf{B}_i]$, where $j,\ k < i$.

(.31) $\mathcal{H} \vdash \mathbf{A} \supset \mathbf{B}_j$ by inductive hypothesis
(.32) $\mathcal{H} \vdash \mathbf{A} \supset \centerdot\, \mathbf{B}_j \supset \mathbf{B}_i$ by inductive hypothesis
(.33) $\mathcal{H} \vdash \mathbf{A} \supset \mathbf{B}_i$ Rule P: .31, .32

Case 4. $\mathbf{B}_i$ is inferred from $\mathbf{B}_j$ by Gen, where $j < i$. Thus, $\mathbf{B}_i$ is $\forall \mathbf{x} \mathbf{B}_j$, and $\mathbf{x}$ does not occur free in $\mathcal{H}$ or in $\mathbf{A}$.

(.41) $\mathcal{H} \vdash \mathbf{A} \supset \mathbf{B}_j$ by inductive hypothesis
(.42) $\mathcal{H} \vdash \mathbf{A} \supset \forall \mathbf{x} \mathbf{B}_j$ $\supset\forall$ rule (2115): .41

Case 5. $\mathbf{B}_i$ is inferred from $\mathbf{B}_j$ by the α Rule, where $j < i$.

(.51) $\mathcal{H} \vdash \mathbf{A} \supset \mathbf{B}_j$ by inductive hypothesis
(.52) $\mathcal{H} \vdash \mathbf{A} \supset \mathbf{B}_i$ α Rule

∎

2117 Extended Substitutivity of Implication and Equivalence. Let $\mathbf{M}$, $\mathbf{N}$, and $\mathbf{A}$ be wffs, and let $\mathbf{A}'$ be the result of replacing $\mathbf{M}$ by $\mathbf{N}$ at certain designated occurrences of $\mathbf{M}$ in $\mathbf{A}$. If $\mathcal{H} \vdash \mathbf{A}$, then $\mathcal{H} \vdash \mathbf{A}'$, provided that one of the following conditions is satisfied:

(a) $\vdash \mathbf{M} \supset \mathbf{N}$, and the designated occurrences of $\mathbf{M}$ are all positive in $\mathbf{A}$;
(b) $\vdash \mathbf{N} \supset \mathbf{M}$, and the designated occurrences of $\mathbf{M}$ are all negative in $\mathbf{A}$;
(c) $\vdash \mathbf{M} \equiv \mathbf{N}$.

Proof:

(.1) $\mathcal{H} \vdash \mathbf{A}$ given

Case (a).

(.11) $\vdash \mathbf{M} \supset \mathbf{N}$ given
(.12) $\vdash \mathbf{A} \supset \mathbf{A}'$ 2105(b)
(.13) $\mathcal{H} \vdash \mathbf{A} \supset \mathbf{A}'$ 2113

(.14) $\mathcal{H} \vdash \mathbf{A}'$ MP: .1, .13

Case (*b*).

(.21) $\vdash \mathbf{N} \supset \mathbf{M}$ given
(.22) $\vdash \mathbf{A} \supset \mathbf{A}'$ 2105(d)

(Note that $\mathbf{A}$ is obtained from $\mathbf{A}'$ upon replacing certain negative occurrences of $\mathbf{N}$ in $\mathbf{A}'$ by occurrences of $\mathbf{M}$.)

(.23) $\mathcal{H} \vdash \mathbf{A} \supset \mathbf{A}'$ 2113
(.24) $\mathcal{H} \vdash \mathbf{A}'$ MP: .1, .23

The proof for Case (c) is like the proof of (a), but uses 2106(b) and Rule P. ■

2118 Rule of Substitution (Sub). Let $\mathbf{m}$ be an individual variable and $\mathbf{t}$ be a term, or let $\mathbf{m}$ be a propositional variable and $\mathbf{t}$ be a wff. If $\mathcal{H} \vdash \mathbf{A}$, then $\mathcal{H} \vdash \mathsf{S}^{\mathbf{m}}_{\cdot\mathbf{t}}\mathbf{A}$, provided that $\mathbf{m}$ does not occur free in $\mathcal{H}$ and $\mathbf{t}$ is free for $\mathbf{m}$ in $\mathbf{A}$.

Proof: If $\mathbf{m}$ is an individual variable, use Gen and $\forall$I (2111).

If $\mathbf{m}$ is a propositional variable, we first consider the case in which $\mathcal{H}$ is empty. Let $\mathbf{A}_1, \ldots, \mathbf{A}_k$ be a proof of $\mathbf{A}$. Let $\mathbf{x}_1, \ldots, \mathbf{x}_n$ be the free individual variables of $\mathbf{t}$, and let $\mathbf{u}_1, \ldots, \mathbf{u}_n$ be distinct individual variables which do not occur in any wff of $\{\mathbf{A}_1, \ldots, \mathbf{A}_k, \mathbf{t}\}$. Let $\mathbf{B}$ be $\mathsf{S}^{\mathbf{x}_1 \ldots \mathbf{x}_n}_{\cdot\mathbf{u}_1 \ldots \mathbf{u}_n}\mathbf{t}$. It is easy to establish by induction on i that $\vdash \mathsf{S}^{\mathbf{m}}_{\mathbf{B}}\mathbf{A}_i$ for $1 \leq i \leq k$.

If $\mathbf{A}_i$ is an axiom, then $\mathsf{S}^{\mathbf{m}}_{\mathbf{B}}\mathbf{A}_i$ is too, since no individual variable of $\mathbf{A}_i$ occurs free in $\mathbf{B}$. (We leave it to the reader to verify the details when $\mathbf{A}_i$ is an instance of Axiom Schema 4 or 5.) The arguments for the cases where $\mathbf{A}_i$ is obtained by Modus Ponens or Generalization are routine.

The inductive proof is complete, so $\vdash \mathsf{S}^{\mathbf{m}}_{\mathbf{B}}\mathbf{A}$. Hence by n applications of the Rule of Substitution for individual variables, $\vdash \mathsf{S}^{\mathbf{u}_1}_{\cdot\mathbf{x}_1} \ldots \mathsf{S}^{\mathbf{u}_n}_{\cdot\mathbf{x}_n} \mathsf{S}^{\mathbf{m}}_{\mathbf{B}}\mathbf{A}$. (To see that $\mathbf{x}_i$ is free for $\mathbf{u}_i$, use the fact that $\mathbf{t}$ is free for $\mathbf{m}$ in $\mathbf{A}$.) From the definition of $\mathbf{B}$ and the conditions on $\mathbf{u}_1, \ldots, \mathbf{u}_n$ it can be seen that $\mathsf{S}^{\mathbf{u}_1}_{\cdot\mathbf{x}_1} \ldots \mathsf{S}^{\mathbf{u}_n}_{\cdot\mathbf{x}_n}\mathbf{B} = \mathbf{t}$, so since $\mathbf{u}_1, \ldots, \mathbf{u}_n$ do not occur in $\mathbf{A}$, $\mathsf{S}^{\mathbf{u}_1}_{\cdot\mathbf{x}_1} \ldots \mathsf{S}^{\mathbf{u}_n}_{\cdot\mathbf{x}_n}\mathsf{S}^{\mathbf{m}}_{\mathbf{B}}\mathbf{A} = \mathsf{S}^{\mathbf{m}}_{\mathbf{t}}\mathbf{A}$ (see Exercise X2012), and the theorem is proved for the case where $\mathcal{H}$ is empty.

Now to prove the theorem for the case that $\mathcal{H}$ is not empty, let $\mathcal{H} = \{\mathbf{H}_1, \ldots, \mathbf{H}_j\}$.

$\mathbf{H}_1, \ldots, \mathbf{H}_j \vdash \mathbf{A}$ given
$\vdash \mathbf{H}_1 \supset \centerdot\, \mathbf{H}_2 \supset \centerdot\, \mathbf{H}_j \supset \mathbf{A}$ Deduction Theorem

$\vdash \mathbf{H}_1 \supset \centerdot\, \mathbf{H}_2 \supset \centerdot \ldots \supset \centerdot\, \mathbf{H}_j \supset \mathsf{S}^{\mathbf{m}}_{\mathbf{t}}\mathbf{A}$

by substitution, since $\mathbf{m}$ does not occur in $\mathcal{H}$.

$\mathcal{H} \vdash \mathsf{S}^{\mathbf{m}}_{\mathbf{t}}\mathbf{A}$ 2113, MP

■

2119. $\vdash \sim \exists\mathbf{x}\mathbf{A} \equiv \forall\mathbf{x} \sim \mathbf{A}$ by Rule P and the Definition of $\exists$

ILLUSTRATION: Nobody votes if and only if everyone abstains.

2120. $\vdash \sim \forall\mathbf{x}\mathbf{A} \equiv \exists\mathbf{x} \sim \mathbf{A}$

by Rule P, Substitutivity of Equivalence, and the definition of $\exists$

ILLUSTRATION: Not everyone votes votes if and only if someone abstains.

Note that 2119 and 2120 are generalized forms of DeMorgan's Laws.

2121. $\vdash \forall\mathbf{x}\forall\mathbf{y}\mathbf{A} \equiv \forall\mathbf{y}\forall\mathbf{x}\mathbf{A}$

Proof:

(.1) $\forall\mathbf{x}\forall\mathbf{y}\mathbf{A} \vdash \mathbf{A}$ ∀I (2111)
(.2) $\forall\mathbf{x}\forall\mathbf{y}\mathbf{A} \vdash \forall\mathbf{y}\forall\mathbf{x}\mathbf{A}$ Gen: .1
(.3) $\vdash \forall\mathbf{x}\forall\mathbf{y}\mathbf{A} \supset \forall\mathbf{y}\forall\mathbf{x}\mathbf{A}$ Deduction Theorem
(.4) $\vdash \forall\mathbf{y}\forall\mathbf{x}\mathbf{A} \supset \forall\mathbf{x}\forall\mathbf{y}\mathbf{A}$ Theorem Schema 2121.3
Then use Rule P. ■

ILLUSTRATION:

$[Raa \wedge Rab \wedge Rac] \wedge [Rba \wedge Rbb \wedge Rbc] \wedge [Rca \wedge Rcb \wedge Rcc] \equiv$
$\centerdot\, [Raa \wedge Rba \wedge Rca] \wedge [Rab \wedge Rbb \wedge Rcb] \wedge [Rac \wedge Rbc \wedge Rcc]$

2122. $\vdash \exists\mathbf{x}\exists\mathbf{y}\mathbf{A} \equiv \exists\mathbf{y}\exists\mathbf{x}\mathbf{A}$ by 2121, Rule P, and 2117

2123. $\vdash \left(\mathsf{S}^{\mathbf{x}}_{\cdot\mathbf{t}}\mathbf{A}\right) \supset \exists\mathbf{x}\mathbf{A}$ provided that $\mathbf{t}$ is a term free for $\mathbf{x}$ in $\mathbf{A}$.

Proof:

(.1) $\vdash \forall\mathbf{x} \sim \mathbf{A} \supset \sim \mathsf{S}^{\mathbf{x}}_{\cdot\mathbf{t}}\mathbf{A}$ Axiom 4
(.2) $\vdash \mathsf{S}^{\mathbf{x}}_{\cdot\mathbf{t}}\mathbf{A} \supset \sim \forall\mathbf{x} \sim \mathbf{A}$ Rule P: .1

■

EXAMPLE: $\vdash P^3xxf^1x \supset \exists yP^3yxf^1y$.

2124. $\vdash \mathbf{A} \supset \exists\mathbf{xA}$ 2123

2125. $\vdash \forall\mathbf{xA} \supset \exists\mathbf{xA}$ Rule P: 2102, 2124

2126. Rule of Existential Generalization ($\exists$ **Gen**). If $\mathcal{H} \vdash \mathsf{S}^{\mathbf{x}}_{\mathbf{t}}\mathbf{A}$, where $\mathbf{t}$ is a term free for $\mathbf{x}$ in $\mathbf{A}$, then $\mathcal{H} \vdash \exists\mathbf{xA}$.

Proof: By 2123 and MP. ■

EXAMPLE: $\vdash \forall xPxy \supset Pyy$ by Axiom 4, so $\vdash \exists z \,\centerdot\, \forall xPxz \supset Pzy$ by 2126.

EXAMPLE: If $\mathcal{H} \vdash 3+2=2+3$, then $\mathcal{H} \vdash \exists x \,\centerdot\, 3+x=2+3$

REMARK. To illustrate the need for the condition that $\mathbf{t}$ be free for $\mathbf{x}$ in $\mathbf{A}$, let $\mathbf{A}$ be $\forall y\ y < x$, $\mathbf{x}$ be x, and $\mathbf{t}$ be $y+1$. Then $\mathsf{S}^{\mathbf{x}}_{\mathbf{t}}\mathbf{A}$ is $\forall y\ y < y+1$, and $\exists\mathbf{xA}$ is $\exists x\forall y\ y < x$. The latter does not follow from the former by Existential Generalization, since $y+1$ is not free for x in $\forall y\ y < x$.

REMARK. We cannot infer $\exists yPyyx$ from $Pxyx$ by Existential Generalization. For example, we cannot infer $\exists y\,[y+y=5]$ (which is a false statement about integers) from $5+y=5$ (which is a true statement about integers when y is assigned the value 0).

2127. $\vdash \forall\mathbf{xA} \equiv \mathbf{A}$ if $\mathbf{x}$ is not free in $\mathbf{A}$.

Proof:

(.1) $\vdash \mathbf{A} \supset \mathbf{A}$ Rule P
(.2) $\vdash \mathbf{A} \supset \forall\mathbf{xA}$ $\supset\forall$ Rule (2103): .1
(.3) $\vdash \forall\mathbf{xA} \equiv \mathbf{A}$ Rule P: .2, 2102

■

2128. $\vdash \exists\mathbf{xA} \equiv \mathbf{A}$ if $\mathbf{x}$ is not free in $\mathbf{A}$.

Proof: By 2127, Rule P, and the definition of $\exists$. ■

2129. $\vdash \forall\mathbf{x}[\mathbf{A} \wedge \mathbf{B}] \equiv \,\centerdot\, \forall\mathbf{xA} \wedge \forall\mathbf{xB}$

Proof:

(.1) $\forall\mathbf{x}[\mathbf{A} \wedge \mathbf{B}] \vdash \mathbf{A} \wedge \mathbf{B}$ $\forall$I (2111): hyp

(.2) $\forall \mathbf{x}[\mathbf{A} \wedge \mathbf{B}] \vdash \mathbf{A}$	Rule P: .1
(.3) $\forall \mathbf{x}[\mathbf{A} \wedge \mathbf{B}] \vdash \forall \mathbf{x}\mathbf{A}$	Gen: .2
(.4) $\forall \mathbf{x}[\mathbf{A} \wedge \mathbf{B}] \vdash \mathbf{B}$	Rule P: .1
(.5) $\forall \mathbf{x}[\mathbf{A} \wedge \mathbf{B}] \vdash \forall \mathbf{x}\mathbf{B}$	Gen: .4
(.6) $\forall \mathbf{x}[\mathbf{A} \wedge \mathbf{B}] \vdash \forall \mathbf{x}\mathbf{A} \wedge \forall \mathbf{x}\mathbf{B}$	Rule P: .3, .5
(.7) $\vdash \forall \mathbf{x}[\mathbf{A} \wedge \mathbf{B}] \supset \centerdot \forall \mathbf{x}\mathbf{A} \wedge \forall \mathbf{x}\mathbf{B}$	Deduction Theorem (2116): .6
(.8) $\forall \mathbf{x}\mathbf{A} \wedge \forall \mathbf{x}\mathbf{B} \vdash \forall \mathbf{x}\mathbf{A}$	Rule P: hyp
(.9) $\forall \mathbf{x}\mathbf{A} \wedge \forall \mathbf{x}\mathbf{B} \vdash \mathbf{A}$	$\forall$I: .8
(.10) $\forall \mathbf{x}\mathbf{A} \wedge \forall \mathbf{x}\mathbf{B} \vdash \forall \mathbf{x}\mathbf{B}$	Rule P: hyp
(.11) $\forall \mathbf{x}\mathbf{A} \wedge \forall \mathbf{x}\mathbf{B} \vdash \mathbf{B}$	$\forall$I: .10
(.12) $\forall \mathbf{x}\mathbf{A} \wedge \forall \mathbf{x}\mathbf{B} \vdash \mathbf{A} \wedge \mathbf{B}$	Rule P: .9, .11
(.13) $\forall \mathbf{x}\mathbf{A} \wedge \forall \mathbf{x}\mathbf{B} \vdash \forall \mathbf{x}[\mathbf{A} \wedge \mathbf{B}]$	Gen: .12
(.14) $\vdash \forall \mathbf{x}\mathbf{A} \wedge \forall \mathbf{x}\mathbf{B} \supset \forall \mathbf{x} \centerdot \mathbf{A} \wedge \mathbf{B}$	Deduction Theorem: .13

Then apply Rule P to .7 and .14. ■

ILLUSTRATION:

$$[Pa \wedge Qa] \wedge [Pb \wedge Qb] \wedge [Pc \wedge Qc] \\ \equiv \centerdot [Pa \wedge Pb \wedge Pc] \wedge [Qa \wedge Qb \wedge Qc].$$

2130. $\vdash \exists \mathbf{x}[\mathbf{A} \vee \mathbf{B}] \equiv \centerdot \exists \mathbf{x}\mathbf{A} \vee \exists \mathbf{x}\mathbf{B}$

Proof:

(.1) $\vdash \forall \mathbf{x}[\sim \mathbf{A} \wedge \sim \mathbf{B}] \equiv \centerdot \forall \mathbf{x} \sim \mathbf{A} \wedge \forall \mathbf{x} \sim \mathbf{B}$	2129
(.2) $\vdash \sim \mathbf{A} \wedge \sim \mathbf{B} \equiv \sim \centerdot \mathbf{A} \vee \mathbf{B}$	Rule P
(.3) $\vdash \forall \mathbf{x} \sim [\mathbf{A} \vee \mathbf{B}] \equiv \centerdot \forall \mathbf{x} \sim \mathbf{A} \wedge \forall \mathbf{x} \sim \mathbf{B}$	2117: .1, .2
(.4) $\vdash \exists \mathbf{x}[\mathbf{A} \vee \mathbf{B}] \equiv \centerdot \exists \mathbf{x}\mathbf{A} \vee \exists \mathbf{x}\mathbf{B}$	Rule P, def of $\exists$: .3

■

REMARK. 2129 and 2130 show that a universal quantifier can be distributed over a conjunction, and an existential quantifier can be distributed over a disjunction.

2131. $\vdash \exists \mathbf{x}[\mathbf{A} \vee \mathbf{B}] \equiv \centerdot \mathbf{A} \vee \exists \mathbf{x}\mathbf{B}$ if $\mathbf{x}$ is not free in $\mathbf{A}$.
$\vdash \exists \mathbf{x}[\mathbf{B} \vee \mathbf{A}] \equiv \centerdot \exists \mathbf{x}\mathbf{B} \vee \mathbf{A}$ if $\mathbf{x}$ is not free in $\mathbf{A}$.

Proof: by 2130, 2128, and Rule P. ■

2132. $\vdash \forall \mathbf{x}\mathbf{A} \lor \forall \mathbf{x}\mathbf{B} \supset \forall \mathbf{x} \centerdot \mathbf{A} \lor \mathbf{B}$.

Proof:

(.1) $\vdash \forall \mathbf{x}\mathbf{A} \supset \mathbf{A}$	2102
(.2) $\vdash \forall \mathbf{x}\mathbf{B} \supset \mathbf{B}$	2102
(.3) $\vdash \forall \mathbf{x}\mathbf{A} \lor \forall \mathbf{x}\mathbf{B} \supset \centerdot \mathbf{A} \lor \mathbf{B}$	Rule P: .1, .2
(.4) $\vdash \forall \mathbf{x}\mathbf{A} \lor \forall \mathbf{x}\mathbf{B} \supset \forall \mathbf{x} \centerdot \mathbf{A} \lor \mathbf{B}$	2103: .3

∎

ILLUSTRATION:

$$[Pa \land Pb \land Pc] \lor [Qa \land Qb \land Qc] \supset \centerdot [Pa \lor Qa] \land [Pb \lor Qb] \land [Pc \lor Qc].$$

Note that the converse need not be true.

2133. $\vdash \forall \mathbf{x}[\mathbf{A} \lor \mathbf{B}] \equiv \centerdot \mathbf{A} \lor \forall \mathbf{x}\mathbf{B}$ if $\mathbf{x}$ is not free in $\mathbf{A}$.
$\vdash \forall \mathbf{x}[\mathbf{B} \lor \mathbf{A}] \equiv \centerdot \forall \mathbf{x}\mathbf{B} \lor \mathbf{A}$ if $\mathbf{x}$ is not free in $\mathbf{A}$.

Proof:

(.1) $\vdash \mathbf{A} \lor \forall \mathbf{x}\mathbf{B} \supset \forall \mathbf{x} \centerdot \mathbf{A} \lor \mathbf{B}$	Rule P: 2127, 2132
(.2) $\vdash \forall \mathbf{x}[\mathbf{A} \lor \mathbf{B}] \equiv \centerdot \mathbf{A} \lor \forall \mathbf{x}\mathbf{B}$	Rule P: .1, Axiom 5
(.3) $\vdash \forall \mathbf{x}[\mathbf{B} \lor \mathbf{A}] \equiv \centerdot \forall \mathbf{x}\mathbf{B} \lor \mathbf{A}$	Rule P, 2117

∎

2134. $\vdash \forall \mathbf{x}[\mathbf{B} \supset \mathbf{A}] \equiv \centerdot \exists \mathbf{x}\mathbf{B} \supset \mathbf{A}$ if $\mathbf{x}$ is not free in $\mathbf{A}$.

Proof:

(.1) $\vdash \forall \mathbf{x}[\sim \mathbf{B} \lor \mathbf{A}] \equiv \centerdot \forall \mathbf{x} \sim \mathbf{B} \lor \mathbf{A}$	2133
(.2) $\vdash \forall \mathbf{x}[\mathbf{B} \supset \mathbf{A}] \equiv \centerdot \exists \mathbf{x}\mathbf{B} \supset \mathbf{A}$	Rule P: .1

∎

2135. Existential Rule ($\exists$ Rule). If $\mathcal{H}$, $\mathbf{B} \vdash \mathbf{A}$ and if $\mathbf{x}$ is not free in $\mathbf{A}$ or in $\mathcal{H}$, then $\mathcal{H}$, $\exists \mathbf{x}\mathbf{B} \vdash \mathbf{A}$.

Proof:

(.1) $\mathcal{H}, \mathbf{B} \vdash \mathbf{A}$	given
(.2) $\mathcal{H} \vdash \mathbf{B} \supset \mathbf{A}$	Deduction Theorem: .1
(.3) $\mathcal{H} \vdash \forall \mathbf{x} \centerdot \mathbf{B} \supset \mathbf{A}$	Gen: .2
(.4) $\mathcal{H} \vdash \exists \mathbf{x}\mathbf{B} \supset \mathbf{A}$	Rule P: 2134, .3

(.5) $\mathcal{H}$, $\exists\mathbf{x}\mathbf{B} \vdash \exists\mathbf{x}\mathbf{B} \supset \mathbf{A}$ — 2113: .4
(.6) $\mathcal{H}$, $\exists\mathbf{x}\mathbf{B} \vdash \mathbf{A}$ — MP: hyp, .5

■

Frequently in a mathematical argument one establishes (under appropriate assumptions) that there is an element with some property, and then one says "choose such an element, and let it be called y". One then continues the argument, assuming that y has the property in question. The following derived rule of inference (not to be confused with the Axiom of Choice, which is much stronger and cannot really be expressed in first-order logic unless set-theoretic postulates are assumed) provides a way of formalizing such an argument.

2136. Rule C.[4] If $\mathcal{H} \vdash \exists\mathbf{x}\mathbf{B}$ and $\mathcal{H}$, $\mathsf{S}^{\mathbf{x}}_{\mathbf{y}}\mathbf{B} \vdash \mathbf{A}$, where $\mathbf{y}$ is an individual variable which is free for $\mathbf{x}$ in $\mathbf{B}$ and which is not free in $\mathcal{H}$, $\exists\mathbf{x}\mathbf{B}$, or in $\mathbf{A}$, then $\mathcal{H} \vdash \mathbf{A}$. (Note that $\mathbf{y}$ may be $\mathbf{x}$.)

Proof:

(.1) $\mathcal{H}$, $\mathsf{S}^{\mathbf{x}}_{\mathbf{y}}\mathbf{B} \vdash \mathbf{A}$	given
(.2) $\mathcal{H}$, $\exists\mathbf{y}\mathsf{S}^{\mathbf{x}}_{\mathbf{y}}\mathbf{B} \vdash \mathbf{A}$	2135: .1
(.3) $\mathcal{H} \vdash \exists\mathbf{y}\mathsf{S}^{\mathbf{x}}_{\mathbf{y}}\mathbf{B} \supset \mathbf{A}$	Deduction Theorem: .2
(.4) $\mathcal{H} \vdash \exists\mathbf{x}\mathbf{B} \supset \mathbf{A}$	α Rule (unless $\mathbf{y}$ is $\mathbf{x}$)
(.5) $\mathcal{H} \vdash \exists\mathbf{x}\mathbf{B}$	given
(.6) $\mathcal{H} \vdash \mathbf{A}$	MP: .4, .5

■

In a proof where $\exists\mathbf{x}\mathbf{B}$ has been established, the next step is often the introduction of $\mathsf{S}^{\mathbf{x}}_{\mathbf{y}}\mathbf{B}$ as a new hypothesis, which will eventually be eliminated by Rule C. To make the proof as easy to comprehend as possible, we shall often provide the explanation "choose y" or "$\exists$I" (existential instantiation) for the new hypothesis, rather than simply indicating that it is a hypothesis.

To illustrate the ideas mentioned above, we consider the problem of proving $\forall x\forall z\forall w \,\blacksquare\, z + x = w + x \supset z = w$ in the theory $\mathcal{G}$ mentioned near the beginning of this section. One can start by assuming $z + x = w + x$ as hypothesis. With the help of postulate (G6) one can prove $\exists y \,\blacksquare\, x + y = 0$, so one chooses such a y by assuming (*) $x + y = 0$ as an additional hypothesis. With the help of (*) and various postulates of $\mathcal{G}$ one can derive

[4] This rule differs from the Rule C of [Rosser 1953], although both rules provide ways of handling acts of choice in an argument.

$$\begin{aligned}(z+x)+y &= (w+x)+y,\\ z+(x+y) &= w+(x+y),\\ z+0 &= w+0\\ z &= w\end{aligned}$$

(Exercise X2105 asks the reader to fill in the details of this proof.) The conclusion $z = w$ says nothing about y, and depends only on the hypothesis $z + x = w + x$, so at this stage hypothesis (*) can be dropped. Of course, in informal mathematics one keeps track of one's hypotheses in an informal way, and just forgets about those that are no longer relevant. In formal logic, however, one needs an explicit rule to eliminate a hypothesis, and this is precisely the purpose of Rule C.

One often finds that the same wff occurs in several lines of a proof, especially if it occurs as a hypothesis. Therefore, for the sake of brevity, it is often convenient to use the number of a line as an abbreviation for the wff which is asserted in that line, when this wff occurs elsewhere in the proof.

Henceforth when a line of a proof is inferred from the immediately preceding line, we shall often omit explicit mention of the preceding line from our justification of the inferred line.

These conventions will be illustrated in the proofs of the next two theorems, which provide examples of applications of Rule C.

2137. $\vdash \exists \mathbf{x} \forall \mathbf{z} \mathbf{A} \supset \forall \mathbf{z} \exists \mathbf{x} \mathbf{A}$.

Proof: If $\mathbf{x}$ and $\mathbf{z}$ are the same variable, then this wff is equivalent via 2128, 2127, and Rule P to an instance of 2125. Therefore we need consider only the case where $\mathbf{x}$ and $\mathbf{z}$ are distinct.

(.1)	$.1 \vdash \exists \mathbf{x} \forall \mathbf{z} \mathbf{A}$	hyp
(.2)	$.1,\ .2 \vdash \forall \mathbf{z} \mathbf{A}$	choose $\mathbf{x}$
(.3)	$.1,\ .2 \vdash \mathbf{A}$	$\forall$I
(.4)	$.1,\ .2 \vdash \exists \mathbf{x} \mathbf{A}$	$\exists$ Gen
(.5)	$.1,\ .2 \vdash \forall \mathbf{z} \exists \mathbf{x} \mathbf{A}$	Gen
(.6)	$.1 \vdash .5$	Rule C: .1, .5
(.7)	$\vdash \exists \mathbf{x} \forall \mathbf{x} \mathbf{A} \supset \forall \mathbf{z} \exists \mathbf{z} \mathbf{A}$	Deduction Theorem

■

ILLUSTRATION:

$$[[Raa \wedge Rab \wedge Rac] \vee [Rba \wedge Rbb \wedge Rbc] \vee [Rca \wedge Rcb \wedge Rcc]] \\ \supset \blacksquare\ [Raa \vee Rba \vee Rca] \wedge [Rab \vee Rbb \vee Rcb] \wedge [Rac \vee Rbc \vee Rcc]$$

Note that the converse need not be true.

2138. $\vdash \exists \mathbf{x}[\mathbf{A} \wedge \mathbf{B}] \supset \blacksquare\, \exists \mathbf{x}\mathbf{A} \wedge \exists \mathbf{x}\mathbf{B}$

Proof:

(.1) $.1 \vdash \exists \mathbf{x}\ .2$	hyp
(.2) $.1,\ .2 \vdash \mathbf{A} \wedge \mathbf{B}$	choose $\mathbf{x}$
(.3) $.1,\ .2 \vdash \mathbf{A}$	Rule P
(.4) $.1,\ .2 \vdash \exists \mathbf{x}\mathbf{A}$	$\exists$ Gen
(.5) $.1,\ .2 \vdash \mathbf{B}$	Rule P: .2
(.6) $.1,\ .2 \vdash \exists \mathbf{x}\mathbf{B}$	$\exists$ Gen
(.7) $.1,\ .2 \vdash \exists \mathbf{x}\mathbf{A} \wedge \exists \mathbf{x}\mathbf{B}$	Rule P: .4, .6
(.8) $.1 \vdash .7$	Rule C: .1, .7
(.9) $\vdash \exists \mathbf{x}[\mathbf{A} \wedge \mathbf{B}] \supset \blacksquare\, \exists \mathbf{x}\mathbf{A} \wedge \exists \mathbf{x}\mathbf{B}$	2116

■

ILLUSTRATION:

$$[[Pa \wedge Qa] \vee [Pb \wedge Qb] \vee [Pc \wedge Qc]] \\ \supset \blacksquare\ [Pa \vee Pb \vee Pc] \wedge [Qa \vee Qb \vee Qc].$$

Note that the converse need not be true.

REMARK. Theorem 2138 is essentially the contrapositive of Theorem 2132 and can be derived from 2132 in a manner analogous to the way 2130 was derived from 2129.

Consider the following attempt at a proof:

(1) $1 \vdash \exists y Pxy$	hyp
(2) $1,\ 2 \vdash Pxx$	choose x
(3) $1,\ 2 \vdash \exists v Pvv$	$\exists$ Gen
(4) $1 \vdash 3$	Rule C: 1, 3 (incorrect)
(5) $\vdash \exists y Pxy \supset \exists v Pvv$	Deduction Theorem

From an intuitive point of view, the essential error is made in line (2), where the element y, whose existence is asserted in (1), is unjustifiably identified with x. However, one can assume anything one likes as a hypothesis, so from a formal point of view line (2) is correct (though useless). The difficulty arises in line (4), when an attempt is made to eliminate (2) from the list of hypotheses. Note how the restrictions in the statement of Rule C block this incorrect inference. We certainly do not wish to be able to derive (5), since the same reasoning would enable us to infer that $\exists y\,[8 < y] \supset \exists v\,[v < v]$ is a true statement about real numbers.

Sometimes it is suggested in connection with Rule C that after deriving $\exists \mathbf{x}\mathbf{B}$ from a set $\mathcal{H}$ of hypotheses, one should simply assert $\mathsf{S}^{\mathbf{x}}_{\cdot\mathbf{y}}\mathbf{B}$ (where $\mathbf{y}$ is a new variable), but not list it as a hypothesis, since it will eventually be deleted from the list of hypotheses via Rule C. Unfortunately, this can lead to a variety of incorrect inferences, such as the following three:

Example A:

(.1) $.1 \vdash \exists \mathbf{x}\mathbf{A}$	hyp
(.2) $.1 \vdash \mathbf{A}$	choose $\mathbf{x}$
(.3) $.1 \vdash \forall \mathbf{x}\mathbf{A}$	Gen
(.4) $\vdash \exists \mathbf{x}\mathbf{A} \supset \forall \mathbf{x}\mathbf{A}$	Deduction Theorem

Example B:

(.1) $.1 \vdash \forall \mathbf{z}\exists \mathbf{x}\mathbf{A}$	hyp
(.2) $.1 \vdash \exists \mathbf{x}\mathbf{A}$	$\forall$I
(.3) $.1 \vdash \mathbf{A}$	choose $\mathbf{x}$
(.4) $.1 \vdash \forall \mathbf{z}\mathbf{A}$	Gen
(.5) $.1 \vdash \exists \mathbf{x}\forall \mathbf{z}\mathbf{A}$	$\exists$ Gen
(.6) $\vdash \forall \mathbf{z}\exists \mathbf{x}\mathbf{A} \supset \exists \mathbf{x}\forall \mathbf{z}\mathbf{A}$	Deduction Theorem

Example C:

(.1) $.1 \vdash \exists x P x \wedge \exists x Q x$	hyp
(.2) $.1 \vdash \exists x P x$	Rule P
(.3) $.1 \vdash P x$	Choose x
(.4) $.1 \vdash \exists x Q x$	Rule P: .1
(.5) $.1 \vdash Q y$	choose y
(.6) $.1 \vdash P x \wedge Q y$	Rule P: .3, .5
(.7) $.1 \vdash P x \wedge Q x$	Sub (2118)
(.8) $.1 \vdash \exists x \,\blacksquare\, P x \wedge Q x$	$\exists$ Gen
(.9) $\vdash \exists x P x \wedge \exists x Q x \supset \exists x \,\blacksquare\, P x \wedge Q x$	2116

Thus one would be led to believe that the converses of 2125, 2137, and 2138 are provable. Actually, we shall see in §23 that they are not. We leave it to the reader to analyze the errors in the examples above.

The proofs of the next three metatheorems require only the Deduction Theorem and Rule P.

2139 Rule of Cases. If $\mathcal{H} \vdash \mathbf{A} \vee \mathbf{B}$ and $\mathcal{H}, \mathbf{A} \vdash \mathbf{C}$ and $\mathcal{H}, \mathbf{B} \vdash \mathbf{C}$, then $\mathcal{H} \vdash \mathbf{C}$.

2140 Corollary. If $\mathcal{H}, \mathbf{A} \vdash \mathbf{C}$ and $\mathcal{H}, \sim \mathbf{A} \vdash \mathbf{C}$ then $\mathcal{H} \vdash \mathbf{C}$.

2141 Indirect Proof (IP). If $\mathcal{H}, \sim \mathbf{A} \vdash \mathbf{B}$, and $\mathcal{H}, \sim \mathbf{A} \vdash \sim \mathbf{B}$, then $\mathcal{H} \vdash \mathbf{A}$. Also, if $\mathcal{H}, \sim \mathbf{A} \vdash \mathbf{B} \wedge \sim \mathbf{B}$, then $\mathcal{H} \vdash \mathbf{A}$.

In Figure 2.2 we provide some rather obvious advice about general strategies which may be used in proving theorems of first-order logic. Of course these strategies may not always provide the most efficient proof of a wff, but in the absence of a better idea they can usually be made to work. When other methods fail to produce a proof, the method of Indirect Proof (2141) can always be forced to work. (See §32.)

We remark that negations can be pushed in via De Morgan's Laws, 2119, and 2120, so they are not considered in Figure 2.2. Of the three strategies presented for proving an implication, the last one (proof by contradiction) is most general. For if $\mathbf{A} \vdash \mathbf{B}$, then $\mathbf{A}, \sim \mathbf{B} \vdash \mathbf{B} \wedge \sim \mathbf{B}$. Similarly, if $\sim \mathbf{B} \vdash \sim \mathbf{A}$, then $\mathbf{A}, \sim \mathbf{B} \vdash \mathbf{A} \wedge \sim \mathbf{A}$. Moreover, if one takes both $\mathbf{A}$ and $\sim\mathbf{B}$ as hypotheses, one has "more to work with" than if one assumes only one of these wffs.

It will be observed that ingenuity is required in the application of the rules above primarily in the cases of Universal Instantiation and Existential Generalization. Actually, the fundamental problems involved in applying these two rules appropriately are the same. For if one wishes to prove $\exists \mathbf{x} \mathbf{A}$, instead of proving $\mathsf{S}^{\mathbf{x}}_{\mathbf{t}} \mathbf{A}$ and applying Existential Generalization, one may assume $\sim \exists \mathbf{x} \mathbf{A}$ as hypothesis, obtain $\forall \mathbf{x} \sim \mathbf{A}$ by 2119, then $\sim \mathsf{S}^{\mathbf{x}}_{\mathbf{t}} \mathbf{A}$ by Universal Instantiation, and derive a contradiction. There is also a fundamental similarity between existential instantiation and Universal Generalization. If one wishes to prove $\forall \mathbf{x} \mathbf{A}$, instead of proving $\mathbf{A}$ and applying Universal Generalization, one may assume $\sim \forall \mathbf{x} \mathbf{A}$ as hypothesis, obtain $\exists \mathbf{x} \sim \mathbf{A}$ by 2120, then $\sim\mathbf{A}$ by existential instantiation, and derive a contradiction. In either case, one must assure (by an alphabetic change of bound variable, if necessary) that the variable $\mathbf{x}$ is not free in any hypothesis.

To Prove	*Do*
$\mathbf{A} \supset \mathbf{B}$	Assume **A**, prove **B**, and use the Deduction Theorem.
$\mathbf{A} \supset \mathbf{B}$	Assume $\sim\mathbf{B}$, and use the Deduction Theorem and Contrapositive Law $\sim \mathbf{B} \supset \sim \mathbf{A} \supset . \mathbf{A} \supset \mathbf{B}$.
$\mathbf{A} \supset \mathbf{B}$	Assume **A** and $\sim \mathbf{B}$, prove $\mathbf{C} \wedge \sim\mathbf{C}$ (for any wff **C**), and use the Deduction Theorem and Rule P.
$\mathbf{A} \vee \mathbf{B}$	Write this as $\sim \mathbf{A} \supset \mathbf{B}$, and use the strategies for $\supset$.
$\mathbf{A} \wedge \mathbf{B}$	Prove **A** and **B** separately, then use Rule P.
$\forall \mathbf{x}\, \mathbf{A}$	Prove **A**, and use Gen. (If **x** is free in a hypothesis, prove $\mathsf{S}^{\mathbf{x}}_{\mathbf{y}}\mathbf{A}$, for some new variable **y**, then use Gen and the α Rule.)
$\exists \mathbf{x} \mathbf{A}$	Prove $\mathsf{S}^{\mathbf{x}}_{\mathbf{t}}\mathbf{A}$ for some term **t**, then use $\exists$ Gen. (Of course, the crucial problem is to find a term **t** such that this is feasible. For complex cases, see the discussion in §30.)

If one has inferred	*Do*
$\mathbf{A} \wedge \mathbf{B}$	Infer **A** and **B** by Rule P.
$\mathbf{A} \vee \mathbf{B}$	Take **A** and **B** as hypotheses separately, and use the Rule of Cases.
$\mathbf{A} \supset \mathbf{B}$	Write this as $\sim \mathbf{A} \vee \mathbf{B}$, and use the strategy for $\vee$. If **A** has been established, use Modus Ponens.
$\forall \mathbf{x} \mathbf{A}$	Infer $\mathsf{S}^{\mathbf{x}}_{\mathbf{t}}\mathbf{A}$ for an *appropriate* term **t** by $\forall$I. This may have to be done with a number of different terms.
$\exists \mathbf{x} \mathbf{A}$	Existentially instantiate by choosing a new variable **y** (which may be **x** if it has not already occurred free), and taking $\mathsf{S}^{\mathbf{x}}_{\mathbf{y}}$ as hypothesis. Eliminate the new hypothesis later by Rule C.

Figure 2.2: General advice on methods of proof

Sometimes attempts to complete a proof appear to be blocked by the restrictions on Rule C. Suppose one has the following fragment of an attempt to prove $\mathcal{H} \vdash \mathbf{A}$:

$\mathcal{H}, \sim \mathbf{A} \vdash \exists \mathbf{x} \mathbf{B}$

$\mathcal{H}, \sim \mathbf{A}, \mathbf{B} \vdash \mathbf{B}$ choose $\mathbf{x}$

$\mathcal{H}, \sim \mathbf{A}, \mathbf{B} \vdash \sim \mathbf{B}$

It appears that one has now reached the desired contradiction. However, hypothesis **B** must be eliminated, and Rule C cannot be applied if **x** occurs free in **B**. (We may assume that **x** was chosen as a variable distinct from the free variables in $\mathcal{H}$ and **A**, but it normally will occur free in **B**.) Nevertheless, one can complete this proof as follows:

$\mathcal{H}, \sim \mathbf{A}, \mathbf{B} \vdash p \wedge \sim p$ Rule P

$\mathcal{H}, \sim \mathbf{A} \vdash p \wedge \sim p$ Rule C

$\mathcal{H} \vdash \mathbf{A}$ IP

When one is constructing proofs, it is often convenient to work down from the top and up from the bottom. For example, if one wishes to prove $\mathbf{A} \supset \mathbf{B}$, one may start by constructing the following proof outline:

(1) $1 \vdash \mathbf{A}$ hyp
(99) $1 \vdash \mathbf{B}$
(100) $\vdash \mathbf{A} \supset \mathbf{B}$ Ded: 99

One must then fill in some lines between (1) and (99) so as to justify (99). If **B** has the form $\forall \mathbf{x} \mathbf{C}$, and **x** does not occur free in **A**, one might expand this outline as follows:

(1) $1 \vdash \mathbf{A}$ hyp
(98) $1 \vdash \mathbf{C}$
(99) $1 \vdash \forall \mathbf{x} \mathbf{C}$ Gen: 98
(100) $\vdash \mathbf{A} \supset \mathbf{B}$ Ded: 99

Working in this way tends to produce well structured proofs.

More formal discussion of proof procedures will be found in Chapter 3. In particular, the reader will find in §30 a summary of particularly useful derived rules of inference.

EXERCISES

X2100. Show that there is an effective test for determining of any wff of $\mathcal{F}$ whether or not it is tautologous.

X2101. In each case below, decide whether the given inference is a correct application of Rule P.

(a) From $\forall x\,[Gx \supset Hx] \equiv \exists x Hx$ and $\forall x\,[Gx \supset Hx] \equiv \sim \centerdot \forall x Gx \supset \exists x Hx$ infer $[\exists x Hx \supset \sim \forall x Gx]$.
(b) From $[q \supset \forall z Rz]$ and $[\forall x Px \supset \centerdot \sim \forall z Rz \wedge q]$ infer $\sim \forall x Px$.
(c) From $[p \supset \centerdot \forall x Qx \vee Ry]$ and $[\forall x Qx \supset \centerdot \exists x[Rx \wedge Qx] \supset Ry]$ and $[\exists x[Rx \wedge Qx] \vee \sim p]$ infer $[p \supset Ry]$.
(d) From $[[q \wedge r \wedge p] \supset \centerdot s \supset \forall x Px]$ infer $[p \supset \centerdot q \supset \centerdot [r \supset s] \supset \forall x Px]$.

In exercises X2102–X2105, prove the given theorems of $\mathcal{G}$, the elementary theory of abelian groups which was presented near the beginning of this section.

X2102. $\forall x \forall y \centerdot x = y \supset y = x$.

X2103. $\forall x \forall y \centerdot x = y \supset \centerdot z = x \supset z = y$.

X2104. $\forall x \forall y \forall z \centerdot x = y \wedge y = z \supset x = z$.

X2105. $\forall x \forall z \forall w \centerdot z + x = w + x \supset z = w$.

Prove the following wffs, using the derived rules of inference discussed in this section. Present your proofs in the manner of the proofs of theorem schemata 2137 and 2138.[5]

X2106. $\forall x\,[Rx \supset Px] \wedge \forall x\,[\sim Qx \supset Rx] \supset \forall x \centerdot Px \vee Qx$.

X2107. $Rab \wedge \forall x \forall y\,[Rxy \supset \centerdot Ryx \wedge Qxy] \wedge \forall u \forall v\,[Quv \supset Quu] \supset \centerdot Qaa \wedge Qbb$.

X2108. $\forall x \exists y \centerdot Px \supset Py$.

X2109. $\exists x\,[p \wedge Qx] \equiv \centerdot p \wedge \exists x Qx$.

X2110. $\exists x Rx \wedge \forall y\,[Ry \supset \exists z Qyz] \wedge \forall x \forall y\,[Qxy \supset Qxx]$
$\supset \exists x \exists y \centerdot Qxy \wedge Ry$.

---

[5]See the end of the Preface for information about the ETPS program, which facilitates work on exercises such as these.

X2111. $\forall x\,[\exists yPxy \supset \forall yQxy] \wedge \forall z\exists yPzy \supset \forall y\forall xQxy.$

X2112. $\exists v\forall xPxv \wedge \forall x\,[Sx \supset \exists yQyx] \wedge \forall x\forall y\,[Pxy \supset {\sim} Qxy] \supset \exists u \sim Su.$

X2113. $\forall y\exists wRyw \wedge \exists z\forall x\,[Px \supset {\sim} Rzx] \supset \exists x \sim Px.$

X2114. $\forall xRxb \wedge \forall y\,[\exists zRyz \supset Ray] \supset \exists u\forall vRuv.$

X2115. $\forall x\,[\exists yPxy \supset \forall zPzz] \wedge \forall u\exists v\,[Puv \vee {\bullet}\, Mu \wedge Qf^2uv] \wedge \forall w\,[Qw \supset {\sim} Mg^1w] \supset \forall u\exists v \,{\bullet}\, Pg^1uv \wedge Puu.$

X2116. $\forall x\exists y\,[Px \supset {\bullet}\, Rxg^1h^1y \wedge Py] \wedge \forall w\,[Pw \supset {\bullet}\, Pg^1w \wedge Ph^1w] \supset \forall x \,{\bullet}\, Px \supset \exists y \,{\bullet}\, Rxy \wedge Py.$

X2117. $\forall u\forall v\,[Ruu \equiv Ruv] \wedge \forall w\forall z\,[Rww \equiv Rzw] \supset {\bullet}\, \exists xRxx \supset \forall yRyy.$

X2118. $\forall x\,[[p \wedge Qx] \vee {\bullet} \sim p \wedge Rx] \supset {\bullet}\, \forall xQx \vee \forall xRx.$
Hint: If it is true every day that the roof was fixed and the rug stays dry, or that the roof was not fixed and the rug gets wet, then the rug stays dry every day or the rug gets wet every day.

X2119. $\exists y\forall x \,{\bullet}\, Py \supset Px.$

X2120. $\forall u\forall v\forall w\,[Puv \vee Pvw] \supset \exists x\forall yPxy.$

X2121. $\exists v\forall y\exists z \,{\bullet}\, [Payh^1y \vee Pvyf^1y] \supset Pvyz.$

X2122. $[\exists xRxx \supset \forall yRyy] \supset \exists u\forall v \,{\bullet}\, Ruu \supset Rvv.$

X2123. $\exists y\,[Py \supset Qx] \supset \exists y \,{\bullet}\, Py \supset Qy.$

X2124. $\exists x\,[Px \supset Qx] \equiv {\bullet}\, \forall xPx \supset \exists xQx.$

X2125. $\exists x\forall y\,[Px \equiv Py] \equiv {\bullet}\, \exists xPx \equiv \forall yPy.$

X2126. $\forall x\,[Px \equiv \exists yPy] \equiv {\bullet}\, \forall xPx \equiv \exists yPy.$

X2127. $\exists x\forall y\,[Py \equiv Px] \supset {\bullet}\, \forall xPx \vee \forall x \sim Px.$

X2128. $\forall x\,[Px \equiv \forall yPy] \equiv {\bullet}\, \exists xPx \equiv \forall yPy.$

X2129. $[\exists x\forall y\,[Px \equiv Py] \equiv {\bullet}\, \exists xQx \equiv \forall yPy] \equiv {\bullet}\, \exists x\forall y\,[Qx \equiv Qy] \equiv {\bullet}\, \exists xPx \equiv \forall yQy.$

X2130. $\forall xPx \supset {\bullet} \sim \exists yQy \vee \exists z \,{\bullet}\, Pz \supset Qz.$

X2131. $\forall xPx \supset \exists y \,{\bullet}\, \forall x\forall zQxyz \supset {\sim} \forall z \,{\bullet}\, Pz \wedge {\sim} Qyyz.$

X2132. $\forall w \sim Rww \supset \exists x \exists y \,\blacksquare\, \sim Rxy \wedge \,\blacksquare\, Qyx \supset \forall z Qzz.$

X2133. $\forall x\,[\exists y Qxy \supset Px] \wedge \forall v \exists u Quv \wedge \forall w \forall z\,[Qwz \supset \,\blacksquare\, Qzw \vee Qzz]$ $\supset \forall z\, Pz.$

X2134. $\forall z \exists x\,[\forall y Pxy \vee Qxz] \supset \forall y \exists x]\,[Pxy \vee Qxy].$

X2135. $\exists x \forall y\,[Px \wedge Qy \supset \,\blacksquare\, Qx \vee Py].$

X2136. $\exists x \exists y \forall u\,[Pxyz \supset Puxx].$

X2137. $\exists x \forall y\,[Px \supset \,\blacksquare\, Qx \vee Py].$

X2138. $\forall x \exists y\, Fxy \wedge \exists x \forall \epsilon \exists n \forall w\,[Snw \supset Dwx\epsilon] \wedge$
$\forall \epsilon \exists \delta \forall x_1 \forall x_2\,[Dx_1 x_2 \delta \supset \forall y_1 \forall y_2 \,\blacksquare\, Fx_1 y_1 \wedge Fx_2 y_2 \supset Dy_1 y_2 \epsilon]$
$\supset \exists y \forall \epsilon \exists m \forall w \,\blacksquare\, Smw \supset \forall z \,\blacksquare\, Fwz \supset Dzy\epsilon.$

(Here δ, ϵ, n, and m are used as individual variables.)

X2139. $(\exists x \epsilon S)\, Px \equiv \sim (\forall x \epsilon S) \sim Px.$

X2140. $(\forall x \epsilon S)\, Px \equiv \sim (\exists x \epsilon S) \sim Px.$

X2141. Can $\exists x\, \forall y\, Pyx$ be inferred from $\forall y\, Py\, fy$ by the rule of Existential Generalization? (Here f is a monadic function constant.) Explain exactly how this inference does or does not conform to this rule.

In exercises X2142–X2143, you are given alleged proofs to evaluate. Decide whether each line of the proof is correctly inferred from the specified line(s) by the specified rule of inference. Treat each line of the proof separately; when dealing with one line, you need not be concerned with the question of whether other lines were obtained correctly.

X2142.

(1)	$1 \vdash Px \wedge \forall x \exists y\, Rxyz$	hyp
(2)	$1 \vdash Px$	Rule P: 1
(3)	$1 \vdash Pu$	Sub: 2
(4)	$1 \vdash \exists y\, Ruyz$	UI: 1
(5)	$1 \vdash \forall z\, \exists y\, Ruyz$	UGen: 4
(6)	$1 \vdash \exists y\, Ruyu$	UI: 5
(7)	$1, 7 \vdash Ruzu$	hyp
(8)	$1, 7 \vdash \exists v\, Rvvu$	EGen: 7
(9)	$1 \vdash \exists v\, Rvvu$	Rule C: 6, 8
(10)	$1, 7 \vdash \exists x\, Rxzu$	EGen: 7

(11) $1 \vdash \forall y \sim Ruyz \supset \sim Ruuu$ Rule P: 4
(12) $1 \vdash \exists y\, Ruyy$ UI: 5

X2143.

(1) $1 \vdash \forall x\,[Px \supset \sim\!\centerdot\, Qxy \;\vee\; Rxy] \supset\centerdot \sim Rxy \supset Qxy$ hyp
(2) $1 \vdash [Px \supset \sim\!\centerdot\, Qxy \;\vee\; Rxy] \supset\centerdot \sim Rxy \supset Qxy$ ∀I: 1
(3) $1 \vdash \sim Qxy \supset Rxy$ Rule P: 2
(4) $1 \vdash \forall y\centerdot \sim Qxy \supset Rxy$ Gen: 3
(5) $1 \vdash \exists z \forall y\centerdot \sim Qxy \supset Rzy$ ∃ Gen: 4
(6) $1, 6 \vdash \forall y\centerdot \sim Qxy \supset Rzy$ choose z
(7) $1, 6 \vdash \sim Qxy \supset Rzx$ ∀I: 6
(8) $1, 6 \vdash \exists x\centerdot \sim Qxx \supset Rzx$ ∃ Gen: 7
(9) $1 \vdash \exists x\centerdot \sim Qxx \supset Rzx$ Rule C: 5, 8
(10) $\vdash \forall x\,[Px \supset \sim\!\centerdot\, Qxy \;\vee\; Rxy] \supset\centerdot \sim Rxy \supset Qxy \supset \exists x\centerdot \sim Qxx \supset Rzx$
Deduction Theorem: 9

X2144. Let $\mathcal{F}^*$ be the logistic system obtained by adding to the system $\mathcal{F}$ the additional axiom schema $\exists\mathbf{x}\mathbf{A} \supset \forall\mathbf{x}\mathbf{A}$. Prove that $\mathcal{F}^*$ is absolutely consistent.

X2145. Let $\mathcal{F}^{**}$ be the logistic system obtained by adding to the system $\mathcal{F}$ the additional axiom schema $\forall\mathbf{x}\exists\mathbf{y}\mathbf{A} \supset \exists\mathbf{y}\forall\mathbf{x}\mathbf{A}$. Is $\mathcal{F}^{**}$ is absolutely consistent?

X2146. Express the following statements in symbolic form, and prove that the negation of statement A implies statement B.

(A) Everyone who bought a car also bought a trailer or a boat without a trailer.

(B) Someone bought a car but no boat and no trailer.

§22. Prenex Normal Form

In this section we wish to show that each wff is equivalent to a wff in which no quantifier is in the scope of a propositional connective. However, we wish to consider existential as well as universal quantifiers, and ignore the negations which occur as parts of existential quantifiers. To be precise about this, we shall introduce a language $\mathcal{E}$ in which existential as well as universal quantifiers are primitive, and formulate certain definitions in terms of wffs of $\mathcal{E}$.

DEFINITION. The primitive symbols of $\mathcal{E}$ are those of $\mathcal{F}$, plus the symbol $\exists$. The formation rules of $\mathcal{E}$ are those of $\mathcal{F}$, plus the following:

(g) If $\mathbf{B}$ is a wff of $\mathcal{E}$ and $\mathbf{x}$ is an individual variable, then $\exists\mathbf{xB}$ is a wff of $\mathcal{E}$.

Clearly each wff of $\mathcal{F}$ is a wff of $\mathcal{E}$. Also, every wff of $\mathcal{E}$ is an abbreviation for a wff of $\mathcal{F}$, under the convention that $\exists\mathbf{xB}$ is an abbreviation for $\sim\forall\mathbf{x}\sim\mathbf{B}$. Thus, $\exists\mathbf{xB}$ and $\sim\forall\mathbf{x}\sim\mathbf{B}$ are distinct wffs of $\mathcal{E}$, but are abbreviations for the same wff of $\mathcal{F}$. In this section, when we assert of a wff $\mathbf{A}$ of $\mathcal{E}$ that $\vdash\mathbf{A}$, we shall mean that the wff of $\mathcal{F}$ for which $\mathbf{A}$ is an abbreviation is a theorem of $\mathcal{F}$.

DEFINITION. An occurrence of a quantifier $\forall\mathbf{x}$ or $\exists\mathbf{x}$ is *vacuous* iff its variable $\mathbf{x}$ has no free occurrence in its scope. Thus, if $\mathbf{x}$ is not free in $\mathbf{B}$, then the exhibited quantifier in $\forall\mathbf{xB}$ is vacuous.

DEFINITION. A wff $\mathbf{G}$ of $\mathcal{E}$ is in *prenex normal form* iff no vacuous quantifier occurs in $\mathbf{G}$ and no quantifier in $\mathbf{G}$ occurs in the scope of a propositional connective of $\mathbf{G}$. A wff $\mathbf{A}$ of $\mathcal{F}$ is in prenex normal form iff there is a wff of $\mathcal{E}$ which is in prenex normal form and which is an abbreviation for $\mathbf{A}$. Thus, a wff in prenex normal form is either quantifier-free, or is of the form $\exists\!\!\!/_1\mathbf{x}_1\ldots\exists\!\!\!/_n\mathbf{x}_n\mathbf{M}$, where $\mathbf{M}$ is a quantifier-free wff and for each i, $\exists\!\!\!/_i\mathbf{x}_i$ is $\forall\mathbf{x}_i$ or $\exists\mathbf{x}_i$. Here $\mathbf{M}$ is called the *matrix* of the wff, and the sequence $\exists\!\!\!/_1\mathbf{x}_1\ldots\exists\!\!\!/_n\mathbf{x}_n$ of quantifiers is called the *prefix*. If $\mathbf{G}$ is a wff in prenex normal form such that $\vdash\mathbf{C}\equiv\mathbf{G}$, we shall say that $\mathbf{G}$ is *a prenex normal form* of $\mathbf{C}$.

DEFINITION. A wff $\mathbf{D}$ is *rectified* iff $\mathbf{D}$ contains no vacuous quantifier, no variable occurs both bound and free in $\mathbf{D}$, and no two quantifier-occurrences in $\mathbf{D}$ bind the same variable. Clearly a wff in prenex normal form must be rectified.

For any rectified wff $\mathbf{D}$ of $\mathcal{E}$, we define *the prenex normal form* of $\mathbf{D}$ as follows. Let $(\forall\mathbf{x})'$ be $\exists\mathbf{x}$ and $(\exists\mathbf{x})'$ be $\forall\mathbf{x}$. Let the quantifiers in $\mathbf{D}$ be $\exists\!\!\!/_1\mathbf{x}_1,\ldots,\exists\!\!\!/_n\mathbf{x}_n$, in the left-to-right order in which they occur in $\mathbf{D}$ (so $n\geq 0$ and each $\exists\!\!\!/_i$ is $\forall$ or $\exists$). We say that $\exists\!\!\!/_i\mathbf{x}_i$ occurs *positively* [*negatively*] in $\mathbf{D}$ iff $\exists\!\!\!/_i\mathbf{x}_i\mathbf{C}_i$, where $\mathbf{C}_i$ is the scope of the quantifier $\exists\!\!\!/_i\mathbf{x}_i$, is a positive [negative] wf part of $\mathbf{D}$. *The prenex normal form* of $\mathbf{D}$ is $\mathbf{Z}_1\mathbf{x}_1\ldots\mathbf{Z}_n\mathbf{x}_n\mathbf{M}$, where $\mathbf{M}$ is the wff obtained from $\mathbf{D}$ by erasing all quantifiers of $\mathbf{D}$, and for $1\leq i\leq n$, $\mathbf{Z}_i\mathbf{x}_i$ is $\exists\!\!\!/_i\mathbf{x}_i$ if $\exists\!\!\!/_i\mathbf{x}_i$ occurs positively in $\mathbf{D}$, but $\mathbf{Z}_i\mathbf{x}_i$ is $(\exists\!\!\!/_i\mathbf{x}_i)'$ if $\exists\!\!\!/_i\mathbf{x}_i$ occurs negatively in $\mathbf{D}$. Examples will be given after we prove the following fundamental theorem.

2200 Prenex Normal Form Theorem. Every wff has a prenex normal form. If **D** is a rectified wff and **G** is the prenex normal form of **D**, then $\vdash \mathbf{D} \equiv \mathbf{G}$.

Proof: For any wff **D** of $\mathcal{E}$, let #**D** be the sum, over all occurrences of quantifiers **D**, of the number of occurrences of propositional connectives in whose scope lies the given quantifier. We first show by induction on #**D** that if **D** is any rectified wff of $\mathcal{E}$ and **G** is the prenex normal form of **D**, then $\vdash \mathbf{D} \equiv \mathbf{G}$.

If $\#\mathbf{D} = 0$, clearly **G** is **D**, so $\vdash \mathbf{D} \equiv \mathbf{G}$ by Rule P.

If $\#\mathbf{D} > 0$, some quantifier of **D** is in the scope of some propositional connective. The left-most such quantifier of **D** is immediately in the scope of some propositional connective. Let **M** be the wf part of **D** consisting of that connective and its scope. (Thus, **M** is the smallest wf part of **D** which properly includes the left-most quantifier of **D** which is in the scope of some propositional connective in **D**.) For each of the possible forms of **M**, let **N** be the wff indicated below:

M	N
$\sim \forall \mathbf{x}\mathbf{B}$	$\exists \mathbf{x} \sim \mathbf{B}$
$\sim \exists \mathbf{x}\mathbf{B}$	$\forall \mathbf{x} \sim \mathbf{B}$
$\mathbf{A} \lor \forall \mathbf{x}\mathbf{B}$	$\forall \mathbf{x}[\mathbf{A} \lor \mathbf{B}]$
$\forall \mathbf{x}\mathbf{B} \lor \mathbf{A}$	$\forall \mathbf{x}[\mathbf{B} \lor \mathbf{A}]$
$\mathbf{A} \lor \exists \mathbf{x}\mathbf{B}$	$\exists \mathbf{x}[\mathbf{A} \lor \mathbf{B}]$
$\exists \mathbf{x}\mathbf{B} \lor \mathbf{A}$	$\exists \mathbf{x}[\mathbf{B} \lor \mathbf{A}]$

(Note that in each case **x** is not free in **A**, since **D** is rectified.) Let **E** be the wff obtained when **M** is replaced by **N** in **D**. $\vdash \mathbf{M} \equiv \mathbf{N}$ by 2120, 2119, 2133, 2131, and Rule P, so $\vdash \mathbf{D} \equiv \mathbf{E}$ by the Substitutivity of Equivalence. Clearly **E** is a rectified wff, and the bound variables of **E** have their quantifiers in the same left-to-right order in **E** as in **D** (since if **M** is $\mathbf{A} \lor \forall \mathbf{x}\mathbf{B}$ or $\mathbf{A} \lor \exists \mathbf{x}\mathbf{B}$, then **A** must be quantifier-free). Hence it is easy to see that the prenex normal form **G** of **D** is also the prenex normal form of **E**. Also, $\#\mathbf{E} = \#\mathbf{D} - 1$, so by inductive hypothesis, $\vdash \mathbf{E} \equiv \mathbf{G}$. Hence $\vdash \mathbf{D} \equiv \mathbf{G}$ by Rule P. This completes the inductive proof.

Now for any wff **C** of $\mathcal{F}$, let **D** be a rectified wff obtained from **C** by deleting any vacuous quantifiers and by making any necessary alphabetic changes of bound variables. $\vdash \mathbf{C} \equiv \mathbf{D}$ by 2127, 2109, and the Substitutivity of Equivalence (2106). Hence if **G** is the prenex normal form of **D**, $\vdash \mathbf{C} \equiv \mathbf{G}$, so **C** has a prenex normal form. ∎

EXAMPLE: Find a prenex normal form of $\forall x \exists y Pxy \equiv \exists z Qzx$.
Eliminating the $\equiv$ sign we obtain

$$[\forall x \exists y Pxy \supset \exists z Qzx] \wedge [\exists z Qzx \supset \forall x \exists y Pxy],$$

which can be rectified to

$$[\forall x_2 \exists y Px_2 y \supset \exists z Qzx] \wedge [\exists z_1 Qz_1 x \supset \forall x_1 \exists y_1 Px_1 y_1].$$

In this wff the quantifiers $\forall x_2$, $\exists y$, and $\exists z_1$ occur negatively, and the others occur positively, so a prenex normal form of this wff is

$$\exists x_2 \forall y \exists z \forall z_1 \forall x_1 \exists y_1 [[Px_2 y \supset Qzx] \wedge \centerdot Qz_1 x \supset Px_1 y_1].$$

REMARK. When students are confronted with an example like the one above, they are sometimes tempted to change its *free* variables. However, there is no justification for doing this, and the resulting wff its usually not equivalent to the original one. Note Exercise X2312 below.

REMARK. Sometimes by using theorem schemata 2129 and 2130 one can reduce the number of quantifiers in a wff. Let us apply this idea to find an alternative prenex normal form of the wff dealt with in the example above. We leave it to the reader to explain why each wff below is equivalent to the preceding one:

$$\begin{array}{l}
[\forall x_2 \exists y Px_2 y \supset \exists z Qzx] \wedge [\exists z_1 Qz_1 x \supset \forall x_1 \exists y_1 Px_1 y_1] \\
[\exists x_2 \forall y \sim Px_2 y \vee \exists z Qzx] \wedge [\forall z_1 \sim Qz_1 x \vee \forall x_1 \exists y_1 Px_1 y_1] \\
[\exists z \forall y \sim Pzy \vee \exists z Qzx] \wedge [\forall y \sim Qyx \vee \forall x_1 \exists y_1 Px_1 y_1] \\
\exists z [\forall y \sim Pzy \vee Qzx] \wedge \forall y \forall x_1 \exists y_1 [\sim Qyx \vee Px_1 y_1] \\
\exists z \centerdot \forall y [\sim Pzy \vee Qzx] \wedge \forall y \forall x_1 \exists y_1 [\sim Qyx \vee Px_1 y_1] \\
\exists z \forall y \centerdot [Pzy \supset Qzx] \wedge \forall x_1 \exists y_1 [Qyx \supset Px_1 y_1] \\
\exists z \forall y \forall x_1 \exists y_1 \centerdot [Pzy \supset Qzx] \wedge \centerdot Qyx \supset Px_1 y_1.
\end{array}$$

EXERCISES

Find *the* prenex normal form of each of the rectified wffs below:

X2200. $\sim \forall x Rxu \wedge \exists y \centerdot \sim \forall z Ryz \supset Quy$

X2201. $\sim \centerdot \exists x Qxy \supset \sim \forall z \centerdot \forall u Puyz \supset \sim \exists v\, Rv$

Find prenex normal forms for the wffs below:

X2202. $\forall x Px \equiv \exists y\, Qy$

X2203. $\forall x \sim \exists y \forall u Pxy \supset \sim \centerdot \exists z Pxy \supset \forall y Puy$

X2204. $\forall x\, [Pxu \vee \exists y \centerdot Qxy \wedge \sim \forall u Ruz] \supset \forall z Qzu$

X2205. $\forall w \exists x\, [Pwx \supset \forall y Pyx] \supset \forall x Qxw$

X2206. $\sim [[\forall u Puz \wedge \sim \exists x \centerdot Qux \vee \forall z \sim Rxz] \supset \exists z Quz]$

X2207. $\forall y\, [Py \wedge \exists y Qy] \supset \centerdot \forall x Rxy \supset \exists z Rzx$

X2208. $\forall x Px \supset \exists x \centerdot \forall z Qxz \supset \exists y \forall x Rxyz$

X2209. $\forall z\, [\sim \exists x\, [Px \wedge \forall y Rxyz] \vee \forall z \sim \centerdot \exists y Qyz \supset Px]$

For each of the wffs below, find a prenex normal form in which the number of occurrences of quantifiers is minimized.

X2210. $\forall u \exists v \sim \exists w \exists x \forall y \forall z \centerdot [Pwz \vee Qy] \supset \sim Rux$

X2211. $\sim [\sim \forall x\, [Qx \vee \sim \forall y Ry] \vee \forall z Qz \vee \sim \forall w Rw]$

X2212. $\forall x Px \supset \exists x \centerdot \forall z Qxz \supset \exists y \forall x Rxyz$

X2213. $\forall z\, [\sim \exists x\, [Px \wedge \forall y\, Rxyz] \vee \forall z \sim \centerdot \exists y Qyz \supset Px]$

X2214. $\forall x Px \equiv \exists y \forall z Qxyz$

Prove the following wffs:

X2215. $\exists x \forall y \exists z \centerdot Pzx \wedge Pxy \supset Pyy$

X2216. $\exists x \forall y \centerdot Qx \supset \centerdot Px \vee Qy$

X2217. $\exists x \exists y \forall z\, [Gxy \supset [Gzx \wedge [Hxy \supset Hzz]]]$

§23. Semantics of $\mathcal{F}$

DEFINITION. An *interpretation* $\mathcal{M} = \langle \mathcal{D}, \mathcal{J} \rangle$ of $\mathcal{F}$ consists of a non-empty set $\mathcal{D}$ called the *universe* or the *domain of individuals* and (if there are constants in the formulation of $\mathcal{F}$ under consideration) a mapping $\mathcal{J}$ defined on the constants of $\mathcal{F}$ satisfying the conditions below. Given an interpretation $\mathcal{M}$, an *assignment* into $\mathcal{M}$ is a function φ defined on the variables of $\mathcal{F}$ and satisfying the conditions below.

(a) If $\mathbf{x}$ is an individual variable [constant], then $\varphi\mathbf{x}$ [$\mathcal{J}\mathbf{x}$] is a member of $\mathcal{D}$.
(b) If $\mathbf{f}^n$ is an n-ary function variable [constant], then $\varphi\mathbf{f}^n$ [$\mathcal{J}\mathbf{f}^n$] is a function from $\mathcal{D}^n$ into $\mathcal{D}$, where $\mathcal{D}^n$ is the set of ordered n-tuples of elements in $\mathcal{D}$.
(c) If $\mathbf{p}$ is a propositional variable [constant], then $\varphi\mathbf{p}$ [$\mathcal{J}\mathbf{p}$] is a truth value.
(d) If $\mathbf{P}^n$ is an n-ary predicate variable [constant], then $\varphi\mathbf{P}^n$ [$\mathcal{J}\mathbf{P}^n$] is a function from $\mathcal{D}^n$ into the set $\{\mathsf{T},\ \mathsf{F}\}$ of truth values.

DEFINITIONS. Let φ and ψ be assignments, $\mathbf{x}$ be a variable (of any type), and $\mathcal{S}$ be a set of variables. We say that φ and ψ *agree off* $\mathbf{x}$ iff $\varphi\mathbf{y} = \psi\mathbf{y}$ for all variables $\mathbf{y}$ (of any type) distinct from $\mathbf{x}$. We say that φ and ψ *agree on* the variables in $\mathcal{S}$ iff $\varphi\mathbf{y} = \psi\mathbf{y}$ for all variables $\mathbf{y}$ in $\mathcal{S}$.

We shall next define the semantical values (meanings) of certain expressions, so that the values of terms are individuals, and the values of wffs are truth values.

DEFINITION. Given an interpretation $\mathcal{M} = \langle\mathcal{D},\ \mathcal{J}\rangle$ and an assignment φ, we define $\mathcal{V}^{\mathcal{M}}_{\varphi}\mathbf{Z}$, the *value of* $\mathbf{Z}$ *with respect to* φ *in* $\mathcal{M}$, for each variable, constant, term, or wff $\mathbf{Z}$, as follows by induction on the construction of $\mathbf{Z}$:

(1) $\mathcal{V}^{\mathcal{M}}_{\varphi}\mathbf{Z} = \varphi\mathbf{Z}$ if $\mathbf{Z}$ is a variable.
(2) $\mathcal{V}^{\mathcal{M}}_{\varphi}\mathbf{Z} = \mathcal{J}\mathbf{Z}$ if $\mathbf{Z}$ is a constant.
(3) If $\mathbf{F}^n$ is an n-ary function or predicate variable or constant, and $\mathbf{t}_1, \ldots, \mathbf{t}_n$ are terms $\mathcal{V}^{\mathcal{M}}_{\varphi}(\mathbf{F}^n\mathbf{t}_1, \ldots, \mathbf{t}_n) = \left(\mathcal{V}^{\mathcal{M}}_{\varphi}\mathbf{F}^n\right)\left(\mathcal{V}^{\mathcal{M}}_{\varphi}\mathbf{t}_1, \ldots, \mathcal{V}^{\mathcal{M}}_{\varphi}\mathbf{t}_n\right)$, i.e., the value of the function $\mathcal{V}^{\mathcal{M}}_{\varphi}\mathbf{F}^n$ on the indicated arguments.
(4) $\mathcal{V}^{\mathcal{M}}_{\varphi} \sim \mathbf{A} = \mathsf{F}$ if $\mathcal{V}^{\mathcal{M}}_{\varphi}\mathbf{A} = \mathsf{T}$.
$= \mathsf{T}$ if $\mathcal{V}^{\mathcal{M}}_{\varphi}\mathbf{A} = \mathsf{F}$.
(5) $\mathcal{V}^{\mathcal{M}}_{\varphi}[\mathbf{A} \vee \mathbf{B}] = \mathsf{T}$ if $\mathcal{V}^{\mathcal{M}}_{\varphi}\mathbf{A} = \mathsf{T}$ or $\mathcal{V}^{\mathcal{M}}_{\varphi}\mathbf{B} = \mathsf{T}$.
$= \mathsf{F}$ if $\mathcal{V}^{\mathcal{M}}_{\varphi}\mathbf{A} = \mathsf{F}$ and $\mathcal{V}^{\mathcal{M}}_{\varphi}\mathbf{B} = \mathsf{F}$.
(6) $\mathcal{V}^{\mathcal{M}}_{\varphi}\forall\mathbf{x}\mathbf{A} = \mathsf{T}$ if $\mathcal{V}^{\mathcal{M}}_{\psi}\mathbf{A} = \mathsf{T}$ for every assignment ψ which agrees with φ off $\mathbf{x}$.
$= \mathsf{F}$ if $\mathcal{V}^{\mathcal{M}}_{\psi}\mathbf{A} = \mathsf{F}$ for some assignment ψ which agrees with φ off $\mathbf{x}$.

EXAMPLE: Let us consider $\mathcal{V}^{\mathcal{M}}\mathcal{M}_{\varphi}\forall x\, Pyx$, where $\mathcal{M} = \langle \mathbb{N}, \mathcal{J} \rangle$, $\mathbb{N}$ is the set of natural numbers, and for all n and m in $\mathbb{N}$, $(\mathcal{J}P)\, nm = \mathsf{T}$ iff $n \leq m$ (so that Pyx means $y \leq x$). Suppose $\varphi y = 0$, and let ψ be an assignment which agrees with φ off x. Then $\psi y = \varphi y = 0$, and $\mathcal{V}_{\psi}^{\mathcal{M}} Pyx = (\mathcal{J}P)\,(\psi y)\,(\psi x) = (\mathcal{J}P)\, 0\, (\psi x) = \mathsf{T}$, since $0 \leq \psi x$ (no matter what natural number ψx is). Therefore $\mathcal{V}_{\varphi}^{\mathcal{M}} \forall x Pyx = \mathsf{T}$ when $\varphi y = 0$. Next, suppose $\varphi y = 5$. Let ψ be the assignment which agrees with φ off x, with $\psi x = 4$. Then $\mathcal{V}_{\psi}^{\mathcal{M}} Pyx = (\mathcal{J}P)\,(\psi y)\,(\psi x) = (\mathcal{J}P)\, 5\, 4 = \mathsf{F}$ (since it is false that $5 \leq 4$), so $\mathcal{V}_{\varphi}^{\mathcal{M}} \forall x Pyx = \mathsf{F}$ in this case.

It is important that we have a rigorous way of defining semantical concepts, but the reader should not let the formalism obscure the basic ideas. The example above can be summarized informally by saying that $\forall x\, y \leq x$ is true in the natural numbers when y is 0, but false when y is 5. It is obvious that some value must be assigned to the free variable y before one can consider whether the wff is true or false.

REMARK. We shall sometimes write $\mathcal{V}_{\varphi}^{\mathcal{M}}$ as $\mathcal{V}_{\varphi}$, $\mathcal{V}^{\mathcal{M}}$, or $\mathcal{V}$, when the interpretation or assignment is clear from the context, or irrelevant.

For convenience, we shall on occasion use "$\sim$" and "$\vee$" in our meta-language to denote the truth functions negation and disjunction, respectively. The context will reveal when these symbols are being used as symbols of the object language $\mathcal{F}$, and when they are being used as notations in the meta-language. Thus, (4) might be written $\mathcal{V}_{\varphi} \sim \mathbf{A} = \sim \mathcal{V}_{\varphi}\mathbf{A}$ and (5) might be written $\mathcal{V}_{\varphi}\, [\mathbf{A} \vee \mathbf{B}] = (\mathcal{V}_{\varphi}\mathbf{A}) \vee (\mathcal{V}_{\varphi}\mathbf{B})$.

2300 Proposition. Let $\mathcal{M}$ be an interpretation, $\mathbf{Z}$ a term or wff, and φ and ψ assignments into $\mathcal{M}$ which agree on all variables which occur free in $\mathbf{Z}$. Then $\mathcal{V}_{\varphi}^{\mathcal{M}}\mathbf{Z} = \mathcal{V}_{\psi}^{\mathcal{M}}\mathbf{Z}$.

The proof is by induction on the construction of $\mathbf{Z}$. We leave the details to the reader.

DEFINITIONS. Let $\mathbf{A}$ be a wff, $\mathcal{M}$ be an interpretation, and φ be an assignment into $\mathcal{M}$.

(1) φ *satisfies* $\mathbf{A}$ in $\mathcal{M}$ iff $\mathcal{V}_{\varphi}^{\mathcal{M}}\mathbf{A} = \mathsf{T}$.
(2) $\mathbf{A}$ is *satisfiable in* $\mathcal{M}$ iff there is an assignment φ which satisfies $\mathbf{A}$ in $\mathcal{M}$.
(3) $\mathbf{A}$ is *satisfiable* iff there is an interpretation in which $\mathbf{A}$ is satisfiable.

(4) A set $\mathcal{G}$ of wffs is (*simultaneously*) *satisfiable* iff there is an interpretation $\mathcal{M}$ and an assignment φ such that for all $\mathbf{B} \in \mathcal{G}$, $\mathcal{V}_\varphi^\mathcal{M}\mathbf{B} = \mathsf{T}$.
(5) $\mathbf{A}$ is *valid* in $\mathcal{M}$ iff for every assignment φ into $\mathcal{M}$, $\mathcal{V}_\varphi^\mathcal{M}\mathbf{A} = \mathsf{T}$.
(6) $\mathbf{A}$ is *valid* iff $\mathbf{A}$ is valid in every interpretation.
(7) A sentence $\mathbf{A}$ is *true* in $\mathcal{M}$ iff $\mathcal{V}^\mathcal{M}\mathbf{A} = \mathsf{T}$ and *false* in $\mathcal{M}$ iff $\mathcal{V}^\mathcal{M}\mathbf{A} = \mathsf{F}$.
(8) $\mathbf{A}$ is *contradictory*, or *unsatisfiable*, iff for every interpretation $\mathcal{M}$ and every assignment φ into $\mathcal{M}$, $\mathcal{V}_\varphi^\mathcal{M}\mathbf{A} = \mathsf{F}$.
(9) A set of wffs is *unsatisfiable* iff it is not simultaneously satisfiable.
(10) A *model* for a set of wffs is an interpretation in which each of the wffs is valid. A *model* for a first order theory is a model for the set of postulates of the theory. (Normally we speak of models for sets of *sentences*.)
(11) A *countermodel* for a sentence is an interpretation in which the sentence if false. (It is a counterexample to the conjecture that the sentence is valid.)

NOTATION. We write $\mathcal{M} \models_\varphi \mathbf{A}$, $\mathcal{M} \models \mathbf{A}$, and $\models \mathbf{A}$ to indicate that φ satisfies $\mathbf{A}$ in $\mathcal{M}$, $\mathbf{A}$ is valid in $\mathcal{M}$, and $\mathbf{A}$ is valid, respectively. The notation $\mathcal{M} \models \mathbf{A}\,[\varphi]$ is often used instead of $\mathcal{M} \models_\varphi \mathbf{A}$.

REMARK. We shall show that the theorems of $\mathcal{F}$ are precisely the valid wffs.

EXAMPLES: We add to $\mathcal{F}$ an individual constant 1, a binary function constant $+^2$, and predicate constants E^1, O^1, $<^2$, $=^2$. We shall let $(\mathbf{s} + \mathbf{t})$, $\mathbf{s} < \mathbf{t}$, and $\mathbf{s} = \mathbf{t}$ serve as abbreviations for $+^2\mathbf{st}$, $<^2\mathbf{st}$, and $=^2\mathbf{st}$, respectively. We consider two interpretations, $\mathbb{N}$ and $\mathbb{R}$. The domain of individuals is the set $\{0, 1, 2, 3, \ldots\}$ of natural numbers for $\mathbb{N}$, and the set of real numbers for $\mathbb{R}$. In both interpretations the constants 1, $+$, E, O, $<$, $=$ denote the number one, the addition function, the set of even integers, the set of odd integers, and the strict ordering and equality relations, respectively.

(a) $x + y = 1$ is satisfiable but not valid in $\mathbb{N}$ and in $\mathbb{R}$.
(b) $\exists x \exists y\,[x + y = 1]$ is true in $\mathbb{N}$ and in $\mathbb{R}$.
(c) $\forall x \forall y\,[x + y = 1]$ is false in $\mathbb{N}$ and in $\mathbb{R}$.
(d) $\forall x \exists y\,[x + y = 1]$ is false in $\mathbb{N}$, true in $\mathbb{R}$. Thus, it is satisfiable but not valid.
(e) $\exists y \forall x\,[x + y = 1]$ is false in $\mathbb{N}$ and in $\mathbb{R}$.
(f) $\forall x \exists y\,[x + y = 1] \supset \exists y \forall x\,[x + y = 1]$ is true in $\mathbb{N}$ but false in $\mathbb{R}$, and so is not valid.
(g) $\exists x O x \supset \forall x O x$ is false in $\mathbb{N}$, and so is not valid.
(i) $\forall x \exists y\, x < y$ is true in $\mathbb{N}$.

(j) $\exists y\, y < y$ is false in $\mathbb{N}$.
(k) $\forall x \exists y\, x < y \supset \exists y\, y < y$ is false in $\mathbb{N}$, and so is not valid. This illustrates the need for the restriction on Axiom 4.
(l) $\forall x\, [x = y \supset y = x]$ is valid in $\mathbb{N}$.
(m) $\mathcal{V}_{\varphi}^{\mathbb{N}}\, [x = y \supset \forall x \,\centerdot\, y = x] = \mathsf{F}$ when $\varphi x = \varphi y$, so this wff is not valid in $\mathbb{N}$.
(n) $\forall x\, [x = y \supset y = x] \supset [x = y \supset \forall x \,\centerdot\, y = x]$ is not valid in $\mathbb{N}$, hence not valid. This illustrates the need for the restriction on Axiom 5. Note, however, that this wff is satisfied in $\mathbb{N}$ by any assignment φ such that $\varphi x \neq \varphi y$.

2301 Proposition. Let $\mathbf{A}$ be a wff and $\mathcal{M}$ be an interpretation.

(a) $\mathbf{A}$ is valid [in $\mathcal{M}$] iff $\sim\mathbf{A}$ is not satisfiable [in $\mathcal{M}$].
(b) $\mathbf{A}$ is satisfiable [in $\mathcal{M}$] iff $\sim\mathbf{A}$ is not valid [in $\mathcal{M}$].
(c) $\mathbf{A}$ is valid [in $\mathcal{M}$] iff $\forall \mathbf{x} \mathbf{A}$ is valid [in $\mathcal{M}$].
(d) $\mathbf{A}$ is satisfiable [in $\mathcal{M}$] iff $\exists \mathbf{x} \mathbf{A}$ is satisfiable [in $\mathcal{M}$].

We leave the proofs of these statements to the reader. However, the proof of (a) is essentially conveyed by the following abbreviated argument:

$$\begin{aligned} \mathbf{A} \text{ is valid} &\Leftrightarrow \forall \mathcal{M}\, \forall \varphi\, \mathcal{M} \models_{\varphi} \mathbf{A} \\ &\Leftrightarrow \sim \exists \mathcal{M}\, \exists \varphi \sim (\mathcal{M} \models_{\varphi} \mathbf{A}) \\ &\Leftrightarrow \sim \exists \mathcal{M}\, \exists \varphi\, \mathcal{M} \models_{\varphi} \sim \mathbf{A} \\ &\Leftrightarrow \sim \mathbf{A} \text{ is not satisfiable.} \end{aligned}$$

DEFINITIONS. For any wff $\mathbf{B}$ of which the free variables are precisely $\mathbf{x}_1, \ldots, \mathbf{x}_n$, we define the *universal closure* $\mathbf{A}$ of $\mathbf{B}$ to be $\forall \mathbf{x}_1 \ldots \forall \mathbf{x}_n \mathbf{B}$, and the *existential closure* $\mathbf{E}$ of $\mathbf{B}$ to be $\exists \mathbf{x}_1 \ldots \exists \mathbf{x}_n \mathbf{B}$. Here $\mathbf{A}$ and $\mathbf{E}$ are closed wffs, and if $\mathbf{B}$ is a closed wff, then $\mathbf{A} = \mathbf{B} = \mathbf{E}$. Clearly $\vdash \mathbf{A} \supset \mathbf{B}$ and $\vdash \mathbf{B} \supset \mathbf{E}$, and $\vdash \mathbf{B}$ iff $\vdash \mathbf{A}$. Also, for any interpretation $\mathcal{M}$, $\mathbf{B}$ is valid [in $\mathcal{M}$] iff $\mathbf{A}$ is valid [in $\mathcal{M}$], and $\mathbf{B}$ is satisfiable [in $\mathcal{M}$] iff $\mathbf{E}$ is satisfiable [in $\mathcal{M}$].

If θ is any proper substitution for free occurrences of variables and $\mathbf{A}$ is a wff, we say that θ is *free on* $\mathbf{A}$ iff for each free variable $\mathbf{m}$ of $\mathbf{A}$, $\theta\mathbf{m}$ is free for $\mathbf{m}$ in $\mathbf{A}$. Note that if $\mathbf{t}$ is free for $\mathbf{x}$ in $\mathbf{A}$, then $\overset{\centerdot}{\mathsf{S}}{}^{\mathbf{x}}_{\mathbf{t}}$ is free on $\mathbf{A}$.

Given an interpretation $\mathcal{M}$, an assignment φ into $\mathcal{M}$, and a proper substitution θ for free occurrences of variables, we define $\varphi \circ \theta$ to be that assignment into $\mathcal{M}$ such that for each variable $\mathbf{m}$, $(\varphi \circ \theta)\, \mathbf{m} = \mathcal{V}_{\varphi}\, (\theta \mathbf{m})$.

2302 Substitution-Value Theorem. If $\mathbf{A}$ is a variable, constant, term, or wff, and θ is a proper substitution for free occurrences of variables such that θ is free on $\mathbf{A}$, and φ is any assignment, then $\mathcal{V}_{\varphi}\theta\mathbf{A} = \mathcal{V}_{\varphi\circ\theta}\mathbf{A}$.

REMARK. Given an interpretation and an assignment φ, $\mathcal{V}_\varphi \mathbf{m} = \varphi \mathbf{m}$ for each variable $\mathbf{m}$, so $\mathcal{V}_\varphi$ is an extension of φ. Therefore, if $\mathbf{K}$ is a constant, term, or wff, we could write $\varphi\mathbf{K}$ as an abbreviation for $\mathcal{V}_\varphi\mathbf{K}$. In this notation the Substitution-Value Theorem says that $\varphi(\theta\mathbf{A}) = (\varphi \circ \theta)\,\mathbf{A}$.

Corollary. If $\mathbf{t}$ is a term free for the individual variable $\mathbf{x}$ in the wff $\mathbf{A}$, φ is any assignment, and ψ is that assignment which agrees with φ off $\mathbf{x}$, while $\psi\mathbf{x} = \mathcal{V}_\varphi \mathbf{t}$, then $\mathcal{V}_\varphi \mathsf{S}^{\mathbf{x}}_{\mathbf{t}}\mathbf{A} = \mathcal{V}_\psi \mathbf{A}$.

Proof: We prove 2302 simultaneously for all proper substitutions θ and assignments φ by induction on the construction of $\mathbf{A}$.

Case 1. $\mathbf{A}$ is a variable. $\mathcal{V}_\varphi(\theta\mathbf{A}) = (\varphi \circ \theta)\,\mathbf{A} = \mathcal{V}_{\varphi\circ\theta}\mathbf{A}$ by definition.

Case 2. $\mathbf{A}$ is a constant. $\mathcal{V}_\varphi\theta\mathbf{A} = \mathcal{V}_\varphi\mathbf{A} = \mathcal{V}\mathbf{A} = \mathcal{V}_{\varphi\circ\theta}\mathbf{A}$.

Case 3. $\mathbf{A}$ is $\mathbf{F}^n\mathbf{t}_1\ldots\mathbf{t}_n$, where $\mathbf{F}^n$ is a function or predicate variable or constant, and $\mathbf{t}_1,\ldots,\mathbf{t}_n$ are terms.

$$\begin{aligned}\mathcal{V}_\varphi(\theta\mathbf{A}) &= \mathcal{V}_\varphi((\theta\mathbf{F}^n)(\theta\mathbf{t}_1)\ldots(\theta\mathbf{t}_n)) = (\mathcal{V}_\varphi\theta\mathbf{F}^n)(\mathcal{V}_\varphi\theta\mathbf{t}_1)\ldots(\mathcal{V}_\varphi\theta\mathbf{t}_n)\\ &= (\mathcal{V}_{\varphi\circ\theta}\mathbf{F}^n)(\mathcal{V}_{\varphi\circ\theta}\mathbf{t}_1)\ldots(\mathcal{V}_{\varphi\circ\theta}\mathbf{t}_n)\ \text{(by inductive hypothesis)}\\ &= \mathcal{V}_{\varphi\circ\theta}\mathbf{F}^n\mathbf{t}_1\ldots\mathbf{t}_n = \mathcal{V}_{\varphi\circ\theta}\mathbf{A}.\end{aligned}$$

Case 4. $\mathbf{A}$ is $\sim\mathbf{B}$.

$$\begin{aligned}\mathcal{V}_\varphi(\theta \sim \mathbf{B}) &= \mathcal{V}_\varphi(\sim \theta\mathbf{B}) = \sim \mathcal{V}_\varphi(\theta\mathbf{B})\\ &= \sim \mathcal{V}_{\varphi\circ\theta}\mathbf{B}\ \text{(by inductive hypothesis)}\\ &= \mathcal{V}_{\varphi\circ\theta} \sim \mathbf{B}\end{aligned}$$

Case 5. $\mathbf{A}$ is $[\mathbf{B} \lor \mathbf{C}]$

$$\begin{aligned}\mathcal{V}_\varphi\theta[\mathbf{B} \lor \mathbf{C}] &= \mathcal{V}_\varphi[(\theta\mathbf{B}) \lor (\theta\mathbf{C})] = \mathcal{V}_\varphi(\theta\mathbf{B}) \lor \mathcal{V}_\varphi(\theta\mathbf{C})\\ &= \mathcal{V}_{\varphi\circ\theta}\mathbf{B} \lor \mathcal{V}_{\varphi\circ\theta}\mathbf{C}\ \text{(by inductive hypothesis)}\\ &= \mathcal{V}_{\varphi\circ\theta}[\mathbf{B} \lor \mathbf{C}]\end{aligned}$$

Case 6. $\mathbf{A}$ is $\forall\mathbf{y}\mathbf{B}$.

Note that if $\mathbf{x}$ is any free variable of $\mathbf{B}$ other than $\mathbf{y}$, then $\mathbf{y}$ is not free in $\theta\mathbf{x}$, since θ is free on $\forall\mathbf{y}\mathbf{B}$. Let θ' be the substitution which agrees with θ off $\mathbf{y}$, while $\theta'\mathbf{y} = \mathbf{y}$. Thus $\theta\mathbf{A} = \forall\mathbf{y}\theta'\mathbf{B}$. Clearly θ' is free on $\mathbf{B}$, so by inductive hypothesis, $\mathcal{V}_{\varphi'}\theta'\mathbf{B} = \mathcal{V}_{\varphi'\circ\theta'}\mathbf{B}$ for any assignment φ'.

Subcase (6a). $\mathcal{V}_\varphi(\theta\mathbf{A}) = \mathsf{F}$.

Thus there is an assignment φ' which agrees with φ off $\mathbf{y}$ such that $\mathcal{V}_{\varphi'}\theta'\mathbf{B} = \mathsf{F}$. (We wish to show that $\mathcal{V}_{\varphi\circ\theta}\forall\mathbf{y}\mathbf{B} = \mathsf{F}$. For this purpose we need an assignment ψ which agrees with $\varphi \circ \theta$ off $\mathbf{y}$ such that $\mathcal{V}_{\psi}\mathbf{B} = \mathsf{F}$.[6]) Let ψ be that assignment which agrees with $\varphi \circ \theta$ off $\mathbf{y}$, while $\psi\mathbf{y} = (\varphi' \circ \theta')\,\mathbf{y}$. If $\mathbf{x}$ is any free variable of $\mathbf{B}$ other than $\mathbf{y}$, then $\mathcal{V}_{\varphi}\theta\mathbf{x} = \mathcal{V}_{\varphi'}\theta\mathbf{x}$ by 2300. Hence $\psi\mathbf{x} = (\varphi \circ \theta)\,\mathbf{x} = \mathcal{V}_{\varphi}\theta\mathbf{x} = \mathcal{V}_{\varphi'}\theta\mathbf{x} = \mathcal{V}_{\varphi'}\theta'\mathbf{x} = (\varphi' \circ \theta')\,\mathbf{x}$, so ψ agrees with $(\varphi' \circ \theta')$ on all free variables of $\mathbf{B}$. Thus by 2300, $\mathcal{V}_{\psi}\mathbf{B} = \mathcal{V}_{\varphi'\circ\theta'}\mathbf{B} = \mathcal{V}_{\varphi'}\theta'\mathbf{B} = \mathsf{F}$, so $\mathcal{V}_{\varphi\circ\theta}\mathbf{A} = \mathcal{V}_{\varphi\circ\theta}\forall\mathbf{y}\,\mathbf{B} = \mathsf{F} = \mathcal{V}_{\varphi}\theta\mathbf{A}$.

Subcase (*6b*). $\mathcal{V}_{\varphi\circ\theta}\mathbf{A} = \mathsf{F}$.

Thus there is an assignment ψ which agrees with $\varphi \circ \theta$ off $\mathbf{y}$ such that $\mathcal{V}_{\psi}\mathbf{B} = \mathsf{F}$. Let φ' be that assignment which agrees with φ off $\mathbf{y}$, while $\varphi'\mathbf{y} = \psi\mathbf{y}$. If $\mathbf{x}$ is any free variable of $\mathbf{B}$ other than $\mathbf{y}$, then $\mathcal{V}_{\varphi}\theta\mathbf{x} = \mathcal{V}_{\varphi'}\theta\mathbf{x}$ by 2300, so $\psi\mathbf{x} = (\varphi \circ \theta)\,\mathbf{x} = \mathcal{V}_{\varphi}\theta\mathbf{x} = \mathcal{V}_{\varphi'}\,(\theta\mathbf{x}) = \mathcal{V}_{\varphi'}\,(\theta'\mathbf{x}) = (\varphi' \circ \theta')\,\mathbf{x}$. Also, $\psi\mathbf{y} = \varphi'\mathbf{y} = \mathcal{V}_{\varphi'}\theta'\mathbf{y} = (\varphi' \circ \theta')\,\mathbf{y}$, so ψ agrees with $\varphi' \circ \theta'$ on all free variables of $\mathbf{B}$. Hence by 2300, $\mathcal{V}_{\varphi'}\theta'\mathbf{B} = \mathcal{V}_{\varphi'\circ\theta'}\mathbf{B} = \mathcal{V}_{\psi}\mathbf{B} = \mathsf{F}$, so $\mathcal{V}_{\varphi}\theta\mathbf{A} = \mathcal{V}_{\varphi}\forall\mathbf{y}\theta'\mathbf{B} = \mathsf{F} = \mathcal{V}_{\varphi\circ\theta}\mathbf{A}$.

If neither subcase (6a) nor (6b) applies, then $\mathcal{V}_{\varphi}\theta\mathbf{A}$ and $\mathcal{V}_{\varphi\circ\theta}\mathbf{A}$ must both be true, so the proof is complete. ■

2303 Soundness Theorem.

(a) Every theorem of $\mathcal{F}$ is valid.
(b) If $\mathcal{T}$ is a first-order theory and $\mathcal{M}$ is a model of $\mathcal{T}$, then every theorem of $\mathcal{T}$ is valid in $\mathcal{M}$.
(c) If $\mathcal{G}$ is a set of wffs and φ simultaneously satisfies $\mathcal{G}$ (in some interpretation) and $\mathcal{G} \vdash \mathbf{B}$, then φ satisfies $\mathbf{B}$.

Proof: We first show that if $\mathbf{D}$ is any axiom of $\mathcal{F}$, then $\models \mathbf{D}$. This is evident if $\mathbf{D}$ is an instance of any of Axiom Schemata 1-3. To deal with Axiom Schemata 4 and 5, let any interpretation, and an arbitrary assignment φ, be given; we shall show that $\mathcal{V}_{\varphi}\mathbf{D} = \mathsf{T}$.

Suppose $\mathbf{D}$ is an instance of Axiom Schema 4, so $\mathbf{D}$ has the form $[\forall\mathbf{x}\mathbf{A} \supset \mathsf{S}^{\mathbf{x}}_{\mathbf{t}}\mathbf{A}]$, where $\mathbf{t}$ is a term free for $\mathbf{x}$ in $\mathbf{A}$. If $\mathcal{V}_{\varphi}\forall\mathbf{x}\,\mathbf{A} = \mathsf{F}$, clearly $\mathcal{V}_{\varphi}\mathbf{D} = \mathsf{T}$. On the other hand, if $\mathcal{V}_{\varphi}\forall\mathbf{x}\,\mathbf{A} = \mathsf{T}$, then $\mathcal{V}_{\psi}\mathbf{A} = \mathsf{T}$ for any

[6] It might be natural to try letting $\psi = \varphi' \circ \theta'$. One could then attempt to show that $\varphi' \circ \theta'$ agrees with $\varphi \circ \theta$ off $\mathbf{y}$ by arguing that for any variable $\mathbf{z}$ distinct from $\mathbf{y}$, $(\varphi' \circ \theta')\,\mathbf{z} = \mathcal{V}_{\varphi'}\theta'\mathbf{z} = \mathcal{V}_{\varphi'}\theta\mathbf{z} \stackrel{?}{=} \mathcal{V}_{\varphi}\theta\mathbf{z} = (\varphi \circ \theta)\,\mathbf{z}$. However, the step marked $\stackrel{?}{=}$ would not be correct, since $\mathbf{y}$ might be free in $\theta\mathbf{z}$.

assignment ψ which agrees with φ off $\mathbf{x}$. In particular, if $\psi = \varphi \circ \mathsf{S}^{\mathbf{x}}_{\cdot\mathbf{t}}$, then by the Substitution-Value Theorem $\mathcal{V}_\varphi \mathsf{S}^{\mathbf{x}}_{\mathbf{t}} \mathbf{A} = \mathcal{V}_\psi \mathbf{A} = \mathsf{T}$, so $\mathcal{V}_\varphi \mathbf{D} = \mathsf{T}$.

Suppose $\mathbf{D}$ is an instance of Axiom Schema 5, and so has the form $[\forall \mathbf{x}\, [\mathbf{A} \lor \mathbf{B}] \supset {\scriptstyle\blacksquare} \mathbf{A} \lor \forall \mathbf{x} \mathbf{B}]$, where $\mathbf{x}$ is not free in $\mathbf{A}$. If $\mathcal{V}_\varphi \forall \mathbf{x}\, [\mathbf{A} \lor \mathbf{B}] = \mathsf{F}$ or $\mathcal{V}_\varphi \mathbf{A} = \mathsf{T}$, then $\mathcal{V}_\varphi \mathbf{D} = \mathsf{T}$. Suppose $\mathcal{V}_\varphi \forall \mathbf{x}\, [\mathbf{A} \lor \mathbf{B}] = \mathsf{T}$ and $\mathcal{V}_\varphi \mathbf{A} = \mathsf{F}$, and let ψ be any assignment which agrees with φ off $\mathbf{x}$. Then $\mathcal{V}_\psi\, [\mathbf{A} \lor \mathbf{B}] = \mathsf{T}$, but $\mathcal{V}_\psi \mathbf{A} = \mathcal{V}_\varphi \mathbf{A} = \mathsf{F}$ by 2300, so $\mathcal{V}_\psi \mathbf{B} = \mathsf{T}$. Hence $\mathcal{V}_\varphi \forall \mathbf{x} \mathbf{B} = \mathsf{T}$, so $\mathcal{V}_\varphi \mathbf{D} = \mathsf{T}$ in this case also.

It is easy to see that if $\mathbf{C}$ and $\mathbf{C} \supset \mathbf{D}$ are valid [in an interpretation], then $\mathbf{D}$ is valid [in that interpretation], and that if $\mathbf{C}$ is valid [in an interpretation], then $\forall \mathbf{x} \mathbf{C}$ is also valid [in that interpretation]. Thus every wff in a proof [in a first-order theory] must be valid [in every model of that theory]. This proves (a) and (b).

To prove (c), if $\mathcal{G} \vdash \mathbf{B}$ then there is some finite subset $\{\mathbf{A}_1, \ldots, \mathbf{A}_n\}$ of $\mathcal{G}$ such that $\mathbf{A}_1, \ldots, \mathbf{A}_n \vdash \mathbf{B}$. (The case where the subset is empty is trivial, using (a).) Hence by the Deduction Theorem, $\vdash \mathbf{A}_1 \supset {\scriptstyle\blacksquare} \ldots \supset {\scriptstyle\blacksquare} \mathbf{A}_n \supset \mathbf{B}$, so by (a) $\mathcal{V}_\varphi\, [\mathbf{A}_1 \supset {\scriptstyle\blacksquare} \ldots \supset {\scriptstyle\blacksquare} \mathbf{A}_n \supset \mathbf{B}] = \mathsf{T}$. However, we are given that $\mathcal{V}_\varphi \mathbf{A}_1 = \ldots = \mathcal{V}_\varphi \mathbf{A}_n = \mathsf{T}$, so $\mathcal{V}_\varphi \mathbf{B} = \mathsf{T}$.

REMARK. Note that the wff $[Px \supset \forall x Px]$ is not valid, and hence is not a theorem. Of course, if $\mathbf{A}$ is any wff such that $\vdash \mathbf{A}$, then $\vdash \forall \mathbf{x} \mathbf{A}$, so in this case $\vdash \mathbf{A} \supset \forall \mathbf{x} \mathbf{A}$ by Rule P.

2304 Theorem. Every set of sentences which has a model is consistent.

Proof: Suppose $\mathcal{S}$ is a set of sentences which has a model $\mathcal{M}$ but $\mathcal{S}$ is inconsistent. Then if $\mathbf{A}$ is any wff, $\mathcal{S} \vdash \mathbf{A} \land \sim \mathbf{A}$, so $\mathcal{M} \models \mathbf{A} \land \sim \mathbf{A}$ by the Soundness Theorem (2303). This is impossible, so $\mathcal{S}$ must be consistent. ∎

EXERCISES

X2300. Show that $\mathcal{M} \models_\varphi \exists \mathbf{x} \mathbf{A}$ iff there is an assignment ψ which agrees with φ off $\mathbf{x}$ such that $\mathcal{M} \models_\psi \mathbf{A}$.

X2301.

(a) Find a wff $\mathbf{A}$ such that $[\mathbf{A} \supset \forall \mathbf{x} \mathbf{A}]$ is valid.
(b) Find a wff $\mathbf{B}$ such that $[\mathbf{B} \supset \forall \mathbf{x} \mathbf{B}]$ is not valid.
(c) Prove that if $\mathbf{C}$ is any wff and $\mathcal{M}$ is an interpretation, then $[\mathbf{C} \supset \forall \mathbf{x} \mathbf{C}]$ is satisfiable in $\mathcal{M}$.

X2302. Is it true that for every wff $\mathbf{A}$ of $\mathcal{F}$, $\models \mathbf{A}$ iff $\models \mathbf{A} \supset \forall \mathbf{x} \mathbf{A}$?

X2303. Are there any valid wffs of the form $\forall \mathbf{x} [\mathbf{A} \lor \mathbf{B}] \supset \centerdot \mathbf{A} \lor \forall \mathbf{x} \mathbf{B}$ with $\mathbf{x}$ free in $\mathbf{A}$?

X2304. Are wffs of the following forms necessarily valid? satisfiable?

(a) $\exists x [\mathbf{M} \equiv \mathbf{N}] \supset \centerdot \exists x \mathbf{M} \equiv \exists x \mathbf{N}$
(b) $\forall x [\mathbf{M} \equiv \mathbf{N}] \equiv \centerdot \forall x \mathbf{M} \equiv \forall x \mathbf{N}$

X2305. Let $A(x)$ be the wff $[Px \supset \centerdot Pa \land Pb]$. Show that $\models \exists x A(x)$, but there is no term $\mathbf{t}$ such that $\models A(\mathbf{t})$.

X2306. Let the wffs A, B, and C be defined as follows:

$$A \text{ is } \exists z [\exists x \forall y Pxyz \supset \exists w Qwz]$$
$$B \text{ is } \exists z [\forall y \exists x Pxyz \supset \exists w Qwz]$$
$$C \text{ is } \forall z \forall y \exists x Pxyz \supset \exists z \exists w Qwz$$

For each ordered pair $(\mathbf{D}, \mathbf{E})$ of distinct wffs, where $\mathbf{D}$ and $\mathbf{E}$ are taken from the set $\{A, B, C\}$, decide whether or not $[\mathbf{D} \supset \mathbf{E}]$ is valid.

X2307. Find a model for the sentence $\forall x [Px \equiv \sim Pfx]$. (Here P and f are constants.)

X2308. For any wff A of $\mathcal{F}$. let A' be defined as follows: $\mathbf{A}' = \mathbf{A}$ if $\mathbf{A}$ is atomic; $(\sim \mathbf{B})' = \sim \mathbf{B}'$; $[\mathbf{B} \lor \mathbf{C}]' = [\mathbf{B}' \lor \mathbf{C}']$; $(\forall \mathbf{x} \mathbf{B})' = \exists \mathbf{x} \mathbf{B}'$. Is it true that $\vdash \mathbf{A} \supset \mathbf{A}'$ for every wff $\mathbf{A}$ of $\mathcal{F}$?

In X2309-X2311, prove or refute the stated conjectures.

X2309. If $\mathbf{A}$ and $\mathbf{B}$ are satisfiable sentences, then the first-order theory with $\{\mathbf{A}, \mathbf{B}\}$ as its set of postulates is consistent.

X2310. If $\mathbf{M}$, $\mathbf{N}$, $\mathbf{C}$, and $\mathbf{C}'$ are wffs of $\mathcal{F}$ such that $\mathbf{C}'$ is obtained from $\mathbf{C}$ by replacing $\mathbf{M}$ by $\mathbf{N}$ at one positive occurrence of $\mathbf{M}$ in $\mathbf{C}$, and if $\vdash \mathbf{M} \lor \mathbf{N}$, then $\vdash \mathbf{C} \lor \mathbf{C}'$.

X2311. If $\mathbf{M}$, $\mathbf{N}$, $\mathbf{C}$, and $\mathbf{C}'$ are wffs of $\mathcal{F}$ such that $\mathbf{C}'$ is obtained from $\mathbf{C}$ by replacing $\mathbf{M}$ by $\mathbf{N}$ at one positive occurrence of $\mathbf{M}$ in $\mathbf{C}$, and if $\vdash \mathbf{M} \land \mathbf{N}$, then $\vdash \mathbf{C} \land \mathbf{C}'$.

Decide which of the wffs below are valid. Prove each valid wff and give a countermodel for each wff which is not valid.

X2312. $\exists y Qyw \equiv \exists z Qzx$

X2313. $[\exists xPx \supset \forall xQx] \supset \forall x[Px \supset Qx]$

X2314. $\forall x[Px \supset Qx] \supset [\exists xPx \supset \forall xQx]$

X2315. $\forall x[Px \supset Qx] \equiv \,\centerdot\, \forall xPx \supset \exists xQx$

X2316. $\forall x[Px \supset Qx] \equiv \,\centerdot\, \forall xPx \supset \forall xQx$

X2317. $\forall x[Qx \supset Rx] \supset \,\centerdot\, \exists x[Px \wedge Qx] \supset \exists x \,\centerdot\, Px \wedge Rx$

X2318. $\exists x\forall y \,\centerdot\, Px \equiv Py$

X2319. $\exists y\forall yRxy \supset \forall x\exists yRxy$

X2320. $\exists x\forall yRxy \equiv \forall y\exists xRxy$

X2321. $\exists y\forall x[Px \equiv \,\centerdot\, Py \wedge Qx] \supset \exists z \,\centerdot\, Pz \supset Qz$

X2322. $\forall x\forall y \,\centerdot\, Pxy \supset \exists zPzz$

X2323. $\forall y \,\centerdot\, Pyy \supset \exists zPzy$

X2324. $\exists x\forall y[Pxy \wedge Qx] \supset \,\centerdot\, \exists yQy \wedge \exists zPzz$

§24. Independence

2400 Lemma. If $\vdash \mathbf{A}$, then $\vee$ occurs in $\mathbf{A}$.

Proof: It is straightforward to establish by induction on the construction of $\mathbf{A}$ that if $\mathbf{A}$ is any wff in which $\vee$ does not occur, then $\mathbf{A}$ is satisfiable in some interpretation with one individual, and $\sim\mathbf{A}$ is satisfiable in some interpretation with one individual. Hence if $\vee$ does not occur in $\mathbf{A}$, by 2301a $\mathbf{A}$ is not valid, and so by 2303 is not a theorem. ■

2401 Theorem. The rules of inference and axiom schemata of $\mathcal{F}$ are independent.

Proof: To see that Modus Ponens is independent, note that every theorem of $\mathcal{F}$ which can be proved without using Modus Ponens is obtained from an axiom by zero or more applications of Generalization. Clearly $p \vee \sim p$ (a theorem by 2100) cannot be so obtained, so Modus Ponens is independent.

To see that Generalization is independent, modify the definition of the valuation function $\mathcal{V}$ of §23 so that $\mathcal{V}_{\varphi}^{\mathcal{M}}\forall\mathbf{x}\mathbf{A}$ is always F. With this modified definition, the axioms are valid and Modus Ponens preserves validity, so any theorem provable without using Generalization if valid in the modified sense. However, $\forall x\,[p \lor p \supset p]$, which is a theorem of $\mathcal{F}$ by Axiom 1 and Generalization, is not valid in the modified sense, and so cannot be proved without using Generalization. Thus Generalization is an independent rule of inference.

The independence of each of Axiom Schemata 1, 2, and 3 of $\mathcal{F}$ is readily established by a modification of the method used in §13 to establish the independence of the corresponding Axiom Schema of $\mathcal{P}$. For any wff **A** of $\mathcal{F}$, let **A**$'$ be the formula obtained by deleting all quantifiers and terms from **A**. (Thus, if **A** is the axiom $\sim \forall x P^2 x f^1 y \lor P^2 g^2 y f^1 x f^1 y$, then **A**$'$ is $\sim P^2 \lor P^2$.) We may regard **A**$'$ as a wff of some system of propositional calculus by including the predicate letters of $\mathcal{F}$ among the propositional variables of that system. To prove the independence of Axiom Schema i (where $i = 1,\ 2$, or 3) of $\mathcal{F}$, define a wff **A** of $\mathcal{F}$ to have property $\mathcal{R}_i$ iff **A**$'$ is special in the sense of §13 when $\lor$ and $\sim$ are interpreted according to Figure 1.9 $+i$ in §13. It is then readily verified that Modus Ponens and Generalization preserve property $\mathcal{R}_i$, and that each axiom of $\mathcal{F}$ which is not an instance of Axiom Schema i has property $\mathcal{R}_i$. (Note that if **A** is an instance of Axiom Schema 4 or 5, then **A**$'$ has the form $\sim \mathbf{B} \lor \mathbf{B}$.) However, as in §13 we can find an instance of Axiom Schema i which does not have property $\mathcal{R}_i$, so this wff cannot be derived from the other axiom schemata of $\mathcal{F}$. Hence Axiom Schema i is independent for $i = 1,\ 2$, and 3.

To prove the independence of Axiom Schema 4, for each wff **A**, let **A**$'$ be the result of replacing $\forall\mathbf{x}$ by $\exists\mathbf{x}$ (i.e., $\sim \forall\mathbf{x} \sim$) everywhere in **A**. Say that **A** has property $\mathcal{R}_4$ iff $\models$ **A**$'$. It is easy to see that all instances of Axiom Schemata 1, 2, 3, and 5 have property $\mathcal{R}_4$, and that Modus Ponens and Generalization preserve property $\mathcal{R}_4$. However, the instance $[\forall x P^1 x \supset P^1 x]$ of Axiom Schema 4 does not have property $\mathcal{R}_4$, since $\mathcal{V}_\varphi^\mathcal{M}\,[\exists x P^1 x \supset P^1 x] = \mathsf{F}$ when $\mathcal{M}$ is an interpretation with the integers as domain of individuals, φP^1 is the property of being positive, and $\varphi x = -2$. Thus Axiom Schema 4 is independent. (This argument suggests that one of the reasons for including Axiom Schema 4 among the axioms of $\mathcal{F}$ is to prevent $\forall\mathbf{x}$ from being soundly interpreted as meaning "there exists **x** such that".)

To prove the independence of Axiom Schema 5, let **A**$'$ be the wff obtained from **A** by replacing each wf part of **A** of the form $\forall\mathbf{x}\mathbf{B}$, where **B** contains no occurrences of $\lor$, by the wff $[p \land \sim p]$. (This may be done sequentially, treating the quantifiers of **A** in the left-to-right order of their occurrences in **A**.) Say that **A** has property $\mathcal{R}_5$ iff $\vdash$ **A** and $\models$ **A**$'$. Clearly Generalization preserves property $\mathcal{R}_5$, since if $\vdash$ **A** and $\models$ **A**$'$, then $\vdash \forall\mathbf{x}\,\mathbf{A}$ and $\models \forall\mathbf{x}(\mathbf{A}')$ (by 2301c), but **A** contains $\lor$ by 2400, so $(\forall\mathbf{x}\,\mathbf{A})'$ is $\forall\mathbf{x}(\mathbf{A}')$. Modus Ponens also preserves property $\mathcal{R}_5$, since $[\mathbf{A} \supset \mathbf{B}]'$ is $[\mathbf{A}' \supset \mathbf{B}']$.

All instances of Axiom Schemata 1-3 clearly have property $\mathcal{R}_5$. It can be seen that $[\forall \mathbf{x}\mathbf{A} \supset \mathsf{S}^{\mathbf{x}}_{\mathbf{t}} \mathbf{A}]'$ is $[[p \wedge \sim p] \supset \mathsf{S}^{\mathbf{x}}_{\mathbf{t}} (\mathbf{A}')]$ if $\vee$ does not occur in $\mathbf{A}$, and otherwise $\forall \mathbf{x} (\mathbf{A}') \supset \mathsf{S}^{\mathbf{x}}_{\mathbf{t}} (\mathbf{A}')$; moreover, if $\mathbf{t}$ is free for $\mathbf{x}$ in $\mathbf{A}$, it is also free for $\mathbf{x}$ in $\mathbf{A}'$. Hence all instances of Axiom Schema 4 have property $\mathcal{R}_5$. However, if $\mathbf{A}$ is the instance $[\forall x [r \vee Q^1 x] \supset . r \vee \forall x\, Q^1 x]$ of Axiom Schema 5, then $\mathcal{V}_\varphi \mathbf{A}' = \mathcal{V}_\varphi [\forall x [r \vee Q_1 x] \supset . r \vee . p \wedge \sim p] = \mathsf{F}$ whenever $\varphi r = \mathsf{F}$ and φQ^1 is the predicate which is true of all individuals (in a given interpretation). Thus, $\mathbf{A}$ does not have property $\mathcal{R}_5$, so Axiom Schema 5 is independent. ■

We can now show, as promised in the remark following metatheorem 2113, that in any proof of $\forall y Py \supset \forall x Px$, there must be an application of the rule of Generalization in which x is the variable generalized upon. For this purpose, modify the definition of $\mathcal{V}_\varphi$ in §23 so that for any wff $\mathbf{A}$, $\mathcal{V}_\varphi \forall x \mathbf{A} = \mathsf{F}$. However, if $\mathbf{y}$ is any individual variable other than x, $\mathcal{V}_\varphi \forall \mathbf{y} \mathbf{A}$ is defined as before. With this modified definition, it can easily be checked that the axioms remain valid, Modus Ponens preserves validity, and Generalization with respect to any variable other than x preserves validity. However, $\forall y Py \supset \forall x Px$ is not valid under this definition, as can be seen by assigning P as value the set of all individuals in some interpretation, so that $\mathcal{V}_\varphi \forall y Py = \mathsf{T}$. Hence any proof of $\forall y Py \supset \forall x Px$ in $\mathcal{F}$ must contain a generalization upon x.

EXERCISES

X2400. Show that if $\vdash \mathbf{A}$, then $\sim$ occurs in $\mathbf{A}$.

X2401. Give a careful inductive definition of $\mathbf{A}'$ in the proof of the independence of Axiom Schema 5. Then give a careful inductive proof that $(\mathsf{S}^{\mathbf{x}}_{\mathbf{t}} \mathbf{A})' = \mathsf{S}^{\mathbf{x}}_{\mathbf{t}} (\mathbf{A}')$ for any term $\mathbf{t}$.

§25. Abstract Consistency and Completeness

We start this section with a number of incidental definitions and results which will be useful later.

DEFINITION. A *literal* is a wff which is atomic, or is the negation of an atomic wff.

2500 Proposition. Every wff of $\mathcal{F}$ is a literal or is of one of the forms $[\mathbf{A} \lor \mathbf{B}]$, $\forall \mathbf{xA}$, $\sim\sim \mathbf{A}$, $\sim [\mathbf{A} \lor \mathbf{B}]$, $\sim \forall \mathbf{xA}$.

Proof: Clearly every wff which is not atomic must have one of the forms $\sim \mathbf{C}$, $[\mathbf{A} \lor \mathbf{B}]$, or $\forall \mathbf{xA}$. If it has the form $\sim \mathbf{C}$, then $\mathbf{C}$ must also be atomic or have one of these forms. ■

Recall that a set $\mathcal{G}$ of wffs is consistent iff there is a wff $\mathbf{A}$ such that not $\mathcal{G} \vdash \mathbf{A}$.

DEFINITION. The set $\mathcal{H}$ of wffs is *consistent with* $\mathcal{G}$ iff $\mathcal{G} \cup \mathcal{H}$ is consistent. We say that $\mathbf{A}$ is *consistent with* $\mathcal{G}$ iff $\mathcal{G} \cup \{\mathbf{A}\}$ is consistent.

2501 Lemma. Let $\mathcal{G}$ be any set of wffs. $\{\mathbf{A}_1, \ldots, \mathbf{A}_n\}$ is inconsistent with $\mathcal{G}$ iff $\mathcal{G} \vdash \sim \mathbf{A}_1 \lor \ldots \lor \sim \mathbf{A}_n$. The wff $\mathbf{A}$ is inconsistent with $\mathcal{G}$ iff $\mathcal{G} \vdash \sim \mathbf{A}$.

Proof: If $\{\mathbf{A}_1, \ldots, \mathbf{A}_n\}$ is inconsistent with $\mathcal{G}$, then $\mathcal{G}, \mathbf{A}_1, \ldots, \mathbf{A}_n \vdash \sim \mathbf{A}_1$ so $\mathcal{G} \vdash \mathbf{A}_1 \supset . \ldots \supset . \mathbf{A}_n \supset \sim \mathbf{A}_1$ by the Deduction Theorem. Hence $\mathcal{G} \vdash \sim \mathbf{A}_1 \lor \ldots \lor \sim \mathbf{A}_n$ by Rule P.

If $\mathcal{G} \vdash \sim \mathbf{A}_1 \lor \ldots \lor \sim \mathbf{A}_n$ then $\mathcal{G}, \mathbf{A}_1, \ldots, \mathbf{A}_n \vdash \sim \mathbf{A}_1 \lor \ldots \lor \sim \mathbf{A}_n$ by 2113. Also, $\mathcal{G}, \mathbf{A}_1, \ldots, \mathbf{A}_n \vdash \mathbf{A}_i$ for $1 \leq i \leq n$ so if $\mathbf{B}$ is any wff, $\mathcal{G}, \mathbf{A}_1, \ldots, \mathbf{A}_n \vdash \mathbf{B}$ by Rule P. Hence $\{\mathbf{A}_1, \ldots, \mathbf{A}_n\}$ is inconsistent with $\mathcal{G}$. ■

Note that the proof above applies when $n = 1$.

To simplify terminology, until further notice we shall restrict our attention to formulations of $\mathcal{F}$ in which there are no variables except individual variables. Thus all predicate and function symbols will be constants, and all closed wffs will be sentences. Various formulations of $\mathcal{F}$ will differ from each other only by having different sets of constants. It will be obvious how to transfer results about these formulations of $\mathcal{F}$ to formulations with other types of variables.

DEFINITIONS. We shall refer to the set of wffs of a logistic system $\mathcal{F}$ as the *language* of $\mathcal{F}$, and denote this set by $\mathcal{L}(\mathcal{F})$. If $\mathcal{F}^1$ and $\mathcal{F}^2$ are logistic systems, we shall say that $\mathcal{F}^2$ is an *expansion* of $\mathcal{F}^1$ iff $\mathcal{L}(\mathcal{F}^1) \subseteq \mathcal{L}(\mathcal{F}^2)$, and a *proper expansion* iff $\mathcal{L}(\mathcal{F}^1) \subset \mathcal{L}(\mathcal{F}^2)$. For the present we are concerned only with formulations of the first-order system $\mathcal{F}$, so $\mathcal{F}^2$ in an expansion of $\mathcal{F}^1$ iff every constant of $\mathcal{F}^1$ is a constant of $\mathcal{F}^2$. Naturally, when one expands a formulation of $\mathcal{F}$ by adding new constants, one also adds new instances of the logical axiom schemata to the list of axioms.

If $\mathcal{F}^1$ and $\mathcal{F}^2$ are first-order theories, we shall say that $\mathcal{F}^2$ is an *extension* of $\mathcal{F}^1$ iff $\mathcal{L}(\mathcal{F}^1) \subseteq \mathcal{L}(\mathcal{F}^2)$ and every theorem of $\mathcal{F}^1$ is a theorem of $\mathcal{F}^2$. The latter condition is satisfied iff every postulate of $\mathcal{F}^1$ is a theorem of $\mathcal{F}^2$. Of course, postulates of $\mathcal{F}^1$ need not be postulates of $\mathcal{F}^2$.

Let $\mathcal{F}^1$ and $\mathcal{F}^2$ be first order theories. We say that $\mathcal{F}^2$ is a *conservative extension* of $\mathcal{F}^1$ iff for every wff $\mathbf{C}$ of $\mathcal{L}(\mathcal{F}^1)$, $\vdash_{\mathcal{F}^1} \mathbf{C}$ iff $\vdash_{\mathcal{F}^2} \mathbf{C}$. Clearly a conservative extension is an extension.

Logistic systems $\mathcal{F}^1$ and $\mathcal{F}^2$ are *equivalent* iff each is an extension of the other. In this case, they have the same wffs and the same theorems, and each is a conservative extension of the other.

An example of a conservative extension might be obtained by letting $\mathcal{F}^1$ be a formalization of the mathematics used in a certain scientific theory, and creating $\mathcal{F}^2$ by adding to $\mathcal{F}^1$ constants to serve as names various scientific concepts, and adding axioms to express the definitions and laws of the scientific theory. One would expect that if a theorem of $\mathcal{F}^2$ had purely mathematical content (i.e., was a wff of $\mathcal{F}^1$), its proof would not require the assumptions about the physical world expressed by the laws of the scientific theory, or the definitions of the concepts of that theory, so it should be provable in $\mathcal{F}^1$. That is, $\mathcal{F}^2$ should be a conservative extension of $\mathcal{F}^1$.

2502 Proposition. Let $\mathcal{F}^1$ and $\mathcal{F}^2$ be first order theories, with $\mathcal{F}^2$ a conservative extension of $\mathcal{F}^1$. Then $\mathcal{F}^2$ is consistent iff $\mathcal{F}^1$ is consistent.

Proof: Left to the reader.

If $\mathcal{G}$ is a set of sentences of $\mathcal{F}$, we shall denote by $\mathcal{F} \cup \mathcal{G}$ the first order theory with the set $\mathcal{G}$ of postulates. (Thus the set of axioms of this theory is the union of $\mathcal{G}$ with the set of axioms $\mathcal{F}$.)

2503 Proposition. Let $\mathcal{F}^1$ and $\mathcal{F}^2$ be formulations of $\mathcal{F}$, so that $\mathcal{F}^2$ is an expansion of $\mathcal{F}^1$ obtained by adding additional individual constants to the set of constants of $\mathcal{F}^1$. Let $\mathcal{G}$ be any set of sentences of $\mathcal{F}^1$. Then

(1) $\mathcal{F}^2 \cup \mathcal{G}$ is a conservative extension of $\mathcal{F}^1 \cup \mathcal{G}$.
(2) $\mathcal{G}$ is consistent in $\mathcal{F}^1$ iff it is consistent in $\mathcal{F}^2$.

Proof: Clearly $\mathcal{F}^2 \cup \mathcal{G}$ is an extension of $\mathcal{F}^1 \cup \mathcal{G}$. If $\mathcal{P}$ is a proof in $\mathcal{F}^2 \cup \mathcal{G}$ of a wff $\mathbf{C}$ of $\mathcal{F}^1$, then one can replace the individual constants of $\mathcal{L}(\mathcal{F}^2) - \mathcal{L}(\mathcal{F}^1)$ in $\mathcal{P}$ by new individual variables to obtain a proof of $\mathbf{C}$ in $\mathcal{F}^1 \cup \mathcal{G}$.

Clearly $\mathcal{G}$ is consistent in $\mathcal{F}$ iff $\mathcal{F} \cup \mathcal{G}$ is consistent, so (2) follows from (1) and 2502. ■

We say that a set is *countable* iff it is finite or denumerably infinite. Since zero is finite, a countable set may be empty.

Customarily we use formulations of $\mathcal{F}$ with countably many constants. Indeed, one never actually writes down more than finitely many wffs, so a countably infinite set of constants provides an inexhaustible supply of new constants for those who wish to use first-order logic as a vehicle for formalizing theories and proving theorems. Nevertheless, for certain model-theoretic purposes it is useful to consider, as objects of mathematical study, formulations of first-order logic with uncountably many constants. Therefore we shall impose no restrictions on the cardinality of the set of constants. We always assume that there are denumerably many individual variables.

We denote the cardinality of any set or ordinal number $\mathbf{S}$ by card($\mathbf{S}$). (The cardinality of an ordinal is the cardinality of the set of ordinals which precede it.) Since every formula is a finite sequence of primitive symbols, it is easy to see that card($\mathcal{L}(\mathcal{F})$) is the cardinality of the set of primitive symbols of $\mathcal{F}$. Formulations of $\mathcal{F}$ with countably many constants have $\aleph_0$ wffs, and are called countable formulations. (Indeed, one can enumerate the primitive symbols of a countable formulation of $\mathcal{F}$, so that each primitive symbol $\mathbf{K}$ is assigned a positive integer '$\mathbf{K}$'. Then to each formula $\mathbf{K}_0 \ldots \mathbf{K}_n$ one can assign the number $\Pi_{i=0}^{n} \mathbf{P}_i^{`\mathbf{K}_i'}$, where $\mathbf{P}_i$ is the ith odd prime (as defined for Theorem 6522 in §65). Then each formula is assigned a number, and distinct formulas have distinct numbers.)

The reader who does not wish to be bothered with cardinality considerations may focus entirely on countable formulations of $\mathcal{F}$. Indeed, this is advisable on first encounter with the material in this section. We phrase the proofs so as to apply to formulations of $\mathcal{F}$ of arbitrary cardinality because very little extra effort is required to achieve this generality.

An interpretation $\langle \mathcal{D}, \mathcal{J} \rangle$ for $\mathcal{F}$ is said to be *frugal* iff card($\mathcal{D}$) $\leq$ card($\mathcal{L}(\mathcal{F})$). If $\mathcal{F}$ is countable, this means that the domain of individuals is countable, and a frugal interpretation [model] is called a countable interpretation [model].

We shall show in Theorem 2508 that every consistent set of sentences has a frugal model. Note that if $\mathcal{F}$ has uncountably many constants, a consistent set of sentences of $\mathcal{F}$ need not have a *countable* model. For example, consider the set $\mathcal{S}$ which contains (1) the sentence $\forall x\, Exx$ and (2) all sentences of the form $\sim E\mathbf{cd}$, where $\mathbf{c}$ and $\mathbf{d}$ are distinct individual constants. In any model of $\mathcal{S}$, the predicate constant E denotes a reflexive relation, such as equality. Therefore, it is easy to see that if $\mathcal{M}$ is any model for $\mathcal{S}$, and $\mathbf{c}$ and $\mathbf{d}$ are distinct constants, then $\mathcal{V}^{\mathcal{M}}\mathbf{c} \neq \mathcal{V}^{\mathcal{M}}\mathbf{d}$. For if $\mathcal{V}^{\mathcal{M}}\mathbf{c} = \mathcal{V}^{\mathcal{M}}\mathbf{d}$, and we let $\psi x = \mathcal{V}^{\mathcal{M}}\mathbf{c}$, then by (1) $\mathsf{T} = \mathcal{V}^{\mathcal{M}}_{\psi}Exx = (\mathcal{V}^{\mathcal{M}}E)(\psi x)(\psi x) = (\mathcal{V}^{\mathcal{M}}E)(\mathcal{V}^{\mathcal{M}}\mathbf{c})(\mathcal{V}^{\mathcal{M}}\mathbf{d}) = \mathcal{V}^{\mathcal{M}}E\mathbf{cd}$, contradicting (2). Thus, any model of $\mathcal{S}$ must have at least as many individuals as there are individual constants in $\mathcal{F}$.

Our main objective in this section is to prove Theorem 2508 and its consequences, including Gödel's Completeness Theorem (2510). Clearly, the concept of consistency is the fundamental syntactic notion in Theorem 2508. It was noticed by Smullyan [Smullyan, 1963] that the proof of Theorem 2508 uses only certain properties of the notion of consistency, and that by treating consistency in an abstract way, with very little extra effort one can obtain a generalization (2507 Smullyan's Unifying Principle) of Theorem 2508 which yields a variety of important metatheorems. The power of Smullyan's Unifying Principle is a consequence of the fact that various logical questions can be formulated as questions about consistency relative to special logical systems. Our only use of Smullyan's Unifying Principle in this chapter will be in the proof of Theorem of 2508, but other examples of its usefulness will be found in later chapters. In §25A we give a direct proof of Theorem 2508 for countable formulations of $\mathcal{F}$. The reader who is not interested in the more general treatment below may now proceed to read §25A, and then go directly to Theorem 2509.

We shall formulate an abstract notion of consistency by considering various properties possessed by the class[7] of consistent sets of sentences. Two such properties are defined below:

DEFINITION. Let Γ be a class of sets.

(1) Γ is *closed under subsets* iff for all sets $\mathcal{S}_1$ and $\mathcal{S}_2$, if $\mathcal{S}_1 \subseteq \mathcal{S}_2$ and $\mathcal{S}_2 \in \Gamma$, then $\mathcal{S}_1 \in \Gamma$.
(2) Γ is of *finite character* iff for every set $\mathcal{S}$, $\mathcal{S} \in \Gamma$ iff every finite subset of $\mathcal{S}$ is a member of Γ.

2504 Lemma. If Γ is a class of sets which is of finite character, then Γ is closed under subsets.

Proof: Suppose $\mathcal{S}_1 \subseteq \mathcal{S}_2$ and $\mathcal{S}_2 \in \Gamma$. Every finite subset of $\mathcal{S}_1$ is a subset of $\mathcal{S}_2$, and so is in Γ, so $\mathcal{S}_1 \in \Gamma$. ∎

[7] We use the word *class* as a synonym for *collection* or *set*, without any of the connotations associated with the word in axiomatic set theory.

A *variable-free* term is a term in which no variables occur.

DEFINITION. We shall say that Γ is an *abstract consistency class* iff Γ is a class of sets of sentences, Γ is closed under subsets, and for all sets $\mathcal{S}$ in Γ, the following conditions hold for all wffs **A** and **B**:

(a) If **A** is atomic, then $\mathbf{A} \notin \mathcal{S}$ or $\sim \mathbf{A} \notin \mathcal{S}$.
(b) If $\sim\sim \mathbf{A} \in \mathcal{S}$, then $\mathcal{S} \cup \{\mathbf{A}\} \in \Gamma$.
(c) If $[\mathbf{A} \lor \mathbf{B}] \in \mathcal{S}$, then $\mathcal{S} \cup \{\mathbf{A}\} \in \Gamma$ or $\mathcal{S} \cup \{\mathbf{B}\} \in \Gamma$.
(d) If $\sim [\mathbf{A} \lor \mathbf{B}] \in \mathcal{S}$, then $\mathcal{S} \cup \{\sim \mathbf{A}, \sim \mathbf{B}\} \in \Gamma$.
(e) If $\forall \mathbf{xA} \in \mathcal{S}$, then for each variable-free term **t**, $\mathcal{S} \cup \left\{\mathsf{S}^{\mathbf{x}}_{\cdot\mathbf{t}}\mathbf{A}\right\} \in \Gamma$.
(f) If $\sim \forall \mathbf{xA} \in \mathcal{S}$, and **c** is any individual constant which does not occur in any sentence of $\mathcal{S}$, then $\mathcal{S} \cup \left\{\sim \mathsf{S}^{\mathbf{x}}_{\cdot\mathbf{c}}\mathbf{A}\right\} \in \Gamma$.

It is easy to see that the class of sets of sentences which have models is an abstract consistency class. Another example, which justifies the terminology, is provided by the following proposition.

2505 Proposition. $\{\mathcal{S} \mid \mathcal{S}$ is a consistent set of sentences of $\mathcal{F}\}$ is an abstract consistency class.

Proof: The class is closed under subsets, since if a set $\mathcal{S}$ is consistent, then every subset of $\mathcal{S}$ is consistent. Clearly if $\mathcal{S}$ is consistent, no wff and its negation can be in $\mathcal{S}$. This establishes (a).

Note that if $\vdash \mathbf{C} \supset \mathbf{D}_1 \land \ldots \land \mathbf{D}_n$ (where $n \geq 1$) and $\mathbf{C} \in \mathcal{S}$ and $\mathcal{S}$ is consistent, then $\mathcal{S} \cup \{\mathbf{D}_1, \ldots, \mathbf{D}_n\}$ must be consistent. For if $\mathcal{S} \cup \{\mathbf{D}_1, \ldots, \mathbf{D}_n\}$ is inconsistent, then $\mathcal{S} \vdash \sim \mathbf{D}_1 \lor \ldots \lor \sim \mathbf{D}_n$ by Lemma 2501, so $\mathcal{S} \vdash \sim \mathbf{C}$ by Rule P, so $\mathcal{S}$ is inconsistent. This establishes (b), (d), and (e).

If $\mathcal{S}$ is consistent but $\mathcal{S} \cup \{\mathbf{A}\}$ and $\mathcal{S} \cup \{\mathbf{B}\}$ are both inconsistent, then $\mathcal{S} \vdash \sim \mathbf{A}$ and $\mathcal{S} \vdash \sim \mathbf{B}$ by 2501, so $\mathcal{S} \vdash \sim [\mathbf{A} \lor \mathbf{B}]$ by Rule P, so $[\mathbf{A} \lor \mathbf{B}] \notin \mathcal{S}$. This establishes (c).

To establish (f), suppose $\mathcal{S}$ is consistent but $\mathcal{S} \cup \left\{\sim \mathsf{S}^{\mathbf{x}}_{\cdot\mathbf{c}}\mathbf{A}\right\}$ is inconsistent, where **c** is an individual constant which does not occur in $\mathcal{S}$ or **A**. Then $\mathcal{S} \vdash \mathsf{S}^{\mathbf{x}}_{\cdot\mathbf{c}}\mathbf{A}$ by 2501 and Rule P, so there is a finite subset $\mathcal{H}$ of $\mathcal{S}$ such that $\mathcal{H} \vdash \mathsf{S}^{\mathbf{x}}_{\cdot\mathbf{c}}\mathbf{A}$. Given a proof $\mathcal{P}$ of $\mathsf{S}^{\mathbf{x}}_{\cdot\mathbf{c}}\mathbf{A}$ from $\mathcal{H}$, let **y** be an individual variable which does not occur in $\mathcal{P}$ or in $\mathcal{H}$. Since **c** does not occur in $\mathcal{H}$ or **A**, it can be seen that $\mathsf{S}^{\mathbf{c}}_{\mathbf{y}}\mathcal{P}$ is a proof of $\mathsf{S}^{\mathbf{x}}_{\cdot\mathbf{y}}\mathbf{A}$ from $\mathcal{H}$, so $\mathcal{H} \vdash \mathsf{S}^{\mathbf{x}}_{\cdot\mathbf{y}}\mathbf{A}$. Hence $\mathcal{H} \vdash \forall \mathbf{y}\mathsf{S}^{\mathbf{x}}_{\cdot\mathbf{y}}\mathbf{A}$ by Generalization, and $\mathcal{H} \vdash \forall \mathbf{xA}$ by the α Rule, so $\mathcal{S} \vdash \sim\sim \forall \mathbf{xA}$, so $\sim \forall \mathbf{xA} \notin \mathcal{S}$. ∎

2506 Proposition. Let Γ be an abstract consistency class. Let $\Delta = \{\mathcal{S} \mid$ every finite subset of $\mathcal{S}$ is in $\Gamma\}$. Then $\Gamma \subseteq \Delta$ and Δ is an abstract consistency class of finite character.

Proof:[8] To see that $\Gamma \subseteq \Delta$, suppose $\mathcal{S} \in \Gamma$. Then every finite subset of $\mathcal{S}$ is in Γ since Γ is closed under subsets, so $\mathcal{S} \in \Delta$.

Next, let us show that Δ is of finite character. Suppose $\mathcal{S} \in \Delta$ and $\mathcal{H}$ is any finite subset of $\mathcal{S}$. Then by the definition of Δ, all finite subsets of $\mathcal{H}$ are in Γ, so $\mathcal{H} \in \Delta$. Thus all finite subsets of $\mathcal{S}$ are in Δ whenever $\mathcal{S}$ is in Δ. On the other hand, suppose all finite subsets of $\mathcal{S}$ are in Δ. Then by the definition of Δ the finite subsets of $\mathcal{S}$ are also in Γ, so $\mathcal{S} \in \Delta$. Thus Δ is of finite character.

Now we show that Δ is an abstract consistency class. By Lemma 2504 it is closed under subsets. Let $\mathcal{S} \in \Delta$.

(a) Suppose there is an atom $\mathbf{A}$ such that $\{\mathbf{A}, \sim \mathbf{A}\} \subseteq \mathcal{S}$. Then $\{\mathbf{A}, \sim \mathbf{A}\} \in \Gamma$, contradicting (a) for Γ.

(b), (d), (e), (f) We handle these conditions simultaneously, since each has the form "If $\mathbf{C} \in \mathcal{S}$, then $\mathcal{S} \cup \mathcal{C} \in \Delta$", where $\mathbf{C}$ is a wff and $\mathcal{C}$ is an appropriate finite set of wffs; for example in case (e) $\mathbf{C}$ is $\forall \mathbf{x}\mathbf{A}$, and $\mathcal{C}$ is $\left\{\mathsf{S}^{\mathbf{x}}_{\mathbf{t}}\mathbf{A}\right\}$ for some variable-free term $\mathbf{t}$. So suppose $\mathbf{C} \in \mathcal{S}$, and let $\mathcal{H}$ be any finite subset of $\mathcal{S} \cup \mathcal{C}$. Let $\mathcal{S}_1 = (\mathcal{H} - \mathcal{C}) \cup \{\mathbf{C}\}$. $\mathcal{S}_1$ is a finite subset of $\mathcal{S}$, so $\mathcal{S}_1 \in \Gamma$. Since Γ is an abstract consistency class and $\mathbf{C} \in \mathcal{S}_1$, we have $\mathcal{S}_1 \cup \mathcal{C} \in \Gamma$. $\mathcal{H} \subseteq \mathcal{S}_1 \cup \mathcal{C}$ and Γ is closed under subsets, so $\mathcal{H} \in \Gamma$. Thus every finite subset $\mathcal{H}$ of $\mathcal{S} \cup \mathcal{C}$ is in Γ, so $\mathcal{S} \cup \mathcal{C} \in \Delta$. (Of course, when dealing with condition (f), we can use the fact that if an individual constant does not occur in any sentence of $\mathcal{S}$, then it does not occur in any sentence of $\mathcal{S}_1$.)

(c) Suppose $[\mathbf{A} \vee \mathbf{B}] \in \mathcal{S}$ but $\mathcal{S} \cup \{\mathbf{A}\} \notin \Delta$ and $\mathcal{S} \cup \{\mathbf{B}\} \notin \Delta$. Then there are finite subsets $\mathcal{S}_1$ and $\mathcal{S}_2$ of $\mathcal{S}$ such that $\mathcal{S}_1 \cup \{\mathbf{A}\} \notin \Gamma$ and $\mathcal{S}_2 \cup \{\mathbf{B}\} \notin \Gamma$ (since all finite subsets of $\mathcal{S}$ are in Γ). Let $\mathcal{S}_3 = \mathcal{S}_1 \cup \mathcal{S}_2 \cup \{[\mathbf{A} \vee \mathbf{B}]\}$. $\mathcal{S}_3$ is a finite subset of $\mathcal{S}$, so $\mathcal{S}_3 \in \Gamma$. Since Γ is an abstract consistency class and $[\mathbf{A} \vee \mathbf{B}] \in \mathcal{S}_3$, we have $\mathcal{S}_3 \cup \{\mathbf{A}\} \in \Gamma$ or $\mathcal{S}_3 \cup \{\mathbf{B}\} \in \Gamma$. Γ is closed under subsets, so $\mathcal{S}_1 \cup \{\mathbf{A}\} \in \Gamma$ or $\mathcal{S}_2 \cup \{\mathbf{B}\} \in \Gamma$. This is a contradiction, so we conclude that if $[\mathbf{A} \vee \mathbf{B}] \in \mathcal{S}$, then $\mathcal{S} \cup \{\mathbf{A}\} \in \Delta$ or $\mathcal{S} \cup \{\mathbf{B}\} \in \Delta$. ∎

[8] Note that if Γ is of finite character then $\Delta = \Gamma$, and this proposition is vacuous. Since many abstract consistency classes can easily be seen to be of finite character, some readers may wish to postpone reading the proof of this proposition.

DEFINITION. Let $\mathcal{S}$ be a set of wffs of $\mathcal{F}$. $\mathcal{S}$ is *pure* iff no wff of $\mathcal{S}$ contains an individual constant. $\mathcal{S}$ is *sufficiently pure* in $\mathcal{F}$ iff there is a set of cardinality $\text{card}(\mathcal{L}(\mathcal{F}))$ of individual constants of $\mathcal{F}$ which do not occur in any wff of $\mathcal{S}$.

2507 Smullyan's Unifying Principle [Smullyan, 1963], [Smullyan, 1968, Chapter VI]. If Γ is an abstract consistency class and $\mathcal{S}$ is a sufficiently pure set of sentences and $\mathcal{S} \in \Gamma$, then $\mathcal{S}$ has a frugal model.

Proof: The theorem is trivial if $\mathcal{S}$ is empty, so we may assume $\mathcal{S}$ is non-empty. As in Proposition 2506, let Δ be an abstract consistency class of finite character such that $\Gamma \subseteq \Delta$. (To make the argument seem less abstract, it is suggested that the reader pretend that "$\mathcal{H} \in \Delta$" means "$\mathcal{H}$ is consistent" when reading the proof for the first time.) Let $K = \text{card}\,(\mathcal{L}\,(\mathcal{F}))$, and let κ be the initial ordinal of cardinality K. Well-order the sentences of $\mathcal{F}$, and for each ordinal $\tau < \kappa$, let $\mathbf{S}_\tau$ be the τ-th sentence in this well-ordering.

Let $\mathcal{C}$ be a well-ordered set of cardinality K of individual constants which do not occur in sentences of $\mathcal{S}$. By transfinite induction, for each ordinal $\tau \leq \kappa$ we define a set $\mathcal{S}_\tau$ of sentences so that $\mathcal{S}_\sigma \subseteq \mathcal{S}_\tau$ whenever $\sigma \leq \tau$, and for each τ there is a finite cardinal n such that at most $n + \text{card}(\tau)$ constants from $\mathcal{C}$ occur in the sentences of $\mathcal{S}_\tau$. $\mathcal{S}_0 = \mathcal{S}$. If τ is a limit ordinal, $\mathcal{S}_\tau = \bigcup_{\sigma<\tau} \mathcal{S}_\sigma$. The definition for successor ordinals $\tau + 1$ is by cases:

(1) If $\mathcal{S}_\tau \cup \{\mathbf{S}_\tau\} \notin \Delta$, then $\mathcal{S}_{\tau+1} = \mathcal{S}_\tau$.
(2) If $\mathcal{S}_\tau \cup \{\mathbf{S}_\tau\} \in \Delta$, and $\mathbf{S}_\tau$ is not of the form $\sim \forall \mathbf{xA}$, then $\mathcal{S}_{\tau+1} = \mathcal{S}_\tau \cup \{\mathbf{S}_\tau\}$.
(3) If $\mathcal{S}_\tau \cup \{\mathbf{S}_\tau\} \in \Delta$ and $\mathbf{S}_\tau$ is $\sim\forall\mathbf{x}\,\mathbf{A}$, then $\mathcal{S}_{\tau+1} = \mathcal{S}_\tau \cup \left\{\sim\forall\mathbf{x}\,\mathbf{A}, \sim \mathsf{S}^{\mathbf{x}}_{\mathbf{c}}\mathbf{A}\right\}$, where $\mathbf{c}$ is the first individual constant from $\mathcal{C}$ which does not occur in any sentence of $\mathcal{S}_\tau \cup \{\mathbf{S}_\tau\}$. (Note that there must be such a constant.)

It is readily seen by induction that the condition on the number of constants from $\mathcal{C}$ in $\mathcal{S}_\tau$ is satisfied.

We next show by induction that $\mathcal{S}_\tau \in \Delta$ for each $\tau \leq \kappa$. $\mathcal{S}_0 = \mathcal{S} \in \Gamma \subseteq \Delta$. If τ is a limit ordinal, let $\mathcal{H}$ be any finite subset of $\mathcal{S}_\tau$. Then there is a $\sigma < \tau$ such that $\mathcal{H} \subseteq \mathcal{S}_\sigma$, and $\mathcal{S}_\sigma \in \Delta$ by inductive hypothesis, so $\mathcal{H} \in \Delta$ since Δ is closed under subsets. Thus every finite subset of $\mathcal{S}_\tau$ is in Δ, so $\mathcal{S}_\tau \in \Delta$. Also, if $\mathcal{S}_\tau \in \Delta$, then $\mathcal{S}_{\tau+1} \in \Delta$ in all cases.

Let $\mathcal{U} = \mathcal{S}_\kappa$. Clearly $\mathcal{U}$ is a set of sentences and $\mathcal{S} \subseteq \mathcal{U}$. We assert that $\mathcal{U}$ has the following properties for all wffs $\mathbf{A}$ and $\mathbf{B}$:

(a) There is no atomic sentence $\mathbf{A}$ such that $\mathbf{A} \in \mathcal{U}$ and $\sim \mathbf{A} \in \mathcal{U}$.
(b) If $\sim\sim \mathbf{A} \in \mathcal{U}$, then $\mathbf{A} \in \mathcal{U}$.
(c) If $[\mathbf{A} \vee \mathbf{B}] \in \mathcal{U}$, then $\mathbf{A} \in \mathcal{U}$ or $\mathbf{B} \in \mathcal{U}$.
(d) If $\sim [\mathbf{A} \vee \mathbf{B}] \in \mathcal{U}$, then $\sim \mathbf{A} \in \mathcal{U}$ and $\sim \mathbf{B} \in \mathcal{U}$.
(e) If $\forall \mathbf{x}\mathbf{A} \in \mathcal{U}$, then $\mathsf{S}^{\mathbf{x}}_{\mathbf{t}}\mathbf{A} \in \mathcal{U}$ for each variable-free term $\mathbf{t}$.
(f) If $\sim \forall \mathbf{x}\mathbf{A} \in \mathcal{U}$, then there is an individual constant $\mathbf{c}$ such that $\sim \mathsf{S}^{\mathbf{x}}_{\mathbf{c}}\mathbf{A} \in \mathcal{U}$.

(a) is true since $\mathcal{U} \in \Delta$. Next note that if $\mathbf{D}$ is a sentence such that $\mathcal{U} \cup \{\mathbf{D}\} \in \Delta$, then $\mathbf{D} \in \mathcal{U}$. For there is an ordinal τ such that $\mathbf{S}_\tau = \mathbf{D}$, and $\mathcal{S}_\tau \cup \{\mathbf{D}\} \subseteq \mathcal{U} \cup \{\mathbf{D}\} \in \Delta$, so $\mathcal{S}_\tau \cup \{\mathbf{D}\} \in \Delta$ since Δ is closed under subsets. Hence $\mathbf{D} \in \mathcal{S}_{\tau+1} \subseteq \mathcal{U}$, so $\mathbf{D} \in \mathcal{U}$. Now to prove (b), suppose $\sim\sim \mathbf{A} \in \mathcal{U}$. Then $\mathcal{U} \cup \{\mathbf{A}\} \in \Delta$ (since Δ is an abstract consistency class), so $\mathbf{A} \in \mathcal{U}$. Similar arguments establish (d) and (e). To prove (c), suppose $[\mathbf{A} \vee \mathbf{B}] \in \mathcal{U}$. Then $\mathcal{U} \cup \{\mathbf{A}\} \in \Delta$ or $\mathcal{U} \cup \{\mathbf{B}\} \in \Delta$, so $\mathbf{A} \in \mathcal{U}$ or $\mathbf{B} \in \mathcal{U}$. To prove (f), suppose $\sim \forall \mathbf{x}\mathbf{A} \in \mathcal{U}$. Choose τ such that $\mathbf{S}_\tau = \sim \forall \mathbf{x}\mathbf{A}$. Then $\mathcal{S}_\tau \cup \{\sim \forall \mathbf{x}\mathbf{A}\} \subseteq \mathcal{U} \in \Delta$, so $\mathcal{S}_\tau \cup \{\sim \forall \mathbf{x}\mathbf{A}\} \in \Delta$. Thus $\mathcal{S}_{\tau+1} = \mathcal{S}_\tau \cup \left\{\sim \forall \mathbf{x}\mathbf{A},\ \sim \mathsf{S}^{\mathbf{x}}_{\mathbf{c}}\mathbf{A}\right\}$ for some individual constant $\mathbf{c}$, so $\sim \mathsf{S}^{\mathbf{x}}_{\mathbf{c}}\mathbf{A} \in \mathcal{S}_{\tau+1} \subseteq \mathcal{U}$.

We next define an interpretation $\mathcal{M} = \langle \mathcal{D}, \mathcal{J} \rangle$. Let $\mathcal{D}$ be the set of variable-free terms of $\mathcal{F}$. Thus, $\mathcal{C} \subseteq \mathcal{D}$, so $\text{card}(\mathcal{D}) = \text{card}(\mathcal{L}(\mathcal{F}))$, and the interpretation will be frugal. For each function constant $\mathbf{f}^n$, if $\mathbf{t}_1, \ldots, \mathbf{t}_n$ are in $\mathcal{D}$, let $(\mathcal{J}\mathbf{f}^n)\mathbf{t}_1, \ldots, \mathbf{t}_n = \mathbf{f}^n\mathbf{t}_1 \ldots \mathbf{t}_n$, and if $\mathbf{a}$ is an individual constant, let $\mathcal{J}\mathbf{a} = \mathbf{a}$. (Thus variable-free terms denote themselves.) For each predicate constant $\mathbf{P}^n$ of $\mathcal{F}$, if $\mathbf{t}_1, \ldots, \mathbf{t}_n$ are in $\mathcal{D}$, let $(\mathcal{J}\mathbf{P}^n)\mathbf{t}_1, \ldots, \mathbf{t}_n = \mathsf{T}$ iff $\mathbf{P}^n\mathbf{t}_1 \ldots \mathbf{t}_n \in \mathcal{U}$. Similarly, if $\mathbf{p}$ is a propositional constant, $\mathcal{J}\mathbf{p} = \mathsf{T}$ iff $\mathbf{p} \in \mathcal{U}$.

Clearly $\mathcal{V}^{\mathcal{M}}\mathbf{t} = \mathbf{t}$ for each variable-free term $\mathbf{t}$, and for each atomic sentence $\mathbf{E}$, $\mathcal{M} \models \mathbf{E}$ iff $\mathbf{E} \in \mathcal{U}$. We show that by induction on the number of occurrences of $\sim$, $\vee$, and $\forall$ in $\mathbf{E}$ that if $\mathbf{E} \in \mathcal{U}$, then $\mathcal{M} \models \mathbf{E}$. (Note Exercise X2500). We consider the following cases:

(a) $\mathbf{E}$ is $\sim\mathbf{A}$ for some atomic sentence $\mathbf{A}$. Since $\mathbf{E} \in \mathcal{U}$, $\mathbf{A}$ cannot be in $\mathcal{U}$, so $\mathcal{V}^{\mathcal{M}}\mathbf{A} = \mathsf{F}$. Hence $\mathcal{M} \models \mathbf{E}$.
(b) $\mathbf{E}$ is $\sim\sim \mathbf{A}$. Then $\mathbf{A} \in \mathcal{U}$, so $\mathcal{M} \models \mathbf{A}$ by inductive hypothesis, so $\mathcal{M} \models \mathbf{E}$.
(c) $\mathbf{E}$ is $[\mathbf{A} \vee \mathbf{B}]$. Then $\mathbf{A} \in \mathcal{U}$ or $\mathbf{B} \in \mathcal{U}$, so by inductive hypothesis $\mathcal{M} \models \mathbf{A}$ or $\mathcal{M} \models \mathbf{B}$. Hence $\mathcal{M} \models \mathbf{E}$ in either case.

(d) $\mathbf{E}$ is $\sim[\mathbf{A} \vee \mathbf{B}]$. Then $\sim \mathbf{A} \in \mathcal{U}$ and $\sim \mathbf{B} \in \mathcal{U}$ so $\mathcal{M} \models \sim \mathbf{A}$ and $\mathcal{M} \models \sim \mathbf{B}$ by inductive hypothesis, so $\mathcal{V}^{\mathcal{M}}\mathbf{A} = \mathsf{F} = \mathcal{V}^{\mathcal{M}}\mathbf{B}$, so $\mathcal{V}^{\mathcal{M}}[\mathbf{A} \vee \mathbf{B}] = \mathsf{F}$ and $\mathcal{M} \models \mathbf{E}$.

(e) $\mathbf{E}$ is $\forall\mathbf{x}\mathbf{A}$. Let ψ be any assignment, and let $\mathbf{t} = \psi\mathbf{x}$. Then $\mathsf{S}^{\mathbf{x}}_{\cdot\mathbf{t}}\mathbf{A} \in \mathcal{U}$, so by inductive hypothesis and the Substitution-Value Theorem (2301), $\mathsf{T} = \mathcal{V}^{\mathcal{M}}\mathsf{S}^{\mathbf{x}}_{\cdot\mathbf{t}}\mathbf{A} = \mathcal{V}^{\mathcal{M}}_{\psi}\mathbf{A}$. Hence $\mathcal{M} \models \mathbf{E}$.

(f) $\mathbf{E}$ is $\sim \forall\mathbf{x}\mathbf{A}$. Then there is an individual constant $\mathbf{c}$ (which is, of course, a variable-free term) such that $\sim \mathsf{S}^{\mathbf{x}}_{\cdot\mathbf{c}}\mathbf{A} \in \mathcal{U}$, so $\mathcal{M} \models \sim \mathsf{S}^{\mathbf{x}}_{\cdot\mathbf{c}}\mathbf{A}$ by inductive hypothesis, so $\mathcal{V}^{\mathcal{M}}\mathsf{S}^{\mathbf{x}}_{\cdot\mathbf{c}}\mathbf{A} = \mathsf{F}$. $\models \forall\mathbf{x}\mathbf{A} \supset \mathsf{S}^{\mathbf{x}}_{\cdot\mathbf{c}}\mathbf{A}$ by 2303, so $\mathcal{V}^{\mathcal{M}}\forall\mathbf{x}\mathbf{A} = \mathsf{F}$, so $\mathcal{M} \models \mathbf{E}$.

Since $\mathcal{S} \subseteq \mathcal{U}$, every member of $\mathcal{S}$ is valid in $\mathcal{M}$, so we see that $\mathcal{M}$ is a frugal model of $\mathcal{S}$. ■

REMARK. It is natural to ask whether the purity condition in Theorem 2507 is really necessary. To see that it is, consider a formulation of $\mathcal{F}$ with k individual constants. Let p be a fixed positive integer greater than k, and let $\mathcal{S}$ be the set of all sentences of the form $\sim \forall\mathbf{x}_1 \ldots \forall\mathbf{x}_m\mathbf{B}(\mathbf{x}_1, \ldots, \mathbf{x}_m, \mathbf{c}_1, \ldots, \mathbf{c}_n)$, where $n \geq 0$, $m + n \geq p$, $\mathbf{x}_1, \ldots, \mathbf{x}_m$ are distinct variables, $\mathbf{c}_1, \ldots, \mathbf{c}_n$ are distinct individual constants, and $\mathbf{B}(\mathbf{x}_1, \ldots, \mathbf{x}_m, \mathbf{c}_1, \ldots \mathbf{c}_n)$ is a quantifier-free wff in which $\mathbf{x}_1, \ldots, \mathbf{x}_m, \mathbf{c}_1, \ldots, \mathbf{c}_n$ all occur. Notice that $m \geq p - k > 0$, since $m + n \geq p$ and $k \geq n$. Let Γ be the class of all subsets of $\mathcal{S}$.

Note that if $\sim \forall\mathbf{x}\mathbf{A} \in \mathcal{S}$, and $\mathbf{c}$ is any individual constant which does not occur in $\mathbf{A}$, then $\sim \mathsf{S}^{\mathbf{x}}_{\cdot\mathbf{c}}\mathbf{A} \in \mathcal{S}$. (It is true vacuously in the case of a sentence such as $\sim \forall\mathbf{x}_1\mathbf{B}(\mathbf{x}_1, \mathbf{c}_1, \ldots, \mathbf{c}_k)$, since there is no individual constant which does not occur in this sentence.) Therefore Γ is an abstract consistency class, and $\mathcal{S} \in \Gamma$. However, $\mathcal{S}$ has no model, since it contains unsatisfiable sentences of the form $\sim \forall\mathbf{x}_1 \ldots \forall\mathbf{x}_m[\mathbf{C} \vee \sim \mathbf{C}]$.

Note that if $p \leq k$, the counterexample fails, so when p is held fixed there is a sense in which one can make Γ fail to be an abstract consistency class by adding at least $p - k$ new individual constants to the language. However, fewer than $p - k$ will not suffice.

REMARK. For many applications of Smullyan's Unifying Principle one can simply add constants to the language without causing any essential disruption of the situation, and thus satisfy the purity condition. In such cases we shall often assume without discussion that this is done, and omit explicit mention of the purity condition.

2508 Theorem. Every consistent set of sentences has a frugal model.[9]

Proof: Let $\mathcal{S}$ be a consistent set of sentences of $\mathcal{F}$. Expand $\mathcal{F}$ by adding to it card($\mathcal{L}(\mathcal{F})$) individual constants; let the expanded formulation of $\mathcal{F}$ be called $\mathcal{F}^+$. $\mathcal{S}$ is consistent in $\mathcal{F}^+$ by 2503. Thus $\mathcal{S}$ has a frugal model by 2505 and 2507. ∎

2509 Generalized Completeness Theorem. If $\mathcal{S}$ is a set of sentences and $\mathbf{A}$ is valid in all frugal models of $\mathcal{S}$, then $\mathcal{S} \vdash \mathbf{A}$.

Proof: Let $\overline{\mathbf{A}}$ be the universal closure of $\mathbf{A}$. Suppose $\mathbf{A}$ is valid in all frugal models of $\mathcal{S}$. Then $\overline{\mathbf{A}}$ is too, so $\mathcal{S} \cup \{\sim \overline{\mathbf{A}}\}$ has no frugal model. Therefore $\mathcal{S} \cup \{\sim \overline{\mathbf{A}}\}$ is inconsistent by 2508, so $\mathcal{S} \vdash \sim\sim \overline{\mathbf{A}}$ by 2501, so $\mathcal{S} \vdash \mathbf{A}$ by Rule P and ∀I. ∎

It should be noted that Theorem 2508 and the Generalized Completeness Theorem (2509) are equivalent in the sense that each can easily be derived from the other. To prove 2508 from 2509, suppose that $\mathcal{S}$ is a set of sentences which has no frugal model. Then $[p \wedge \sim p]$ is valid in all frugal models of $\mathcal{S}$, so $\mathcal{S} \vdash p \wedge \sim p$ by 2509, so $\mathcal{S}$ is inconsistent.

2510 Gödel's Completeness Theorem [Gödel, 1930]. Every valid wff of $\mathcal{F}$ is a theorem.

Proof: Let $\mathcal{S} = \emptyset$ in 2509.

REMARK. Since we have shown that the axiom schemata and rules of inference of $\mathcal{F}$ are independent, it is clear that each of them must be used in some way in any proof of the Completeness Theorem. The reader may wish to note where each of them has been used in the proof above.

2511 Löwenheim's Theorem [Löwenheim, 1915]. If $\mathbf{A}$ is valid in every countable interpretation, then $\mathbf{A}$ is valid.

Proof: We may regard $\mathbf{A}$ as a wff of a countable formulation of $\mathcal{F}$. If $\mathbf{A}$ is valid in every countable interpretation (model of $\emptyset$), then $\vdash \mathbf{A}$ by 2509, so $\models \mathbf{A}$ by 2303. ∎

NOTE: By 2303 and 2510, the theorems of $\mathcal{F}$ are precisely the valid wffs.

[9] This theorem was proved in [Gödel, 1930] but the proof given here (which is mostly embodied in the proof of Theorem 2507) is based on the approach introduced in [Henkin, 1949].

Thus we have two radically different characterizations of this set of wffs, one syntactic, and one semantic. The next theorem also shows the equivalence of a syntactic and a semantic notion.

2512 Theorem. A set of sentences is consistent iff it has a model.

Proof: by 2304 and 2508. ∎

2513 Löwenheim-Skolem Theorem [Skolem, 1920]. Every set of sentences which has a model has a frugal model.

Proof: By 2512 and 2508. ∎

The Löwenheim-Skolem theorem has a rather puzzling consequence which is known as Skolem's Paradox. Set Theory can be axiomatized within first-order logic, and this is customarily done so that one can prove the existence of infinite sets, the existence of the power set $\mathcal{P}(S)$ (collection of all subsets) of each set S, and Cantor's Theorem (which is discussed in a type-theoretic formulation in exercise X5304) that $\mathcal{P}(S)$ always has a greater cardinality than S. Of course, this means that one can prove within presumably consistent formulations of Axiomatic Set Theory the existence of uncountably infinite sets, although 2512 and 2513 imply that such a theory must have a countable model. The paradox is best understood when one has a clear notion of a non-standard model, so we postpone further discussion of it to §55 (following theorem 5503).

2514 Compactness Theorem . A set $\mathcal{S}$ of sentences has a model iff every finite subset of $\mathcal{S}$ has a model.

Proof: Clearly any model for $\mathcal{S}$ is a model for each subset of $\mathcal{S}$. If $\mathcal{S}$ has no model, then by 2508 $\mathcal{S}$ is inconsistent, so if $\mathbf{A}$ is any wff, $\mathcal{S} \vdash \mathbf{A} \land \sim \mathbf{A}$. Hence there is a finite subset $\mathcal{H}$ of $\mathcal{S}$ such that $\mathcal{H} \vdash \mathbf{A} \land \sim \mathbf{A}$. Thus $\mathcal{H}$ is inconsistent, and so by 2512 has no model. ∎

APPLICATION: An interesting application of the Compactness Theorem permits an extension of the celebrated Four Color Theorem [Appel and Haken, 1976] of graph theory from finite to infinite graphs.

A simple *graph* consists of a set V of *vertices* and a set of E of *edges*, which are unordered pairs of vertices. Vertices u and v are said to be *adjacent* iff $(u, v) \in E$. The graph is *finite* iff V is finite, and it is *planar* iff it can be embedded in the plane so that its edges (represented as lines) never cross.

(We refer the reader to a text on graph theory for more details.) Let C be a set of four colors, which we shall refer to as the first, second, third, and fourth color. A 4-*coloring* of a graph is an assignment of colors (i.e., members of C) to vertices so that each vertex is assigned exactly one color, and adjacent vertices never receive the same color. The Four Color Theorem says that every planar graph has a 4-coloring. We shall show that once one knows that all finite planar graphs have 4-colorings, one also knows this for infinite planar graphs.

Let an infinite planar graph $\mathcal{G}$ be given. We shall introduce a first-order theory $\mathcal{F}_{\mathcal{G}}$ to describe certain aspects of $\mathcal{G}$. For each vertex of $\mathcal{G}$ we shall have an individual constant to denote that vertex. We shall have four monadic predicate constants C_1, C_2, C_3, and C_4, and $C_i\mathbf{a}$ will mean that vertex $\mathbf{a}$ has color i. We shall use the notation $\bigvee_{j=1}^{4} C_j x$ as an abbreviation for $[C_1 x \lor C_2 x \lor C_3 x \lor C_4 x]$, and we shall use obvious extensions of this notation.

Let $\mathcal{S}$ be the set which contains the following sentences:

$$\forall x \bigvee_{j=1}^{4} C_j x$$

$$\forall x \bigwedge_{1 \le j < k \le 4} \sim [C_j x \land C_k x]$$

$$\bigwedge_{j=1}^{4} \sim [C_j \mathbf{a} \land C_j \mathbf{b}],$$

where $\mathbf{a}$ and $\mathbf{b}$ are any constants denoting adjacent vertices.

In the intended interpretation, the vertices of $\mathcal{G}$ are the individuals and the sentences of $\mathcal{S}$ express the statements that every vertex has a color, that every vertex has only one color, and that adjacent vertices do not have the same color. Of course, $\mathcal{S}$ will be infinite if $\mathcal{G}$ has infinitely many edges.

Let $\mathcal{S}'$ be any finite subset of $\mathcal{S}$, and let $\mathcal{G}'$ be the subgraph of $\mathcal{G}$ which contains those vertices whose names occur in sentences of $\mathcal{S}'$. Since $\mathcal{G}'$ is finite and planar, it has a 4-coloring, and if one interprets the constants of $\mathcal{F}_{\mathcal{G}}$ as indicated above with respect to this coloring, the sentences of $\mathcal{S}'$ will be true in the interpretation where the domain of individuals is the set of vertices of $\mathcal{G}'$. Thus every finite subset of $\mathcal{S}$ has a model, so by the Compactness Theorem, $\mathcal{S}$ has a model. If one assigns colors to vertices so that a vertex with name $\mathbf{a}$ is given the unique color j such that $C_j\mathbf{a}$ is true in this model, one obtains a 4-coloring of the entire graph $\mathcal{G}$.

APPLICATION: Another important application of the Compactness Theorem leads to nonstandard number theory and nonstandard analysis. Let $\mathcal{T}$ be a first-order theory which is a formalization of the theory of natural numbers. Thus, $\mathcal{T}$ may have constants denoting certain natural numbers, certain functions of natural numbers (such as successor, addition, and multiplication), and certain properties of natural numbers and relations between natural numbers. Also, $\mathcal{T}$ will have postulates (perhaps Peano's Postulates[10] and definitions of the functions and relations denoted by constants of $\mathcal{T}$) which are true in a model $\mathcal{N}$ which has the set of natural numbers as its domain of individuals. Let us assume that each natural number n is denoted by some variable-free term $\overline{n}$, and that the equality relation is definable by some wff of $\mathcal{T}$. Let $\mathcal{T}^*$ be the expansion and extension of $\mathcal{T}$ obtained by adding a new individual constant c and the set of postulates $\{\sim c = \overline{n} \mid n \text{ is a natural number}\}$. It is easy to see that every finite subset of the postulates of $\mathcal{T}^*$ has a model (simply use $\mathcal{N}$ with an appropriate interpretation for c), so by the Compactness Theorem $\mathcal{T}^*$ has a model $\mathcal{N}^*$, which is called a *nonstandard model for number theory.* Since the denotation of c in $\mathcal{N}^*$ is not any finite number, $\mathcal{N}^*$ may be said to contain "infinite" numbers. Of course, $\mathcal{N}^*$ is a model of $\mathcal{T}$, so we see that all formalizations of number theory have nonstandard models .

We may even abstractly consider a theory $\mathcal{T}$ which has constants to denote all functions and relations on the set of natural numbers, whose postulates are all the sentences of $\mathcal{T}$ which are true in $\mathcal{N}$. (It will be seen that such a set of postulates cannot be effectively specified, but that is irrelevant for our present purposes.) Then every true statement about natural numbers expressible in $\mathcal{T}$ will also be true in the nonstandard model $\mathcal{N}^*$. This rather paradoxical situation may be regarded as revealing an inadequacy in the expressive power of first order logic.

Similar considerations apply to formalizations of other infinite mathematical systems. For another example, let $\mathcal{T}$ be a first-order theory for the real numbers, and let $\mathcal{T}^*$ be obtained by adding to $\mathcal{T}$ a new constant c and the set of postulates $\{0 < c \wedge c < \overline{r} \mid r \text{ is a positive real number}\}$. $\mathcal{T}^*$ will have a model in which there are numbers, called *infinitesimals*, which are smaller than all standard real numbers, but greater than zero.

---

[10]Properly speaking, the Induction Axiom cannot be expressed in first-order logic (unless on uses some device such as embedding two-sorted logic or axiomatic set theory into first-order logic). Nevertheless, its first-order consequences can be expressed by the infinite set of wffs of the form $[\mathbf{A}(\overline{0}) \wedge \forall \mathbf{x}[\mathbf{A}(\mathbf{x}) \supset \mathbf{A}(\overline{s}\mathbf{x})] \supset \forall \mathbf{x}\, \mathbf{A}(\mathbf{x})]$, where $\overline{0}$ is a constant denoting the number zero, $\overline{s}$ is a function constant denoting the successor function, and $\mathbf{A}(\mathbf{x})$ is a wff in which $\mathbf{x}$ is free.

We can now easily extend 2503 by using a semantic, rather than syntactic, proof.

2515 Proposition. Let $\mathcal{F}^1$ and $\mathcal{F}^2$ be formulations of $\mathcal{F}$, so that $\mathcal{F}^2$ is an expansion of $\mathcal{F}^1$ obtained by adding additional constants (of any type) to the set of constants of $\mathcal{F}^1$. Let $\mathcal{G}$ be any set of sentences of $\mathcal{F}^1$. Then

(1) $\mathcal{F}^2 \cup \mathcal{G}$ is a conservative extension of $\mathcal{F}^1 \cup \mathcal{G}$.
(2) $\mathcal{G}$ is consistent in $\mathcal{F}^1$ iff it is consistent in $\mathcal{F}^2$.

Proof: Let $\mathbf{C}$ be any wff of $\mathcal{F}^1$. Clearly if $\vdash_{\mathcal{F}^1 \cup \mathcal{G}} \mathbf{C}$, then $\vdash_{\mathcal{F}^2 \cup \mathcal{G}} \mathbf{C}$. Suppose $\vdash_{\mathcal{F}^2 \cup \mathcal{G}} \mathbf{C}$, and let $\mathcal{M}$ be any model of $\mathcal{F}^1 \cup \mathcal{G}$. $\mathcal{M}$ can be extended to a model $\mathcal{M}'$ of $\mathcal{F}^2 \cup \mathcal{G}$ simply by providing arbitrary interpretations for the new constants of $\mathcal{F}^2$. $\mathcal{M}' \models \mathbf{C}$ by 2303, so $\mathcal{M} \models \mathbf{C}$, since the truth value of a wff is not affected by constants which do not occur in it. Thus $\mathbf{C}$ is valid in all models of $\mathcal{F}^1 \cup \mathcal{G}$, so $\vdash_{\mathcal{F}^1 \cup \mathcal{G}} \mathbf{C}$ by 2509. This proves (1).

(2) follows from (1) and 2502. ∎

§25A. Supplement: Simplified Completeness Proof

In this section we provide a direct proof of Theorem 2508 for countable formulations of $\mathcal{F}$, bypassing Theorems 2504-2507 and the associated definitions.

2508A Theorem. Every consistent set of sentences of a countable formulation of $\mathcal{F}$ has a countable model.

Proof: Let $\mathcal{S}$ be a consistent set of sentences of $\mathcal{F}$. The theorem is trivial if $\mathcal{S}$ is empty, so we may assume $\mathcal{S}$ is non-empty. Expand $\mathcal{F}$ by adding to it countably many individual constants; let the expanded formulation of $\mathcal{F}$ be called $\mathcal{F}^+$. Enumerate the sentences of $\mathcal{F}^+$, and for each index τ, let $\mathbf{S}_\tau$ be the τ-th sentence in this enumeration.

Let $\mathcal{C}$ be a countable set of individual constants which do not occur in sentences of $\mathcal{S}$. By induction, for each index τ we define a set $\mathcal{S}_\tau$ of sentences so that $\mathcal{S}_\sigma \subseteq \mathcal{S}_\tau$ whenever $\sigma \leq \tau$, and for each τ only finitely many constants from $\mathcal{C}$ occur in the sentences of $\mathcal{S}_\tau$. $\mathcal{S}_0 = \mathcal{S}$. The definition for successor indices $\tau + 1$ is by cases:

(1) If $\mathcal{S}_\tau \cup \{\mathbf{S}_\tau\}$ is not consistent, then $\mathcal{S}_{\tau+1} = \mathcal{S}_\tau$.

(2) If $\mathcal{S}_\tau \cup \{\mathbf{S}_\tau\}$ is consistent, and $\mathbf{S}_\tau$ is not of the form $\sim \forall \mathbf{x}\mathbf{A}$, then $\mathcal{S}_{\tau+1} = \mathcal{S}_\tau \cup \{\mathbf{S}_\tau\}$.

(3) If $\mathcal{S}_\tau \cup \{\mathbf{S}_\tau\}$ is consistent and $\mathbf{S}_\tau$ is $\sim\forall\mathbf{x}\mathbf{A}$, then $\mathcal{S}_{\tau+1} = \mathcal{S}_\tau \cup \left\{\sim \forall\mathbf{x}\mathbf{A},\ \sim \mathsf{S}^{\mathbf{x}}_{\mathbf{c}}\mathbf{A}\right\}$, where $\mathbf{c}$ is the first individual constant from $\mathcal{C}$ which does not occur in any sentence of $\mathcal{S}_\tau \cup \{\mathbf{S}_\tau\}$. (Note that there must be such a constant.)

It is readily seen by induction that only finitely many constants from $\mathcal{C}$ occur in the sentences of each $\mathcal{S}_\tau$.

We next show by induction that $\mathcal{S}_\tau$ is consistent for each τ. $\mathcal{S}_0 = \mathcal{S}$, which is consistent in $\mathcal{F}^+$ by Theorem 2503. It is clear that if $\mathcal{S}_\tau$ is consistent, then $\mathcal{S}_{\tau+1}$ is consistent in cases (1) and (2); we must show that this is also true in case (3). Suppose $\mathcal{S}_\tau \cup \{\sim \forall\mathbf{x}\mathbf{A}\}$ is consistent but $\mathcal{S}_\tau \cup \{\sim \forall\mathbf{x}\mathbf{A},\ \sim \mathsf{S}^{\mathbf{x}}_{\mathbf{c}}\mathbf{A}\}$ is inconsistent, where $\mathbf{c}$ is an individual constant which does not occur in $\mathcal{S}_\tau$ or $\mathbf{A}$. Then $\mathcal{S}_\tau \cup \{\sim \forall\mathbf{x}\mathbf{A}\} \vdash \mathsf{S}^{\mathbf{x}}_{\mathbf{c}}\mathbf{A}$ by Theorem 2501 and Rule P, so there is a finite subset $\mathcal{H}$ of $\mathcal{S}_\tau \cup \{\sim \forall\mathbf{x}\mathbf{A}\}$ such that $\mathcal{H} \vdash \mathsf{S}^{\mathbf{x}}_{\mathbf{c}}\mathbf{A}$. Given a proof $\mathcal{P}$ of $\mathsf{S}^{\mathbf{x}}_{\mathbf{c}}\mathbf{A}$ from $\mathcal{H}$, let $\mathbf{y}$ be an individual variable which does not occur in $\mathcal{P}$ or in $\mathcal{H}$. Since $\mathbf{c}$ does not occur in $\mathcal{H}$ or $\mathbf{A}$, it can be seen that $\mathsf{S}^{\mathbf{c}}_{\mathbf{y}}\mathcal{P}$ is a proof of $\mathsf{S}^{\mathbf{x}}_{\mathbf{y}}\mathbf{A}$ from $\mathcal{H}$, so $\mathcal{H} \vdash \mathsf{S}^{\mathbf{x}}_{\mathbf{y}}\mathbf{A}$. Hence, $\mathcal{H} \vdash \forall\mathbf{y}\mathsf{S}^{\mathbf{x}}_{\mathbf{y}}\mathbf{A}$ by Generalization, and $\mathcal{H} \vdash \forall\mathbf{x}\mathbf{A}$ by the α Rule, so $\mathcal{S}_\tau \cup \{\sim \forall\mathbf{x}\mathbf{A}\} \vdash \forall\mathbf{x}\mathbf{A}$, so $\mathcal{S}_\tau \cup \{\sim \forall\mathbf{x}\mathbf{A}\}$ is inconsistent. This contradiction proves our assertion.

Let $\mathcal{U} = \bigcup_{\tau<\omega} \mathcal{S}_\tau$. Clearly $\mathcal{U}$ is a set of sentences and $\mathcal{S} \subseteq \mathcal{U}$. We show that $\mathcal{U}$ is consistent. Let $\mathcal{H}$ be any finite subset of $\mathcal{U}$. Then there is a τ such that $\mathcal{H} \subseteq \mathcal{S}_\tau$, and $\mathcal{S}_\tau$ is consistent, so $\mathcal{H}$ is also consistent. Thus every finite subset of $\mathcal{U}$ is consistent, so $\mathcal{U}$ is consistent.

Note that (*) if $\mathbf{D}$ is a sentence such that $\mathcal{U} \cup \{\mathbf{D}\}$ is consistent, then $\mathbf{D} \in \mathcal{U}$. For there is an index τ such that $\mathbf{S}_\tau = \mathbf{D}$, and $\mathcal{S}_\tau \cup \{\mathbf{D}\} \subseteq \mathcal{U} \cup \{\mathbf{D}\}$, which is consistent; thus $\mathcal{S}_\tau \cup \{\mathbf{D}\}$ is consistent. Hence $\mathbf{D} \in \mathcal{S}_{\tau+1} \subseteq \mathcal{U}$, so $\mathbf{D} \in \mathcal{U}$. From (*) it follows that (**) if $\mathcal{U} \vdash \mathbf{D}$, then $\mathbf{D} \in \mathcal{U}$.

We assert that $\mathcal{U}$ has the following properties for all wffs $\mathbf{A}$ and $\mathbf{B}$:

(a) There is no atomic sentence $\mathbf{A}$ such that $\mathbf{A} \in \mathcal{U}$ and $\sim \mathbf{A} \in \mathcal{U}$.

(b) If $\sim\sim \mathbf{A} \in \mathcal{U}$, then $\mathbf{A} \in \mathcal{U}$.

(c) If $[\mathbf{A} \lor \mathbf{B}] \in \mathcal{U}$, then $\mathbf{A} \in \mathcal{U}$ or $\mathbf{B} \in \mathcal{U}$.

(d) If $\sim [\mathbf{A} \lor \mathbf{B}] \in \mathcal{U}$, then $\sim \mathbf{A} \in \mathcal{U}$ and $\sim \mathbf{B} \in \mathcal{U}$.

(e) If $\forall\mathbf{x}\mathbf{A} \in \mathcal{U}$, then $\mathsf{S}^{\mathbf{x}}_{\mathbf{t}}\mathbf{A} \in \mathcal{U}$ for each variable-free term $\mathbf{t}$.

(f) If $\sim \forall\mathbf{x}\mathbf{A} \in \mathcal{U}$, then there is an individual constant $\mathbf{c}$ such that $\sim \mathsf{S}^{\mathbf{x}}_{\mathbf{c}}\mathbf{A} \in \mathcal{U}$.

(a) is true since $\mathcal{U}$ is consistent, (b), (d) and (e) follow from (**). To prove (c), suppose $[\mathbf{A} \vee \mathbf{B}] \in \mathcal{U}$. Then $\mathcal{U} \cup \{\mathbf{A}\}$ is consistent or $\mathcal{U} \cup \{B\}$ is consistent, so $\mathbf{A} \in \mathcal{U}$ or $\mathbf{B} \in \mathcal{U}$ by (*). To prove (f), suppose $\sim \forall \mathbf{xA} \in \mathcal{U}$. Choose τ such that $\mathbf{S}_\tau = \sim \forall \mathbf{xA}$. Then $\mathcal{S}_\tau \cup \{\sim \forall \mathbf{xA}\} \subseteq \mathcal{U}$, which is consistent, so $\mathcal{S}_\tau \cup \{\sim \forall \mathbf{xA}\}$ is consistent. Thus $\mathcal{S}_{\tau+1} = \mathcal{S}_\tau \cup \left\{\sim \forall \mathbf{xA},\ \sim \mathsf{S}^{\mathbf{x}}_{\mathbf{c}}\mathbf{A}\right\}$ for some individual constant $\mathbf{c}$, so $\sim \mathsf{S}^{\mathbf{x}}_{\mathbf{c}}\mathbf{A} \in \mathcal{S}_{\tau+1} \subseteq \mathcal{U}$.

We next define an interpretation $\mathcal{M} = \langle \mathcal{D}, \mathcal{J} \rangle$. Let $\mathcal{D}$ be the set of variable-free terms of $\mathcal{F}^+$. Clearly $\mathcal{D}$ is countable. For each function constant $\mathbf{f}^n$, if $\mathbf{t}_1, \ldots, \mathbf{t}_n$ are in $\mathcal{D}$, let $(\mathcal{J}\mathbf{f}^n)\,\mathbf{t}_1, \ldots, \mathbf{t}_n = \mathbf{f}^n\mathbf{t}_1 \ldots \mathbf{t}_n$, and if $\mathbf{a}$ is an individual constant, let $\mathcal{J}\,\mathbf{a} = \mathbf{a}$. (Thus variable-free terms denote themselves.) For each predicate constant $\mathbf{P}^n$ of $\mathcal{F}^+$, if $\mathbf{t}_1, \ldots, \mathbf{t}_n$ are in $\mathcal{D}$, let $(\mathcal{J}\mathbf{P}^n)\,\mathbf{t}_1 \ldots \mathbf{t}_n = \mathsf{T}$ iff $\mathbf{P}^n\mathbf{t}_1 \ldots \mathbf{t}_n \in \mathcal{U}$. Similarly, if $\mathbf{p}$ is a propositional constant, $\mathcal{J}\mathbf{p} = \mathsf{T}$ iff $\mathbf{p} \in \mathcal{U}$.

Clearly $\mathcal{V}^{\mathcal{M}}\mathbf{t} = \mathbf{t}$ for each variable-free term $\mathbf{t}$, and for each atomic sentence $\mathbf{E}$, $\mathcal{M} \models \mathbf{E}$ iff $\mathbf{E} \in \mathcal{U}$.

We show by induction on the number of occurrences of $\sim$, $\vee$, and $\forall$ in $\mathbf{E}$ that if $\mathbf{E} \in \mathcal{U}$, then $\mathcal{M} \models \mathbf{E}$. We suppose that $E \in \mathcal{U}$, and consider the following cases:

(a) $\mathbf{E}$ is $\sim\mathbf{A}$ for some atomic sentence $\mathbf{A}$. Since $\mathbf{E} \in \mathcal{U}$, $\mathbf{A}$ cannot be in $\mathcal{U}$, so $\mathcal{V}^{\mathcal{M}}\mathbf{A} = \mathsf{F}$. Hence $\mathcal{M} \models \mathbf{E}$.

(b) $\mathbf{E}$ is $\sim\sim\mathbf{A}$. Then $\mathbf{A} \in \mathcal{U}$, so $\mathcal{M} \models \mathbf{A}$ by inductive hypothesis, so $\mathcal{M} \models \mathbf{E}$.

(c) $\mathbf{E}$ is $[\mathbf{A} \vee \mathbf{B}]$. Then $\mathbf{A} \in \mathcal{U}$ or $\mathbf{B} \in \mathcal{U}$, so by inductive hypothesis $\mathcal{M} \models \mathbf{A}$ or $\mathcal{M} \models \mathbf{B}$. Hence $\mathcal{M} \models \mathbf{E}$ in either case.

(d) $\mathbf{E}$ is $\sim[\mathbf{A} \vee \mathbf{B}]$. Then $\sim\mathbf{A} \in \mathcal{U}$ and $\sim\mathbf{B} \in \mathcal{U}$ so $\mathcal{M} \models \sim\mathbf{A}$ and $\mathcal{M} \models \sim\mathbf{B}$ by inductive hypothesis, so $\mathcal{V}^{\mathcal{M}}\mathbf{A} = \mathsf{F} = \mathcal{V}^{\mathcal{M}}B$, so $\mathcal{V}^{\mathcal{M}}[\mathbf{A} \vee \mathbf{B}] = \mathsf{F}$ and $\mathcal{M} \models \mathbf{E}$.

(e) $\mathbf{E}$ is $\forall\mathbf{xA}$. Let ψ be any assignment, and let $\mathbf{t} = \psi\mathbf{x}$. Then $\mathsf{S}^{\mathbf{x}}_{\mathbf{t}}\mathbf{A} \in \mathcal{U}$, so by inductive hypothesis and the Substitution-Value Theorem (2302), $\mathsf{T} = \mathcal{V}^{\mathcal{M}}\mathsf{S}^{\mathbf{x}}_{\mathbf{t}}\mathbf{A} = \mathcal{V}^{\mathcal{M}}_{\psi}\mathbf{A}$. Hence $\mathcal{M} \models \mathbf{E}$.

(f) $\mathbf{E}$ is $\sim \forall\mathbf{xA}$. Then there is an individual constant $\mathbf{c}$ (which is, of course, a variable-free term) such that $\sim\mathsf{S}^{\mathbf{x}}_{\mathbf{c}}\mathbf{A} \in \mathcal{U}$, so $\mathcal{M} \models \sim \mathsf{S}^{\mathbf{x}}_{\mathbf{c}}\mathbf{A}$ by inductive hypothesis, so $\mathcal{V}^{\mathcal{M}}\mathsf{S}^{\mathbf{x}}_{\mathbf{c}}\mathbf{A} = \mathsf{F}$. $\models \forall\mathbf{xA} \supset \mathsf{S}^{\mathbf{x}}_{\mathbf{c}}\mathbf{A}$ by 2303, so $\mathcal{V}^{\mathcal{M}}\forall\mathbf{xA} = \mathsf{F}$, so $\mathcal{M} \models \mathbf{E}$.

Since $\mathcal{S} \subseteq \mathcal{U}$, every member of $\mathcal{S}$ is valid in $\mathcal{M}$, so we see that $\mathcal{M}$ is a countable model of $\mathcal{S}$. ■

EXERCISES

X2500. In the proof of Theorem 2507, suppose that Γ (and therefore Δ) is the class of consistent sets of sentences. Show that in this case for each sentence **E**:

(1) $\mathbf{E} \in \mathcal{U}$ or $\sim \mathbf{E} \in \mathcal{U}$.
(2) $\mathbf{E} \in \mathcal{U}$ if and only if $\mathcal{M} \models \mathbf{E}$.

X2501. Let Γ be the class of all finite sets of quantifier-free sentences whose conjunctions are not contradictory (i.e., not substitution instances of contradictions of propositional calculus). Show that Γ is an abstract consistency class.

X2502. Show how to derive Löwenheim's Theorem (2511) from the Löwenheim-Skolem Theorem (2513) without using any other theorems from §25.

X2503. A set $\mathcal{H}$ of sentences of $\mathcal{F}$ is said to be *complete* if for every sentence **A** of $\mathcal{F}$, either $\mathcal{H} \vdash \mathbf{A}$ or $\mathcal{H} \vdash \sim\mathbf{A}$. (Do not confuse this use of the word with other uses.) Show that if $\mathcal{G}$ is any consistent set of sentences, then there is a complete, consistent set $\mathcal{H}$ of sentences such that $\mathcal{G} \subseteq \mathcal{H}$.

X2504. We say that **B** is a (*semantic*) *consequence* of **A** if for every interpretation $\mathcal{M}$ and every assignment ϕ, if $\mathcal{M} \models_\phi \mathbf{A}$ then $\mathcal{M} \models_\phi \mathbf{B}$. Consider the following assertions:

(1) If **B** is a consequence of **A**, then **B** is valid in all models of **A**.
(2) If **B** is valid in all models of **A**, then **B** is a consequence of **A**.

Are these assertions true for all wffs **A** and **B**? If so, prove them; if not, provide counterexamples.

X2505. Show how Theorem 2514 can be used to prove that every partial ordering of a set can be extended to a total ordering of the set, once one knows that this is true for finite sets.

§26. Equality

The system $\mathcal{F}^=$ of *first-order logic with equality* is obtained by adding to the primitive symbols of $\mathcal{F}$ a binary predicate constant $=$ (unless this constant is already present)[11] and adding to the axioms of $\mathcal{F}$ the following axioms (where $\mathbf{s} = \mathbf{t}$ is an abbreviation for $= \mathbf{st}$):

Axiom 6. $x = x$ (Reflexivity)

Axiom Schema 7. $\mathbf{x} = \mathbf{y} \supset \centerdot \mathsf{S}^{\mathbf{z}}_{\mathbf{x}} \supset \mathsf{S}^{\mathbf{z}}_{\mathbf{y}}\mathbf{A}$ (Substitutivity)
where **A** is an atomic wff.

A first order theory is a *first-order theory with equality* if it has a binary predicate = such that the wffs above are theorems of the theory.

It can be seen (Exercise X2601) that the elementary theory $\mathcal{G}$ of additive abelian groups which was presented in §21 is an example of such a theory. It can also be seen (Exercise X2602) that the notion of equality is unambiguous in such a theory.

We shall often write $\mathbf{s} \neq \mathbf{t}$ as an abbreviation for $\sim \mathbf{s} = \mathbf{t}$.

In any first order theory with equality, the following are theorems:

2600. $\vdash x = y \supset y = x$

Proof:

(.1) $\vdash x = y \supset \centerdot x = x \supset y = x$	Axiom 7
(.2) $\vdash x = y \supset y = x$	Rule P: .1, Axiom 6

■

2601. $\vdash x = y \supset \centerdot y = z \supset x = z$

Proof:

(.1) $\vdash y = x \supset \centerdot y = z \supset x = z$	Axiom 7
(.2) $\vdash x = y \supset \centerdot y = z \supset x = z$	Rule P: .1, 2600

■

2602. $\vdash \mathbf{x} = \mathbf{y} \supset \centerdot \left(\mathsf{S}^{\mathbf{z}}_{\cdot\mathbf{x}}\mathbf{A}\right) \equiv \mathsf{S}^{\mathbf{z}}_{\cdot\mathbf{y}}\mathbf{A}$, where **x** and **y** are free for **z** in **A**.

Proof: by induction on the construction of **A**

Case 1. **A** is atomic.

(.1) $\vdash \mathbf{x} = y \supset \centerdot \mathsf{S}^{\mathbf{z}}_{\cdot\mathbf{x}}\mathbf{A} \supset \mathsf{S}^{\mathbf{z}}_{\cdot\mathbf{y}}\mathbf{A}$	Axiom 7

[11]Actually, it is not essential that = be a primitive symbol, since in some first-order theories = can be defined. For example, in Axiomatic Set Theory $x = y$ is sometimes regarded as an abbreviation for $\forall z\,[z \in x \equiv z \in y]$ or for $\forall w\,[x \in w \equiv y \in w]$.

(.2) $\vdash \mathbf{y} = \mathbf{x} \supset \centerdot \mathsf{S}^{\mathbf{z}}_{\mathbf{y}}\mathbf{A} \supset \mathsf{S}^{\mathbf{z}}_{\mathbf{x}}\mathbf{A}$ Axiom 7
(.3) $\vdash \mathbf{x} = \mathbf{y} \supset \mathbf{y} = \mathbf{x}$ Sub: 2600
(.4) $\vdash \mathbf{x} = \mathbf{y} \supset \centerdot \left(\mathsf{S}^{\mathbf{z}}_{\mathbf{x}}\mathbf{A}\right) \equiv \mathsf{S}^{\mathbf{z}}_{\mathbf{y}}\mathbf{A}$ Rule P: .1, .2, .3

The cases where **A** has the form $\sim$**B** or [**B** $\vee$ **C**] follow directly by Rule P from the inductive hypotheses.

Case 2. **A** has the form $\forall$**wB**. If **z** is not free in **A**, this instance of 2602 follows by Rule P, so we may assume **z** is free in **A**. Thus, since **x** and **y** are free for **z** in **A**, they are distinct from **w** and free for **z** in **B**.

$\vdash \mathbf{x} = \mathbf{y} \supset \centerdot \left(\mathsf{S}^{\mathbf{z}}_{\mathbf{x}}\mathbf{B}\right) \equiv \mathsf{S}^{\mathbf{z}}_{\mathbf{y}}B$ by inductive hypothesis
$\vdash \mathbf{x} = \mathbf{y} \supset \forall \mathbf{w} \centerdot \left(\mathsf{S}^{\mathbf{z}}_{\mathbf{x}}\mathbf{B}\right) \equiv \mathsf{S}^{\mathbf{z}}_{\mathbf{y}}\mathbf{B}$ $\supset \forall$ Rule (2103)
$\vdash \mathbf{x} = \mathbf{y} \supset \centerdot \forall \mathbf{w} \left(\mathsf{S}^{\mathbf{z}}_{\mathbf{x}}\mathbf{B}\right) \equiv \forall \mathbf{w} \mathsf{S}^{\mathbf{z}}_{\mathbf{y}}\mathbf{B}$ Rule P: 2107

■

2603. $\vdash \mathbf{x} = \mathbf{y} \supset \left(\mathsf{S}^{\mathbf{z}}_{\mathbf{x}}\mathbf{t}\right) = \left(\mathsf{S}^{\mathbf{z}}_{\mathbf{y}}\mathbf{t}\right)$ for any term **t**.

Proof: Let **w** be an individual variable distinct from **x**, **y**, **z** and the variables in **t**.

(.1) $\vdash \mathbf{x} = \mathbf{y} \supset \centerdot \mathbf{w} = \left(\mathsf{S}^{\mathbf{z}}_{\mathbf{x}}\mathbf{t}\right) \supset \mathbf{w} = \mathsf{S}^{\mathbf{z}}_{\mathbf{y}}\mathbf{t}$ Axiom 7 (**A** is $\mathbf{w} = \mathbf{t}$)
(.2) $\vdash \mathbf{x} = \mathbf{y} \supset \centerdot \left(\mathsf{S}^{\mathbf{z}}_{\mathbf{x}}\mathbf{t}\right) = \left(\mathsf{S}^{\mathbf{z}}_{\mathbf{x}}\mathbf{t}\right) \supset \left(\mathsf{S}^{\mathbf{z}}_{\mathbf{x}}\mathbf{t}\right) = \mathsf{S}^{\mathbf{z}}_{\mathbf{y}}\mathbf{t}$ Sub: .1
(.3) $\vdash \mathbf{x} = \mathbf{y} \supset \left(\mathsf{S}^{\mathbf{z}}_{\mathbf{x}}\mathbf{t}\right) = \mathsf{S}^{\mathbf{z}}_{\mathbf{y}}\mathbf{t}$ Sub: Axiom 6; Rule P: .2

■

2604. $\vdash x_1 = y_1 \wedge \ldots \wedge x_n = y_n \supset \mathbf{f}x_1 \ldots x_n = \mathbf{f}y_1 \ldots y_n$ for any n-ary function symbol **f**.

Proof:

(a) $\vdash x_1 = y_1 \supset \mathbf{f}x_1x_2 \ldots x_n = \mathbf{f}y_1x_2 \ldots x_n$ 2603

We show by induction on i that for $1 \leq i \leq n$,

(b$_i$) $\vdash x_1 = y_1 \wedge \ldots \wedge x_n = y_n \supset \mathbf{f}x_1 \ldots x_n = \mathbf{f}y_1 \ldots y_i x_{i+1} \ldots x_n$

The basis (b_1) of the induction follows from (a) by Rule P. For the induction step we assume (b_i) as inductive hypothesis.

(c) $\vdash x_{i+1} = y_{i+1} \supset \mathbf{f}y_1 \ldots y_i x_{i+1} x_{i+2} \ldots x_n = \mathbf{f}y_1 \ldots y_i y_{i+1} x_{i+2} \ldots x_n$ 2603

$(b_{i+1})\ \vdash x_1 = y_1\ \wedge \ldots \wedge\ x_n = y_n\ \supset\ \mathbf{f}x_1 \ldots x_n = \mathbf{f}y_1 \ldots y_{i+1}x_{i+2} \ldots x_n$
Sub: 2601; Rule P: b_i, c

This completes the inductive proof. (b_n) is 2604. ∎

2605. $\vdash x_1 = y_1\ \wedge \ldots \wedge\ x_n = y_n\ \supset \centerdot\ \mathbf{P}x_1 \ldots x_n\ \equiv\ \mathbf{P}y_1 \ldots y_n$ for any n-ary predicate symbol $\mathbf{P}$.

The proof is similar to that of 2604.

2606 Proposition. Let $\mathcal{T}$ be a first-order theory with a binary predicate constant $=$ such that

(a) $\vdash_{\mathcal{T}} x = x$

and for each n-ary function symbol $\mathbf{f}^n$

(b) $\vdash_{\mathcal{T}} x_1 = y_1\ \wedge \ldots \wedge\ x_n = y_n\ \supset\ \mathbf{f}^n x_1 \ldots x_n = \mathbf{f}^n y_1 \ldots y_n$

and for each n-ary predicate symbol $\mathbf{P}^n$

(c) $\vdash_{\mathcal{T}} x_1 = y_1\ \wedge \ldots \wedge\ x_n = y_n\ \supset \centerdot\ \mathbf{P}^n_1 \ldots x_n\ \supset\ \mathbf{P}^n y_1 \ldots y_n$.

Then $\mathcal{T}$ is a first-order theory with equality.

Proof: In this proof we shall let $\vdash \mathbf{A}$ mean $\vdash_{\mathcal{T}} \mathbf{A}$.

We first show by induction on the construction of $\mathbf{t}$ that if $\mathbf{t}$ is any term, then

(d) $\vdash \mathbf{x} = \mathbf{y}\ \supset\ (\mathsf{S}^{\mathbf{z}}_{\mathbf{x}}\mathbf{t}) = \mathsf{S}^{\mathbf{z}}_{\mathbf{y}}\mathbf{t}$. If $\mathbf{t}$ is an individual constant or variable, then (d) is easily proved using (a), Sub, and Rule P. Suppose $\mathbf{t}$ is $\mathbf{f}^n\mathbf{t}_1 \ldots \mathbf{t}_n$. For $1 \le i \le n$, let $\mathbf{t}'_i$ be $\mathsf{S}^{\mathbf{z}}_{\mathbf{x}}\mathbf{t}_i$ and $\mathbf{t}''_i$ be $\mathsf{S}^{\mathbf{z}}_{\mathbf{y}}\mathbf{t}_i$. Then

(e$_i$) $\vdash \mathbf{x} = \mathbf{y}\ \supset\ \mathbf{t}'_i = \mathbf{t}''_i$ — by inductive hypothesis

(f) $\vdash \mathbf{t}'_1 = \mathbf{t}''_1\ \wedge \ldots \wedge\ \mathbf{t}'_n = \mathbf{t}''_n\ \supset\ \mathbf{f}^n\mathbf{t}'_1 \ldots \mathbf{t}'_n = \mathbf{f}^n\mathbf{t}''_1 \ldots \mathbf{t}''_n$ — Sub: (b)

(g) $\vdash \mathbf{x} = \mathbf{y}\ \supset\ \mathbf{f}^n\mathbf{t}'_1 \ldots \mathbf{t}'_n = \mathbf{f}^n\mathbf{t}''_1 \ldots \mathbf{t}''_n$ — Rule P: (e$_1$)-(e$_n$), (f)

This completes the inductive proof of (d).

Now let $\mathbf{A}$ be an atomic wff $\mathbf{P}^n\mathbf{t}_1 \ldots \mathbf{t}_n$.

(h$_i$) $\vdash \mathbf{x} = \mathbf{y}\ \supset\ \left(\mathsf{S}^{\mathbf{z}}_{\mathbf{x}}\mathbf{t}_i\right) = \mathsf{S}^{\mathbf{z}}_{\mathbf{y}}\mathbf{t}_i$ — (d)

(j) $\vdash \mathbf{x} = \mathbf{y}\ \supset \centerdot\ \mathsf{S}^{\mathbf{z}}_{\mathbf{x}}\mathbf{A}\ \supset\ \mathsf{S}^{\mathbf{z}}_{\mathbf{y}}\mathbf{A}$ — Sub: (c); Rule P: (h$_1$)-(h$_n$)

Since (j) is Axiom Schema 7, this completes the proof. ■

DEFINITION. In any first order theory with equality,

$$\exists_1 \mathbf{x} \mathbf{A} \text{ stands for } \exists \mathbf{x} \mathbf{A} \ \wedge \ \forall \mathbf{x} \, \forall \mathbf{y} \left[\mathbf{A} \ \wedge \ \left(\mathsf{S}^{\mathbf{x}}_{\mathbf{y}} \mathbf{A} \right) \ \supset \ \mathbf{x} = \mathbf{y} \right],$$

where $\mathbf{y}$ is the first individual variable distinct from $\mathbf{x}$ and all variables which occur in $\mathbf{A}$. Clearly, $\exists_1 \mathbf{x}\ \mathbf{A}$ means "there exists a unique $\mathbf{x}$ such that $\mathbf{A}$". Alternative possible definitions are suggested by the following theorems of any first-order theory with equality:

2607. $\vdash \exists_1 x\ Px \ \equiv \ \exists x \centerdot Px \ \wedge \ \forall y \centerdot Py \ \supset \ x = y$

2608. $\vdash \exists_1 x Px \ \equiv \ \exists y \forall x \centerdot Px \ \equiv \ x = y$

The proofs of 2607 and 2608 are left to the reader.

It follows from Axiom 6, 2600, and 2601 that in any model for a first-order theory with equality, the interpretation of the equality predicate must be an equivalence relation on the domain of individuals. We next see that one can actually require this relation to the identity relation.

DEFINITION. An interpretation [model] $\langle \mathcal{D}, \mathcal{J} \rangle$ for a first-order language with an equality predicate $=$ is an *equality-interpretation* [*model*] iff $\mathcal{J}=$ is the identity relation on $\mathcal{D}$, i.e., for all elements d and e of $\mathcal{D}$, $(\mathcal{J} =)\, de$ is T iff d is e.

2609 Soundness Theorem for Logic with Equality. If $\mathcal{M}$ is an equality model for a set $\mathcal{S}$ of sentences and $\mathcal{S} \vdash_{\mathcal{F}^=} \mathbf{A}$, then $\mathcal{M} \models \mathbf{A}$.

Proof: Let $\mathcal{F}$ be the formulation of first-order logic with $\mathcal{L}(\mathcal{F}) = \mathcal{L}(\mathcal{F}^=)$, and let $\mathcal{E}$ be the set of universal closures of Axioms 6 and 7. We first show that any equality-model $\mathcal{M}$ for $\mathcal{S}$ is a model for $\mathcal{E}$.

Clearly $\mathcal{M} \models$ Axiom 6. To treat Axiom 7, let ϕ be an assignment into $\mathcal{M}$. If $\varphi\mathbf{x} \neq \varphi\mathbf{y}$, then $\mathcal{V}_\varphi[\mathbf{x} = \mathbf{y}] = \mathsf{F}$, so $\mathcal{M} \models_\varphi$ Axiom 7. If $\varphi\mathbf{x} = \varphi\mathbf{y}$, let ψ be that assignment which agrees with φ off $\mathbf{z}$, while $\psi\mathbf{z} = \varphi\mathbf{x} = \varphi\mathbf{y}$. Then by the Substitution-Value Theorem (2302), $\mathcal{V}_\varphi \mathsf{S}^{\mathbf{z}}_{\mathbf{x}} \mathbf{A} = \mathcal{V}_\psi \mathbf{A} = \mathcal{V}_\psi \mathsf{S}^{\mathbf{z}}_{\mathbf{y}} \mathbf{A}$, so $\mathcal{M} \models_\varphi$ Axiom 7.

Thus $\mathcal{M}$ is a model for $\mathcal{S} \cup \mathcal{E}$. Since $\mathcal{S} \vdash_{\mathcal{F}}= \mathbf{A}$, we have $\mathcal{S} \cup \mathcal{E} \vdash_{\mathcal{F}} \mathbf{A}$, so $\mathcal{M} \models \mathbf{A}$ by the Soundness Theorem for first-order logic (2303). ■

DEFINITION. Interpretations $\mathcal{M}$ and $\mathcal{N}$ of $\mathcal{F}$ are *elementarily equivalent* iff for every sentence $\mathbf{A}$ of $\mathcal{F}$, $\mathcal{M} \models \mathbf{A}$ iff $\mathcal{N} \models \mathbf{A}$.

2610 Proposition. Let $\mathcal{M}$ be any model for Axioms 6 and 7. Then there is an equality-model $\mathcal{N}$ for Axioms 6 and 7 such that the cardinality of $\mathcal{N}$ is $\leq$ the cardinality of $\mathcal{M}$, and $\mathcal{N}$ is elementarily equivalent to $\mathcal{M}$.

Proof: Let $\mathcal{M} = \langle \mathcal{D}, \mathcal{I} \rangle$. Let $\mathcal{E}$ be the set of equivalence classes of $\mathcal{D}$ under the equivalence relation $(\mathcal{I} =)$. Define an interpretation function $\mathcal{J}$ into $\mathcal{E}$ as follows. If $\mathbf{a}$ is an individual constant, $\mathcal{J}\mathbf{a}$ is the equivalence class $[\mathcal{I}\mathbf{a}]$ containing $\mathcal{I}\mathbf{a}$. If $\mathbf{f}^n$ is any n-ary function constant, $\mathbf{P}^n$ is any n-ary predicate constant, and $d_1, \ldots, d_n$ are elements of $\mathcal{D}$, then $(\mathcal{J}\mathbf{f}^n)\,[d_1]\ldots[d_n]$ is $[(\mathcal{I}\mathbf{f}^n)\,d_1\ldots d_n]$ and $(\mathcal{J}\ \mathbf{P}^n)\,[d_1]\ldots[d_n]$ is $(\mathcal{I}\ \mathbf{P}^n)\,d_1\ldots d_n$. It is easily seen that this definition is independent of the particular representative d_i chosen from the equivalence class $[d_i]$, since $\mathcal{M} \models 2604$ and $\mathcal{M} \models 2605$. Hence if $[d_i] = [e_i]$ for $1 \leq i \leq n$, then $[(\mathcal{I}\mathbf{f}^n)\,d_1\ldots d_n]$ is $[(\mathcal{I}\ \mathbf{f}^n)\,e_1\ldots e_n]$ and $(\mathcal{I}\ \mathbf{P}^n)\,d_1\ldots d_n$ is $(\mathcal{I}\ \mathbf{P}^n)\,e_1\ldots e_n$, so the definition of $\mathcal{J}$ is unambiguous.

Let $\mathcal{N}$ be $\langle \mathcal{E}, \mathcal{J} \rangle$. Then $\mathcal{N}$ is an equality-model, since $(\mathcal{J} =)\,[d]\,[e]$ is T iff $(\mathcal{I} =)\,de$ is T iff $[d]$ is $[e]$. Clearly the cardinality of $\mathcal{E}$ is $\leq$ that of $\mathcal{D}$.

For each assignment φ into $\mathcal{M}$, let $*\varphi$ be that assignment into $\mathcal{N}$ such that $(*\varphi)\,\mathbf{x} = [\varphi\mathbf{x}]$ for each individual variable $\mathbf{x}$. It is easily seen by induction on the construction of $\mathbf{t}$ that for any term $\mathbf{t}$ and assignment φ into $\mathcal{M}$, $\mathcal{V}^{\mathcal{N}}_{*\varphi}\mathbf{t} = [\mathcal{V}^{\mathcal{M}}_{\varphi}\mathbf{t}]$. We show by induction on the construction of $\mathbf{A}$ that $\mathcal{V}^{\mathcal{N}}_{*\varphi}\mathbf{A} = \mathcal{V}^{\mathcal{M}}_{\varphi}\mathbf{A}$ for any assignment φ into $\mathcal{M}$ and any wff $\mathbf{A}$. The cases where $\mathbf{A}$ is atomic or of the form $\sim\mathbf{B}$ or $[\mathbf{B} \lor \mathbf{C}]$ are routine, and details are left to the reader.

Case. $\mathbf{A}$ is $\forall\mathbf{x}\mathbf{B}$.

Suppose $\mathcal{V}^{\mathcal{M}}_{\varphi}\mathbf{A} = \mathsf{F}$. Hence there is an assignment ψ into $\mathcal{M}$ which agrees with φ off $\mathbf{x}$ such that $\mathcal{V}^{\mathcal{M}}_{\psi}\mathbf{B} = \mathsf{F}$. For any variable $\mathbf{y}$ distinct from $\mathbf{x}$, $(*\psi)\,\mathbf{y} = [\psi\mathbf{y}] = [\varphi\mathbf{y}] = (*\varphi)\,\mathbf{y}$, so $*\psi$ agrees with $*\varphi$ off $\mathbf{x}$. By inductive hypothesis $\mathcal{V}^{\mathcal{N}}_{*\psi}\mathbf{B} = \mathcal{V}^{\mathcal{M}}_{\psi}\mathbf{B} = \mathsf{F}$, so $\mathcal{V}^{\mathcal{N}}_{*\varphi}\mathbf{A} = \mathsf{F}$.

Suppose $\mathcal{V}^{\mathcal{N}}_{*\varphi}\mathbf{A} = \mathsf{F}$. Then there is an assignment τ into $\mathcal{N}$ which agrees with $*\varphi$ off $\mathbf{x}$ such that $\mathcal{V}^{\mathcal{N}}_{\tau}\mathbf{B} = \mathsf{F}$. Let ψ be an assignment into $\mathcal{M}$ which agrees with φ off $\mathbf{x}$ such that $\psi\mathbf{x}$ is a member of the equivalence class $\tau\mathbf{x}$. Then $(*\psi)\,\mathbf{x} = [\psi\mathbf{x}] = \tau\mathbf{x}$, and for any variable $\mathbf{y}$ distinct from $\mathbf{x}$, $(*\psi)\,\mathbf{y} = [\psi\mathbf{y}] = [\varphi\mathbf{y}] = (*\varphi)\,\mathbf{y} = \tau\mathbf{y}$, so $*\psi = \tau$. By inductive hypothesis $\mathcal{V}^{\mathcal{M}}_{\psi}\mathbf{B} = \mathcal{V}^{\mathcal{N}}_{*\psi}\mathbf{B} = \mathcal{V}^{\mathcal{N}}_{\tau}\mathbf{B} = \mathsf{F}$, so $\mathcal{V}^{\mathcal{M}}_{\varphi}\mathbf{A} = \mathsf{F}$.

This completes the inductive proof. It immediately follows that $\mathcal{N} \models \mathbf{A}$ iff $\mathcal{M} \models \mathbf{A}$ for any sentence $\mathbf{A}$. ∎

2611 Theorem. Let $\mathcal{G}$ be a set of sentences consistent in $\mathcal{F}^=$. Then $\mathcal{G}$ has a frugal equality-model.

Proof: Let $\mathcal{E}$ be the set of universal closures of Axioms 6 and 7 of $\mathcal{F}^=$. Clearly $\mathcal{G} \cup \mathcal{E}$ is consistent in first-order logic, so $\mathcal{G} \cup \mathcal{E}$ has a frugal model by 2508. This is a model for Axioms 6 and 7, so by 2610 one obtains a frugal equality-model for $\mathcal{G}$. ■

2612 Extended Completeness and Soundness Theorem. Let $\mathcal{S}$ be a set of sentences and $\mathbf{A}$ be a wff of $\mathcal{F}^=$. Then $\mathcal{S} \vdash_{\mathcal{F}^=} \mathbf{A}$ iff $\mathbf{A}$ is valid in all frugal equality-models of $\mathcal{S}$.

Proof: Let $\overline{\mathbf{A}}$ be the universal closure of $\mathbf{A}$. If not $\mathcal{S} \vdash_{\mathcal{F}^=} \mathbf{A}$, then $\mathcal{S} \cup \{\sim \overline{\mathbf{A}}\}$ is consistent in $\mathcal{F}^=$, and has a frugal equality-model by 2611, so $\mathbf{A}$ is not valid in all frugal equality-models of $\mathcal{S}$.

Theorem 2609 completes the argument. ■

2613 Löwenheim-Skolem-Tarski Theorem. If $\mathcal{S}$ is a set of sentences of $\mathcal{F}^=$ which has an infinite equality-model, and K is any cardinal number such that $K \geq \text{card}(\mathcal{L}(\mathcal{F}^=))$, then $\mathcal{S}$ has an equality-model of cardinality K.

Proof: Let $\mathcal{M}$ be an infinite equality-model for $\mathcal{S}$. Let $\mathcal{C}$ be a set of new individual constants of cardinality K, and let $\mathcal{F}^{=+}$ be the expansion of $\mathcal{F}^=$ obtained by adding these constants to the primitive symbols of $\mathcal{F}^=$. Let $\mathcal{H}$ be the set of all sentences of the form $\mathbf{c} \neq \mathbf{d}$, where $\mathbf{c}$ and $\mathbf{d}$ are distinct constants in $\mathcal{C}$. Every finite subset of $\mathcal{S} \cup \mathcal{H}$ has an equality-model, since $\mathcal{M}$ is infinite and the finitely many constants of $\mathcal{C}$ which actually occur in the sentences of this finite subset can be interpreted as distinct individuals of $\mathcal{M}$. Hence every finite subset of $\mathcal{S} \cup \mathcal{H}$ is consistent (in $\mathcal{F}^{=+}$), so $\mathcal{S} \cup \mathcal{H}$ is consistent. Thus by 2611, $\mathcal{S} \cup \mathcal{H}$ has an equality-model $\mathcal{N}$ of cardinality $\leq K$. Since every sentence of $\mathcal{H}$ is true in $\mathcal{N}$, the cardinality of $\mathcal{N}$ cannot be less than K, so it is precisely K. ■

We next show that the problem of provability in $\mathcal{F}^=$ can be reduced to the problem of provability in $\mathcal{F}$.

2614 Proposition. Suppose $\mathcal{F}$ is a formulation of first-order logic such that $\mathcal{L}(\mathcal{F}) = \mathcal{L}(\mathcal{F}^=)$. For any wff $\mathbf{D}$ of $\mathcal{F}$, let $\mathbf{B}$ be the conjunction of all wffs of the form

$$\forall x_1 \ldots \forall x_n \forall y_1 \ldots \forall y_n [x_1 = y_1 \ \wedge \ldots \wedge \ x_n = y_n \ \supset \ \mathbf{f}x_1 \ldots x_n = \mathbf{f}y_1 \ldots y_n],$$

where **f** is a function symbol which occurs in **D**, and let **C** be the conjunction of all wffs of the form

$$\forall x_1 \ldots \forall x_n \forall y_1 \ldots \forall y_n \, [x_1 = y_1 \ \wedge \ldots \wedge \ x_n = y_n \supset \\ \centerdot \ \mathbf{P} x_1 \ldots x_n \supset \mathbf{P} y_1 \ldots y_n] \, ,$$

where **P** is $=$ or a predicate symbol which occurs in **D**. Then $\vdash_{\mathcal{F}^=} \mathbf{D}$ iff $\vdash_{\mathcal{F}} \forall x \, [x = x] \ \wedge \ \mathbf{B} \ \wedge \ \mathbf{C} \ \supset \ \mathbf{D}$.

Proof: Suppose $\vdash_{\mathcal{F}} \forall x \, [x = x] \ \wedge \ \mathbf{B} \ \wedge \ \mathbf{C} \ \supset \ \mathbf{D}$. Then this wff is also a theorem of $\mathcal{F}^=$, and $\vdash_{\mathcal{F}^=} \forall x \, [x = x] \ \wedge \ \mathbf{B} \ \wedge \ \mathbf{C}$ by Axiom 6, 2604, and 2605, so $\vdash_{\mathcal{F}^=} \mathbf{D}$.

Suppose $\vdash_{\mathcal{F}^=} \mathbf{D}$. Let $\mathcal{T}$ be the first-order theory whose only constants are the binary predicate constant $=$ and the constants which occur in **D**, and whose sole postulate is $\forall x \, [x = x] \ \wedge \ \mathbf{B} \ \wedge \ \mathbf{C}$. Let $\mathcal{M}$ be any model for $\mathcal{T}$. $\mathcal{T}$ is a first-order theory with equality by 2606, so by 2610 $\mathcal{T}$ has an equality-model $\mathcal{N}$ elementarily equivalent to $\mathcal{M}$. One can easily extend $\mathcal{N}$ to an equality-interpretation $\mathcal{N}'$ for $\mathcal{F}^=$ by arbitrarily interpreting the additional constants of $\mathcal{F}^=$, so $\mathcal{N}' \models \mathbf{D}$ by 2609. Thus $\mathcal{N} \models \mathbf{D}$ since the truth of **D** depends only on the constants which occur in it, so $\mathcal{M} \models \mathbf{D}$ since $\mathcal{N}$ and $\mathcal{M}$ are elementarily equivalent. Therefore $\vdash_{\mathcal{T}} \mathbf{D}$ by the Completeness Theorem (2510), so $\vdash_{\mathcal{F}} \forall x \, [x = x] \ \wedge \ \mathbf{B} \ \wedge \ \mathbf{C} \ \supset \ \mathbf{D}$ by the Deduction Theorem. ■

EXERCISES

X2600. Prove that $\mathcal{F}^=$ is a conservative extension of $\mathcal{F}$. Assume $=$ is not a predicate symbol of $\mathcal{F}$.

X2601. Show that the elementary theory of $\mathcal{G}$ of abelian groups which was presented in §21 is a first-order theory with equality.

X2602. Show that if a formulation of $\mathcal{F}$ has two binary predicates $\overset{1}{=}$ and $\overset{2}{=}$, and is a first-order theory with equality with respect to both of them, then $\vdash \forall x \forall y \centerdot x \overset{1}{=} y \ \equiv \ x \overset{2}{=} y$.

X2603. Show that there is no consistent set $\mathcal{S}$ of sentences of $\mathcal{F}^=$ such that all models of $\mathcal{S}$ are equality-models.

Prove the following theorems of $\mathcal{F}^=$:

X2604. Theorem 2607.

X2605. Theorem 2608.

X2606. $\forall x\,[fx = ffx] \;\wedge\; \forall x \forall y\,[fx = fy \;\supset\; x = y] \;\supset\; \forall z \exists x \,\blacksquare\, fx = z$

X2607. $\forall x \forall y\,[fx = fy \;\supset\; x = y] \;\wedge\; \exists x\,[\exists u\,[Pu \;\wedge\; fu = x] \;\wedge\; \exists v \,\blacksquare\, Qv \;\wedge$
$fv = x] \;\supset\; \exists w \,\blacksquare\, Pw \;\wedge\; Qw$

X2608. $\forall x \forall y\,[x = y] \;\supset\; \blacksquare\, \exists x Px \;\equiv\; \forall x Px$

X2609. $\exists x \exists y\, x \neq y \;\wedge\; \exists x \exists y \forall z\,[z = x \;\vee\; z = y]$
$\equiv \exists x \exists y \,\blacksquare\, x \neq y \;\wedge\; \forall z \,\blacksquare\, z = x \;\vee\; z = y$

Chapter 3

Provability and Refutability

We now turn our attention to the practical problems of proving theorems of first-order logic. If one has unlimited resources and patience, one can simply enumerate proofs until one finds a proof of the wff under consideration (provided, of course, that it actually is a theorem). As a practical matter, however, one usually wishes to find a proof as quickly and easily as possible, so various methods have been devised to facilitate efficient searches for proofs. We shall discuss some of these methods in this chapter, along with various results related to them.

A *refutation* of a wff $\mathbf{B}$ is a proof that $\mathbf{B}$ is contradictory; this amounts to a proof of $\sim \mathbf{B}$. Thus, one way to prove a wff $\mathbf{A}$ is to refute $\sim \mathbf{A}$. This method is proving a wff $\mathbf{A}$ is often called the indirect method, and it is surprisingly useful, so a number of the procedures we discuss below are refutation procedures rather than proof procedures. Since a refutation of $\sim \mathbf{A}$ provides a proof of $\mathbf{A}$, we shall sometimes speak of refutation procedures as proof procedures.

We say that we refute a set $\mathcal{S}$ of sentences when we drive a contradiction from $\mathcal{S}$. Naturally, such a refutation shows that some finite conjunction of members of $\mathcal{S}$ is contradictory.

§30. Natural Deduction

Proofs of theorems of $\mathcal{F}$ (which is known as a *Hilbert-style* system).are rarely written out in full, since they tend to be long, awkward, and unpleasant to read. In practice, one uses proofs from hypotheses and derived rules of inference of $\mathcal{F}$. In this section we introduce a system $\mathcal{N}$ of *natural deduction*,[1] in which it is possible to give fairly natural and well structured proofs, and express the forms of arguments which arise in mathematical practice. Of course, the rules of inference of $\mathcal{N}$ are all derived rules of inference of $\mathcal{F}$, and this section may be regarded simply as a summary of those derived rules of inference of $\mathcal{F}$ which are most useful for writing our proofs.

Later in this chapter we discuss a variety of methods and logical tools for establishing the validity of wffs of first-order logic. Generally, proofs can be translated from one format into another (though real insight may be required to design algorithms to do this), and once one has found the essential ingredients of a proof, one may wish to express them in a proof which is reasonably comprehensible. The system $\mathcal{N}$ provides one standard for what such a proof should look like.

The exact choice of primitive connectives and quantifiers for this section does not matter very much, but it is natural to take at least $\sim$, $\vee$, $\wedge$, $\supset$, and f (falsehood) as primitive connectives, and both $\forall$ and $\exists$ as primitive quantifiers. f plays the role of a contradiction in indirect proofs.

Various definitions, such as *free for*, must be adjusted in trivial ways to take account of the fact that $\exists$ is now a primitive quantifier, and we leave this task to the reader. To avoid needless redundancy we shall use $\mathbf{M}$ and $\mathbf{N}$ as syntactical variables for wffs or for the "null formula" (empty disjunction). When $\mathbf{M}$ is null, $\mathbf{M} \vee \mathbf{A}$ and $\mathbf{A} \vee \mathbf{M}$ both stand for $\mathbf{A}$. A null formula standing alone may be regarded as an abbreviation for f. In this section we use $\mathcal{H}$ as a syntactical variable for a finite (possibly empty) *sequence* of wffs, and write $\mathcal{H}, \mathbf{A}$ for the sequence obtained by appending $\mathbf{A}$ to the sequence $\mathcal{H}$. As in §21, we may say that $\mathbf{x}$ is not free in $\mathcal{H}$ when we mean that $\mathbf{x}$ is not free in any wff of $\mathcal{H}$.

A *natural deduction proof* consists of a finite sequence of *lines* of the form $\mathcal{H} \vdash \mathbf{A}$. The members of the sequence $\mathcal{H}$ are called *hypotheses*, and the wff $\mathbf{A}$ is called the *assertion* of the line. Each line must be inferred from zero or more preceding lines by one of the following rules of inference.

Rules of Inference

Hypothesis Rule (Hyp). Infer $\mathcal{H} \vdash \mathbf{A}$ whenever $\mathbf{A}$ is a member of $\mathcal{H}$.

Rule for Expanding or Rearranging Hypotheses. From $\mathcal{H}_1 \vdash \mathbf{A}$ infer $\mathcal{H}_2 \vdash \mathbf{A}$, provided that every wff in $\mathcal{H}_1$ is also in $\mathcal{H}_2$.

[1] Natural deduction was introduced in [Gentzen, 1935]. Also see [Quine, 1950], [Prawitz, 1965], and references cited therein.

Deduction Rule (Ded). From $\mathcal{H},\ \mathbf{A} \vdash \mathbf{B}$ infer $\mathcal{H} \vdash \mathbf{A} \supset \mathbf{B}$.

Rule P. From $\mathcal{H} \vdash \mathbf{A}_1, \ldots,$ and $\mathcal{H} \vdash \mathbf{A}_n$ infer $\mathcal{H} \vdash \mathbf{B}$, provided that $[[\mathbf{A}_1 \wedge \ldots \wedge \mathbf{A}_n] \supset \mathbf{B}]$ is tautologous.

Negation Rule (Neg.). From $\mathcal{H} \vdash \mathbf{A}$, infer $\mathcal{H} \vdash \mathbf{B}$, where $\mathbf{A}$ is $\sim \forall \mathbf{x}\mathbf{C}$, $\sim \exists \mathbf{x}\mathbf{C}$, $\forall \mathbf{x} \sim \mathbf{C}$, or $\exists \mathbf{x} \sim \mathbf{C}$, and $\mathbf{B}$ is $\exists \mathbf{x} \sim \mathbf{C}$, $\forall \mathbf{x} \sim \mathbf{C}$, $\sim \exists \mathbf{x}\mathbf{C}$, or $\sim \forall \mathbf{x}\mathbf{C}$, respectively.

Rule of Indirect Proof (IP). From $\mathcal{H},\ \sim \mathbf{A} \vdash \mathsf{f}$ infer $\mathcal{H} \vdash \mathbf{A}$.

Rule of Cases (Cases). From $\mathcal{H} \vdash \mathbf{A} \vee \mathbf{B}$ and $\mathcal{H},\ \mathbf{A} \vdash \mathbf{C}$ and $\mathcal{H},\ \mathbf{B} \vdash \mathbf{C}$ infer $\mathcal{H} \vdash \mathbf{C}$.

Rule of Alphabetic Change of Bound Variables (α). From $\mathcal{H} \vdash \mathbf{A}$ infer $\mathcal{H} \vdash \mathbf{A}'$, where $\mathbf{A}'$ is obtained from $\mathbf{A}$ upon replacing an occurrence of $\forall \mathbf{x}\mathbf{C}$ $[\exists \mathbf{x}\mathbf{C}]$ in $\mathbf{A}$ by an occurrence of $\forall \mathbf{y} \mathsf{S}^{\mathbf{x}}_{\mathbf{y}} \mathbf{C}$ $[\exists \mathbf{y} \mathsf{S}^{\mathbf{x}}_{\mathbf{y}} \mathbf{C}]$, where $\mathbf{y}$ is not free in $\mathbf{C}$ and $\mathbf{y}$ is free for $\mathbf{x}$ in $\mathbf{C}$.

Universal Generalization ($\forall$**G**). From $\mathcal{H} \vdash \mathbf{M} \vee \mathbf{A} \vee \mathbf{N}$ infer $\mathcal{H} \vdash \mathbf{M} \vee \forall \mathbf{x}\mathbf{A} \vee \mathbf{N}$, provided that $\mathbf{x}$ is not free in $\mathcal{H}$, $\mathbf{M}$, or $\mathbf{N}$.

Existential Generalization ($\exists$**G**). Let $\mathbf{A}(\mathbf{x})$ be a wff and let $\mathbf{t}$ be a term which is free for $\mathbf{x}$ in $\mathbf{A}(\mathbf{x})$. ($\mathbf{t}$ may occur in $\mathbf{A}(\mathbf{x})$.) From $\mathcal{H} \vdash \mathbf{M} \vee \mathbf{A}(\mathbf{t}) \vee \mathbf{N}$ infer $\mathcal{H} \vdash \mathbf{M} \vee \exists \mathbf{x}\mathbf{A}(\mathbf{x}) \vee \mathbf{N}$.

Universal Instantiation ($\forall$**I**). From $\mathcal{H} \vdash \forall \mathbf{x}\mathbf{A}(\mathbf{x})$ infer $\mathcal{H} \vdash \mathbf{A}(\mathbf{t})$, provided that $\mathbf{t}$ is a term free for $\mathbf{x}$ in $\mathbf{A}(\mathbf{x})$.

Rule C. From $\mathcal{H} \vdash \exists \mathbf{x}\mathbf{B}(\mathbf{x})$ and $\mathcal{H},\ \mathbf{B}(\mathbf{y}) \vdash \mathbf{A}$ infer $\mathcal{H} \vdash \mathbf{A}$, where $\mathbf{y}$ is an individual variable which is free for $\mathbf{x}$ in $\mathbf{B}(\mathbf{x})$ and which is not free in $\mathcal{H}$, $\exists \mathbf{x}\mathbf{B}(\mathbf{x})$ or in $\mathbf{A}$.

Note that the system $\mathcal{N}$ has no axioms. We remark that the Rule for Expanding or Rearranging Hypotheses is often used tacitly and without explicit mention in combination with other rules. Rule $\forall$G really has four forms:

from $\mathcal{H} \vdash \mathbf{A}$	infer $\mathcal{H} \vdash \forall \mathbf{x}\mathbf{A}$;
from $\mathcal{H} \vdash \mathbf{A} \vee \mathbf{C}$	infer $\mathcal{H} \vdash \forall \mathbf{x}\mathbf{A} \vee \mathbf{C}$;
from $\mathcal{H} \vdash \mathbf{B} \vee \mathbf{A}$	infer $\mathcal{H} \vdash \mathbf{B} \vee \forall \mathbf{x}\mathbf{A}$;
from $\mathcal{H} \vdash \mathbf{B} \vee \mathbf{A} \vee \mathbf{C}$	infer $\mathcal{H} \vdash \mathbf{B} \vee \forall \mathbf{x}\mathbf{A} \vee \mathbf{C}$.

(In each case, $\mathbf{x}$ must not be free in $\mathcal{H}$, $\mathbf{B}$, or $\mathbf{C}$.) Our use of the null formula simply enables us to compress these four statements into one. Rule P can actually be restricted to certain special cases of Rule P (such as certain rules of the system $\mathcal{G}$ of §31), but we shall not discuss that here. Naturally, if one wishes to prove theorems involving $=$, one should add to the rules above certain derived rules of inference of the system discussed in §26. Also, in some contexts it would be natural to include a rule permitting one to infer any previously proved theorem.

The soundness and completeness of $\mathcal{N}$ follow from the corresponding results for $\mathcal{F}$, though a careful proof must deal with the fact that $\exists$ is a primitive symbol of $\mathcal{N}$, but not of $\mathcal{F}$. We leave further consideration of these matters to the reader (Exercises X3003 and X3004).

If $\mathbf{A}(\mathbf{x})$ is a wff in which $\mathbf{x}$ occurs free, and one wishes to prove $\exists\mathbf{x}\mathbf{A}(\mathbf{x})$, a natural approach is to try to find a term $\mathbf{t}$ such that one can prove $\mathbf{A}(\mathbf{t})$, and then derive $\exists\mathbf{x}\mathbf{A}(\mathbf{x})$ by $\exists$G. For example, if $\mathbf{A}(\mathbf{x})$ is $[Qy \supset Q\mathbf{x}]$, one can prove $\mathbf{A}(y)$ (i.e., $[Qy \supset Qy]$), and from this infer $\exists\mathbf{x}\mathbf{A}(\mathbf{x})$ (i.e., $\exists\mathbf{x}[Qy \supset Q\mathbf{x}]$).

Sometimes this approach will not work, as in the case where $\mathbf{A}(\mathbf{x})$ is $[Q\mathbf{x} \supset \centerdot\, Qa \wedge Qb]$. Nevertheless, in this case one can prove $\mathbf{A}(a) \vee \mathbf{A}(b)$ (i.e., $[Qa \supset \centerdot\, Qa \wedge Qb] \vee [Qb \supset \centerdot\, Qa \wedge Qb]$, which is tautologous), from which one can infer $\exists\mathbf{x}\mathbf{A}(\mathbf{x}) \vee \exists\mathbf{x}\mathbf{A}(\mathbf{x})$ and hence infer $\exists\mathbf{x}\mathbf{A}(\mathbf{x})$ by Rule P. Thus, a natural generalization of the approach to proving $\exists\mathbf{x}\mathbf{A}(\mathbf{x})$ mentioned above is to find terms $\mathbf{t}_1, \ldots, \mathbf{t}_n$ such that one can prove $\mathbf{A}(\mathbf{t}_1) \vee \ldots \vee \mathbf{A}(\mathbf{t}_n)$ and from this infer $\exists\mathbf{x}\mathbf{A}(\mathbf{x}) \vee \ldots \vee \exists\mathbf{x}\mathbf{A}(\mathbf{x})$, and hence $\exists\mathbf{x}\mathbf{A}(\mathbf{x})$.

However, there are cases when even this generalized approach does not quite work. Consider the problem of giving a direct natural deduction proof of $\exists x\forall y \centerdot\, Px \supset Py$. One proof is:

$\vdash [Px \supset Py] \vee \centerdot\, Py \supset Pz$	Rule P
$\vdash [Px \supset Py] \vee \forall z \centerdot\, Py \supset Pz$	$\forall$G
$\vdash [Px \supset Py] \vee \exists x\forall z \centerdot\, Px \supset Pz$	$\exists$G
$\vdash \forall y[Px \supset Py] \vee \exists x\forall z \centerdot\, Px \supset Pz$	$\forall$G
$\vdash \exists x\forall y[Px \supset Py] \vee \exists x\forall z \centerdot\, Px \supset Pz$	$\exists$G
$\vdash \exists x\forall y[Px \supset Py] \vee \exists x\forall y \centerdot\, Px \supset Py$	α
$\vdash \exists x\forall y \centerdot\, Px \supset Py$	Rule P

Note that in this proof we do not have terms $\mathbf{t}_1$ and $\mathbf{t}_2$ in which y is not free such that we prove $\forall y[P\mathbf{t}_1 \supset Py] \vee \forall y[P\mathbf{t}_2 \supset Py]$. Could there be such terms? Consider an interpretation $\mathcal{M} = \langle \mathcal{D}, \mathcal{J} \rangle$, where

$$\mathcal{D} = \{a, b\};$$

$$\mathcal{J}\mathbf{c} = a \text{ for all constants } \mathbf{c};$$

for each function symbol $\mathbf{f}^n$

$$(\mathcal{J}\mathbf{f}^n)d_1 \ldots d_n = a \text{ for all } d_1, \ldots, d_n \text{ in } \mathcal{D};$$

$$(\mathcal{J}\ P)a = \mathsf{T}, \text{ and } (\mathcal{J}\ P)b = \mathsf{F}.$$

Let $\varphi\mathbf{x} = a$ for all individual variables $\mathbf{x}$; then $\mathcal{V}_\varphi^\mathcal{M}\mathbf{t} = a$ for each term $\mathbf{t}$. Thus if y does not occur in a term $\mathbf{t}$, and ψ is the assignment which agrees with φ off y, while $\psi y = b$, then

$$\mathcal{V}_\psi[P\mathbf{t} \supset Py] = \mathsf{F}, \text{ so } \mathcal{V}_\varphi^\mathcal{M}\forall y[P\mathbf{t} \supset Py] = \mathsf{F}.$$

Thus one cannot prove any disjunction of the form $\bigvee_{i=1}^n \forall y[P\mathbf{t}_i \supset Py]$, where the terms $\mathbf{t}_i$ do not contain y.

EXERCISES

Prove the following theorems in $\mathcal{N}$:

X3000. $\sim \forall\mathbf{x}\mathbf{A} \equiv \exists\mathbf{x} \sim \mathbf{A}$.

X3001. $\sim \exists\mathbf{x}\mathbf{A} \equiv \forall\mathbf{x} \sim \mathbf{A}$.

X3002. $\forall u\forall v\forall w[Puv \lor Pvw] \supset \exists x\forall yPxy$. (*Hint:* recall the advice given near the end of §21 about proving theorems of the form $\mathbf{A} \supset \mathbf{B}$.)

X3003. Prove that the system $\mathcal{N}$ is sound in the sense of 2303.

X3004. Prove that the system $\mathcal{N}$ is complete in the sense of 2509.

X3005. Add rules of inference for $=$ to $\mathcal{N}$, and prove that the resulting system is sound and complete in the sense of 2609 and 2612.

X3006. Are any of the rules of inference of $\mathcal{N}$ dependent (non-independent) in the sense of §13?

X3007. Find a system $\mathcal{N}'$ which can be obtained from $\mathcal{N}$ by deleting certain rules of inference, and whose rules of inference are all independent. Prove the independence of these rules.

§31. Gentzen's Theorem

One of the difficulties of analyzing possible proofs of theorems of $\mathcal{F}$ is that the form of a theorem gives few clues concerning what a proof of it might look like. If one seeks to analyze the possible proofs of a theorem **B**, one must consider the possibility that it was obtained by Modus Ponens from theorems **A** and **A** $\supset$ **B**, but it is hard to guess what **A** might have been. In this section we consider a system $\mathcal{G}$ whose rules of inference are chosen so that one can search in a fairly systematic way for a proof of any specified theorem. The system $\mathcal{G}$ also has the "subformula property", i.e., every wff which occurs in a proof of a theorem bears some resemblance to some subformula of the theorem. $\mathcal{G}$ is based on ideas introduced by Gentzen [Gentzen, 1935], and is closely related to a simplification of Gentzen's sequent calculus due to Schütte [Schütte, 1950].

The wffs of $\mathcal{G}$ are the same as those of $\mathcal{F}$. As in §30, we use **M** and **N** as syntactical variables for wffs which may be null.

The occurrences of wf parts of a wff which are in the scope of no negations or quantifiers of that wff are called *disjunctive components* of that wff. For example, the disjunctive components of $[[\forall x Fx \vee Gy] \vee [[\sim p \vee \sim Gy] \vee q]]$ are the entire wff, the first occurrence of Gy, and the sole occurrences in the wff of $\forall x\ Fx$, $[\forall x Fx \ \vee \ Gy]$, $\sim p$, $\sim Gy$, $[\sim p \ \vee \ \sim Gy]$, q, and $[[\sim p \vee \ \sim Gy] \ \vee \ q]$.

Now we can state the axioms and rules of inference of $\mathcal{G}$:

Axiom. $\sim$ **A** $\vee$ **A**, where **A** is any atomic wff.

Rules of Inference

Disjunction Rules ($\vee$). Replace a disjunctive component **D** of a wff by **E**, where **D** is [[**A** $\vee$**B**] $\vee$**C**] and **E** is [**A** $\vee$ [**B** $\vee$ **C**]], or **D** is [**A** $\vee$ [**B** $\vee$ **C**]] and **E** is [[**A** $\vee$**B**] $\vee$ **C**], or **D** is [**A** $\vee$ **B**] and **E** is [**B** $\vee$ **A**].

Weakening (Weaken). From **B** infer **B** $\vee$ **A**.

Simplification (Simp). From **M** $\vee$ [**A** $\vee$ **A**] infer **M** $\vee$ **A**.

Negation Introduction ($\sim$). From **M** $\vee$ **A** infer **M** $\vee$ $\sim\sim$ **A**.

Conjunction Introduction($\wedge$). From **M** $\vee$ $\sim$ **A** and **M** $\vee$ $\sim$ **B** infer **M** $\vee$ $\sim$ [**A** $\vee$ **B**].

Alphabetic Change of Bound Variables (α). (See 2110 for a statement of this rule.)

Universal Generalization ($\forall$G). From $\mathbf{M} \lor \mathbf{A}$ infer $\mathbf{M} \lor \forall\mathbf{x}\mathbf{A}$, provided $\mathbf{x}$ is not free in $\mathbf{M}$.

Existential Generalization ($\exists$G). From $\mathbf{M} \lor \sim \mathsf{S}^{\mathbf{x}}_{\mathbf{t}}\mathbf{A}$ infer $\mathbf{M} \lor \sim \forall\mathbf{x}\mathbf{A}$, where $\mathbf{t}$ is a term free for $\mathbf{x}$ in $\mathbf{A}$.

We shall write $\vdash_{\mathcal{G}} \mathbf{A}$ to indicate that $\mathbf{A}$ is a theorem of $\mathcal{G}$.

The appropriateness of the names for rules $\land$ and $\exists$G should become clear to the reader who does exercises X3100 and X3101. The $\lor$ rules permit one to move the disjunctive components of a wff around so that the component which is to be modified by a rule of inference is on the right. The need for them could be eliminated by stating the rules of inference in a more general (but less perspicuous) way. We shall often combine applications of the $\lor$ rules with those of other rules, and omit mention of the $\lor$ rules.

EXAMPLE: The following proof in $\mathcal{G}$ illustrates a number of features of the system, including the need for the Simplification rule. The reader may find it enlightening to prove the theorem for himself before looking at the proof below:

(1)	$\vdash \sim Pxy \lor Pxy$	Axiom
(2)	$\vdash \sim Pxy \lor \bullet\, Pxy \lor Pyz$	Weaken
(3)	$\vdash \sim Pyz \lor Pyz$	Axiom
(4)	$\vdash \sim Pyz \lor \bullet\, Pxy \lor Pyz$	Weaken
(5)	$\vdash \sim [Pxy \lor Pyz] \lor \bullet\, Pxy \lor Pyz$	$\land$: 2,4
(6)	$\vdash \sim \forall u\forall v\forall w[Puv \lor Pvw] \lor \bullet\, Pxy \lor Pyz$	$\exists$G (thrice)
(7)	$\vdash \sim \forall u\forall v\forall w[Puv \lor Pvw] \lor \bullet\, Pxy \lor \forall z\, Pyz$	$\forall$G
(8)	$\vdash \sim \forall u\forall v\forall w[Puv \lor Pvw] \lor \bullet\, Pxy \lor \exists x\forall z\, Pxz$	X3101
(9)	$\vdash \sim \forall u\forall v\forall w[Puv \lor Pvw] \lor \bullet\, \forall y\, Pxy \lor \exists x\forall z\, Pxz$	$\forall$G
(10)	$\vdash \sim \forall u\forall v\forall w[Puv \lor Pvw] \lor \bullet\, \exists x\forall y\, Pxy \lor \exists x\forall z\, Pxz$	X3101
(11)	$\vdash \sim \forall u\forall v\forall w[Puv \lor Pvw] \lor \bullet\, \exists x\forall y\, Pxy \lor \exists x\forall y\, Pxy$	α
(12)	$\vdash \sim \forall u\forall v\forall w[Puv \lor Pvw] \lor \exists x\forall y\, Pxy$	Simp

It can be shown that without loss of the completeness of $\mathcal{G}$ (which we establish below), the Simplification rule can be restricted to the form: from $\mathbf{M} \lor \sim \forall\mathbf{x}\mathbf{A} \lor \sim \forall\mathbf{x}\mathbf{A}$ to infer $\mathbf{M} \lor \sim \forall\mathbf{x}\mathbf{A}$. Also, one need use it only when the occurrences of $\sim \forall\mathbf{x}\mathbf{A}$ have been introduced by rule $\exists$G. Therefore the rules Simp and $\exists$G can be combined into the single rule:

$\exists$G-Simp. From $\mathbf{M} \lor \sim \forall\mathbf{x}\mathbf{A} \lor \sim \mathsf{S}^{\mathbf{x}}_{\mathbf{t}}\mathbf{A}$ infer $\mathbf{M} \lor \sim \forall\mathbf{x}\mathbf{A}$, where $\mathbf{t}$ is free for $\mathbf{x}$ in $\mathbf{A}$.

Note exercise X3102.

3100 Lemma. If $\vdash_{\mathcal{G}} \mathbf{M} \vee \sim\sim \mathbf{E}$ then $\vdash_{\mathcal{G}} \mathbf{M} \vee \mathbf{E}$.

Proof: Let $\mathbf{P}_1, \ldots, \mathbf{P}_m$ be a proof in $\mathcal{G}$ of $\mathbf{M} \vee \sim\sim \mathbf{E}$. We prove by induction on i that if $\mathbf{Q}_i$ is the result of replacing zero or more disjunctive components of $\mathbf{P}_i$ of the form $\sim\sim \mathbf{D}$ by components of the form $\mathbf{D}$, then $\vdash_{\mathcal{G}} \mathbf{Q}_i$.

Clearly no axiom contains double negations. Suppose $\mathbf{P}_i$ is inferred by some rule, which we shall call rule R, from $\mathbf{P}_j$ [and $\mathbf{P}_k$], where $j < i$ [and $k < i$]. If R is not Weakening or $\sim$, we may apply the inductive hypotheses to $\mathbf{P}_j$ [and to $\mathbf{P}_k$] to obtain a wff $\mathbf{Q}_j$ [and a wff $\mathbf{Q}_k$] from which $\mathbf{Q}_i$ may be inferred by rule R. If R is $\sim$ or Weakening, we proceed in the same way, except that we may omit the application of rule $\sim$, and we may modify the application of weakening to infer $\mathbf{B} \vee \mathbf{D}$ instead of $\mathbf{B} \vee \sim\sim \mathbf{D}$ from $\mathbf{B}$. ■

3101 Theorem. For every wff $\mathbf{A}$ of $\mathcal{F}$, $\vdash_{\mathcal{F}} \mathbf{A}$ iff $\vdash_{\mathcal{G}} \mathbf{A}$.

Proof: It is easy to see that every theorem of $\mathcal{G}$ is a theorem of $\mathcal{F}$, since the axioms of $\mathcal{G}$ are all theorems of $\mathcal{F}$, and the rules of inference of $\mathcal{G}$ are all derived rules of inference of $\mathcal{F}$.

For the proof in the other direction, we let Γ be the class of all finite sets $\{\mathbf{C}_1, \ldots, \mathbf{C}_n\}$ of sentences such that not $\vdash_{\mathcal{G}} \sim \mathbf{C}_1 \vee \ldots \vee \sim \mathbf{C}_n$, and use Smullyan's Unifying Principle. By virtue of rules $\vee$, Weaken, and Simp, it does not matter in what order the $\mathbf{C}_i$ are listed, and $\{\mathbf{C}_1, \ldots, \mathbf{C}_n\} \cup \{\mathbf{D}_1, \ldots, \mathbf{D}_m\} \in \Gamma$ iff not $\vdash_{\mathcal{G}} \sim \mathbf{C}_1 \vee \ldots \vee \sim \mathbf{C}_n \vee \sim \mathbf{D}_1 \vee \ldots \vee \sim \mathbf{D}_m$ whether or not some $\mathbf{C}_i$ and $\mathbf{D}_j$ are the same.

Next we verify that Γ is an abstract consistency class. Because $\mathcal{G}$ has the Weakening rule, Γ is clearly closed under subsets.

(a) If $\mathbf{A}$ is atomic, then $\vdash_{\mathcal{G}} \mathbf{M} \vee \sim\sim \mathbf{A} \vee \sim \mathbf{A}$ for any $\mathbf{M}$, so clearly (a) is satisfied.

Conditions (b) and (d)-(f) all have the form: if $\mathcal{S} \in \Gamma$ and $\mathbf{D} \in \mathcal{S}$, then $\mathcal{S} \cup \{\mathbf{E}_1, \ldots, \mathbf{E}_n\} \in \Gamma$ (where $1 \leq n \leq 2$). We shall consider the condition in the equivalent (more or less contrapositive) form: assume $\mathbf{D} \in \mathcal{S}$; if $\mathcal{S} \cup \{\mathbf{E}_1, \ldots, \mathbf{E}_n\} \notin \Gamma$, then $\mathcal{S} \notin \Gamma$. We may write $\mathcal{S}$ as $\{\mathbf{C}_1, \ldots, \mathbf{C}_k, \mathbf{D}\}$ (where $k \geq 0$) and $\sim \mathbf{C}_1 \vee \ldots \vee \sim \mathbf{C}_k$ as $\mathbf{M}$. Then in each case the condition takes the form: if $\vdash_{\mathcal{G}} \mathbf{M} \vee \sim \mathbf{D} \vee \sim \mathbf{E}_1 \vee \ldots \vee \sim \mathbf{E}_n$, then $\vdash_{\mathcal{G}} \mathbf{M} \vee \sim \mathbf{D}$. Condition (c) can be handled in a similar way. The detailed statements to be verified are the following:

(b) If $\vdash_{\mathcal{G}} \mathbf{M} \lor {\sim\sim\sim} \mathbf{A} \lor {\sim} \mathbf{A}$ then $\vdash_{\mathcal{G}} \mathbf{M} \lor {\sim\sim\sim} \mathbf{A}$. Rules $\sim$ and Simp assure this.

(c) If $\vdash_{\mathcal{G}} \mathbf{M} \lor {\sim} [\mathbf{A} \lor \mathbf{B}] \lor {\sim} \mathbf{A}$ and $\vdash_{\mathcal{G}} \mathbf{M} \lor {\sim} [\mathbf{A} \lor \mathbf{B}] \lor {\sim} \mathbf{B}$ then $\vdash_{\mathcal{G}} \mathbf{M} \lor {\sim} [\mathbf{A} \lor \mathbf{B}]$. Rules $\land$ and Simp assure this.

(d) If $\vdash_{\mathcal{G}} \mathbf{M} \lor {\sim\sim} [\mathbf{A} \lor \mathbf{B}] \lor {\sim\sim} \mathbf{A} \lor {\sim\sim} \mathbf{B}$, then $\vdash_{\mathcal{G}} \mathbf{M} \lor {\sim\sim} [\mathbf{A} \lor \mathbf{B}]$. Lemma 3100, $\sim$, and Simp assure this.

(e) If $\vdash_{\mathcal{G}} \mathbf{M} \lor {\sim} \forall \mathbf{x} \mathbf{A} \lor {\sim} \overset{\bullet}{\mathsf{S}}{}^{\mathbf{x}}_{\mathbf{t}} \mathbf{A}$ for some variable-free term $\mathbf{t}$, then $\vdash_{\mathcal{G}} \mathbf{M} \lor {\sim} \forall \mathbf{x} \mathbf{A}$. Rules $\exists$G and Simp assure this.

(f) If (∗) $\vdash_{\mathcal{G}} \mathbf{M} \lor {\sim\sim} \forall \mathbf{x} \mathbf{A} \lor {\sim\sim} \overset{\bullet}{\mathsf{S}}{}^{\mathbf{x}}_{\mathbf{c}} \mathbf{A}$ and $\mathbf{c}$ is an individual constant which does not occur in $\mathbf{M}$ or $\mathbf{A}$, then $\vdash_{\mathcal{G}} \mathbf{M} \lor {\sim\sim} \forall \mathbf{x} \mathbf{A}$.

To verify (f), note that we can replace $\mathbf{c}$ everywhere in the proof of (∗) by some new variable $\mathbf{y}$ which does not occur in the proof to obtain

$\vdash_{\mathcal{G}} \mathbf{M} \lor {\sim\sim} \forall \mathbf{x} \mathbf{A} \lor {\sim\sim} \overset{\bullet}{\mathsf{S}}{}^{\mathbf{x}}_{\mathbf{y}} \mathbf{A}$	
$\vdash_{\mathcal{G}} \mathbf{M} \lor \forall \mathbf{x} \mathbf{A} \lor \forall \mathbf{y} \overset{\bullet}{\mathsf{S}}{}^{\mathbf{x}}_{\mathbf{y}} \mathbf{A}$	3100 and $\forall$G
$\vdash_{\mathcal{G}} \mathbf{M} \lor \forall \mathbf{x} \mathbf{A} \lor \forall \mathbf{x} \mathbf{A}$	α
$\vdash_{\mathcal{G}} \mathbf{M} \lor \forall \mathbf{x} \mathbf{A}$	Simp
$\vdash_{\mathcal{G}} \mathbf{M} \lor {\sim\sim} \forall \mathbf{x} \mathbf{A}$	$\sim$

Thus Γ is an abstract consistency class. Now let $\mathbf{A}$ be any theorem of $\mathcal{F}$. If $\mathbf{A}$ is not a sentence, let $\dot{\mathbf{A}}$ be the result of replacing the free variables $\mathbf{x}_1 \ldots \mathbf{x}_n$ of $\mathbf{A}$ by distinct new constants $\mathbf{c}_1 \ldots \mathbf{c}_n$ at all their free occurrences in $\mathbf{A}$. $\vdash_{\mathcal{F}} \dot{\mathbf{A}}$ by 2118, so $\left\{\sim \dot{\mathbf{A}}\right\}$ has no model by 2303 and 2301, so $\left\{\sim \dot{\mathbf{A}}\right\} \notin \Gamma$ by Smullyan's Unifying Principle (2507), so $\vdash_{\mathcal{G}} {\sim\sim} \dot{\mathbf{A}}$, so $\vdash_{\mathcal{G}} \dot{\mathbf{A}}$ by 3100.

Now it remains only to see that a proof of $\dot{\mathbf{A}}$ in $\mathcal{G}$ can be transformed into a proof of $\mathbf{A}$. Let $\mathbf{y}_1, \ldots, \mathbf{y}_n$ be distinct variables which do not occur in the given proof of $\dot{\mathbf{A}}$. Replace all occurrences of the $\mathbf{x}_i$ in the given proof of $\dot{\mathbf{A}}$ by occurrences of the $\mathbf{y}_i$. Then replace all occurrences of the $\mathbf{c}_i$ in the resulting proof by occurrences of the $\mathbf{x}_i$. It is easy to see that one thus obtains a proof in $\mathcal{G}$ of a wff which differs from $\mathbf{A}$ at most by certain alphabetic changes of bound variables, and by applying Rule α one can then obtain a proof in $\mathcal{G}$ of $\mathbf{A}$. ∎

3102 Corollary. If $\vdash_{\mathcal{G}} \mathbf{M} \lor \mathbf{A}$ and $\vdash_{\mathcal{G}}, {\sim} \mathbf{A} \lor \mathbf{N}$, then $\vdash_{\mathcal{G}} \mathbf{M} \lor \mathbf{N}$.

Proof: By 3101 and Rule P (2100) for $\mathcal{F}$. ∎

REMARK. The rule of inference of 3102 is known as the *Cut* rule. Note that it is a generalization of Modus Ponens, since **M** may be null. We prefer not to call Cut a derived rule of inference of $\mathcal{G}$ at this stage, since we have not shown how to actually construct a proof of $\mathbf{M} \lor \mathbf{N}$ from given proofs of $\mathbf{M} \lor \mathbf{A}$ and $\sim \mathbf{A} \lor \mathbf{N}$.

It is now easy to describe an alternative approach to 3101 and 3102 which provides a purely syntactic proof of these theorems, and which parallels Gentzen's original argument. Let $\mathcal{G}^+$ be the result of adding Cut to the primitive rules of inference of $\mathcal{G}$. It is easy to see that $\mathcal{G}^+$ and $\mathcal{F}$ have the same theorems, since the axioms of $\mathcal{F}$ are all theorems of $\mathcal{G}$ (Exercise X3104), and the rules of inference of $\mathcal{F}$ are all rules of $\mathcal{G}^+$. Then show that by manipulating proofs one can eliminate all applications of Cut from all proofs in $\mathcal{G}^+$. This would establish the *Cut-Elimination Theorem* for $\mathcal{G}^+$, and show that $\mathcal{G}$ and $\mathcal{G}^+$ are equivalent systems. Gentzen's Hauptsatz (Main Theorem) was the Cut-Elimination Theorem for this system.

$\mathcal{G}$ is called a *Cut-free* system. As we shall mention in §35, Herbrand also discussed a Cut-free system, and part of Herbrand's Theorem is closely related to Gentzen's Hauptsatz.

EXERCISES

X3100. Give a direct constructive proof that if $\vdash_{\mathcal{G}} \mathbf{M} \lor \mathbf{A}$ and $\vdash_{\mathcal{G}} \mathbf{M} \lor \mathbf{B}$, then $\vdash_{\mathcal{G}} \mathbf{M} \lor [\mathbf{A} \land \mathbf{B}]$, where $\land$ as is defined in §12.

X3101. Give a direct constructive proof that if **t** is free for **x** in **C** and $\vdash_{\mathcal{G}} \mathbf{M} \lor \overset{\bullet}{\mathsf{S}}{}^{\mathbf{x}}_{\mathbf{t}}\mathbf{C}$, then $\vdash_{\mathcal{G}} \mathbf{M} \lor \exists\mathbf{x}\mathbf{C}$, where $\exists$ is defined as in §20.

X3102. Let $\mathcal{G}'$ be the system obtained from $\mathcal{G}$ by deleting rules Simp and $\exists$G, and adding rule $\exists$G-Simp. Show how to prove $\sim \forall u \forall v \forall w[Puv \lor Pvw] \lor \exists x \forall y\, Pxy$ in $\mathcal{G}'$.

X3103. Show that $\vdash_{\mathcal{G}} \sim \mathbf{A} \lor \mathbf{A}$ for every wff **A**.

X3104. Show that the axioms of $\mathcal{F}$ are all theorems of $\mathcal{G}$.

§32. Semantic Tableaux

In this section we introduce a method for refuting sets of sentences which was introduced by Beth [Beth, 1959], Hintikka [Hintikka, 1955], and Schütte [Schütte, 1956]. In Chapters 1 and 2 we dealt with proofs which were finite sequences of wffs, but in this section we shall arrange the occurrences of wffs

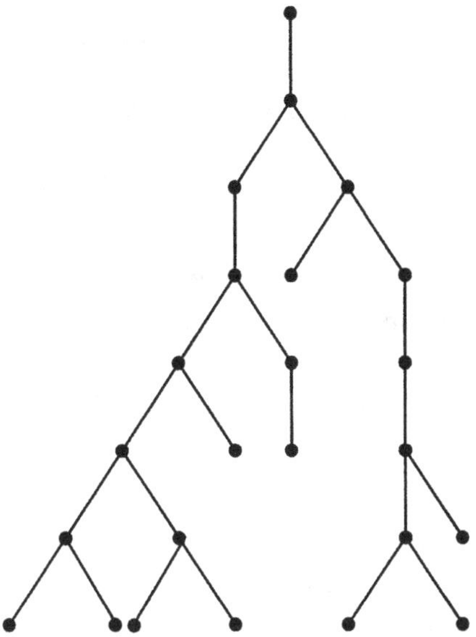

Figure 3.1: A tree

in a refutation as nodes of a tree, and call the resulting structure a *semantic tableau*.

The trees we shall use will be *downward dyadic* trees, meaning that we picture them as having their branches extending downward, and at most two nodes are immediately below any node. An example of such a tree is shown in Figure 3.1, where the nodes are depicted as large dots.

A (*dyadic*) *tree of sentences* is a downward dyadic tree with each node of which is associated some sentence of $\mathcal{F}$. (We do not use sentences for nodes, since a sentence may have several occurrences in a tree of sentences.) (An occurrence of) a wff **A** in such a tree is *below* (an occurrence of) a wff **B** iff the node with which **A** is associated is below the node with which **B** is associated.

DEFINITION. A *semantic tableau* for a finite set $\mathcal{S}$ of sentences is a tree of sentences which can be constructed by the following inductive rules:

(1) **Initial Hypothesis (Hyp).** A tree of sentences which has just one node, whose sentence is in $\mathcal{S}$, is a semantic tableau for $\mathcal{S}$.

(2) If $\mathcal{U}$ is a semantic tableau for $\mathcal{S}$, **G** is a sentence of $\mathcal{U}$, and $\mathcal{U}'$ is obtained from $\mathcal{U}$ by appending to a leaf of $\mathcal{U}$ a new node with sentence **H** so that **H** is below **G** in $\mathcal{U}'$, then U' is a semantic tableau for $\mathcal{S}$, provided that **H** is obtained from **G** by one of the following rules of inference:

(2a) **Hypothesis (Hyp).** From **G** infer **H**, where **H** is any sentence in $\mathcal{S}$.

(2b) **Double-Negation Elimination ($\sim\sim$).** From $\sim\sim$ **A** infer **A**.

(2c) **Conjunction Elimination to the Left ($\wedge\ \ell$).** From $\sim$ [**A** $\vee$ **B**] infer $\sim$ **A**.

(2d) **Conjunction Elimination to the Right ($\wedge\ r$).** From $\sim$ [**A** $\vee$ **B**] infer $\sim$ **B**.

(2e) **Universal Instantiation ($\forall$I).** From $\forall$**xA** infer $\mathbf{S}^{\mathbf{x}}_{\mathbf{t}}\mathbf{A}$, where t is any variable-free term.

(3) **Existential Instantiation ($\exists$I).** If $\mathcal{U}$ is a semantic tableau for $\mathcal{S}$, $\sim \forall$**xA** is a sentence of $\mathcal{U}$, and $\mathcal{U}'$ is obtained from $\mathcal{U}$ by appending to a leaf of $\mathcal{U}$ a new node with sentence $\sim \mathbf{S}^{\mathbf{x}}_{\mathbf{c}}\mathbf{A}$, so that $\sim \mathbf{S}^{\mathbf{x}}_{\mathbf{c}}\mathbf{A}$ is below $\sim \forall$**xA** in $\mathcal{U}'$ and **c** is an individual constant which does not occur in any sentence of $\mathcal{U}$ or $\mathcal{S}$, then $\mathcal{U}'$ is a semantic tableau for $\mathcal{S}$.

(4) **Disjunction Elimination ($\vee$).** If $\mathcal{U}$ is a semantic tableau for $\mathcal{S}$, [**A** $\vee$ **B**] is a sentence of $\mathcal{U}$ with node n, n' is a leaf of $\mathcal{U}$ which is n or is below n, and $\mathcal{U}'$ is obtained from $\mathcal{U}$ by appending below n' two new nodes with sentences **A** and **B**, respectively, then $\mathcal{U}'$ is a semantic tableau for $\mathcal{S}$.

A branch in a semantic tableau is *closed* iff it contains a sentence and its negation. We write $\Box$ under the terminal leaf of a closed branch. A semantic tableau is *closed* iff all of its branches are closed.

An example of a semantic tableau is displayed in Figure 3.2. Only the sentences associated with the nodes are actually written, with their positions indicating which sentences are below which other sentences. The vertical lines separate nodes which are on different branches of the tree.

$$\exists v\, \forall x\, Pxv \;\wedge\; \forall x[Sx \;\supset\; \exists y\, Qyx] \;\wedge\; \forall x\, \forall y[Pxy \supset \sim Qxy] \supset \exists u \sim Su$$

is proved by constructing a closed semantic tableau for sentences (1)-(4) in Figure 3.2. (The conjunction of (1)-(4) is equivalent to the negation of the sentence to be proved.)

(1)	$\exists v \forall x\, Pxv$	hyp
(2)	$\forall x\, [Sx \supset \exists y\, Qyx]$	hyp
(3)	$\forall x \forall y\, [Pxy \supset \sim Qxy]$	hyp
(4)	$\forall u Su$	hyp
(5)	$\sim\sim \forall x\, Pxa$	$\exists$I : 1
(6)	$\forall x\, Pxa$	$\sim\sim$: 5
(7)	$\sim Sa \lor \exists y\, Qya$	$\forall$I : 2

Left branch:

(8)	$\sim Sa$	$\lor$: 7
(10)	Sa	$\forall$I : 4
	$\Box$	8, 10

Right branch:

(9)	$\exists y\, Qya$	$\lor$: 7
(11)	$\sim\sim Qba$	$\exists$I : 9
(12)	Qba	$\sim\sim$: 11
(13)	$\forall y\, [Pby \supset \sim Qby]$	$\forall$I : 3
(14)	$\sim Pba \lor \sim Qba$	$\forall$I : 13

Right branch, left sub-branch:

(15)	$\sim Pba$	$\lor$: 14
(17)	Pba	$\forall$I : 6
	$\Box$	15, 17

Right branch, right sub-branch:

(16)	$\sim Qba$	$\lor$: 14
	$\Box$	12, 16

Figure 3.2: A semantic tableau

One can speed up the construction of semantic tableaux by applying several rules of inference as one step, and this process can be systematized by formulating, justifying, and applying derived rules of inference. For example, the inference from (1) to (6) in Figure 3.2 can be justified by a slight generalization of rule $\exists$I. Similarly, $\land\, l$ and $\land\, r$ can be generalized to permit inferring **A** and **B** from [**A** $\land$ **B**] (which is defined as in §10, of course). These generalized rules are used in the abbreviated tableau in Figure 3.3. There is a sense in which modus ponens is a derived rule of inference, since a branch containing **A** and [$\sim$ **A** $\lor$ **B**] can be split into a closed branch containing $\sim$ **A**, and another branch containing **B**, as illustrated in Figure 3.4. We leave it to the reader to pursue this subject further (Exercise X3200).

3200 Theorem. Let $\mathcal{S}$ be a set of sentences of $\mathcal{F}$. $\mathcal{S}$ has no model iff $\mathcal{S}$ has a closed semantic tableau.

Proof: It is easy to see that if $\mathcal{S}$ is a set of sentences which has a model and $\mathcal{U}$ is a semantic tableau for $\mathcal{S}$, then $\mathcal{U}$ has a branch $\mathcal{B}$ such that the set of sentences associated with (nodes of) $\mathcal{B}$ has a model. This is seen by induction on the construction of $\mathcal{U}$, considering each of the rules (1)-(4) in turn, and we leave the details to the reader. Clearly, if $\mathcal{B}$ is a branch of

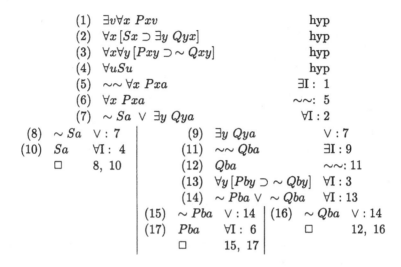

(1)	$\forall u \forall v \forall w [Puv \lor Pvw] \land \forall x \exists y \sim Pxy$	*hyp*
(2)	$\forall u \forall v \forall w [Puv \lor Pvw]$	$\land l : 1$
(3)	$\forall x \exists y \sim Pxy$	$\land r : 1$
(4)	$\exists y \sim Pay$	$\forall \mathrm{I} : 3$
(5)	$\sim Pab$	$\exists \mathrm{I} : 4$
(6)	$\exists y \sim Pby$	$\forall \mathrm{I} : 3$
(7)	$\sim Pbc$	$\exists \mathrm{I} : 6$
(8)	$Pab \lor Pbc$	$\forall \mathrm{I}(thrice) : 2$

(9) Pab $\lor : 8$	(10) Pbc $\lor : 8$
$\square$ 5, 9	$\square$ 7, 10

Figure 3.3: An abbreviated tableau

$\mathcal{U}$ which has a model (in the sense above), then $\mathcal{B}$ cannot be closed, so $\mathcal{U}$ cannot be closed. Consequently, if $\mathcal{S}$ has a closed tableau, then $\mathcal{S}$ can have no model.

To establish that every set of sentences $\mathcal{S}$ which has no model has a closed tableau, we use Smullyan's Unifying Principle (2507). Let $\Gamma = \{ \mathcal{S} \mid \mathcal{S}$ is a set of sentences which has no closed tableau $\}$.

Clearly, if a finite set $\mathcal{S}$ of sentences has a closed tableau, every finite superset of $\mathcal{S}$ has a closed tableau. From this it easily follows that Γ is closed under subsets.

It is straightforward to establish that Γ is an abstract consistency class by checking conditions (a)-(f) of the definition in §25. Note that when $\mathcal{G}$ is a set of sentences, $\mathcal{G} \notin \Gamma$ means that $\mathcal{G}$ has a closed tableau. Thus to check condition (e) (in contrapositive form), it suffices to show that if, for some variable-free term $\mathbf{t}$, $\mathcal{S} \cup \left\{\mathsf{S}^{\mathbf{x}}_{\mathbf{t}}\mathbf{A}\right\}$ has a closed tableau $\mathcal{Q}$, and $\forall \mathbf{x} \mathbf{A} \in \mathcal{S}$, then $\mathcal{S}$ also has a closed tableau $\mathcal{Q}'$. Here $\mathcal{Q}'$ can be constructed from $\mathcal{Q}$ by placing a new node, with sentence $\forall \mathbf{x} \mathbf{A}$, above the origin of $\mathcal{Q}$, so that the

$\vdots$

(6) $\mathbf{A}$
(7) $\sim \mathbf{A} \lor \mathbf{B}$

(8) $\sim \mathbf{A}$ $\lor : 7$	(9) $\mathbf{B}$ $\lor : 7$
$\square$ 6, 8	

Figure 3.4: Modus ponens in a tableau

new node becomes the origin of $\mathcal{Q}'$. We leave it to the reader to check the remaining conditions.

Now let $\mathcal{S}$ be a set of sentences which has no model. By the Compactness Theorem (2514), $\mathcal{S}$ has a finite subset $\mathcal{S}_0$ which has no model. By adjusting the language, if necessary, we may assume that $\mathcal{S}_0$ is a set of sentences in a countable language with infinitely many individual constants. Therefore, by Smullyan's Unifying Principle, $\mathcal{S}_0 \notin \Gamma$, so $\mathcal{S}_0$, and hence $\mathcal{S}$, has a closed tableau. ∎

Thus the method of semantic tableaux provides a sound and complete refutation procedure for first-order logic. While it is in some ways a simple-minded and inefficient procedure, constructing a closed tableau for a sentence $\sim$ **A** often seems easier than directly finding a proof for **A** in the system $\mathcal{F}$ of Chapter 2.

Indeed, a semantic tableau can be constructed in a very systematic way. For this purpose let us combine $\wedge\, l$ and $\wedge\, r$ into the single rule $\wedge$: From $\sim [\mathbf{A} \vee \mathbf{B}]$ infer $\sim \mathbf{A}$ and $\sim \mathbf{B}$. Now let **C** be any sentence in a semantic tableau. If **C** is a literal, no rule of inference can be applied to **C**. If **C** is a non-literal, exactly one rule of inference can be applied to **C**, and unless **C** has the form $\forall \mathbf{x}\mathbf{A}$, there is no point in applying that rule to **C** more than once. The rule ∀I may have to be applied more than once to a wff of the form $\forall \mathbf{x}\mathbf{A}$, so that **x** is instantiated with a number of different terms, as in Figure 3.3. However, the choice of these terms can be restricted as described in Exercise X3203. When function symbols are not present, this restricts the choice of terms to the finite set of individual constants which have already occurred on the current branch of the tableau (unless that set is empty). Of course, the set of available constants keeps expanding with applications of ∃I.

EXERCISES

X3200. Formulate, justify, and illustrate some derived rules of inference for semantic tableaux.

X3201. Use the method of semantic tableau to establish the validity of the wffs in exercises X2106- X2113.

X3202. Show by a purely syntactic argument (as opposed to the semantic one given in the proof of Theorem 3200) that if a set $\mathcal{S}$ of sentences has a closed semantic tableau, then $\mathcal{S}$ is inconsistent. Indeed, give an explicit method for transforming a closed tableau for $\mathcal{S}$ into a refutation of $\mathcal{S}$ in the system of §21.

X3203. Show that if a set of sentences has a closed semantic tableau, then it has one satisfying the following condition: Let $\mathsf{S}^{\mathbf{x}}_{\mathbf{t}}\mathbf{A}$ be any wff inferred from a wff of the form $\forall\mathbf{x}\mathbf{A}$ in the tableau, and let $\mathcal{B}$ be the set of sentences which are above $\mathsf{S}^{\mathbf{x}}_{\mathbf{t}}\mathbf{A}$ in the tableau. If at least one individual constant occurs in some sentence of $\mathcal{B}$, then every individual and function constant which occurs in $\mathbf{t}$ also occurs in some sentence of $\mathcal{B}$.

§33. Skolemization

It will be recalled that when a wff is put into prenex normal form, its positive (universal) quantifiers come out as universal quantifiers, and its negative (universal) quantifiers come out as existential quantifiers. Thus, when one speaks of wffs of $\mathcal{F}$ (without abbreviations), so that only universal quantifiers occur, one may regard positively occurring quantifiers as essentially universal, and negatively occurring quantifiers as essentially existential.

DEFINITION. A wff $\mathbf{A}$ of $\mathcal{F}$ is *universal* iff every occurrence of a quantifier in $\mathbf{A}$ is positive in $\mathbf{A}$.

We now wish to associate with each sentence $\mathbf{B}$ of a countable formulation of $\mathcal{F}$ a universal sentence $\bigstar\mathbf{B}$ of an expansion $\mathcal{F}^\star$ of $\mathcal{F}$. We let $\mathcal{F}^\star$ be the result of adding to $\mathcal{F}$ a denumerable sequence $\overline{g}^n_1, \overline{g}^n_2, \ldots, \overline{g}^n_j, \ldots$ of new n-ary function constants for each natural number $n \geq 0$. (When $n = 0$, $\overline{g}^n_j$ is an individual constant.) We may refer to these constants as *existential constants*. We fix an enumeration $\mathbf{B}_1, \mathbf{B}_2, \ldots, \mathbf{B}_k, \ldots$ of the sentences of $\mathcal{F}$, and define $\bigstar\mathbf{B}_1, \bigstar\mathbf{B}_2, \ldots, \bigstar\mathbf{B}_k, \ldots$ in order. When, in the definition of $\bigstar\mathbf{B}_k$, we refer to the first "new n-ary existential constant", we mean the first of the constants $\overline{g}^n_j$ which has not been used in the definition of $\bigstar\mathbf{B}_1, \ldots, \bigstar\mathbf{B}_1, \ldots, \bigstar\mathbf{B}_{k-1}$, or used previously in the definition of $\bigstar\mathbf{B}_k$.

To define $\bigstar\mathbf{B}$ for a given sentence $\mathbf{B}$ of $\mathcal{F}$, first let $\bigstar^0\mathbf{B}$ be a rectified sentence (see §22) obtained from $\mathbf{B}$ by deleting vacuous quantifiers and making alphabetic changes of bound variables; thus $\vdash \mathbf{B} \equiv \bigstar^0\mathbf{B}$. Let m be the number of negative occurrences of quantifiers in $\bigstar^0\mathbf{B}$. We inductively define $\bigstar^i\mathbf{B}$, for $0 < i \leq m$, and let $\bigstar\mathbf{B}$ be $\bigstar^m\mathbf{B}$. Given $\bigstar^i\mathbf{B}$, let $\forall\mathbf{y}\mathbf{D}$ be the first (in left-to-right order) negative occurrence of a quantifier with its scope in $\bigstar^i\mathbf{B}$, and let $\mathbf{z}_1, \ldots, \mathbf{z}_n$ (where $n \geq 0$) be the free variables of $\forall\mathbf{y}\ \mathbf{D}$. Let $\overline{\mathbf{g}}^n$ be the first new n-ary existential constant, and let $\bigstar^{i+1}\mathbf{B}$ be the result of replacing $\forall\mathbf{y}\ \mathbf{D}$ by $\mathsf{S}^{\mathbf{y}}_{\overline{\mathbf{g}}^n\mathbf{z}_1\ldots\mathbf{z}_n}\mathbf{D}$ in $\bigstar^i\mathbf{B}$. (Naturally, if $n = 0$, $\forall\mathbf{y}\ \mathbf{D}$

is replaced by $\mathsf{S}^{\mathbf{y}}_{\cdot\overline{\mathbf{g}}^0}\mathbf{D}$.)

This completes the definition of $\bigstar\mathbf{B}$. Clearly $\bigstar^i\mathbf{B}$ contains $m-i$ negative occurrences of quantifiers, so $\bigstar\mathbf{B}$ is a universal sentence.

We shall refer to $\bigstar\mathbf{B}$ as the *Skolemized form*[2], or the *functional form (for satisfiability)* of **B**. Since Skolem introduced [Skolem, 1928], and Herbrand extensively exploited [Herbrand, 1930], the idea of replacing existential quantifiers by function symbols, the existential constants are sometimes referred to *Skolem functors* or *Herbrand functors.*

EXAMPLE 1: If B is $\forall z\ \exists y\ Pzy$, $\bigstar B$ is $\forall z \sim\sim Pz\overline{g}^1z$.

EXAMPLE 2: (We shall introduce a^0, g^1, and h^1 as existential constants. For greater legibility we shall write a^0 as a, g^1z as (gz), and h^1z as (hz).)

$$\begin{aligned}
B &= \exists x\ \forall z[Pzx \equiv \forall u\forall v \,\centerdot\, Pzu \supset \,\centerdot\, Puv \supset \sim Pvz]\\
\bigstar^0 B &= \exists x\ \forall z[[Pzx \supset \forall u\forall v \,\centerdot\, Pzu \supset \,\centerdot\, Puv \supset \sim Pvz]\\
&\quad \wedge \centerdot\, \forall y\forall w[Pzy \supset \,\centerdot\, Pyw \supset \sim Pwz] \supset Pzx]\\
\bigstar^1 B &= \sim\sim \forall z[[Pza \supset \forall u\forall v \,\centerdot\, Pzu \supset \,\centerdot\, Puv \supset \sim Pvz]\\
&\quad \wedge \centerdot\, \forall y\forall w[Pzy \supset \,\centerdot\, Pyw \supset \sim Pwz] \supset Pza]\\
\bigstar^2 B &= \sim\sim \forall z[[Pza \supset \forall u\forall v \,\centerdot\, Pzu \supset \,\centerdot\, Puv \supset \sim Pvz]\\
&\quad \wedge \centerdot\, \forall w[Pz(gz) \supset \,\centerdot\, P(gz)w \supset \sim Pwz] \supset Pza]\\
\bigstar B &= \sim\sim \forall z[[Pza \supset \forall u\forall v \,\centerdot\, Pzu \supset \,\centerdot\, Puv \supset \sim Pvz]\\
&\quad \wedge \centerdot\, [Pz(gz) \supset \,\centerdot\, P(gz)(hz) \supset \sim P(hz)z] \supset Pza]
\end{aligned}$$

It should be observed that if the negative quantifier $\forall w$ had been eliminated before $\forall y$, then w would have been replaced first by (h^2zy), and then by $(h^2z(g^1z))$ when $\forall y$ was eliminated. The simplicity of (h^1z) as compared with $(h^2z(g^1z))$ is the chief reason for eliminating the negative quantifiers in left-to-right order.

3300 Lemma. For any sentence **B** of $\mathcal{F}$, $\vdash_{\mathcal{F}^\star} (\bigstar\mathbf{B}) \supset \mathbf{B}$.

Proof: In the definition of $\bigstar^{i+1}\mathbf{B}$, $\vdash \forall\mathbf{y}\mathbf{D} \supset \mathsf{S}^{\mathbf{x}}_{\cdot\overline{g}^n\mathbf{z}_1\ldots\mathbf{z}_n}\mathbf{D}$, since the fact that $\bigstar^i\mathbf{B}$ is rectified assures that $\overline{g}^n\mathbf{z}_1\ldots\mathbf{z}_n$ is free for **y** in **D**. Hence $\vdash \bigstar^{i+1}\mathbf{B} \supset \bigstar^i\mathbf{B}$ for $0 \leq i < m$ by the Substitutivity of Implication (2105), so $\vdash (\bigstar\mathbf{B}) \supset \mathbf{B}$ by Rule P. ∎

[2]This is to be distinguished from the *Skolem normal form* of a wff, in which the sentence is in prenex normal form, has at least one existential quantifier, and all existential quantifiers precede all universal quantifiers; see [Church, 1956, pp. 224-227]

3301 Theorem. Let $\mathcal{G}$ be a set of sentences of $\mathcal{F}$ and let $\mathcal{G}^\star = \{\bigstar\mathbf{B} \mid \mathbf{B} \in \mathcal{G}\}$. Then $\mathcal{G}^\star$ has a model iff $\mathcal{G}$ has a model, and $\mathcal{F}^\star \cup \mathcal{G}^\star$ is a conservative extension $\mathcal{F} \cup \mathcal{G}$.

EXAMPLE: Let $\mathcal{G}$ be the set $\{G1, \ldots, G7\}$ of postulates for the elementary theory of additive abelian groups discussed near the beginning of §21. Here $\mathcal{G}^\star$ is the result of replacing $G6$ by $\bigstar G6$, which is (modulo deletion of a double negation) $\forall x \centerdot x + gx = 0$. Traditionally one writes gx as $-x$, and abbreviates $\forall x \centerdot x + (-x) = 0$ as $\forall x \centerdot x - x = 0$.

We could have used the single postulate $\exists z \centerdot \forall x[x + z = x] \wedge \forall x\ \exists y[x + y = z]$ in place of $G5$ and $G6$ in $\mathcal{G}$. The Skolemized form of this sentence could be written as $\forall x[x + 0 = x] \wedge \forall x[x - x = 0]$.

REMARK. Since $\mathcal{F}^\star \cup \mathcal{G}^\star$ is a conservative extension of $\mathcal{F} \cup \mathcal{G}$, one knows that one can use the convenient notation introduced by the Skolem functions without changing the mathematical content of the theory.

Proof: By Lemma 3300 and the Soundness Theorem (2303), every model of $\mathcal{G}^\star$ is a model of $\mathcal{G}$, and $\mathcal{F}^\star \cup \mathcal{G}^\star$ is an extension of $\mathcal{F} \cup \mathcal{G}$.

If $\mathcal{G}$ has a model, it has a denumerable model by the Löwenheim-Skolem Theorem (2513), so let $\mathcal{M} = \langle \mathcal{D}, \mathcal{J} \rangle$ be any denumerable model of the first order theory $\mathcal{F} \cup \mathcal{G}$. We shall extend $\mathcal{M}$ to a model $\mathcal{M}^\star = \langle \mathcal{D}, \mathcal{J}^\star \rangle$ of $\mathcal{F}^\star \cup \mathcal{G}^\star$ with the same domain of individuals such that for any sentence $\mathbf{B}$ of $\mathcal{F}$, $\mathcal{M} \models \mathbf{B}$ iff $\mathcal{M}^\star \models \bigstar\mathbf{B}$. Since the domain of individuals of $\mathcal{M}$ is in one-one correspondence with the natural numbers, we may assume it is well-ordered. (This will enable us to avoid any appeal to the Axiom of Choice in the argument below.) Let 0 denote the least individual of $\mathcal{D}$ under this ordering.

For each existential constant $\overline{\mathbf{g}}^n$ of $\mathcal{F}^\star$ there is a unique sentence $\mathbf{B}$ of $\mathcal{F}$ such that $\overline{\mathbf{g}}^n$ occurs in $\bigstar\mathbf{B}$. We extend $\mathcal{J}$ to $\mathcal{J}^\star$ in an inductive fashion by considering the existential constants in the order in which they were used in defining the sequence $\bigstar\mathbf{B}_1, \bigstar\mathbf{B}_2, \ldots, \bigstar\mathbf{B}_k, \ldots$, where $\mathbf{B}_k$ is the kth sentence of $\mathcal{F}$. As we extend $\mathcal{J}$ to $\mathcal{J}^\star$ we show that if $\mathbf{B}$ is any sentence of $\mathcal{F}$ and $\bigstar^i\mathbf{B}$ is defined and $\mathcal{J}$ has been extended to an interpretation function $\mathcal{J}'$ whose domain contains all existential constants of $\bigstar^i\mathbf{B}$, then $\mathcal{V}^{\langle \mathcal{D}, \mathcal{J}' \rangle}\mathbf{B} = \mathcal{V}^{\langle \mathcal{D}, \mathcal{J} \rangle}\mathbf{B}$.

Suppose $\bigstar^{i+1}\mathbf{B}$ was obtained by replacing a wf part $\forall \mathbf{y}\ \mathbf{D}$ of $\bigstar^i\mathbf{B}$ by $\mathsf{S}^{\mathbf{y}}_{\centerdot\overline{\mathbf{g}}^n\mathbf{z}_1\ldots\mathbf{z}_n}\mathbf{D}$, where $\mathbf{z}_1, \ldots, \mathbf{z}_n$ are the free variables of $\forall \mathbf{y}\ \mathbf{D}$. We may suppose that $\mathcal{J}$ has already been extended to $\mathcal{J}'$, which is already defined on all constants of $\bigstar^i\mathbf{B}$, so $\mathcal{V}^{\langle \mathcal{D}, \mathcal{J}' \rangle}\bigstar^i\mathbf{B} = \mathcal{V}^{\langle \mathcal{D}, \mathcal{J} \rangle}\mathbf{B}$, but $\mathcal{J}'\overline{\mathbf{g}}^n$ is not defined. We extend $\mathcal{J}'$ to a function $\mathcal{J}''$ which also provides an interpretation for $\mathbf{\overline{g}}^n$. Let $d_1, \ldots, d_n$ be arbitrary elements of $\mathcal{D}$; we must define $(\mathcal{J}''\mathbf{\overline{g}}_n)(d_1, \ldots, d_n)$. Let φ be an assignment such that $\varphi\mathbf{z}_j = d_j$ for $1 \leq j \leq n$. Let $(\mathcal{J}''\mathbf{\overline{g}}^n)(d_1, \ldots, d_n)$ be the least member d of $\mathcal{D}$ such that $\mathcal{V}_\psi^{\langle \mathcal{D}, \mathcal{J}' \rangle}\mathbf{D} = \mathsf{F}$, where ψ agrees with φ off $\mathbf{y}$ and $\psi\mathbf{y} = d$, if such an element exists; if not, let $d = 0$. This completes the definition of $\mathcal{J}''\overline{g}^n$.

We next show that for any assignment φ,

$$\mathcal{V}_\varphi^{\langle \mathcal{D}, \mathcal{J}'' \rangle}\forall \mathbf{y}\ \mathbf{D} = \mathcal{V}_\varphi^{\langle \mathcal{D}, \mathcal{J}'' \rangle}\mathsf{S}^{\mathbf{y}}_{\cdot\mathbf{\overline{g}}^n\mathbf{z}_1\ldots\mathbf{z}_n}\mathbf{D}.$$

If $\mathcal{V}_\varphi\mathsf{S}^{\mathbf{y}}_{\cdot\mathbf{\overline{g}}\, n\mathbf{z}\ldots\mathbf{z}_n}\mathbf{D} = \mathsf{F}$, then $\mathcal{V}_\varphi\forall\mathbf{y}\ \mathbf{D} = \mathsf{F}$, since $\vdash \forall\mathbf{y}\ \mathbf{D} \supset \mathsf{S}^{\mathbf{y}}_{\cdot\mathbf{\overline{g}}\, n\mathbf{z}_1\ldots\mathbf{z}_n}\mathbf{D}$, and every theorem is valid. Suppose $\mathcal{V}_\varphi\forall\mathbf{y}\ \mathbf{D} = \mathsf{F}$. Then there is an assignment ψ which agrees with φ off $\mathbf{y}$ such that $\mathcal{V}_\psi\mathbf{D} = \mathsf{F}$; let ψ be such an assignment with the least possible value for $\psi\mathbf{y}$. Then $\mathcal{V}_\varphi\mathbf{\overline{g}}^n\mathbf{z}_1\ldots\mathbf{z}_n = (\mathcal{J}''\mathbf{\overline{g}}^n)(\varphi\mathbf{z}_1, \ldots, \varphi\mathbf{z}_n) = \psi\mathbf{y}$ by the definition of $\mathcal{J}''\mathbf{\overline{g}}^n$ (since $\mathcal{V}_\psi^{\langle \mathcal{D}, \mathcal{J}' \rangle}\mathbf{D} = \mathcal{V}_\psi^{\langle \mathcal{D}\, \mathcal{J}'' \rangle}\mathbf{D}$), so $\mathcal{V}_\varphi\mathsf{S}^{\mathbf{y}}_{\cdot\mathbf{\overline{g}}^n\mathbf{z}_1\ldots\mathbf{z}_n}\mathbf{D} = \mathcal{V}_\psi\mathbf{D} = \mathsf{F}$ by the Substitution-Value Theorem (2302).

It is now clear that $\langle \mathcal{D}, \mathcal{J}'' \rangle \models \forall\mathbf{z}_1 \ldots \forall\mathbf{z}_n \centerdot \forall\mathbf{y}\ \mathbf{D} \equiv \mathsf{S}^{\mathbf{y}}_{\cdot\mathbf{\overline{g}}^n\mathbf{z}_1\ldots\mathbf{z}_n}\mathbf{D}$, but by the Substitutivity of Equivalence (2106), $\vdash \forall\mathbf{z}_1 \ldots \forall\mathbf{z}_n[\forall\mathbf{y}\ \mathbf{D} \equiv \mathsf{S}^{\mathbf{y}}_{\cdot\mathbf{\overline{g}}^n\mathbf{z}_1\ldots\mathbf{z}_n}\mathbf{D}] \supset \centerdot \bigstar^i\mathbf{B} \equiv \bigstar^{i+1}\mathbf{B}$. Since every theorem is valid, $\langle \mathcal{D}, \mathcal{J}'' \rangle \models \bigstar^i\mathbf{B} \equiv \bigstar^{i+1}\mathbf{B}$. Since $\mathbf{\overline{g}}^n$ does not occur in $\bigstar^i\mathbf{B}$, it is easy to see that $\mathcal{V}^{\langle \mathcal{D}, \mathcal{J}'' \rangle}\bigstar^{i+1}\mathbf{B} = \mathcal{V}^{\langle \mathcal{D}, \mathcal{J}'' \rangle}\bigstar^i\mathbf{B} = \mathcal{V}^{\langle \mathcal{D}, \mathcal{J}' \rangle}\bigstar^i\mathbf{B} = \mathcal{V}^{\langle \mathcal{D}, \mathcal{J} \rangle}\mathbf{B}$, and the induction is complete.

Naturally when $\bigstar^i\mathbf{B}$ is $\bigstar\mathbf{B}$, no further extension of $\mathcal{J}'$ to $\mathcal{J}^\star$ affects the value of $\mathbf{B}$, so $\mathcal{V}^{\mathcal{M}^\star}\bigstar\mathbf{B} = \mathcal{V}^{\langle \mathcal{D}, \mathcal{J}' \rangle}\bigstar\mathbf{B} = \mathcal{V}^{\mathcal{M}}\mathbf{B}$. Consequently, for any denumerable model $\mathcal{M}$ of $\mathcal{F} \cup \mathcal{G}$, $\mathcal{M}^\star$ is a model of $\mathcal{F}^\star \cup \mathcal{G}^\star$.

To complete the proof of the theorem, suppose $\mathbf{C}$ is any wff of $\mathcal{F}$ such that $\vdash_{\mathcal{F}^\star\cup\mathcal{G}^\star} \mathbf{C}$. Let $\mathcal{M}$ be any denumerable model of $\mathcal{F} \cup \mathcal{G}$. Then $\mathcal{M}^\star$ is a model of $\mathcal{F}^\star \cup \mathcal{G}^\star$, so $\mathcal{M}^\star \models \mathbf{C}$ by the Soundness Theorem (2303b), so $\mathcal{M} \models \mathbf{C}$. Thus $\mathbf{C}$ is valid in all denumerable models of $\mathcal{F} \cup \mathcal{G}$, so $\vdash_{\mathcal{F}\cup\mathcal{G}} \mathbf{C}$ by the Generalized Completeness Theorem (2509). ■

3302 Corollary. A sentence $\mathbf{B}$ of $\mathcal{F}$ is satisfiable iff $\bigstar\mathbf{B}$ is satisfiable.

Proof: Let $\mathcal{G} = \{\mathbf{B}\}$.

EXAMPLE: Let $B = \forall x[\exists y\ Pxy \supset \exists y\ Pxy]$, where P is a binary predicate constant. B is valid, hence satisfiable. Hence the sentence $\forall x[\exists y\, Pxy \supset Px(gx)]$ is also satisfiable.

3303 Corollary. Let $\mathbf{A}$ be any wff, and let $\overline{\mathbf{A}}$ be its universal closure. $\mathbf{A}$ is valid iff $\bigstar \sim \overline{\mathbf{A}}$ is unsatisfiable.

Proof: The following conditions are equivalent:

(a) $\mathbf{A}$ is valid.
(b) $\overline{\mathbf{A}}$ is valid. (by 2301 c)
(c) $\sim \overline{\mathbf{A}}$ is unsatisfiable. (2301 a)
(d) $\bigstar \sim \overline{\mathbf{A}}$ is unsatisfiable. (by 3302)

■

In addition to the method of Skolemization $\bigstar$ which is discussed above, there are two other methods of performing Skolemization which have been discussed in the literature. We shall call them $\mathcal{S}_1$ and $\mathcal{S}_2$. In each method one replaces wf parts $\exists \mathbf{y}\ \mathbf{M}(\mathbf{y})$ of a rectified sentence $\mathbf{C}$ by $\mathbf{M}(\mathbf{g}\mathbf{x}^1 \ldots \mathbf{x}^n)$ where $n \geq 0$, $\mathbf{g}$ is a new function (or individual) constant, and the arguments $\mathbf{x}^1, \ldots, \mathbf{x}^n$ of $\mathbf{g}$ are certain variables. In method $\mathcal{S}_1$, the arguments of $\mathbf{g}$ are the variables $\mathbf{x}^i$ such that there is an occurrence of $\forall \mathbf{x}^i$ in $\mathbf{C}$ with $\exists \mathbf{y}\ \mathbf{M}$ in its scope. In method $\mathcal{S}_2$, the arguments of $\mathbf{g}$ are the variables $\mathbf{x}^i$ such that there is an occurrence of $\forall \mathbf{x}^i$ in $\mathbf{C}$ with the minimal scope of $\exists \mathbf{y}\ \mathbf{M}$ in its minimal scope. Of course, in method $\bigstar$, the arguments of $\mathbf{g}$ are the free variables of $\exists \mathbf{y} \mathbf{M}$. For each method, the result of eliminating all existential quantifiers from $\mathbf{C}$ in this way is called the Skolemization $\mathcal{S}(\mathbf{C})$ of $\mathbf{C}$, and it can be shown that $\mathbf{C}$ has a model if and only if $\mathcal{S}(\mathbf{C})$ has a model. We shall write $\mathcal{S}(\mathbf{C})$ as $\mathcal{S}_1(\mathbf{C})$, $\mathcal{S}_2(\mathbf{C})$, or $\bigstar(\mathbf{C})$ when we need to specify which Skolemization method is used. To see that the three methods can give different results, let C be the sentence

$$\forall x \exists w \forall u [\exists z\ Pwz \ \wedge\ [Mxuw \ \vee\ \exists y\ Quy]].$$

$$\mathcal{S}_1(C) \text{ is } \forall x \forall u [P(fx)(gxu) \ \wedge\ [Mxu(fx) \ \vee\ Qu(hxu)]].$$

$$\mathcal{S}_2(C) \text{ is } \forall x \forall u [P(fx)(gx) \ \wedge\ [Mxu(fx) \ \vee\ Qu(hxu)]].$$

$$\bigstar(C) \text{ is } \forall x \forall u [P(fx)(gx) \ \wedge\ [Mxu(fx) \ \vee\ Qu(hu)]].$$

The most commonly used method of Skolemization is $\mathcal{S}_1$, which was introduced by Skolem [Skolem, 1928] (see pp. 517-518 of [van Heijenoort, 1967], though he applied it only to wffs in prenex normal form. In [Herbrand, 1930], Herbrand used both methods $\mathcal{S}_1$ and $\mathcal{S}_2$ (see pp. 138, 151, and 154-154 of [Herbrand, 1971] or pp. 533 and 542-544 of [van Heijenoort, 1967]), and applied them to wffs which need not be in prenex normal form.

Herbrand used $\mathcal{S}_1$ in defining Property C of wffs, and $\mathcal{S}_2$ in defining Property B. Method ★ is used in [Andrews, 1981]. When applied to formulas in antiprenex (miniscope) form, $\mathcal{S}_2$ seems to represent a compromise between $\mathcal{S}_1$ and ★. As illustrated above, in some cases ★ gives the Skolem functions fewer arguments than $\mathcal{S}_1$ or $\mathcal{S}_2$, since $\mathbf{x}^i$ may not occur free in $\exists \mathbf{y}\ \mathbf{M}$ even though $\exists \mathbf{y}\ \mathbf{M}$ is in the scope of $\forall \mathbf{x}^i$. Therefore ★(**C**) may be simpler than $\mathcal{S}_1(\mathbf{C})$ or $\mathcal{S}_2(\mathbf{C})$.

§34. Refutations of Universal Sentences

In this section we extend the Ground Resolution Theorem (1600) to first-order logic. The proof procedure whose completeness is established by Theorem 3401 below has many of the elements of the Resolution method for automated theorem proving introduced in [Robinson, 1965]. However, the Resolution method has an additional very important element, the use of a unification algorithm to generate the substitutions which are needed to prepare for applications of the Simplification and Cut rules discussed below. Such an algorithm will be discussed in §36.

When one uses the semantic tableau method (see §32), one can introduce many unnecessary nodes by performing universal instantiations which are not actually needed for the proof. Of course, when one has a universally quantified formula to instantiate, one may not yet have enough information to decide intelligently which instantiation terms to use. One approach to this problem is to simply instantiate universal quantifiers with variables, and make appropriate substitutions for these variables later. However, if one is not careful how one handles existential instantiations in this context, one can destroy the soundness of the refutation procedure. A way to avoid this problem is to Skolemize the sentence to be refuted before starting the refutation. When this is done, one simply needs to refute a universal sentence, and existential instantiations never need be made.

Another feature of the semantic tableau method which is sometimes disadvantageous is the division into cases (two branches of the tree) whenever one encounters a disjunction. When one encounters many disjunctions, and does not know which wffs are actually needed for the proof, this procedure can lead to a very redundant tree with a large number of branches. An alternative to this method of handling disjunctions is to use the Cut rule (see §31).

In this section we discuss a method for refuting sets of universal sentences which incorporates these ideas. We also consider extending the preprocessing so as to eliminate conjunctions as well as essentially existential quantifiers.

For this section, it is convenient to use a formulation of first-order logic which we call $\mathcal{F}_{\mathsf{f}}$, whose primitive symbols are those of $\mathcal{F}$ plus a 0-ary propositional connective f which is false in all interpretations. (With appropriate minor modifications, the theorem below also applies to formulations of $\mathcal{F}$ which do not contain f.) It is readily verified that Smullyan's Unifying Principle (2507) applies to $\mathcal{F}_{\mathsf{f}}$ if one modifies condition (a) in the definition of abstract consistency class to read:

(a') $\mathsf{f} \notin \mathcal{S}$, and if $\mathbf{A}$ is atomic, then $\mathbf{A} \notin \mathcal{S}$ or $\sim \mathbf{A} \notin \mathcal{S}$.

As in §31, occurrences of wf parts of a wff $\mathbf{A}$ of $\mathcal{F}_{\mathsf{f}}$ which are in the scope of no negations or quantifiers of $\mathbf{A}$ are called disjunctive components of $\mathbf{A}$. We shall use $\Box$ as a notation for the empty disjunction or "null formula". (Technically, this is a slightly different usage of $\Box$ from that in §16, though the two usages are closely related; they simply occur in slightly different contexts.) As in §30, we use $\mathbf{M}$ and $\mathbf{N}$ as syntactical variables which may take as values $\Box$ as well as wffs, and we regard $\mathbf{A} \lor \Box$ and $\Box \lor \mathbf{A}$ as abbreviations for $\mathbf{A}$. (For example, we may regard $\sim\sim \mathbf{B}$ as having the form $\mathbf{M} \lor \sim\sim \mathbf{B}$.) $\Box$ standing alone may be regarded as an abbreviation for f.

DEFINITION. An *$\mathcal{R}$-derivation of* $\mathbf{G}$ *from* a set $\mathcal{S}$ of universal sentences is a finite sequence $\mathbf{B}_1, \ldots, \mathbf{B}_k$ of wffs such that $\mathbf{B}_k = \mathbf{G}$, and for each i ($1 \le i \le k$), $\mathbf{B}_i \in \mathcal{S}$ or $\mathbf{B}_i$ is inferred from preceding members of the sequence by one of the following rules of inference:

($\mathcal{R}1$) **Disjunction Rules.** To replace a disjunctive component $\mathbf{D}$ of a wff by $\mathbf{E}$, where

$\mathbf{D}$ is $[[\mathbf{A} \lor \mathbf{B}] \lor \mathbf{C}]$ and $\mathbf{E}$ is $[\mathbf{A} \lor [\mathbf{B} \lor \mathbf{C}]]$, or
$\mathbf{D}$ is $[\mathbf{A} \lor [\mathbf{B} \lor \mathbf{C}]]$ and $\mathbf{E}$ is $[[\mathbf{A} \lor \mathbf{B}] \lor \mathbf{C}]$, or
$\mathbf{D}$ is $[\mathbf{A} \lor \mathbf{B}]$ and $\mathbf{E}$ is $[\mathbf{B} \lor \mathbf{A}]$.

($\mathcal{R}2$) **Simplification.** From $\mathbf{M} \lor \mathbf{A} \lor \mathbf{A}$ to infer $\mathbf{M} \lor \mathbf{A}$.

($\mathcal{R}3$) **Substitution.** From $\mathbf{A}$ to infer $\mathsf{S}^{\mathbf{x}_1 \ldots \mathbf{x}_n}_{\mathbf{t}_1 \ldots \mathbf{t}_n} \mathbf{A}$, where $\mathbf{x}_1, \ldots, \mathbf{x}_n$ are distinct individual variables and $\mathbf{t}_1, \ldots, \mathbf{t}_n$ are terms such that $\mathbf{t}_i$ is free for $\mathbf{x}_i$ in $\mathbf{A}$ for $1 \le i \le n$.

($\mathcal{R}4$) **Cut.** From $\mathbf{M} \vee \mathbf{A}$ and $\mathbf{N} \vee \sim \mathbf{A}$ to infer $\mathbf{M} \vee \mathbf{N}$.

($\mathcal{R}5$) **Negation Elimination.** From $\mathbf{M} \vee \sim\sim \mathbf{A}$ to infer $\mathbf{M} \vee \mathbf{A}$.

($\mathcal{R}6$) **Conjunction Elimination.** From $\mathbf{M} \vee \sim [\mathbf{A} \vee \mathbf{B}]$ to infer $\mathbf{M} \vee \sim \mathbf{A}$ and $\mathbf{M} \vee \sim \mathbf{B}$.

($\mathcal{R}7$) **Universal Instantiation.** From $\mathbf{M} \vee \forall \mathbf{x} \mathbf{A}$ to infer $\mathbf{M} \vee \mathbf{A}$.

REMARK. It might be supposed that the α Rule is also needed for a complete refutation system, to assure that $\mathbf{x}$ is distinct from the free variables of $\mathbf{M}$ in $\mathcal{R}7$ so that after applying $\mathcal{R}7$ one can substitute for the free occurrences of $\mathbf{x}$ in $\mathbf{A}$ without involving those in $\mathbf{M}$. However, one can use $\mathcal{R}3$ to substitute new variables for the free variables of $\mathbf{M} \vee \forall \mathbf{x} \mathbf{A}$ before applying $\mathcal{R}7$.

DEFINITION. We shall write $\mathcal{S} \vdash_{\mathcal{R}} \mathbf{E}$ to indicate that there is an $\mathcal{R}$-derivation of $\mathbf{E}$ from $\mathcal{S}$. An $\mathcal{R}$-derivation of $\Box$ from $\mathcal{S}$ will be called an *$\mathcal{R}$-refutation* of $\mathcal{S}$. If $\mathcal{S} \vdash_{\mathcal{R}} \Box$ we shall say that $\mathcal{S}$ is *$\mathcal{R}$-refutable.*

3400 Theorem. Let $\mathcal{S}$ be any set of universal sentences. $\mathcal{S}$ is $\mathcal{R}$-refutable iff $\mathcal{S}$ is unsatisfiable.

Proof: Each of the rules $\mathcal{R}1$ - $\mathcal{R}7$ preserves validity in any model, so if $\mathcal{S}$ is a set of sentences and $\mathcal{M}$ is a model for $\mathcal{S}$ and $\mathcal{S} \vdash_{\mathcal{R}} \mathbf{E}$, then $\mathcal{M} \models \mathbf{E}$. Consequently, if $\mathcal{S}$ has a model, $\mathcal{S}$ cannot be $\mathcal{R}$-refutable.

To prove the completeness of the method, let Γ be the class of all finite sets $\mathcal{S}$ of universal sentences which are not $\mathcal{R}$-refutable. Note that when $\mathcal{S}$ is any finite set of universal sentences, $\mathcal{S} \notin \Gamma$ means that $\mathcal{S} \vdash_{\mathcal{R}} \Box$. We shall show that Γ is an abstract consistency class. Now suppose that $\mathcal{S}$ is an unsatisfiable set of universal sentences. By the Compactness Theorem (2514) $\mathcal{S}$ has an unsatisfiable finite subset $\mathcal{S}_0$, and $\mathcal{S}_0$ is sufficiently pure, so by Smullyan's Unifying Principle (2507) $\mathcal{S}_0 \notin \Gamma$, so $\mathcal{S}_0 \vdash_{\mathcal{R}} \Box$, so $\mathcal{S} \vdash_{\mathcal{R}} \Box$.

Next we verify that Γ is an abstract consistency class.

Clearly Γ is closed under subsets.

(a′) If $\Box \in \mathcal{S}$, then $\mathcal{S} \vdash_{\mathcal{R}} \Box$, and if $\{\mathbf{A}, \sim \mathbf{A}\} \subseteq \mathcal{S}$, then $\mathcal{S} \vdash_{\mathcal{R}} \Box$ by $\mathcal{R}4$, so if $\mathcal{S} \in \Gamma$, then condition (a′) is satisfied.

We check conditions (b)-(f) by the indirect method. In each case we assume that $\mathcal{S} \in \Gamma$ and that the stated condition is false, and derive the conclusion that $\mathcal{S} \vdash_{\mathcal{R}} \Box$, contradicting the assumption that $\mathcal{S} \in \Gamma$.

(b) We assume $\sim\sim \mathbf{A} \in \mathcal{S}$ and $\mathcal{S} \cup \{\mathbf{A}\} \vdash_{\mathcal{R}} \Box$. Then $\mathcal{S} \vdash_{\mathcal{R}} \Box$ by $\mathcal{R}5$.

(c) We assume that $[\mathbf{A} \lor \mathbf{B}] \in \mathcal{S}$ and $\mathcal{S} \cup \{\mathbf{A}\} \vdash_{\mathcal{R}} \Box$ and $\mathcal{S} \cup \{\mathbf{B}\} \vdash_{\mathcal{R}} \Box$. Let $\mathbf{C}_1, \ldots, \mathbf{C}_n$ be a refutation of $\mathcal{S} \cup \{\mathbf{B}\}$, and let $\mathbf{E}_1, \ldots, \mathbf{E}_m$ be a refutation of $\mathcal{S} \cup \{\mathbf{A}\}$. It is easy to establish by induction on i that for each i $(1 \leq i \leq n)$, $\mathcal{S} \vdash_{\mathcal{R}} \mathbf{C}_i$ or $\mathcal{S} \vdash_{\mathcal{R}} \mathbf{A} \lor \mathbf{C}_i$. Simply consider how $\mathbf{C}_i$ was inferred in the refutation of $\mathcal{S} \cup \{\mathbf{B}\}$. In case $\mathbf{C}_i$ is $\mathbf{B}$, $\mathbf{A} \lor \mathbf{C}_i$ is $\mathbf{A} \lor \mathbf{B}$, which is in $\mathcal{S}$. In case $\mathbf{C}_i$ was inferred by substitution, note that $\mathbf{A}$ is a sentence. In case $\mathbf{C}_i$ was inferred by Cut ($\mathcal{R}4$), rules $\mathcal{R}1$ and $\mathcal{R}2$ can be used (if necessary) to infer $\mathbf{A} \lor [\mathbf{M} \lor \mathbf{N}]$ from $[\mathbf{A} \lor \mathbf{M}] \lor [\mathbf{A} \lor \mathbf{N}]$. Hence $\mathcal{S} \vdash_{\mathcal{R}} \Box$ or $\mathcal{S} \vdash_{\mathcal{R}} \mathbf{A}$. In the latter case it is easy to establish that $\mathcal{S} \vdash_{\mathcal{R}} \mathbf{E}_j$ for $1 \leq j \leq m$ by induction on j, so $\mathcal{S} \vdash_{\mathcal{R}} \Box$ in this case also.

(d) We assume that $\sim [\mathbf{A} \lor \mathbf{B}] \in \mathcal{S}$ and $\mathcal{S} \cup \{\sim \mathbf{A}, \sim \mathbf{B}\} \vdash_{\mathcal{R}} \Box$. Then $\mathcal{S} \vdash_{\mathcal{R}} \Box$ by $\mathcal{R}6$.

(e) We assume that $\forall \mathbf{x} \mathbf{A} \in \mathcal{S}$ (so $\forall \mathbf{x} \mathbf{A}$ is a sentence) and that there is a variable-free term $\mathbf{t}$ such that $\mathcal{S} \cup \left\{ \mathsf{S}^{\mathbf{x}}_{\mathbf{t}} \mathbf{A} \right\} \vdash_{\mathcal{R}} \Box$. Then $\mathcal{S} \vdash_{\mathcal{R}} \Box$ by $\mathcal{R}7$ and $\mathcal{R}3$.

(f) The condition is vacuously satisfied. Since $\mathcal{S} \in \Gamma$, $\mathcal{S}$ is a set of universal sentences, so it is impossible that $\sim \forall \mathbf{x} \mathbf{A} \in \mathcal{S}$. ■

3401 Theorem. Let $\mathbf{A}$ be any wff, let $\overline{\mathbf{A}}$ be its universal closure, and let $\mathbf{M}$ be the quantifier-free wff obtained by deleting all quantifiers from $\bigstar \sim \overline{\mathbf{A}}$. Let $\mathbf{N}_1 \land \ldots \land \mathbf{N}_k$ be a conjunctive normal form of $\mathbf{M}$ (so each $\mathbf{N}_i$ is a disjunction of literals).

The following conditions are equivalent:

(a) $\models \mathbf{A}$.

(b) $\{\mathbf{N}_1, \ldots, \mathbf{N}_k\}$ has no model.

(c) $\Box$ can be derived from $\{\mathbf{N}_1, \ldots, \mathbf{N}_k\}$ by using rules $\mathcal{R}1$, $\mathcal{R}2$, $\mathcal{R}3$, and $\mathcal{R}4$.

Proof: Note that $\overline{\mathbf{A}}$ is a sentence, and $\bigstar \sim \overline{\mathbf{A}}$ is a rectified universal sentence, so it has $\forall \mathbf{x}_1 \ldots \forall \mathbf{x}_n \mathbf{M}$ as a prenex normal form, where $\mathbf{x}_1, \ldots, \mathbf{x}_n$ (where $n \geq 0$) are the bound variables of $\bigstar \sim \overline{\mathbf{A}}$. Hence

$\vdash \bigstar \sim \overline{\mathbf{A}} \equiv \forall \mathbf{x}_1 \ldots \forall \mathbf{x}_n \mathbf{M}$.

$\vdash \mathbf{M} \equiv . \mathbf{N}_1 \land \ldots \land \mathbf{N}_k$ by Rule P

$\vdash \forall \mathbf{x}_1 \ldots \forall \mathbf{x}_n \mathbf{M} \equiv \forall \mathbf{x}_1 \ldots \forall \mathbf{x}_n [\mathbf{N}_1 \land \ldots \land \mathbf{N}_k]$

by the Substitutivity of Equivalence (2106)

$\vdash \forall \mathbf{x}_1 \ldots \forall \mathbf{x}_n [\mathbf{N}_1 \land \ldots \land \mathbf{N}_k] \equiv . \forall \mathbf{x}_1 \ldots \forall \mathbf{x}_n \mathbf{N}_1 \land \ldots \land \forall \mathbf{x}_1 \ldots \forall \mathbf{x}_n \mathbf{N}_k$

by 2129 and 2106

$\vdash \bigstar \sim \overline{\mathbf{A}} \equiv . \forall \mathbf{x}_1 \ldots \forall \mathbf{x}_n \mathbf{N}_1 \land \ldots \land \forall \mathbf{x}_1 \ldots \forall \mathbf{x}_n \mathbf{N}_k$ by Rule P.

Hence the following conditions are equivalent:

(a) $\models \mathbf{A}$
(a1) $\bigstar \sim \overline{\mathbf{A}}$ is unsatisfiable. by 3303
(b1) $\forall \mathbf{x}_1 \ldots \forall \mathbf{x}_n \mathbf{N}_1 \wedge \ldots \wedge \forall \mathbf{x}_1 \ldots \forall \mathbf{x}_n \mathbf{N}_k$ is unsatisfiable (by the Theorem above and 2303).
(b2) $\{\forall \mathbf{x}_1 \ldots \forall \mathbf{x}_n \mathbf{N}_1, \ldots, \forall \mathbf{x}_1 \ldots \forall \mathbf{x}_n \mathbf{N}_k\}$ has no model.
(b) $\{\mathbf{N}_1, \ldots, \mathbf{N}_k\}$ has no model.

Thus (a) and (b) are equivalent.

Clearly (c) implies (b), since the rules $\mathcal{R}1$ - $\mathcal{R}4$ preserve validity in any model. To prove the converse, suppose that (b), and hence (b2), is true. Let $\mathcal{S}$ be the set $\{\forall \mathbf{x}_1 \ldots \forall \mathbf{x}_n \mathbf{N}_1, \ldots, \forall \mathbf{x}_1 \ldots \forall \mathbf{x}_n \mathbf{N}_k\}$. By Theorem 3400, $\mathcal{S}$ is $\mathcal{R}$-refutable. It is not hard to see that from any $\mathcal{R}$-refutation of $\mathcal{S}$ one can construct an $\mathcal{R}$-refutation in which all applications of Universal Instantiation ($\mathcal{R}7$) precede all applications of the other rules. In wffs of $\mathcal{S}$ the scope of each negation is an atom, so Rules $\mathcal{R}5$ and $\mathcal{R}6$ can never be used. Thus the part of the refutation of $\mathcal{S}$ following all applications of $\mathcal{R}7$ is a refutation of $\{\mathbf{N}_1, \ldots, \mathbf{N}_k\}$ using only $\mathcal{R}1$ - $\mathcal{R}4$.

EXAMPLE: Let A be $\sim \forall z[Pzx \equiv \forall u \, \forall v \, \centerdot \, Pzu \supset \, \centerdot \, Puv \supset\sim Pvz]$. Then $\sim \overline{A}$ is the wff B in Example 2 of §33, so M is

$$\sim\sim [[\sim Pza \vee \sim Pzu \vee \sim Puv \vee \sim Pvz] \\ \wedge \centerdot \sim [\sim Pz(gz) \vee \sim P(gz)(hz) \vee \sim P(hz)z] \vee Pza].$$

Thus A is valid iff $\{[\sim Pza \vee \sim Pzu \vee \sim Puv \vee \sim Pvz], [Pz(gz) \vee Pza], [P(gz)(hz) \vee Pza], [P(hz)z \vee Pza]\}$ has no model. We derive $\Box$ from this set as follows:

(1)	$\sim Pza \vee \sim Pzu \vee \sim Puv \vee \sim Pvz$	given
(2)	$Pz(gz) \vee Pza$	given
(3)	$P(gz)(hz) \vee Pza$	given
(4)	$P(hz)z \vee Pza$	given
(5)	$\sim Paa$	Sub: 1, $\mathcal{R}2$
(6)	$Pa(ga)$	Sub: 2; Cut: 5
(7)	$P(ga)(ha)$	Sub: 3; Cut: 5
(8)	$P(ha)a$	Sub: 4; Cut: 5
(9)	$\sim P(ha)a \vee \sim P(ha)a \vee \sim Pa(ga) \vee \sim P(ga)(ha)$	Sub: 1
(10)	$\Box$	Cut: 9, 7, 6, 8

Thus A is valid.

If we replace $P\mathbf{xy}$ by $\mathbf{x} \in \mathbf{y}$, then the refutable wff $\sim \forall x A$ is equivalent to $\exists x \forall z \, \centerdot \, z \in x \equiv \sim \exists u \exists v \, \centerdot \, z \in u \wedge u \in v \wedge v \in z$. This wff expresses an extension of Russell's Paradox due to Quine.

EXERCISES

Establish the validity of the wffs below by one of the methods discussed in this section.

X3400. $\forall x[\exists y\ Pxy \supset \forall y\ Qxy] \land \forall z \exists y\ Pzy \supset \forall y \forall x\ Qxy$

X3401. $\exists x \forall y[Py \equiv Px] \supset \centerdot \forall x\ Px \lor \forall x \sim Px$

X3402. $\forall y \exists w\ Ryw \lor \exists z \forall x[Px \supset \sim Rzx] \supset \exists x \sim Px$

X3403. $\forall x \forall y \forall z[Pxy \supset \sim Pyz] \supset \exists y \forall x \sim Pxy$

X3404. $\forall x \forall y \centerdot [\exists z Hxz \supset \forall z Gxz] \land \forall z[Gzz \supset Hzy] \supset \centerdot Hxy \equiv \forall z\ Gxz$

X3405. $\exists x \centerdot [Px \supset Pa] \land [Px \supset Pb]$

X3406. $\forall x \exists y\ Rxy \land \forall u \forall v \forall w[Ruv \land Rvw \supset \centerdot Rwu \land \exists s \centerdot Rsv \land Rvs] \supset \forall z\ Rzz$

X3407. $\forall u \forall v \forall w[Puv \lor Pvw] \supset \exists x \forall y\ Pxy$

Determine whether the wffs below are valid. Justify your answers by appropriate arguments.

X3408. $\forall x \forall y \centerdot \sim Rxy \supset \forall z \centerdot Rzy \supset \sim Rxz$

X3409. $\exists y \forall x \centerdot Px \supset \centerdot Py \lor Qx$

X3410. $\forall z[\exists x\ \forall y\ Pxyz \supset Qz] \supset \centerdot \forall z \forall y \exists x\ Pxyz \supset \exists z\ Qz$

X3411. $\forall y[\forall x\ Pxy \supset \forall u\ Quy] \supset \exists z \centerdot \exists v\ Pvz \supset \exists w \centerdot Pzw \lor Qwz$

§35. Herbrand's Theorem

In this section, which is adapted from [Andrews, 1981][3], we shall prove a fundamental theorem which is a consequence of the results presented by Herbrand in his thesis [Herbrand, 1930]. While this theorem involves some of the most important ideas in Herbrand's thesis, it differs from any theorem Herbrand actually stated, so at the end of this section we shall briefly indicate what Herbrand actually did.

[3]Adapted with permission from *Journal of the ACM*, **28** (1981), 193-214. Copyright 1981 Association for Computing Machinery, Inc.

Herbrand's Theorem is one of the most fundamental tools available for dealing with solvable cases of the decision problem[4] and underlies many approaches to automated theorem proving. An automated theorem proving procedure called the *mating method* based on the ideas in this section is described in [Andrews, 1981]. It is closely related to the *connection method* described in [Bibel, 1987].

In this section we shall use a formulation of first-order logic in which the primitive connectives and quantifiers are $\sim$, $\vee$, $\wedge$, and $\forall$. It is not difficult to see that the results and concepts which we have obtained for $\mathcal{F}$ also apply to this system with trivial modifications. A wff is said to be in *negation normal form* (*nnf*), and is a *negation normal formula* (*nnf*) iff the scope of each occurrence of $\sim$ in it is atomic. A sentence in nnf is called a *negation normal sentence* (*nns*). By the methods of §14 it is easy to find for each universal wff an equivalent nnf which contains no more occurrences of atoms than the original wff. For the sake of simplicity and clarity, we shall restrict our attention to nnfs at various points in this section even though it is not essential to do so, since little real generality is lost by this restriction.

Given a wff $\mathbf{A}$ which we wish to show is valid, let $\overline{\mathbf{A}}$ be its universal closure, and let $\mathbf{D}$ be a negation normal form of $\bigstar \sim \overline{\mathbf{A}}$. Then $\mathbf{D}$ is a universal nns which has no model if and only if $\mathbf{A}$ is valid. Thus we shall concentrate on the problem of refuting universal negation normal sentences.

We first define a *compound instance* (*c-instance*) of a universal nnf $\mathbf{D}$ to be the result of replacing each wf part of $\mathbf{D}$ of the form $\forall \mathbf{x}\mathbf{B}(\mathbf{x})$ by $\mathbf{B}(\mathbf{t}_1) \wedge \ldots \wedge \mathbf{B}(\mathbf{t}_n)$, where $n \geq 1$, and for each $i \leq n$, $\mathbf{t}_i$ is a closed (variable-free) term; different terms $\mathbf{t}_1, \ldots, \mathbf{t}_n$ (and different values of n) may be chosen for different occurrences of quantifiers in $\mathbf{D}$. If $n = 1$ for each occurrence of a quantifier in $\mathbf{D}$, the c-instance is called a *simple* instance.

To avoid any ambiguities, we define the c-instances of $\mathbf{D}$ inductively as follows:

(a) If $\mathbf{D}$ is a literal, $\mathbf{D}$ is the only c-instance of $\mathbf{D}$.
(b) If $\mathbf{D}$ is $[\mathbf{D}_1 * \mathbf{D}_2]$, where $*$ is $\vee$ or $\wedge$, and $\mathbf{H}_i$ is a c-instance of $\mathbf{D}_i$ for $i = 1, 2$, then $[\mathbf{H}_1 * \mathbf{H}_2]$ is a c-instance of $\mathbf{D}$.
(c) If $\mathbf{D}$ is $\forall \mathbf{x}\mathbf{C}(\mathbf{x})$, $n \geq 1$, and for each $i\,(1 \leq i \leq n)$, $\mathbf{t}_i$ is a closed term and $\mathbf{H}_i$ is a c-instance of $\mathbf{C}(\mathbf{t}_i)$, then $[\mathbf{H}_1 \wedge \ldots \wedge \mathbf{H}_n]$ is a c-instance of $\mathbf{D}$.

3500 Lemma. Let $\mathbf{D}$ be a universal nnf, and let $\mathbf{H}$ be a c-instance of $\mathbf{D}$. Then there is a conjunction $\mathbf{M}$ of simple instances of $\mathbf{D}$ such that $\mathbf{M} \equiv \mathbf{H}$ is tautologous; also, $\mathbf{M}$ and $\mathbf{H}$ contain exactly the same atomic wffs.

[4] See, for example, [Dreben and Goldfarb, 1979].

Proof: The proof is by induction on the number of occurrences of $\vee$, $\wedge$, and $\forall$ in $\mathbf{D}$.

Case 1. $\mathbf{D}$ is a literal. Let $\mathbf{M}$ be $\mathbf{H}$.

Case 2. $\mathbf{D}$ is $[\mathbf{D}_1 * \mathbf{D}_2]$, where $*$ is $\wedge$ or $\vee$. Then $\mathbf{H}$ has the form $[\mathbf{H}_1 * \mathbf{H}_2]$, where $\mathbf{H}_i$ is a c-instance of $\mathbf{D}_i$ for $i = 1, 2$. By inductive hypothesis there are conjunctions $\bigwedge_{j=1}^{n} \mathbf{K}_j$ and $\bigwedge_{k=1}^{m} \mathbf{L}_k$ of simple instances of $\mathbf{D}_1$ and $\mathbf{D}_2$, respectively, such that $\models \mathbf{H}_1 \equiv \bigwedge_{j=1}^{n} \mathbf{K}_j$ and $\models \mathbf{H}_2 \equiv \bigwedge_{k=1}^{m} \mathbf{L}_k$. Let $\mathbf{M}$ be $\bigwedge_{j=1}^{n} \bigwedge_{k=1}^{m} [\mathbf{K}_j * \mathbf{L}_k]$. Then $\mathbf{M}$ is a conjunction of simple instances of $\mathbf{D}$, and $\models \mathbf{M} \equiv \mathbf{H}$.

Case 3. $\mathbf{D}$ is $\forall \mathbf{x} \mathbf{C}(\mathbf{x})$. Then $\mathbf{H}$ has the form $\mathbf{H}_1 \wedge \ldots \wedge \mathbf{H}_n$, where each $\mathbf{H}_i$ is a c-instance of $\mathbf{C}(\mathbf{t}_i)$ for some closed term $\mathbf{t}_i$. By inductive hypothesis each $\mathbf{H}_i$ is equivalent to a conjunction $\mathbf{M}_i$ of simple instances of $\mathbf{C}(\mathbf{t}_i)$. Let $\mathbf{M}$ be $\mathbf{M}_1 \wedge \ldots \wedge \mathbf{M}_n$. It is easy to see that $\mathbf{M}$ is a conjunction of simple instances of $\mathbf{D}$, and that $\models \mathbf{M} \equiv \mathbf{H}$. ■

3501 Lemma. Let $\mathbf{G}$ and $\mathbf{H}$ be c-instances of a universal nnf $\mathbf{D}$. Then there is a c-instance $\mathbf{K}$ of $\mathbf{D}$ such that $\models \mathbf{K} \supset [\mathbf{G} \wedge \mathbf{H}]$.

Proof: The proof is by induction on the number of occurrences of $\wedge$, $\vee$ and $\forall$ in $\mathbf{D}$.

Case 1. $\mathbf{D}$ is a literal. Then $\mathbf{G} = \mathbf{D} = \mathbf{H}$, and we let $\mathbf{K} = \mathbf{D}$ also.

Case 2. $\mathbf{D}$ has the form $\mathbf{D}_1 * \mathbf{D}_2$, where $*$ is $\wedge$ or $\vee$. Then $\mathbf{G}$ has the form $\mathbf{G}_1 * \mathbf{G}_2$, and $\mathbf{H}$ has the form $\mathbf{H}_1 * \mathbf{H}_2$, where $\mathbf{G}_i$ and $\mathbf{H}_i$ are c-instances of $\mathbf{D}_i$ such that $\models \mathbf{K}_i \supset [\mathbf{G}_i \wedge \mathbf{H}_i]$. Let $\mathbf{K} = \mathbf{K}_1 * \mathbf{K}_2$; then $\mathbf{K}$ is a c-instance of $\mathbf{D}$ and $\models \mathbf{K} \supset [\mathbf{G} \wedge \mathbf{H}]$.

Case 3. $\mathbf{D}$ has the form $\forall \mathbf{x} \mathbf{C}(\mathbf{x})$. Then $\mathbf{G}$ has the form $\mathbf{E}_1 \wedge \ldots \wedge \mathbf{E}_n$ and $\mathbf{H}$ has the form $\mathbf{E}_{n+1} \wedge \ldots \wedge \mathbf{E}_m$, where for each i there is a closed term $\mathbf{t}_i$ such that $\mathbf{E}_i$ is a c-instance of $\mathbf{C}(\mathbf{t}_i)$. Clearly we can let $\mathbf{K}$ be $\mathbf{G} \wedge \mathbf{H}$. ■

EXAMPLE: Let D be $[\forall x\, Px \vee \forall y\, Qy]$, let G be $[Pa \vee Qb]$, H be $[Pc \vee Qd]$, and K be $[[Pa \wedge Pc] \vee [Qb \wedge Qd]]$. The wffs G, H, and K are c-instances of D, and $\models K \supset [G \wedge H]$. Note, however, that $K \equiv [G \wedge H]$ is not valid.

A *truth assignment* is an assignment of truth values (T or F) to atomic wffs. If $\mathbf{G}$ is a quantifier-free wff and Φ is a truth assignment which assigns truth values to all atomic wf parts of **G**, we let $\mathcal{V}_\Phi\mathbf{G}$ denote the truth value of **G** with respect to Φ; this is computed in the usual way using the truth tables for the propositional connectives. We call **G** a *truth-functional contradiction* (*t-f contradiction*) iff $\mathcal{V}_\Phi\mathbf{G} = \mathsf{F}$ for all such truth assignments Φ. We say that Φ *verifies* **G** iff $\mathcal{V}_\Phi\mathbf{G} = \mathsf{T}$. If $\mathcal{S}$ is a set of quantifier-free wffs and Φ is a truth assignment which verifies each member of $\mathcal{S}$, we say that Φ *truth-functionally satisfies* (*t-f satisfies*) $\mathcal{S}$.

3502 Lemma. Let **D** be a universal nnf. If no *c*-instance of **D** is a t-f contradiction, there is a truth assignment which verifies every *c*-instance of **D**.

Proof: Let $\mathcal{S}$ be the set of all *c*-instances of **D**, and let $\{\mathbf{H}_1, \ldots, \mathbf{H}_n\}$ be a finite subset of $\mathcal{S}$. Using Lemma 3501 we obtain a *c*-instance **K** of **D** such that $\models \mathbf{K} \supset [\mathbf{H}_1 \wedge \ldots \mathbf{H}_n]$. Since **K** is not a t-f contradiction, there is a truth-assignment Ψ which verifies **K**, so Ψ t-f satisfies $\{\mathbf{H}_1, \ldots, \mathbf{H}_n\}$. Thus every finite subset of $\mathcal{S}$ is t-f satisfiable, so by the Compactness Theorem (1501) $\mathcal{S}$ is too. ∎

DEFINITIONS. The *Herbrand universe* $\mathcal{H}(\mathbf{D})$ of a sentence **D** is the set of terms which can be constructed from the individual and function constants in **D** (and from the individual constant *c*, if no other individual constant occurs in **D**). A *compound Herbrand instance* (*cH-instance*) of a universal nns **D** is a *c*-instance of **D** in which all terms are members of $\mathcal{H}(\mathbf{D})$. A *simple Herbrand instance* of **D** is a *cH*-instance which is a simple instance of **D**.

3503 Herbrand's Theorem. Let **A** be a wff, let $\overline{\mathbf{A}}$ be its universal closure, and let **D** be a negation normal form of the Skolemization $\bigstar \sim \overline{\mathbf{A}}$ of $\sim \overline{\mathbf{A}}$. Then **A** is valid if and only if **D** has a compound Herbrand instance which is a truth-functional contradiction.

Proof: We may assume that we are using a formulation of first-order logic in which there are no individual or function constants except those from which the terms in $\mathcal{H}(\mathbf{D})$ are constructed. Thus all *c*-instances of **D** are *cH*-instances.

Clearly $\models$ **A** iff **D** has no model. Thus we shall show that **D** has no model iff **D** has a contradictory *c*-instance.

It is easy to see by induction on the number of occurrences of $\wedge$, $\vee$, and $\forall$ in **D**, that if **D** is any universal nns, and **G** is any *c*-instance of **D**, then $\models \mathbf{D} \supset \mathbf{G}$. Hence if **G** is t-f contradictory, it is false in any model, so **D** is too, so **D** has no model.

For the proof in the other direction, suppose **D** has no *t-f* contradictory *c*-instance. Then by Lemma 3502 there is an assignment Φ of truth values to the atoms which occur in *c*-instances of **D** which verifies every *c*-instance of **D**.

We next construct a model for **D**. We let the domain of individuals of our model be the Herbrand universe of **D**. As in the proof of 2507, we interpret the individual constants and function symbols so that if **P** is any n-ary predicate and $\mathbf{t}_1, \ldots, \mathbf{t}_n$ are individuals, then $\mathbf{Pt}_1, \ldots, \mathbf{t}_n$ is true in the interpretation iff $\Phi(\mathbf{Pt}_1, \ldots, \mathbf{t}_n) = \mathsf{T}$. This specifies an interpretation $\mathcal{M}$ for our system of first-order logic.

Next, we show, by induction on the number of occurrences of $\wedge$, $\vee$, and $\forall$ in **B**, that if **B** is any universal nns such that Φ verifies every *c*-instance of **B**, then $\mathcal{M} \models \mathbf{B}$.

Case 1. **B** is a literal. Then **B** is the only *c*-instance of **B**, and $\mathcal{V}^{\mathcal{M}}\mathbf{B} = \mathcal{V}_{\Phi}\mathbf{B}$ by the interpretation of predicates.

Case 2. **B** has the form $\mathbf{B}_1 \wedge \mathbf{B}_2$. For $i = 1, 2$, let $\mathbf{G}_i$ be any *c*-instance of $\mathbf{B}_i$. Then $\mathbf{G}_1 \wedge \mathbf{G}_2$ is a *c*-instance of **B**, so $\mathcal{V}_{\Phi}[\mathbf{G}_1 \wedge \mathbf{G}_2] = \mathsf{T}$, so $\mathcal{V}_{\Phi}\mathbf{G}_i = \mathsf{T}$. Therefore, $\mathcal{M} \models \mathbf{B}_i$ by inductive hypothesis, so $\mathcal{M} \models \mathbf{B}$.

Case 3. **B** has the form $[\mathbf{B}_1 \vee \mathbf{B}_2]$. Suppose there are *c*-instances $\mathbf{G}_1$ of $\mathbf{B}_1$ and $\mathbf{G}_2$ of $\mathbf{B}_2$ such that $\mathcal{V}_{\Phi}\mathbf{G}_1 = \mathsf{F}$ and $\mathcal{V}_{\Phi}\mathbf{G}_2 = \mathsf{F}$. Then $\mathcal{V}_{\Phi}[\mathbf{G}_1 \vee \mathbf{G}_2] = \mathsf{F}$, which contradicts our assumption about **B**, since $[\mathbf{G}_1 \vee \mathbf{G}_2]$ is a *c*-instance of **B**. Therefore for $i = 1$ or $i = 2$, Φ verifies every *c*-instance of $\mathbf{B}_i$, so $\mathcal{M} \models \mathbf{B}_i$ by inductive hypothesis, so $\mathcal{M} \models \mathbf{B}$.

Case 4. **B** has the form $\forall\mathbf{xC}(\mathbf{x})$. Let **t** be any closed term. Every *c*-instance of $\mathbf{C}(\mathbf{t})$ is a *c*-instance of **B**, and so is verified by Φ, so $\mathcal{M} \models \mathbf{C}(\mathbf{t})$ by inductive hypothesis. Therefore $\mathcal{M} \models \forall\mathbf{xC}(\mathbf{x})$, since every individual of $\mathcal{M}$ is the denotation of some closed term.

(In case the argument in the preceding sentence seems too informal, we can make it more rigorous as follows. Let φ be any assignment, and let ψ be any assignment which agrees with φ off **x**. There is some closed term **t** such that $\psi\mathbf{x} = \mathbf{t} = \mathcal{V}^{\mathcal{M}}\mathbf{t}$, so by the Corollary to the Substitution-Value Theorem (2302), $\mathcal{V}^{\mathcal{M}}_{\psi}\mathbf{C}(\mathbf{x}) = \mathcal{V}^{\mathcal{M}}_{\varphi}\mathsf{S}^{\mathbf{x}}_{\mathbf{t}}\, \mathbf{C}(\mathbf{x}) = \mathcal{V}^{\mathcal{M}}_{\varphi}\mathbf{C}(\mathbf{t}) = \mathsf{T}$. Thus $\mathcal{V}^{\mathcal{M}}_{\varphi}\forall\mathbf{xC}(\mathbf{x}) = \mathsf{T}$, and φ is arbitrary, so $\mathcal{M} \models \forall\mathbf{xC}(\mathbf{x})$.)

This completes the inductive argument, and establishes that $\mathcal{M} \models \mathbf{D}$, so **D** has a model. ■

Alternative formulations of Herbrand's Theorem are provided by the following corollaries:

3504 Corollary. Let $\mathbf{A}$ be a wff, let $\overline{\mathbf{A}}$ be its universal closure, and let $\mathbf{D}$ be a negation normal form of the Skolemization $\bigstar \sim \overline{\mathbf{A}}$ of $\sim \overline{\mathbf{A}}$. Then $\mathbf{A}$ is valid if and only if some conjunction of simple Herbrand instances of $\mathbf{D}$ is a truth-functional contradiction.

Proof: By 3500 every cH-instance of $\mathbf{D}$ is truth-functionally equivalent to a conjunction of simple Herbrand instances, and by 3501 every conjunction of simple Herbrand instances of $\mathbf{D}$ is truth-functionally implied by some cH-instance, so the corollary follows from 3503. ■

3505 Corollary. Let $\mathbf{B}$ be a sentence, and let $\mathbf{D}$ be a negation normal form of the Skolemization $\bigstar\mathbf{B}$ of $\mathbf{B}$. Then $\mathbf{B}$ is satisfiable if and only if the set of simple Herbrand instances of $\mathbf{D}$ is t-f satisfiable.

Proof: Let $\mathbf{A}$ be $\sim \mathbf{B}$. Then $\mathbf{D}$ is a nnf of $\bigstar \sim \overline{\mathbf{A}}$, since $\sim \overline{\mathbf{A}}$ is $\sim\sim \mathbf{B}$. Since $\mathbf{B}$ is satisfiable iff $\mathbf{A}$ is not valid, by 3504 $\mathbf{B}$ is satisfiable iff every conjunction of simple Herbrand instances of $\mathbf{D}$ is t-f satisfiable. The corollary follows from this by the Compactness Theorem (1501). ■

EXAMPLE: Suppose we wish to establish the validity of the following wff:

$$(A)\ \exists x\, \forall y[Px \equiv Py] \supset [\exists x\, Px \equiv \forall y\, Py]$$

We negate it, and eliminate $\equiv$ and $\supset$ to obtain:

$$\begin{aligned}(B)\ &\exists x\, \forall y[(\sim Px \lor Py) \land (\sim Py \lor Px)]\\ &\quad \land ([\exists x\, Px \land \exists y \sim Py] \lor [\forall y\, Py \land \forall x \sim Px])\end{aligned}$$

Next we Skolemize it by the methods of §33 to obtain the following nns:

$$\begin{aligned}(D)\ &\forall y[(\sim Pc \lor Py) \land (\sim Py \lor Pc)]\\ &\quad \land([Pd \land \sim Pe] \lor [\forall z\, Pz \land \forall x \sim Px])\end{aligned}$$

It is convenient to display this in the 2-dimensional format discussed in §14:

$$(D')\quad \left[\begin{array}{c} \forall y \left[\begin{array}{ccc} \sim Pc & \vee & Py \\ \sim Py & \vee & Pc \end{array}\right] \\ \\ \left[\begin{array}{c} Pd \\ \sim Pe \end{array}\right] \vee \left[\begin{array}{c} \forall z\, Pz \\ \forall x \sim Px \end{array}\right] \end{array}\right]$$

The following is a *c*-instance of (D):

$$(E)\quad \left[\begin{array}{ccc} \sim Pc & \vee & Pd \\ \sim Pd & \vee & Pc \\ \sim Pc & \vee & Pe \\ \sim Pe & \vee & Pc \\ \left[\begin{array}{c} Pd \\ \sim Pe \end{array}\right] & \vee & \left[\begin{array}{c} Pc \\ \sim Pd \end{array}\right] \end{array}\right]$$

It can be easily be seen by the methods of §14 (especially 1408) that (E) is contradictory, so by 3503 (A) is valid.

Since a variety of results in logic are sometimes called Herbrand's theorem, we conclude this section with a brief and rather imprecise description of the main results which Herbrand actually presented in this thesis.

Herbrand introduced a cut-free system of first-order logic which we shall call $\mathcal{F}_H$. $\mathcal{F}_H$ had all quantifier-free tautologies as axioms, and had as rules of inference universal and existential generalization, anti-prenex rules for pushing in quantifiers, the rule of alphabetic change of bound variables, and a rule of simplification (replace $\mathbf{D} \vee \mathbf{D}$ by $\mathbf{D}$). Modus ponens was not a rule of inference of $\mathcal{F}_H$.

He also considered a system of first-order logic which did have modus ponens as a primitive rule of inference, and which we may pretend was the system $\mathcal{F}$ of Chapter 2. (Systems of this sort are called *Hilbert-style* systems.)

He also showed how to associate with a wff $\mathbf{A}$ of first-order logic certain quantifier-free wffs which we shall call *Herbrand expansions* of $\mathbf{A}$, which were equivalent to negations of the cH-instances of the wff $\mathbf{D}$ mentioned in 3503. Let us say that a wff $\mathbf{A}$ has the *Herbrand property* iff some Herbrand expansion of $\mathbf{A}$ is tautologous.

Herbrand's Theorem consisted of the following claims about any wff $\mathbf{A}$ of first-order logic:

(1) If $\vdash_{\mathcal{F}_H} \mathbf{A}$, then $\vdash_{\mathcal{F}} \mathbf{A}$.
(2) If $\vdash_{\mathcal{F}} \mathbf{A}$, then $\mathbf{A}$ has the Herbrand property.
(3) If $\mathbf{A}$ has the Herbrand property, then $\vdash_{\mathcal{F}_H} \mathbf{A}$.

Thus Herbrand claimed that the following are equivalent:

(a) $\vdash_{\mathcal{F}_H} \mathbf{A}$
(b) $\vdash_{\mathcal{F}} \mathbf{A}$
(c) **A** has the Herbrand property.

If we add the soundness (2303) and completeness (2510) of $\mathcal{F}$ to the claims above, it is natural to add to this list of equivalent statements the following:

(d) **A** is valid.

Various versions of the theorem that (c) and (d) are equivalent, such as our theorem 3503, are commonly called Herbrand's Theorem in the literature of logic. Notice, however, that Herbrand's version of his theorem was a purely syntactic result. His proofs were purely syntactic too, and it is much more difficult to prove these results by purely syntactic arguments than it is to prove them with the aid of semantic as well as syntactic concepts. Actually, Herbrand's proof of (2) contained a basic error [5] [Dreben *et al.*, 1963], which was finally corrected by Dreben and Denton [Dreben and Denton, 1966].

Note that the full Herbrand theorem implies that modus ponens is a derived rule of inference of the system $\mathcal{F}_H$, a result that is closely related to Gentzen's Cut-Elimination Theorem (3102).

§36. Unification

Most systematic theorem-proving procedures involve making substitutions for variables in such a way that certain expressions become the same. In this case we say that they are *unified.* For example, when using the methods of §34, the purpose of applying the substitution rule is to prepare for applications of Simplification or Cut by unifying certain expressions. Similarly, when using the methods of §35, one must find a contradictory Herbrand

[5]It is reported in [van Heijenoort, 1967, p. 525] that Kurt Gödel discovered an essential gap in Herbrand's proof in the early 1940's, but he never published anything on the subject. In 1962 Andrews discovered a gap in Herbrand's proof while examining it to see if it could be extended to higher-order logic. He communicated this to Dreben, who was an authority on Herbrand's work. After some discussion Dreben became convinced that there was indeed a gap in the proof, and set about repairing it. Instead, he found a counterexample to one of Herbrand's lemmas. An attempt was made to replace that lemma by a weaker lemma which would still suffice to prove Herbrand's Theorem, but Aanderaa found a counterexample to this.

instance of a sentence, and (as one sees from 1408) this requires substituting terms for the quantified variables in such a way that certain expressions become unified.

When trying to prove a theorem by such methods, one often proceeds in steps, and each step involves unifying additional expressions. Since a set of expressions may have many unifiers, it is advisable to use at each stage a unifier which is as general as possible, i.e., one which leaves as much freedom as possible for imposing additional constraints on the terms to be finally substituted for the variables.

Before discussing an example, let us clarify some terminology. In this section we shall use the word *expression* to mean an atomic formula or a term from a formulation of first order logic in which the only variables are individual variables. Also, by a *substitution* we shall mean a substitution for variables. (Review the definitions at the end of §10.) If $\mathcal{W}$ is a set of expressions and θ is a substitution, we define $\theta\mathcal{W}$ to be $\{\theta\mathbf{X} \mid \mathbf{X} \in \mathcal{W}\}$. If $\theta\mathcal{W}$ is a unit set (set with exactly one member), we say that θ *unifies* $\mathcal{W}$, and is a *unifier of* $\mathcal{W}$. This means, of course, that $\theta\mathbf{X} = \theta\mathbf{Y}$ for all $\mathbf{X}$ and $\mathbf{Y}$ in $\mathcal{W}$.

EXAMPLE: Let $A = \overline{Q}^2xy$, $B = \overline{Q}^2z\overline{h}^1z$, $C = \overline{Q}^2x\overline{h}^1\overline{a}$, and $D = \overline{Q}^2\overline{g}^2uvv$, where $\overline{Q}^2$, $\overline{g}^2$, and $\overline{h}^1$ are predicate and function constants, and $\overline{a}$ is an individual constant. (We shall henceforth omit the superscripts.) Suppose we wish to simultaneously unify $\{A,\ B\}$ and $\{C,\ D\}$, i.e., find a substitution θ such that $\theta A = \theta B$ and $\theta C = \theta D$. We might do this by finding a substitution τ which unifies $\{A,\ B\}$ and then finding a substitution σ which unifies $\{\tau C,\ \tau D\}$, and letting $\theta = \sigma \circ \tau$. In this case $\theta A = (\sigma \circ \tau)A = \sigma(\tau A) = \sigma(\tau B) = \theta B$ and $\theta C = \sigma(\tau C) = \sigma(\tau D) = \theta D$. Let $\tau_1 = \mathsf{S}^{x\ y\ z}_{\overline{a}\,\overline{h}\overline{a}\,\overline{a}}$. We might happen to notice that τ_1 unifies $\{A,\ B\}$, and proceed to try to unify $\{\tau_1 C,\ \tau_1 D\}$, which is $\{\overline{Q}\overline{a}\overline{h}\overline{a},\ \overline{Q}\overline{g}uvv\}$, but this set is not unifiable. Let $\tau = \mathsf{S}^{x\ y}_{z\ \overline{h}z}$ and $\sigma = \mathsf{S}^{v\ \ z}_{\overline{h}\overline{a}\ \overline{g}u\overline{h}\overline{a}}$. τ also unifies $\{A,\ B\}$, and $\{\tau C,\ \tau D\}$, which is $\{\overline{Q}z\overline{h}\overline{a},\ \overline{Q}\overline{g}uvv\}$ has a unifier, namely σ. It can be seen that $\sigma \circ \tau = \mathsf{S}^{v\ \ x\ \ y\ \ z}_{\overline{h}\overline{a}\ \overline{g}u\overline{h}\overline{a}\ \overline{h}\overline{g}u\overline{h}\overline{a}\ \overline{g}u\overline{h}\overline{a}}$ (for example, $(\sigma \circ \tau)y = \sigma(\tau y) = \sigma(\overline{h}z) = \overline{h}\sigma z = \overline{h}\overline{g}u\overline{h}\overline{a}$), and that $\sigma \circ \tau$ unifies both $\{A,\ B\}$ and $\{C,\ D\}$. Indeed, $(\sigma \circ \tau)A = \overline{Q}\overline{g}u\overline{h}\overline{a}\overline{h}\overline{g}u\overline{h}\overline{a} = (\sigma \circ \tau)B$ and $(\sigma \circ \tau)C = \overline{Q}\overline{g}u\overline{h}\overline{a}\overline{h}\overline{a} = (\sigma \circ \tau)D$.

Clearly τ is a more general substitution that τ_1. Next we make this notion precise.

DEFINITION. Let τ_0 and τ_1 be substitutions. We write $\tau_0 \leq \tau_1$, and say that τ_0 is *less specified than* τ_1, or that τ_0 is *more general than* τ_1, iff there is a substitution ρ such that $\rho \circ \tau_0 = \tau_1$.

In the example above, $\mathsf{S}^{z}_{\overline{a}} \circ \tau = \tau_1$, so $\tau \leq \tau_1$.

It is easy to see that the relation $\leq$ is reflexive (X3600) and transitive (X3601).

DEFINITION. Let $\mathcal{W}$ be a set of expressions and θ be substitution. θ is a *most general unifier* (*mgu*) of $\mathcal{W}$ iff θ unifies $\mathcal{W}$ and for any unifier τ of $\mathcal{W}$, $\theta \leq \tau$.

DEFINITION. Let τ_1 and τ_2 be substitutions. We write $\tau_1 \approx \tau_2$ iff $\tau_1 \leq \tau_2$ and $\tau_2 \leq \tau_1$.

EXAMPLE: Let $\tau_1 = \mathsf{S}^{w\ \ y}_{hx\ x}$, $\tau_2 = \mathsf{S}^{w\ \ x}_{hy\ y}$, and $\rho = \mathsf{S}^{x\ y}_{y\ x}$. Then $\rho \circ \tau_1 = \tau_2$ and $\rho \circ \tau_2 = \tau_1$, so $\tau_1 \approx \tau_2$.

EXAMPLE: Let $\overline{k}$ be a ternary function constant, let $\tau_1 = \mathsf{S}^{w\ \ \ \ \ \ x\ y\ z}_{\overline{k}xyz\ y\ z\ x}$, $\tau_2 = \mathsf{S}^{w\ \ \ \ \ \ x\ y\ z}_{\overline{k}yzx\ z\ x\ y}$, $\sigma_1 = \mathsf{S}^{x\ y\ z}_{y\ z\ x}$, and $\sigma_2 = \mathsf{S}^{x\ y\ z}_{z\ x\ y}$. Then $\sigma_1 \circ \tau_1 = \tau_2$ and $\sigma_2 \circ \tau_2 = \tau_1$, so $\tau_1 \approx \tau_2$.

It is easy to see that $\approx$ is an equivalence relation (X3602). It can also be seen that substitutions which are equivalent in this sense differ only by alphabetic changes of variables. The next proposition shows that most general unifiers of sets of expressions of first-order logic are essentially unique.

3600 Proposition. If $\mathcal{W}$ is a set of expressions and θ and τ are most general unifiers of $\mathcal{W}$, then $\theta \approx \tau$.

Proof: Under the hypotheses θ and τ both unify $\mathcal{W}$, and $\theta \leq \tau$ and $\tau \leq \theta$, so $\theta \approx \tau$. ■

3601 Unification Theorem [Robinson, 1965]. There is an algorithm which can be applied to an arbitrary nonempty finite set $\mathcal{W}$ of expressions to determine whether $\mathcal{W}$ is unifiable. The algorithm always terminates, and produces a most general unifier of $\mathcal{W}$ if $\mathcal{W}$ is unifiable.

Proof: If $\mathcal{W}$ is a finite set of expressions with at least two members, define the *disagreement set* $D(\mathcal{W})$ of $\mathcal{W}$ as follows. The expressions in $\mathcal{W}$ are finite sequences of symbols. Let n be the first position at which not all expressions in $\mathcal{W}$ have the same symbol, and let $D(\mathcal{W})$ be the set of expressions $\mathbf{B}$ such that there is an expression $\mathbf{A}$ in $\mathcal{W}$ such that $\mathbf{B}$ is a sub-expression of $\mathbf{A}$ (expression which is a consecutive subsequence of $\mathbf{A}$) and the first symbol of $\mathbf{B}$ is the nth symbol of $\mathbf{A}$.

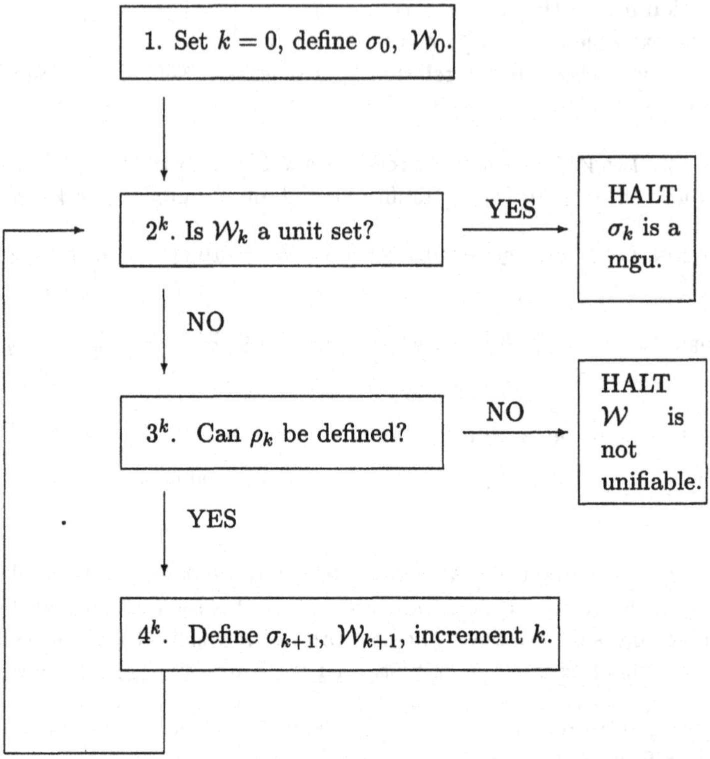

Figure 3.5: Flow chart for the unification algorithm

EXAMPLE: Let $\mathcal{W} = \left\{\overline{Q}^2 x\overline{h}^1\overline{a},\ \overline{Q}^2\overline{g}^2uvv,\ \overline{Q}^2\overline{g}^2\overline{h}^1x\overline{g}^2u\overline{h}^1yx\right\}$. Then $n = 2$ and $D(\mathcal{W}) = \left\{x,\ \overline{g}^2uv,\ \overline{g}^2\overline{h}^1x\overline{g}^2u\overline{h}^1y\right\}$.

We next describe how the algorithm, known as the *Unification Algorithm*, works when applied to a nonempty finite set $\mathcal{W}$ of expressions. (See Figure 3.5.)

Step 1. Set $k = 0$, $\mathcal{W}_0 = \mathcal{W}$, and let σ_0 be the identity substitution. Go to Step 2^0.

Step 2^k. If $\mathcal{W}_k$ is a unit set, halt; σ_k is a most general unifier of $\mathcal{W}$. Otherwise, go to Step 3^k.

Step 3^k. If the disagreement set $D(\mathcal{W}_k)$ contains a variable $\mathbf{v}_k$ and a term $\mathbf{t}_k$ such that $\mathbf{v}_k$ does not occur in $\mathbf{t}_k$, choose such a $\mathbf{v}_k$ and $\mathbf{t}_k$ in some systematic way, let $\rho_k = \mathsf{S}^{\mathbf{v}_k}_{\mathbf{t}_k}$, and go to Step 4^k. Otherwise halt; $\mathcal{W}$ is not unifiable.

Step 4^k. Let $\sigma_{k+1} = \rho_k \circ \sigma_k$ and $\mathcal{W}_{k+1} = \rho_k \mathcal{W}_k$. Set k to $k+1$, and then go to Step 2^k.

It is easy to see by induction on k that for each k, $\mathcal{W}_k = \sigma_k \mathcal{W}$, that $\sigma_{k+1} = \rho_k \circ \rho_{k-1} \circ \ldots \circ \rho_2 \circ \rho_1 \circ \rho_0$, and that none of the variables $\mathbf{v}_0, \ldots, \mathbf{v}_k$ occurs in any expression in $\mathcal{W}_{k+1}$. Since $\mathbf{v}_k$ is chosen from some expression $\mathcal{W}_k$ in Step 3^k, the variables $\mathbf{v}_0$, $\mathbf{v}_1$, $\mathbf{v}_2 \ldots$ are clearly all distinct from one another and all occur in expressions of $\mathcal{W}$. There are only finitely many such variables, so the algorithm clearly terminates.

If it terminates at Step 2^k, then σ_k unifies $\mathcal{W}$. Therefore, if $\mathcal{W}$ is not unifiable, the algorithm cannot terminate at Step 2^k for any k, so it must terminate at some Step 3^k.

Next consider the case where $\mathcal{W}$ is unifiable. Let θ be any unifier of $\mathcal{W}$. We shall first show that whenever σ_k is defined and $\sigma_k \leq \theta$, the algorithm does not terminate at Step 3^k. If $\mathcal{W}_k$ is a unit set, the algorithm terminates at Step 2^k. Otherwise, it reaches Step 3^k, and $D(\mathcal{W}_k)$ has at least two members. Since $\sigma_k \leq \theta$, there is a substitution τ such that $\tau \circ \sigma_k = \theta$. $\tau \mathcal{W}_k = \tau(\sigma_k \mathcal{W}) = (\tau \circ \sigma_k)\mathcal{W} = \theta \mathcal{W}$, which is a unit set, so τ unifies $\mathcal{W}_k$, and hence $D(\mathcal{W}_k)$. If the first symbol of each expression in $D(\mathcal{W}_k)$ were a constant, $\mathbf{D}(\mathcal{W}_k)$ would not be unifiable, since (by the definition of a disagreement set) these constants could not all be the same. Therefore the first symbol of some expression $\mathbf{X}$ in $D(\mathcal{W}_k)$ is a variable, and therefore $\mathbf{X}$ is a variable, which we shall call $\mathbf{v}_k$. Let $\mathbf{t}_k$ be any other expression in $D(\mathcal{W}_k)$. Since $\tau \mathbf{v}_k = \tau \mathbf{t}_k$but $\mathbf{v}_k \neq \mathbf{t}_k$, it is clear that $\mathbf{v}_k$ cannot occur in $\mathbf{t}_k$; otherwise $\tau \mathbf{v}_k$ would be a proper subsequence of $\tau \mathbf{t}_k$, which is $\tau \mathbf{v}_k$. Therefore the algorithm does not terminate at Step 3^k.

Next we show by induction on k that $\sigma_k \leq \theta$ for each k such that σ_k is defined. This is clear for $k = 0$, since $\theta \circ \sigma_0 = \theta$. For the induction step suppose $\sigma_k \leq \theta$ and σ_k is defined. If the algorithm terminates at Step 2^k, then σ_{k+1} is not defined. Otherwise, it proceeds to Step 3^k, where it does not terminate (as shown above), so $\mathbf{v}_k$, $\mathbf{t}_k$, ρ_k, σ_{k+1}, and $\mathcal{W}_{k+1}$ are defined as in Steps 3^k and 4^k. We are given that there is a substitution τ such that $\tau \circ \sigma_k = \theta$, and (as shown above) $\tau \mathbf{v}_k = \tau \mathbf{t}_k$.

Define the substitution τ' so that $\tau' \mathbf{v}_k = \mathbf{v}_k$, and $\tau' \mathbf{v} = \tau \mathbf{v}$ for every variable $\mathbf{v}$ distinct from $\mathbf{v}_k$. We show that $\tau' \circ \rho_k = \tau$ by showing that these substitutions agree on all variables. $(\tau' \circ \rho_k)\mathbf{v}_k = \tau'(\rho_k \mathbf{v}_k) = \tau' \mathbf{t}_k = \tau \mathbf{t}_k$ (since $\mathbf{v}_k$ does not occur in $\mathbf{t}_k$, and τ' agrees with τ on all other variables) $= \tau \mathbf{v}_k$. If $\mathbf{v}$ is a variable distinct from $\mathbf{v}_k$, then $(\tau' \circ \rho_k)\mathbf{v} = \tau'(\rho_k \mathbf{v}) = \tau' \mathbf{v} =$

τv. Thus $\tau' \circ \rho_k = \tau$.

Now $\tau' \circ \sigma_{k+1} = \tau' \circ (\rho_k \circ \sigma_k)$ (by the definition of σ_{k+1}) $= (\tau' \circ \rho_k) \circ \sigma_k = \tau \circ \sigma_k = \theta$, so $\sigma_{k+1} \leq \theta$, as required. This completes the inductive argument.

Since $\sigma_k \leq \theta$ whenever σ_k is defined, and therefore the algorithm cannot terminate at any Step 3^k, it must terminate at some Step 2^K. Since we have shown that $\sigma_K \leq \theta$ for an arbitrary unifier θ of $\mathcal{W}$, σ_K is a most general unifier of $\mathcal{W}$. ■

There is much more to be said about unification. One can perform unification modulo certain theories (such as the associativity or commutativity of certain functions) or in a richer language (such as type theory). One can also seek more efficient unification algorithms. We refer the reader to [Baader and Siekmann, 1994], [Baader and Snyder, 2000], [Degtyarev and Voronkov, 1998], [Gallier and Snyder, 1989], [Gallier *et al.*, 1992], [Huet and Oppen, 1980], [Jouannaud and Kirchner, 1991], [Paterson and Wegman, 1978], [Siekmann, 1984], [Siekmann, 1989], [Snyder, 1991], and [van Vaalen, 1975] for more details and references.

EXERCISES

X3600. Show that the relation $\leq$ is reflexive.

X3601. Show that the relation $\leq$ is transitive.

X3602. Show that $\approx$ is an equivalence relation on the set of substitutions.

Chapter 4

Further Topics in First-Order Logic

§40. Duality

We introduce this subject with a parable about two scholars, named Oren and Nero, who were visiting an archaeologist and were shown a recently discovered tablet, the contents of which are reproduced in Figure 4.1. They soon realized that the figures on the tablet were truth tables, and they set about translating them into more familiar notations. (Before proceeding further, the reader is advised to do this for himself for at least a few of the tables.) Oren produced the translation in Figure 4.2, and Nero produced that in Figure 4.3. As they started to show their translations to the archaeologist, Nero modestly remarked "That was really quite easy, as soon as I realized that $\oplus$ denoted conjunction". "But you're wrong!" exclaimed Oren. "$\oplus$ denoted disjunction!" They argued for some time, and neither was able to persuade the other that he was wrong. All they could agree on was that $\ominus$ denoted negation.

Then the archaeologist showed them an inscription which had been found on another tablet. "We've figured out how to translate most of the language they apparently used in everyday affairs", he said, "but this inscription seems to be in a mixture of two languages. Here's how the inscription looks under the translation we've achieved so far:"

> 'The value of $(\boxplus\ \mathbf{x})\mathbf{A}(\mathbf{x})$ is $\triangledown$ iff there is an n such that the value of $\mathbf{A}(n)$ is $\triangledown$.'

Oren and Nero recognized that this was another fragment of a text on logic,

○	△	▽
△	△	△
▽	△	△

□	▽	△
▽	▽	▽
△	▽	▽

⊕	△	▽
△	△	△
▽	△	▽

⊞	▽	△
▽	▽	▽
△	▽	△

⦸	△	▽
△	△	△
▽	▽	△

⧅	▽	△
▽	▽	▽
△	△	▽

∅	△	▽
△	△	▽
▽	△	△

⧄	▽	△
▽	▽	△
△	▽	▽

⊗	△	▽
△	△	▽
▽	▽	△

⊠	▽	△
▽	▽	△
△	△	▽

⊙	△	▽
△	▽	▽
▽	▽	△

⊡	▽	△
▽	△	△
△	△	▽

	⊖
△	▽
▽	△

Figure 4.1: The tablet

and each soon produced a translation of it. Oren's translation was:

> 'The value of $\forall \mathbf{x}\mathbf{A}(\mathbf{x})$ is falsehood iff there is an n such that the value of $\mathbf{A}(n)$ is falsehood.'

Nero's translation was:

> 'The value of $\exists \mathbf{x}\mathbf{A}(\mathbf{x})$ is truth iff there is an n such that the value of $\mathbf{A}(n)$ is truth.'

Again, they could find no way to resolve their disagreement.

t	T	F
T	T	T
F	T	T

f	F	T
F	F	F
T	F	F

$\vee$	T	F
T	T	T
F	T	F

$\wedge$	F	T
F	F	F
T	F	T

$\subset$	T	F
T	T	T
F	F	T

$\not\supset$	F	T
F	F	F
T	T	F

$\supset$	T	F
T	T	F
F	T	T

$\not\subset$	F	T
F	F	T
T	F	F

$\equiv$	T	F
T	T	F
F	F	T

$\not\equiv$	F	T
F	F	T
T	T	F

$\downarrow$	T	F
T	F	F
F	F	T

$\mid$	F	T
F	T	T
T	T	F

	$\sim$
T	F
F	T

Figure 4.2: Oren's translation of the tablet

f	F	T
F	F	F
T	F	F

t	T	F
T	T	T
F	T	T

$\wedge$	F	T
F	F	F
T	F	T

$\vee$	T	F
T	T	T
F	T	F

$\not\supset$	F	T
F	F	F
T	T	F

$\subset$	T	F
T	T	T
F	F	T

$\not\subset$	F	T
F	F	T
T	F	F

$\supset$	T	F
T	T	F
F	T	T

$\not\equiv$	F	T
F	F	T
T	T	F

$\equiv$	T	F
T	T	F
F	F	T

$\mid$	F	T
F	T	T
T	T	F

$\downarrow$	T	F
T	F	F
F	F	T

	$\sim$
F	T
T	F

Figure 4.3: Nero's translation of the tablet

"There's one more fragment of a tablet you might be interested in", said the archeologist. "We managed to translate the first word on it as '**AXIOMS**', and there's a line under that which looks like this:"

$$(\boxplus p)[p \oplus \ominus p].$$

"That settles it!" exclaimed Oren. "That translates to '$\forall p[p \vee \sim p]$', which is a fine axiom. Clearly my method of translation is correct." "Not at all", replied Nero. "It translates to '$\exists p[p \wedge \sim p]$'. The people who wrote these tablets were known throughout the ancient world as deceitful, treacherous, inveterate liars. Apparently they even axiomatized their lies!"

We leave it to the reader to decide what the moral of this story is.

Clearly there is a symmetry, or *duality*, in logic induced by systematically interchanging truth and falsehood. We shall show how this duality can be used. The ideas in this section apply to any sound and complete formulation of first-order logic. We shall present them in terms of $\mathcal{F}$, but they apply equally well to a system in which both $\forall$ and $\exists$, and all binary propositional connectives, are primitive symbols.

When $\subset$, $\not\supset$, $\not\subset$, $\not\equiv$, $\downarrow$, and $|$ occur in abbreviations of wffs of $\mathcal{F}$, they are to be regarded as having been introduced by the following definitions, which we hereby add to the definitions of $\wedge$, $\supset$, and $\equiv$ in §10, and of $\exists$ in §20:

$[\mathbf{A} \subset \mathbf{B}]$	stands for	$[\mathbf{A} \vee \sim \mathbf{B}]$.
$[\mathbf{A} \not\supset \mathbf{B}]$	stands for	$\sim [\sim \mathbf{A} \vee \mathbf{B}]$.
$[\mathbf{A} \not\subset \mathbf{B}]$	stands for	$\sim [\mathbf{A} \vee \sim \mathbf{B}]$.
$[\mathbf{A} \not\equiv \mathbf{B}]$	stands for	$\sim [\mathbf{A} \equiv \mathbf{B}]$.
$[\mathbf{A} \downarrow \mathbf{B}]$	stands for	$\sim [\mathbf{A} \vee \mathbf{B}]$.
$[\mathbf{A} \mid \mathbf{B}]$	stands for	$\sim [\mathbf{A} \wedge \mathbf{B}]$.

Definition. Let $\mathbf{A}$ be a wff of $\mathcal{F}$. A wff $\mathbf{B}$ is a *dual* of $\mathbf{A}$ iff there is an abbreviation $\mathbf{C}$ for $\mathbf{A}$ such that $\mathbf{B}$ is the result of interchanging $\forall$ with $\exists$, $\vee$ with $\wedge$, $\supset$ with $\not\subset$, $\subset$ with $\not\supset$, $\equiv$ with $\not\equiv$, and $\downarrow$ with $|$ everywhere in $\mathbf{C}$ (See Figure 4.4.) The *principal dual* $\mathbf{A}^d$ of $\mathbf{A}$ is the result of making these interchanges in the unabbreviated form of $\mathbf{A}$.

$\forall$	$\exists$
$\wedge$	$\vee$
$\supset$	$\not\subset$
$\subset$	$\not\supset$
$\equiv$	$\not\equiv$
$\downarrow$	$\mid$

Figure 4.4: Dual Pairs

4000 Proposition. If $\mathbf{B}$ and $\mathbf{D}$ are duals of $\mathbf{A}$, then $\vdash \mathbf{B} \equiv \mathbf{D}$.

Proof: We show that if $\mathbf{B}$ is any dual of $\mathbf{A}$, then $\vdash \mathbf{B} \equiv \mathbf{A}^d$. Clearly this will suffice to prove the proposition. We show, by induction on the construction of $\mathbf{D}$, that if $\mathbf{D}$ is any abbreviation for a wff $\mathbf{D}_0$ of $\mathcal{F}$, and $\mathbf{D}'$ is the result of interchanging $\forall$ with $\exists$, $\vee$ with $\wedge$, $\supset$ with $\not\subset$, etc., in $\mathbf{D}$, then $\vdash \mathbf{D}' \equiv \mathbf{D}_0^d$.

Of course, this is trivially true if $\mathbf{D}$ is atomic. Now consider the following cases:

(a) $\mathbf{D}$ is $\forall\mathbf{xM}$. Then $\mathbf{D}' = \exists\mathbf{xM}'$, but $\vdash \mathbf{M}' \equiv \mathbf{M}_0^d$ by inductive hypothesis, so $\vdash \mathbf{D}' \equiv \exists\mathbf{xM}_0^d$. Also $\mathbf{D}_0^d = (\forall\mathbf{xM}_0)^d = \exists\mathbf{xM}_0^d$, so $\vdash \mathbf{D}' \equiv \mathbf{D}_0^d$.

(b) $\mathbf{D}$ is $\exists\mathbf{xM}$. Then $\mathbf{D}' = \forall\mathbf{xM}'$, but $\vdash \mathbf{M}' \equiv \mathbf{M}_0^d$ by inductive hypothesis, so $\vdash \mathbf{D}' \equiv \forall\mathbf{xM}_0^d$. Also $\mathbf{D}_0^d = (\sim \forall\mathbf{x} \sim \mathbf{M}_0)^d = \sim \exists\mathbf{x} \sim \mathbf{M}_0^d$, so $\vdash \mathbf{D}' \equiv \mathbf{D}_0^d$.

(c) $\mathbf{D}$ is $[\mathbf{M} \vee \mathbf{N}]$. Then $\mathbf{D}' = [\mathbf{M}' \wedge \mathbf{N}']$, but $\vdash \mathbf{M}' \equiv \mathbf{M}_0^d$ and $\vdash \mathbf{N}' \equiv \mathbf{N}_0^d$ by inductive hypothesis, so $\vdash \mathbf{D}' \equiv [\mathbf{M}_0^d \wedge \mathbf{N}_0^d]$. Also $\mathbf{D}_0^d = [\mathbf{M}_0 \vee \mathbf{N}_0]^d = [\mathbf{M}_0^d \wedge \mathbf{N}_0^d]$, so $\vdash \mathbf{D}' \equiv \mathbf{D}_0^d$.

(d) $\mathbf{D}$ is $[\mathbf{M} \wedge \mathbf{N}]$. Then $\mathbf{D}' = [\mathbf{M}' \vee \mathbf{N}']$, but $\vdash \mathbf{M}' \equiv \mathbf{M}_0^d$ and $\vdash \mathbf{N}' \equiv \mathbf{N}_0^d$ by inductive hypothesis, so $\vdash \mathbf{D}' \equiv [\mathbf{M}_0^d \vee \mathbf{N}_0^d]$. Also $\mathbf{D}_0^d = (\sim [\sim \mathbf{M}_0 \vee \sim \mathbf{N}_0])^d = \sim [\sim \mathbf{M}_0^d \wedge \sim \mathbf{N}_0^d] = \sim\sim [\sim\sim \mathbf{M}_0^d \vee \sim\sim \mathbf{N}_0^d]$, so $\vdash \mathbf{D}' \equiv \mathbf{D}_0^d$.

Since the pattern of proof is now clear, we leave it to the reader to check the remaining cases (Exercise X4000). Note that there is some purpose to this exercise, since it should be verified that the duals of the remaining connectives are defined correctly. ■

4001 Corollary. For any wff $\mathbf{A}$ of $\mathcal{F}$, $\vdash \mathbf{A}^{dd} \equiv \mathbf{A}$.

Proof: It is easy to see that $\mathbf{A}$ is a dual of $\mathbf{A}^d$, so $\vdash \mathbf{A}^{dd} \equiv \mathbf{A}$ by 4000. ■

REMARK. Before looking at the next theorem, the reader should test his understanding of the basic idea of duality by trying to figure out what the theorem should say. The theorem has the form $\mathcal{V}_\varphi^\mathcal{M}\mathbf{A}^d = ?$, and essentially says that one can compute $\mathcal{V}_\varphi^\mathcal{M}\mathbf{A}^d$ if one knows $\mathcal{V}_\varphi^\mathcal{N}\mathbf{A}$ for an appropriate interpretation $\mathcal{N}$.

4002 Theorem. Let $\mathcal{M} = \langle \mathcal{D}, \mathcal{J} \rangle$ be any interpretation. Define $\mathcal{J}'$ to agree with $\mathcal{J}$ on all individual and function constants, and for all n-ary predicate constants $\mathbf{P}$, let $(\mathcal{J}'\mathbf{P})d_1 \ldots d_n = \sim (\mathcal{J}\mathbf{P})d_1 \ldots d_n$ for all $d_1, \ldots, d_n \in \mathcal{D}$. Let $\mathcal{M}' = \langle \mathcal{D}, \mathcal{J}' \rangle$. Then for any wff $\mathbf{A}$ and assignment φ, $\mathcal{V}_\varphi^\mathcal{M}\mathbf{A} = \sim \mathcal{V}_\varphi^{\mathcal{M}'}\mathbf{A}$.

NOTES:

(1) In accord with the convention introduced in §25, for simplicity we are

assuming that no propositional, predicate, or function variables occur in wffs of $\mathcal{F}$.

(2) Clearly φ can be regarded as an assignment into $\mathcal{M}'$ as well as into $\mathcal{M}$.

Proof by induction on the construction of **A**:

Case. **A** is an atomic wff $\mathbf{P}\mathbf{t}_1 \ldots \mathbf{t}_n$. It is easy to see by induction that $\mathcal{V}_\varphi^{\mathcal{M}'}\mathbf{t} = \mathcal{V}_\varphi^{\mathcal{M}}\mathbf{t}$ for each term **t**. Let

$$\begin{aligned} d_i &= \mathcal{V}_\varphi^{\mathcal{M}}\mathbf{t}_i = \mathcal{V}_\varphi^{\mathcal{M}'}\mathbf{t}_i \text{ for } i = 1, \ldots, n. \\ \mathcal{V}_\varphi^{\mathcal{M}}\mathbf{A}^d &= \mathcal{V}_\varphi^{\mathcal{M}}\mathbf{A} = (\mathcal{J}\mathbf{P})d_1 \ldots d_n = \sim (\mathcal{J}'\mathbf{P})d_1 \ldots d_n = \sim \mathcal{V}_\varphi^{\mathcal{M}'}\mathbf{A}. \end{aligned}$$

Case. **A** is $\sim$ **B**.

$$\begin{aligned} \mathcal{V}_\varphi^{\mathcal{M}}\mathbf{A}^d &= \mathcal{V}_\varphi^{\mathcal{M}} \sim \mathbf{B}^d = \sim \mathcal{V}_\varphi^{\mathcal{M}}\mathbf{B}^d \\ &= \sim\sim \mathcal{V}_\varphi^{\mathcal{M}'}\mathbf{B} \quad \text{(by inductive hypothesis)} \\ &= \sim \mathcal{V}_\varphi^{\mathcal{M}'} \sim \mathbf{B} = \sim \mathcal{V}_\varphi^{\mathcal{M}'}\mathbf{A}. \end{aligned}$$

Case. **A** is $[\mathbf{B} \lor \mathbf{C}]$.

$$\begin{aligned} \mathcal{V}_\varphi^{\mathcal{M}}\mathbf{A}^d &= \mathcal{V}_\varphi^{\mathcal{M}}[\mathbf{B}^d \land \mathbf{C}^d] = \mathcal{V}_\varphi^{\mathcal{M}} \sim [\sim \mathbf{B}^d \lor \sim \mathbf{C}^d] \\ &= \sim (\sim (\mathcal{V}_\varphi^{\mathcal{M}}\mathbf{B}^d) \lor \sim (\mathcal{V}_\varphi^{\mathcal{M}}\mathbf{C}^d)) \\ &= \sim (\sim\sim (\mathcal{V}_\varphi^{\mathcal{M}'}\mathbf{B}) \lor \sim\sim (\mathcal{V}_\varphi^{\mathcal{M}'}\mathbf{C})) \quad \text{(by inductive hypothesis)} \\ &= \sim \mathcal{V}_\varphi^{\mathcal{M}'}[\mathbf{B} \lor \mathbf{C}] = \sim \mathcal{V}_\varphi^{\mathcal{M}'}\mathbf{A}. \end{aligned}$$

Case. **A** is $\forall\mathbf{x}\mathbf{B}$. We must show that $\mathcal{V}_\varphi^{\mathcal{M}}\exists\mathbf{x}\mathbf{B}^d = \sim \mathcal{V}_\varphi^{\mathcal{M}'}\forall\mathbf{x}\mathbf{B}$. Suppose that $\mathcal{V}_\varphi^{\mathcal{M}}\exists\mathbf{x}\mathbf{B}^d = \mathsf{T}$. Then by X2300 there is an assignment ψ which agrees with φ off **x** such that $\mathcal{V}_\psi^{\mathcal{M}}\mathbf{B}^d = \mathsf{T}$. By inductive hypothesis, $\mathcal{V}_\psi^{\mathcal{M}}\mathbf{B}^d = \sim \mathcal{V}_\psi^{\mathcal{M}'}\mathbf{B}$, so $\mathcal{V}_\psi^{\mathcal{M}'}\mathbf{B} = \mathsf{F}$. Hence $\mathcal{V}_\varphi^{\mathcal{M}'}\forall\mathbf{x}\mathbf{B} = \mathsf{F}$, so $\sim \mathcal{V}_\varphi^{\mathcal{M}'}\forall\mathbf{x}\mathbf{B} = \mathsf{T}$. For the proof in the other direction, suppose that $\sim \mathcal{V}_\varphi^{\mathcal{M}'}\forall\mathbf{x}\mathbf{B} = \mathsf{T}$. Then $\mathcal{V}_\varphi^{\mathcal{M}'}\forall\mathbf{x}\mathbf{B} = \mathsf{F}$, so there is an assignment ψ which agrees with φ off **x** such that $\mathcal{V}_\psi^{\mathcal{M}'}\mathbf{B} = \mathsf{F}$. By inductive hypothesis $\mathcal{V}_\psi^{\mathcal{M}}\mathbf{B}^d = \sim \mathcal{V}_\psi^{\mathcal{M}'}\mathbf{B} = \mathsf{T}$, so $\mathcal{V}_\varphi^{\mathcal{M}}\exists\mathbf{x}\mathbf{B}^d = \mathsf{T}$. Thus $\mathcal{V}_\varphi^{\mathcal{M}}(\forall\mathbf{x}\mathbf{B})^d = \mathcal{V}_\varphi^{\mathcal{M}}\exists\mathbf{x}\mathbf{B}^d = \sim \mathcal{V}_\varphi^{\mathcal{M}'}\forall\mathbf{x}\mathbf{B}$, as required. ■

4003 Corollary. If $\vdash$ **A**, and **B** is any dual of **A**, then $\vdash \sim$ **B**.

Proof: Let $\mathcal{M}$ be any interpretation, and define $\mathcal{M}'$ as in 4002. Let φ be any assignment. $\mathcal{V}_\varphi^\mathcal{M} \sim \mathbf{A}^d = \sim \mathcal{V}_\varphi^\mathcal{M}\mathbf{A}^d = \sim\sim \mathcal{V}_\varphi^{\mathcal{M}'}\mathbf{A}$ (by 4002) $=$ T since $\models \mathbf{A}$. Hence $\models \sim \mathbf{A}^d$, so $\vdash \sim \mathbf{A}^d$ by Gödel's Completeness Theorem (2510). Hence $\vdash \sim \mathbf{B}$ by 4000 and Rule P. ∎

REMARK. 4003 can also be proved by a purely syntactic argument. See Exercise X4002.

4004 Corollary. If $\vdash \mathbf{A} \supset \mathbf{B}$ then $\vdash \mathbf{B}^d \supset \mathbf{A}^d$.

Proof: Suppose $\vdash \mathbf{A} \supset \mathbf{B}$. Then $\vdash \sim [\mathbf{A}^d \not\subset \mathbf{B}^d]$ by 4003, so $\vdash \mathbf{B}^d \supset \mathbf{A}^d$ by Rule P. ∎

EXAMPLE: $\vdash \forall x\, Px \supset Py$ by Axiom 4, so $\vdash Py \supset \exists x Px$ (which is an instance of 2123).

4005 Corollary. If $\vdash \mathbf{A} \equiv \mathbf{B}$, then $\vdash \mathbf{A}^d \equiv \mathbf{B}^d$

Proof: Suppose $\vdash \mathbf{A} \equiv \mathbf{B}$. Then $\vdash \sim [\mathbf{A}^d \not\equiv \mathbf{B}^d]$ by 4003, so $\vdash \mathbf{A}^d \equiv \mathbf{B}^d$ by Rule P. ∎

EXERCISES

X4000. Complete the proof of Proposition 4000.

X4001. Give a semantic proof of 4001 (using Gödel's Completeness Theorem) by using 4002 to prove that $\mathcal{V}_\varphi^\mathcal{M}\mathbf{A}^{dd} = \mathcal{V}_\varphi^\mathcal{M}\mathbf{A}$ for all $\mathcal{M}$ and φ.

X4002. Give a completely syntactic proof of 4003 as follows: Let $\mathbf{A}_1, \ldots, \mathbf{A}_n$ be a proof of **A**; show that $\vdash \sim \mathbf{A}_i^d$ for $1 \leq i \leq n$ by complete induction on i.

X4003. Give examples to illustrate Corollaries 4003, 4004, and 4005.

X4004. Prove or refute the following conjecture about wffs of the system $\mathcal{P}$ of propositional calculus. For any assignment φ of truth values to propositional variables, let $\varphi'\mathbf{p} = \sim \varphi\mathbf{p}$ for all **p**. Suppose **A** is any wff of $\mathcal{P}$ such that $\mathcal{V}_\varphi\mathbf{A} = \sim \mathcal{V}_{\varphi'}\mathbf{A}$ for every assignment φ. Then there is a propositional variable **q** such that $\models \mathbf{A} \equiv \mathbf{q}$ or $\models \mathbf{A} \equiv \sim \mathbf{q}$.

§41. Craig's Interpolation Theorem

As in §34, for this section it is convenient to use a formulation of first-order logic which we call $\mathcal{F}_{\mathsf{f}}$, whose primitive symbols are those of $\mathcal{F}$ plus a 0-ary propositional connective f which is false in all interpretations. (With appropriate minor modifications, the theorem below also applies to formulations of $\mathcal{F}$ which do not contain f.) Note that we regard f as a connective, not a propositional constant. It is readily verified that Smullyan's Unifying Principle (2507) applies to $\mathcal{F}_{\mathsf{f}}$ if one modifies condition (a) in the definition of abstract consistency class to read:

(a′) $\mathsf{f} \notin \mathcal{S}$, and if $\mathbf{A}$ is atomic, then $\mathbf{A} \notin \mathcal{S}$ or $\sim \mathbf{A} \notin \mathcal{S}$.

We slightly extend the terminology introduced in §21 by defining an occurrence of a constant or variable in a wff $\mathbf{A}$ of $\mathcal{F}_{\mathsf{f}}$ to be *positive* [*negative*] in $\mathbf{A}$ iff that occurrence is in the scope of an even [odd] number of occurrences of $\sim$ in $\mathbf{A}$.

DEFINITION. A sentence $\mathbf{J}$ is an *interpolation sentence* for a sentence $[\mathbf{H} \supset \mathbf{K}]$ iff the following conditions are all satisfied:

(1) $\models \mathbf{H} \supset \mathbf{J}$ and $\models \mathbf{J} \supset \mathbf{K}$;
(2) each individual constant which occurs in $\mathbf{J}$ occurs in both $\mathbf{H}$ and $\mathbf{K}$;
(3) each propositional or predicate constant which has a positive [negative] occurrence in $\mathbf{J}$ also has a positive [negative] occurrence in both $\mathbf{H}$ and $\mathbf{K}$.

EXAMPLE: $\mathbf{J}$ is an interpolation sentence for $[\mathbf{M} \wedge \mathbf{J}] \supset [\mathbf{J} \vee \mathbf{N}]$.

REMARK. If $\models \mathbf{K}$, then $\models \mathbf{H} \supset \mathbf{K}$ for any sentence $\mathbf{H}$. Nevertheless, $\mathbf{H} \supset \mathbf{K}$ has an interpolation sentence, namely $\sim \mathsf{f}$. Similarly, if $\models \sim \mathbf{H}$, then f is an interpolation sentence for $\mathbf{H} \supset \mathbf{K}$.

4100 Craig-Lyndon Interpolation Theorem.[1] If $\mathbf{H} \supset \mathbf{K}$ is a valid sentence of $\mathcal{F}_{\mathsf{f}}$, then there is an interpolation sentence for $\mathbf{H} \supset \mathbf{K}$.

Proof: If $\mathcal{S}_1$ and $\mathcal{S}_2$ are sets of sentences, we shall say that a sentence $\mathbf{J}$ is an interpolation sentence for the ordered pair $\langle \mathcal{S}_1, \mathcal{S}_2 \rangle$ iff

(1) $\mathcal{S}_1 \cup \{\sim \mathbf{J}\}$ and $\mathcal{S}_2 \cup \{\mathbf{J}\}$ are unsatisfiable;

[1] [Craig, 1957], [Lyndon, 1959]. Also see [Henkin, 1963a].

(2) each individual constant which occurs in $\mathbf{J}$ occurs in some sentence of $\mathcal{S}_1$ and in some sentence of $\mathcal{S}_2$;

(3) each propositional or predicate constant which has a positive [negative] occurrence in $\mathbf{J}$ also has a positive [negative] occurrence in some sentence of $\mathcal{S}_1$ and a negative [positive] occurrence in some sentence of $\mathcal{S}_2$.

It is readily verified that $\mathbf{J}$ is an interpolation sentence for $\langle\{\mathbf{C}_1,\ldots,\mathbf{C}_m\},\{\sim \mathbf{D}_1,\ldots,\sim \mathbf{D}_n\}\rangle$ iff it is an interpolation sentence for $\mathbf{C}_1 \wedge \ldots \wedge \mathbf{C}_m \supset \mathbf{D}_1 \vee \ldots \vee \mathbf{D}_n$. Also note that :

(*1) $\mathbf{J}$ is an interpolation sentence for $\langle\mathcal{S}_1, \mathcal{S}_2\rangle$ iff $\sim \mathbf{J}$ is an interpolation sentence for $\langle\mathcal{S}_2, \mathcal{S}_1\rangle$.

(*2) If $\mathbf{J}$ is an interpolation sentence for $\langle\mathcal{S}_1, \mathcal{S}_2\rangle$ and $\mathcal{S}_1 \subseteq \mathcal{U}_1$ and $\mathcal{S}_2 \subseteq \mathcal{U}_2$, then $\mathbf{J}$ is an interpolation sentence for $\langle\mathcal{U}_1, \mathcal{U}_2\rangle$.

Let $\Gamma = \{\mathcal{S} \mid \mathcal{S}$ is a set of sentences such that for some sets $\mathcal{S}_1$ and $\mathcal{S}_2$, $\mathcal{S} = \mathcal{S}_1 \cup \mathcal{S}_2$ and $\langle\mathcal{S}_1, \mathcal{S}_2\rangle$ has no interpolation sentence $\}$. Note that if $\mathcal{S}$ is a set of sentences, then $\mathcal{S} \notin \Gamma$ iff for all sets $\mathcal{S}_1$ and $\mathcal{S}_2$ such that $\mathcal{S} = \mathcal{S}_1 \cup \mathcal{S}_2$, $\langle\mathcal{S}_1, \mathcal{S}_2\rangle$ has an interpolation sentence. We shall show that Γ is an abstract consistency class. The theorem will then follow directly, for if $\mathbf{H} \supset \mathbf{K}$ is a valid sentence, then $\{\mathbf{H}, \sim \mathbf{K}\}$ has no model, so $\{\mathbf{H}, \sim \mathbf{K}\} \notin \Gamma$ by Smullyan's Unifying Principle (2507), so $\langle\{\mathbf{H}\}, \{\sim \mathbf{K}\}\rangle$ has an interpolation sentence, so $\mathbf{H} \supset \mathbf{K}$ does too.

Γ is closed under subsets by (*2).

We next check conditions (a′), (b)-(f) in the definition of an abstract consistence class.

(a′) f is an interpolation sentence for $\langle\{\mathsf{f}\},\emptyset\rangle$, so $\{\mathsf{f}\} \notin \Gamma$ by (*1) and (*2). Also, a sentence $\mathbf{A}$ is an interpolation sentence for $\langle\{\mathbf{A}\}, \{\sim \mathbf{A}\}\rangle$, and f is an interpolation sentence for $\langle\{\mathbf{A}, \sim \mathbf{A}\}, \emptyset\rangle$, so $\{\mathbf{A}, \sim \mathbf{A}\} \notin \Gamma$ by (*1) and (*2). Hence (a′) is true since Γ is closed under subsets.

(b) is clear.

We shall check (c)-(f) by the indirect method. In each case (with various choices of the sentence $\mathbf{D}$) we shall assume (among other things) that $\mathbf{D} \in \mathcal{S}$ and $\mathcal{S} \in \Gamma$. Hence we can choose $\mathcal{S}_1$ and $\mathcal{S}_2$ such that $\mathcal{S} = \mathcal{S}_1 \cup \mathcal{S}_2$ and $\langle\mathcal{S}_1, \mathcal{S}_2\rangle$ has no interpolation sentence. Since (by (*1)) $\langle\mathcal{S}_1, \mathcal{S}_2\rangle$ has an interpolation sentence iff $\langle\mathcal{S}_2, \mathcal{S}_1\rangle$ does, without loss of generality we may assume that $\mathbf{D} \in \mathcal{S}_1$. In each case we shall reach a contradiction by showing that $\langle\mathcal{S}_1, \mathcal{S}_2\rangle$ must have an interpolation sentence.

(c) Suppose $[\mathbf{A} \vee \mathbf{B}] \in \mathcal{S}$, $\mathcal{S} \in \Gamma$, but (c_1) $\mathcal{S} \cup \{\mathbf{A}\} \notin \Gamma$ and (c_2) $\mathcal{S} \cup \{\mathbf{B}\} \notin \Gamma$. We are assuming that $[\mathbf{A} \vee \mathbf{B}] \in \mathcal{S}_1$. By ($c_1$) $\langle \mathcal{S}_1 \cup \{\mathbf{A}\}, \mathcal{S}_2 \rangle$ has an interpolation sentence **J**, and by (c_2) $\langle \mathcal{S}_1 \cup \{\mathbf{B}\}, \mathcal{S}_2 \rangle$ has an interpolation sentence **K**. It is easily seen that $[\mathbf{J} \vee \mathbf{K}]$ is an interpolation sentence for $\langle \mathcal{S}_1, \mathcal{S}_2 \rangle$.

(d) Suppose $\sim [\mathbf{A} \vee \mathbf{B}] \in \mathcal{S}$, $\mathcal{S} \in \Gamma$, but (d_1) $\mathcal{S} \cup \{\sim \mathbf{A}, \sim \mathbf{B}\} \notin \Gamma$. We are assuming that $\sim [\mathbf{A} \vee \mathbf{B}] \in \mathcal{S}_1$. By ($d_1$) $\langle \mathcal{S}_1 \cup \{\sim \mathbf{A}, \sim \mathbf{B}\}, \mathcal{S}_2 \rangle$ has an interpolation sentence **J**, and **J** is also an interpolation sentence for $\langle \mathcal{S}_1, \mathcal{S}_2 \rangle$.

(e) Suppose $\forall \mathbf{x} \mathbf{A} \in \mathcal{S}$ and $\mathcal{S} \in \Gamma$, but there is a variable-free term **t** such that $\mathcal{S} \cup \{\mathsf{S}^{\mathbf{x}}_{\mathbf{t}} \mathbf{A}\} \notin \Gamma$. We are assuming that $\forall \mathbf{x} \mathbf{A} \in \mathcal{S}_1$. There is an interpolation sentence **J** for $\langle \mathcal{S}_1 \cup \{\mathsf{S}^{\mathbf{x}}_{\mathbf{t}} \mathbf{A}\}, \mathcal{S}_2 \rangle$. Let $\mathbf{c}_1, \ldots, \mathbf{c}_n$ (where $n \geq 0$) be the individual constants which occur in **t** but not in any wff of $\mathcal{S}_1$, and let $\mathbf{y}_1, \ldots, \mathbf{y}_n$ be distinct individual variables which do not occur in **J**. Let $\mathbf{J}'$ be $\mathsf{S}^{\mathbf{c}_1 \ldots \mathbf{c}_n}_{\mathbf{y}_1 \ldots \mathbf{y}_n} \mathbf{J}$ and let **K** be $\forall \mathbf{y}_1 \ldots \forall \mathbf{y}_n \mathbf{J}'$. (Thus **K** is **J** if $n = 0$.) We assert that **K** is an interpolation sentence for $\langle \mathcal{S}_1, \mathcal{S}_2 \rangle$.

$\mathcal{S}_1 \cup \{\mathsf{S}^{\mathbf{x}}_{\mathbf{t}} \mathbf{A}, \sim \mathbf{J}\}$ is unsatisfiable, hence inconsistent, so:

$\mathcal{S}_1, \mathsf{S}^{\mathbf{x}}_{\mathbf{t}} \mathbf{A} \vdash \mathbf{J}$	by 2501
$\mathcal{S}_1 \vdash \mathsf{S}^{\mathbf{x}}_{\mathbf{t}} \mathbf{A} \supset \mathbf{J}$	Deduction Theorem
$\mathcal{S}_1 \vdash \forall \mathbf{x} \mathbf{A}$	since $\forall \mathbf{x} \mathbf{A} \in \mathcal{S}_1$
$\mathcal{S}_1 \vdash \mathsf{S}^{\mathbf{x}}_{\mathbf{t}} \mathbf{A}$	$\forall$I
$\mathcal{S}_1 \vdash \mathbf{J}$	Rule P
$\mathcal{S}_1 \vdash \mathbf{J}'$	since $\mathbf{c}_1, \ldots, \mathbf{c}_n$ do not occur in any sentence of $\mathcal{S}_1$
$\mathcal{S}_1 \vdash \mathbf{K}$	Gen

Thus $\mathcal{S}_1 \cup \{\sim \mathbf{K}\}$ is inconsistent, hence unsatisfiable. Also, $\vdash \mathbf{K} \supset \mathbf{J}$ and $\mathcal{S}_2 \cup \{\mathbf{J}\}$ is unsatisfiable, so $\mathcal{S}_2 \cup \{\mathbf{K}\}$ is unsatisfiable.

If **d** is any individual constant in **K**, then **d** occurs in **J**, so **d** occurs in some sentence of $\mathcal{S}_2$ and in some sentence of $\mathcal{S}_1 \cup \{\mathsf{S}^{\mathbf{x}}_{\mathbf{t}} \mathbf{A}\}$. Suppose **d** does not occur in any sentence of $\mathcal{S}_1$. Then (since $\forall \mathbf{x} \mathbf{A} \in \mathcal{S}_1$) **d** must occur in **t**, and so is some $\mathbf{c}_i$. However, none of the $\mathbf{c}_i$'s occurs in **K**, so this is impossible; thus **d** must indeed occur in some sentence of $\mathcal{S}_1$. Condition (3) in the definition of an interpolation sentence clearly satisfied, so **K** is an interpolation sentence for $\langle \mathcal{S}_1, \mathcal{S}_2 \rangle$.

(f) Suppose $\sim \forall \mathbf{x} \mathbf{A} \in \mathcal{S}$ and $\mathcal{S} \in \Gamma$ and **c** is an individual constant which does not occur in any sentence of $\mathcal{S}$ and $\mathcal{S} \cup \{\sim \mathsf{S}^{\mathbf{x}}_{\mathbf{c}} \mathbf{A}\} \notin \Gamma$. Let **J** be an interpolation sentence for $\langle \mathcal{S}_1 \cup \{\sim \mathsf{S}^{\mathbf{x}}_{\mathbf{c}} \mathbf{A}\}, \mathcal{S}_2 \rangle$. We assert that **J** is also an interpolation sentence for $\langle \mathcal{S}_1, \mathcal{S}_2 \rangle$. **c** does not occur in any sentence of $\mathcal{S}_2$, so **c** does not occur in **J**. Hence it is easy to see that every individual constant in **J** occurs in some sentence of $\mathcal{S}_1$ as well as some sentence of $\mathcal{S}_2$. Clearly **J** satisfies condition (3).

To complete the proof we simply need to establish that $\mathcal{S}_1 \cup \{\sim \mathbf{J}\}$ is unsatisfiable. Suppose it is satisfiable, and let Δ be the collection of satisfiable sets of sentences of $\mathcal{F}_\mathfrak{f}$. By 2505 and 2512, Δ is an abstract consistency class. $\mathcal{S}_1 \cup \{\sim \mathbf{J}\} \in \Delta$ and $\sim \forall \mathbf{x}\mathbf{A} \in \mathcal{S}_1 \cup \{\sim \mathbf{J}\}$, so by condition (f) in the definition of an abstract consistency class, $\mathcal{S}_1 \cup \{\sim \mathbf{J}\} \cup \{\sim \mathsf{S}^{\mathbf{x}}_{\mathbf{c}}\mathbf{A}\} \in \Delta$, which is to say that this set is satisfiable. This contradicts the fact that **J** is an interpolation sentence for $\langle \mathcal{S}_1 \cup \{\sim \mathsf{S}^{\mathbf{x}}_{\mathbf{c}}\mathbf{A}\},\ \mathcal{S}_2\rangle$. Thus $\mathcal{S}_1 \cup \{\sim \mathbf{J}\}$ is unsatisfiable.

Hence **J** is indeed an interpolation formula for $\langle \mathcal{S}_1,\ \mathcal{S}_2\rangle$. ∎

§42. Beth's Definability Theorem

DEFINITION. Let **H** be a sentence containing a k-ary predicate constant **P**, let **P**′ be a k-ary predicate constant which does not occur in **H**, and let **H**′ be $\mathsf{S}^{\mathbf{P}}_{\mathbf{P}'}\mathbf{H}$.

(1) **H** *defines* **P** *implicitly* iff

$$\vdash \mathbf{H} \wedge \mathbf{H}' \supset \forall x_1 \ldots \forall x_k \centerdot \mathbf{P}x_1 \ldots x_k \equiv \mathbf{P}'x_1 \ldots x_k.$$

(2) The wff **D** is an *explicit definition of* **P** *from* **H** iff

(a) **P** does not occur in **D**;
(b) no variables other than $x_1, \ldots, x_k$ occur free in **D**;
(c) every constant in **D** occurs in **H**;
(d) $\vdash \mathbf{H} \supset \forall x_1 \ldots \forall x_k \centerdot \mathbf{P}x_1 \ldots x_k \equiv \mathbf{D}$.

(3) **P** is *explicitly definable from* **H** iff there is an explicit definition of **P** from **H**.

4200 Beth's Definability Theorem [Beth, 1953]. Let **H** be a sentence containing a k-ary predicate constant **P**. Then **H** defines **P** implicitly iff **P** is explicitly definable from **H**.

Proof: If **P** is explicitly definable from **H** by a wff **D**, then

$\vdash \mathbf{H} \supset \forall x_1 \ldots \forall x_k \centerdot \mathbf{P}x_1 \ldots x_k \equiv \mathbf{D}$

$\vdash \mathbf{H}' \supset \forall x_1 \ldots \forall x_k \centerdot \mathbf{P}'x_1 \ldots x_k \equiv \mathbf{D}$ by substitution

(since **P** does not occur in **D**), so it is easily seen that **H** defines **P** implicitly.

Suppose **H** defines **P** implicitly. Then

$$\vdash \mathbf{H} \wedge \mathbf{H}' \supset \forall x_1 \ldots \forall x_k \centerdot \mathbf{P} x_1 \ldots x_k \equiv \mathbf{P}' x_1 \ldots x_k.$$

By 2515 we may assume that this theorem is provable in a formulation of $\mathcal{F}$ which contains no predicate or function constants other than $\mathbf{P}'$ and the constants which occur in $\mathbf{H}$. Let $\mathbf{c}_1, \ldots, \mathbf{c}_k$ be distinct individual constants of $\mathcal{F}$ which do not occur in $\mathbf{H}$.

$\vdash \mathbf{H} \wedge \mathbf{H}' \supset \centerdot \mathbf{P}\mathbf{c}_1 \ldots \mathbf{c}_k \equiv \mathbf{P}'\mathbf{c}_1 \ldots \mathbf{c}_k$

$\vdash \mathbf{H} \wedge \mathbf{P}\mathbf{c}_1 \ldots \mathbf{c}_k \supset \centerdot \mathbf{H}' \supset \mathbf{P}'\mathbf{c}_1 \ldots \mathbf{c}_k$ Rule P

Let $\mathbf{B}$ be an interpolation sentence (see §41) for the sentence above. By making alphabetical changes of bound variables, if necessary, we may assume that the variables $x_1, \ldots, x_k$ do not occur in $\mathbf{B}$. Let $\mathbf{D}$ be $\mathsf{S}^{\mathbf{c}_1 \ldots \mathbf{c}_k}_{x_1 \ldots x_k} \mathbf{B}$. Note that neither $\mathbf{P}$ nor $\mathbf{P}'$ can occur in $\mathbf{D}$, and every constant in $\mathbf{D}$ must occur in $\mathbf{H}$.

$\vdash \mathbf{H} \wedge \mathbf{P}\mathbf{c}_1 \ldots \mathbf{c}_k \supset \mathbf{B}$	by the definition of $\mathbf{B}$
$\vdash \mathbf{H} \wedge \mathbf{P} x_1 \ldots x_k \supset \mathbf{D}$	by substitution
$\vdash \mathbf{B} \supset \centerdot \mathbf{H}' \supset \mathbf{P}'\mathbf{c}_1 \ldots \mathbf{c}_k$	by the definition of $\mathbf{B}$
$\vdash \mathbf{D} \supset \centerdot \mathbf{H}' \supset \mathbf{P}' x_1 \ldots x_k$	by substitution
$\vdash \mathbf{D} \supset \centerdot \mathbf{H} \supset \mathbf{P} x_1 \ldots x_k$	by substitution
$\vdash \mathbf{H} \supset \centerdot \mathbf{P} x_1 \ldots x_k \equiv \mathbf{D}$	Rule P
$\vdash \mathbf{H} \supset \forall x_1 \ldots \forall x_k \centerdot \mathbf{P} x_1 \ldots x_k \equiv \mathbf{D}$	$\supset \forall$ Rule (2103)

Thus $\mathbf{D}$ is an explicit definition of $\mathbf{P}$ from $\mathbf{H}$. ∎

REMARK. Note that if $\mathbf{P}$ is (implicitly or explicitly) definable from $\mathbf{H}$, then $\mathbf{P}$ has an explicit definition $\mathbf{D}$ in which predicates occur only if they have both positive and negative occurrences in $\mathbf{H}$. This follows from the fact that the interpolation sentence $\mathbf{B}$ constructed in the proof above has this property.

EXERCISE

X4200. Let $A(G, P)$ be the wff

$$\forall x \forall y[Px \supset \centerdot Gxx \vee \sim Gxy] \wedge \forall x[Px \vee \exists z\, Gxz] \wedge \forall x[Gxx \supset Px]$$

(a) Show that $A(G, P)$ defines P implicitly in terms of G. That is, prove $A(G, P_1) \wedge A(G, P_2) \supset \forall w \centerdot P_1 w \equiv P_2 w$.

(b) Illustrate Beth's Theorem on Definability by finding an explicit definition $B(G, w)$ of P in terms of G from $A(G, P)$.

Prove $A(G, P) \supset \forall w \centerdot Pw \equiv B(G, w)$.

Chapter 5

Type Theory

§50. Introduction

So far we have been concerned with first-order logic, and its subsystem propositional calculus, which we might regard as zeroth-order logic. We now wish to discuss higher-order logics.

It seems very natural to extend the system $\mathcal{F}$ of first-order logic by permitting quantification on predicate, propositional, and function variables as well as individual variables. One thus obtains the wffs of a system of *second-order logic*. One might then introduce predicate and function variables of higher type to denote relations and functions whose arguments may be relations and functions of individuals as well as individuals. Thus one would be led to a system of *third-order logic*, and if one permitted quantification with respect to these new variables, one would obtain a system of *fourth-order logic*. This process can be continued indefinitely to obtain logics of arbitrarily high order. Of course, after a while one runs out of different types of letters to use for different types of variables, but this problem can be solved by introducing *type symbols* to indicate the types of variables, and using a letter with type symbol α as subscript for a variable of type α.

To make these ideas precise, we shall briefly describe a system $\mathcal{F}^\omega$ of ω-order logic which has all finite order logics as subsystems. We shall not discuss $\mathcal{F}^\omega$ very extensively, since we prefer to devote most of our attention to an improved formulation of higher-order logic called $\mathcal{Q}_0$ which we shall soon introduce. Nevertheless, $\mathcal{F}^\omega$ provides a convenient starting point for a discussion of higher-order logics. For the sake of simplicity, $\mathcal{F}^\omega$ will have no function symbols. It is well known that assertions about functions can be expressed in terms of relations, since to every n-ary function f there corresponds an $(n+1)$-ary relation R such that for all $x_1, \ldots, x_n$, and y, $Rx_1 \ldots x_n y$ iff $fx_1 \ldots x_n = y$.

A system of logic which includes logics of all finite orders (but no infinite orders) is known as *ω-order logic*, or *finite type theory*. (There are also systems of transfinite type theory, such as that in [Andrews, 1965], but we shall not discuss them here.) Type theory was invented by Bertrand Russell [Russell, 1908]. In [Whitehead and Russell, 1913], which had extraordinary influence on the development of logic in the early twentieth century, he and Alfred North Whitehead demonstrated that various fundamental parts of mathematics could indeed be formalized in this system. Russell was concerned with providing secure foundations for mathematics, and wished to enforce the *vicious-circle principle* that no totality can contain members defined in terms of itself. He therefore originally advocated using what is now called *ramified* type theory,[1] which is still important in various proof-theoretic studies (such as [Feferman, 1964]) of the consistency of restricted portions of mathematics. However, ramified type theory did not prove adequate for formalizing mathematics in general, and we shall devote our attention to a simplified version of it which is known as *simple* type theory.

Additional information about type theory may be obtained from sources such as [Andrews, 2001], [van Benthem and Doets, 1983], [Hatcher, 1968], [Hindley, 1997], [Jacobs, 1999], [Leivant, 1994], [Mendelson, 1987], [Quine, 1963], [Shapiro, 1991], [Wolfram, 1993], and references cited therein.

The *type symbols* of $\mathcal{F}^\omega$ and their *orders* are defined inductively as follows:

(a) $\imath$ is a type symbol (denoting the type of individuals) of order 0.
(b) o is a type symbol (denoting the type of truth values) of order 1.
(c) If $\tau_1, \ldots, \tau_n$ are type symbols (with $n \geq 1$), then $(\tau_1 \ldots \tau_n)$ is a type symbol (denoting the type of n-ary relations with arguments of types $\tau_1, \ldots, \tau_n$, respectively), and its order is 1+ the maximum of the orders of $\tau_1, \ldots, \tau_n$. (One may think of "o" as "( )".)

The primitive symbols of $\mathcal{F}^\omega$ are the improper symbols [,], $\sim$, $\vee$, and $\forall$, a denumerable list of variables of each type, and (optionally) constants of certain types. Variables of type o are called *propositional variables*, and variables of type $\imath$ are called *individual variables*.

The formation rules of $\mathcal{F}^\omega$ are the following:

[1] See [Hazen, 1983] or [Church, 1956, p. 347] for more details.

(a) Every propositional variable or constant is a wff.
(b) If $\mathbf{u}_{\tau_1\ldots\tau_n}, \mathbf{v}^1_{\tau_1}\ldots, \mathbf{v}^n_{\tau_n}$ are variables or constants of the types indicated by their subscripts, then $\mathbf{u}_{\tau_1\ldots\tau_n}\mathbf{v}^1_{\tau_1}\ldots\mathbf{v}^n_{\tau_n}$ is a wff.
(c) If $\mathbf{A}$ and $\mathbf{B}$ are wffs and $\mathbf{u}$ is a variable of any type, then $\sim \mathbf{A}$, $[\mathbf{A} \vee \mathbf{B}]$, and $\forall\mathbf{uA}$ are wffs.

The *order* of a variable or constant is the order of its type symbol. For any positive integer m, the wffs of $2m$th-order logic are the wffs in which no variable or constant of order greater than m occurs, and the wffs of $(2m-1)$th-order logic are the wffs of $2m$th-order logic in which no variable of order m is quantified. (Note that first-order logic, as here defined, is just the result of dropping function symbols from the system $\mathcal{F}$ of §20.)

The definitions of connectives and quantifiers in §10 and §20 can be extended to apply to $\mathcal{F}^\omega$. If $\mathbf{x}_\tau$ and $\mathbf{y}_\tau$ are variables or constants of the same type τ, $\mathbf{x}_\tau = \mathbf{y}_\tau$ is defined to be an abbreviation for

$$\forall z_{(\tau)}[z_{(\tau)}\mathbf{x}_\tau \supset z_{(\tau)}\mathbf{y}_\tau] \tag{*1}$$

where $z_{(\tau)}$ is a variable of type (τ). An alternative possibility is to define $\mathbf{x}_\tau = \mathbf{y}_\tau$ as

$$\forall p_{(\tau\tau)}[\forall z_\tau p_{(\tau\tau)} z_\tau z_\tau \supset p_{(\tau\tau)}\mathbf{x}_\tau\mathbf{y}_\tau]. \tag{*2}$$

(See Exercise X5007.) Note that the two basic properties of equality which we took as axioms in §26 are reflexivity and substitutivity. *1 defines equality in terms of substitutivity, and *2 defines equality in terms of reflexivity (indeed, as the intersection of all reflexive relations). Thus, all the properties of equality follow from either of the basic properties of substitutivity or reflexivity when these are expressed in a sufficiently powerful way using the expressiveness of higher-order logic.

The rules of inference of $\mathcal{F}^\omega$ are:

(1) **Modus Ponens.** From $\mathbf{A}$ and $\mathbf{A} \supset \mathbf{B}$ to infer $\mathbf{B}$.
(2) **Generalization.** From $\mathbf{A}$ to infer $\forall\mathbf{xA}$, where $\mathbf{x}$ is a variable of any type.

The axiom schemata of $\mathcal{F}^\omega$ are:

(1) $\mathbf{A} \vee \mathbf{A} \supset \mathbf{A}$
(2) $\mathbf{A} \supset_\blacksquare \mathbf{B} \vee \mathbf{A}$
(3) $\mathbf{A} \supset \mathbf{B} \supset_\blacksquare \mathbf{C} \vee \mathbf{A} \supset \mathbf{B} \vee \mathbf{C}$

(4) $\forall \mathbf{x}_\tau \mathbf{A} \supset \mathsf{S}^{\mathbf{x}_\tau}_{\mathbf{y}_\tau} \mathbf{A}$, where $\mathbf{y}_\tau$ is a variable or constant of the same type as the variable $\mathbf{x}_\tau$, and $\mathbf{y}_\tau$ is free for $\mathbf{x}_\tau$ in $\mathbf{A}$.
(5) $\forall \mathbf{x}[\mathbf{A} \vee \mathbf{B}] \supset \centerdot \mathbf{A} \vee \forall \mathbf{x}\mathbf{B}$, where $\mathbf{x}$ is any variable not free in $\mathbf{A}$.
(6) **(Comprehension Axioms)**

$\exists \mathbf{u}_o[\mathbf{u}_o \equiv \mathbf{A}]$, where $\mathbf{u}_o$ does not occur free in $\mathbf{A}$.

$\exists \mathbf{u}_{(\tau_1 \ldots \tau_n)} \forall \mathbf{v}^1_{\tau_1} \ldots \forall \mathbf{v}^n_{\tau_n} [\mathbf{u}_{(\tau_1 \ldots \tau_n)} \mathbf{v}^1_{\tau_1} \ldots \mathbf{v}^n_{\tau_n} \equiv \mathbf{A}]$, where $\mathbf{u}_{(\tau_1 \ldots \tau_n)}$ does not occur free in $\mathbf{A}$, and $\mathbf{v}^1_{\tau_1}, \ldots, \mathbf{v}^n_{\tau_n}$ are distinct variables.

(7) **(Axioms of Extensionality)**

$[x_o \equiv y_o] \supset \centerdot x_o = y_o$

$\forall w^1_{\tau_1} \ldots \forall w^n_{\tau_n} [x_{(\tau_1 \ldots \tau_n)} w^1_{\tau_1} \ldots w^n_{\tau_n} \equiv y_{(\tau_1 \ldots \tau_n)} w^1_{\tau_1} \ldots w^n_{\tau_n}] \supset \centerdot x_{(\tau_1 \ldots \tau_n)} = y_{(\tau_1 \ldots \tau_n)}$

The axioms of kth-order logic are those axioms of $\mathcal{F}^\omega$ which are wffs of kth-order logic. Note that Comprehension Axioms first appear in second-order logic, and Axioms of Extensionality first appear in fourth-order logic.

Note the principal features which are added to first-order logic in order to obtain this formulation of higher-order logic:

1. Variables of arbitrarily high orders.

2. Quantification on variables of all types.

3. Comprehension Axioms.

4. Axioms of Extensionality.

One may also wish to assume additional axioms, such as the Axiom Schema of Choice, an Axiom of Infinity, and perhaps even the Continuum Hypothesis, though there is room for argument whether these are axioms of pure logic or of mathematics.

It should be remarked that one can restrict the formation rules of $\mathcal{F}^\omega$ by requiring that $n = 1$ in the definitions above, and obtain a still simpler system of ω-order logic which is nevertheless adequate to express most mathematical ideas, since one can define ordered pairs (and, more generally, n-tuples) in various ways, and then represent an n-ary relation B by the monadic relation P such that $Bx^1 \ldots x^n$ iff $P\langle x_1, \ldots, x_n \rangle$. Thus, $P\langle x^1, \ldots, x^n \rangle$ would be an abbreviation for $\exists u[Pu \wedge u = \langle x^1, \ldots, x^n \rangle]$, where $u = \langle x^1, \ldots, x^n \rangle$ is defined in some appropriate way, and suitable types are assigned to u and P.

If one wishes to thus restrict the language so that all predicates are monadic, one may find it congenial to write $\mathbf{v}_\tau \in \mathbf{u}_{(\tau)}$ rather than $\mathbf{u}_{(\tau)} \mathbf{v}_\tau$, and to say "$\mathbf{v}_\tau$ is in the set $\mathbf{u}_{(\tau)}$" rather than "$\mathbf{v}_\tau$ has the property $\mathbf{u}_{(\tau)}$". Of course, this is a mere notational change. A more radical change is obtained by also dropping the type symbols, so that all variables are of the same type. (Propositional variables are customarily dispensed with.) One thus obtains Naive Axiomatic Set Theory, which can be formalized in first-order logic. $\in$ is regarded as a binary predicate constant, and all atomic wffs are of the form $\mathbf{x} \in \mathbf{y}$, where $\mathbf{x}$ and $\mathbf{y}$ are (individual) variables. $\mathbf{x} \in \mathbf{y}$ is interpreted to mean "$\mathbf{x}$ is a member of the set $\mathbf{y}$", which is of course false if $\mathbf{y}$ is not a set at all. The Comprehension Axiom Schema now takes the simple form $\exists \mathbf{u} \forall \mathbf{v}[\mathbf{v} \in \mathbf{u} \equiv \mathbf{A}]$, where $\mathbf{u}$ is not free in the wff $\mathbf{A}$. This axiom schema asserts that for any wff $\mathbf{A}$ in which $\mathbf{v}$ occurs free, there is a set $\mathbf{u}$ consisting of all those elements $\mathbf{v}$ of which $\mathbf{A}$ is true. Unfortunately, the axiom schema guarantees the existence of too many sets. In particular, $\exists u \forall v[v \in u \equiv \sim v \in v]$, i.e., there is a set whose members are all those elements which are not members of themselves. Choosing such a set u, we conclude that $u \in u \equiv \sim u \in u$, which is a contradiction, so Naive Axiomatic Set Theory is inconsistent! The paradox above was discovered by Bertrand Russell, and is known as Russell's Paradox.

One can restrict the Comprehension Axiom Schema, and take only certain carefully chosen instances of it as axioms, to obtain apparently consistent formulations of Axiomatic Set Theory. Alternatively, one can keep type symbols, so that Russell's paradox (and various other paradoxes) cannot be expressed at all within the formal system. Each approach leads to systems which are adequate for formalizing mathematics, and each approach has advantages for certain purposes. At present, Axiomatic Set Theory is more popular with mathematicians simply because it is much better known. However, most mathematicians do make mental distinctions between different types of mathematical objects, and use different types of letters to denote them, so it might be claimed that type theory provides a more natural formalization of mathematics as it is actually practiced. In addition, explicit syntactic distinctions between expressions denoting intuitively different types of mathematical entities are very useful in computerized systems for exploring and applying mathematics (on a theoretical rather than purely computational level). Of course, type symbols need not be explicitly written as long as one knows the types of the variables and constants.

Therefore we shall turn our attention to finding a formulation of type theory which is as expressive as possible, allowing mathematical ideas to be expressed precisely with a minimum of circumlocutions, and which is as simple and economical as is possible without sacrificing expressiveness. The reader will observe that the formal language we find arises very naturally from a few fundamental design decisions.

Since function symbols are very useful in practice, we shall admit them into the language. Indeed, for all types α and β, we shall have a type $(\alpha\beta)$ of functions from elements of type β to elements of type α.[2] (Of course, we are now giving the type symbol $(\alpha\beta)$ a different meaning from that used for the system $\mathcal{F}^\omega$.)

Since the values of functions can themselves be functions, by using a device of Schönfinkel [Schönfinkel, 1924] we can avoid explicitly admitting into our language functions of more than one argument. For example, if f is a function of two arguments, for each element x of the left domain of f there is a function g (depending on x) such that for each element y of the right domain of f, $gy = fxy$. We may now write $g = fx$, and regard f as a function of a single argument, whose value for any argument x in its domain is a function fx, whose value for any argument y in its domain is fxy.

For a more explicit example, consider the function $+$ which carries any pair of natural numbers to their sum. We may denote this function by $+_{((\sigma\sigma)\sigma)}$, where σ is the type of natural numbers. Given any number x, $[+_{((\sigma\sigma)\sigma)}x]$ is the function which, when applied to any number y, gives the value $[[+_{((\sigma\sigma)\sigma)}x]y]$, which is ordinarily abbreviated as $x + y$. Thus $[+_{((\sigma\sigma)\sigma)}x]$ is the function of one argument as a which adds x to any number. When we think of $+_{((\sigma\sigma)\sigma)}$ as a function of one argument, we see that it maps any number x to the function $[+_{((\sigma\sigma)\sigma)}x]$.

More generally, if f is a function which maps n-tuples $\langle x^1_{\beta_1}, \ldots, x^n_{\beta_n}\rangle$ of elements of types $\beta_1, \ldots, \beta_n$, respectively, to elements of type α, we may assign to f the type $((\ldots(\alpha\beta_n)\ldots\beta_2)\beta_1)$. (We may omit parentheses, using the convention of association to the left, and write this type symbol simply as $(\alpha\beta_n\ldots\beta_1)$.)

A set or property can be represented by a function which maps elements to truth values, so that an element is in the set, or has the property, in question iff the function representing the set or property maps that element to truth. We take o as the type symbol denoting the type of truth values, so we may speak of any function of type $(o\alpha)$ as a *set* of elements of type α. A function of type $((o\alpha)\beta)$ is a binary relation between elements of type β and elements of type α. For example, if σ is the type of the natural numbers, and $<$ is the order relation between natural numbers, $<$ has type $(o\sigma\sigma)$, and for all natural numbers x and y, $<xy$ (which we ordinarily write as $x < y$)

[2]Some authors write $(\beta \to \alpha)$ instead of $(\alpha\beta)$. It is easy to remember the meaning of the notation $(\beta \to \alpha)$, but when it is used in long wffs with many variables, the numerous arrows provide useless visual clutter.

has the value truth iff x is less than y. Of course, $<$ can also be regarded as the function which maps each natural number x to the set $<x$ of all natural numbers y such that x is less than y. Thus we shall regard sets, properties and relations as particular kinds of functions.

In first-order logic we let *terms* denote individuals, while *wffs* expressed statements which took on truth values, i.e., for given interpretations and assignments, wffs denoted truth values. Since we now have many more types under consideration, we need a more uniform terminology, so we shall call expressions which denote elements of type α *wffs of type* α. Thus, statements of type theory will be wffs of type o.

The Comprehension Axiom Schema says that for any wff $\mathbf{A}_o$ of type o in which $\mathbf{u}_{o\beta}$ does not occur free, $\exists \mathbf{u}_{o\beta} \forall \mathbf{v}_\beta [\mathbf{u}_{o\beta} \mathbf{v}_\beta \equiv \mathbf{A}_o]$. We may generalize this to assert the existence of functions as well as sets, and state it as follows: $\exists \mathbf{u}_{\alpha\beta} \forall \mathbf{v}_\beta [\mathbf{u}_{\alpha\beta} \mathbf{v}_\beta = \mathbf{A}_\alpha]$ for any wff $\mathbf{A}_\alpha$ of type α in which $\mathbf{u}_{\alpha\beta}$ is not free. We shall denote the function $\mathbf{u}_{\alpha\beta}$ whose existence is thus asserted by $[\lambda \mathbf{v}_\beta \mathbf{A}_\alpha]$. Thus $\lambda \mathbf{v}_\beta$ is a new variable-binder, like $\forall \mathbf{v}_\beta$ or $\exists \mathbf{v}_\beta$ (but with a quite different meaning, of course); λ is known as an *abstraction operator*. $[\lambda \mathbf{v}_\beta \mathbf{A}_\alpha]$ denotes the function whose value on any argument $\mathbf{v}_\beta$ is $\mathbf{A}_\alpha$, where $\mathbf{v}_\beta$ may occur free in $\mathbf{A}_\alpha$. For example, $[\lambda n_\sigma \bullet n_\sigma^2 + 3]$ denotes the function whose value on any natural number n is $n^2 + 3$. Hence when we apply this function to the number 5 we obtain $[\lambda n_\sigma \bullet n_\sigma^2 + 3]5 = 5^2 + 3 = 28$.

When written in terms of λ-notation, the Comprehension Axiom becomes $\forall \mathbf{v}_\beta \bullet [\lambda \mathbf{v}_\beta \mathbf{A}_\alpha] \mathbf{v}_\beta = \mathbf{A}_\alpha$. From this by universal instantiation we obtain $[\lambda \mathbf{v}_\beta \mathbf{A}_\alpha] \mathbf{B}_\beta = \mathsf{S}^{\mathbf{v}_\beta}_{\mathbf{B}_\beta} \mathbf{A}_\alpha$ (provided $\mathbf{B}_\beta$ is free for $\mathbf{v}_\beta$ in $\mathbf{A}_\alpha$), i.e., the result of applying the function $[\lambda \mathbf{v}_\beta \mathbf{A}_\alpha]$ to the argument $\mathbf{B}_\beta$ is $\mathsf{S}^{\mathbf{v}_\beta}_{\mathbf{B}_\beta} \mathbf{A}_\alpha$. The process of replacing $[\lambda \mathbf{v}_\beta \mathbf{A}_\alpha] \mathbf{B}_\beta$ by $\mathsf{S}^{\mathbf{v}_\beta}_{\mathbf{B}_\beta} \mathbf{A}_\alpha$ (or vice-versa) is known as λ-*conversion*. Of course, when $\mathbf{A}_o$ is a wff of type o, $[\lambda \mathbf{v}_\beta \mathbf{A}_o]$ denotes the set of all elements $\mathbf{v}_\beta$ (of type β) of which $\mathbf{A}_o$ is true; this set may also be denoted by $\{\mathbf{v}_\beta \mid \mathbf{A}_o\}$. In familiar set-theoretic notation, $[\lambda \mathbf{v}_\beta \mathbf{A}_o] \mathbf{B}_\beta = \mathsf{S}^{\mathbf{v}_\beta}_{\mathbf{B}_\beta} \mathbf{A}_o$ would be written $\mathbf{B}_\beta \in \{\mathbf{v}_\beta \mid \mathbf{A}_o\} \equiv \mathsf{S}^{\mathbf{v}_\beta}_{\mathbf{B}_\beta} \mathbf{A}_o$. (By the Axiom of Extensionality for truth values, when $\mathbf{C}_o$ and $\mathbf{D}_o$ are of type o, $\mathbf{C}_o \equiv \mathbf{D}_o$ is equivalent to $\mathbf{C}_o = \mathbf{D}_o$.)

It is interesting to note that the quantifiers and connectives need no longer be regarded as improper symbols, but can be assigned types and can be denoted by constants of these types. The function n_{oo} whose existence is asserted by the comprehension axiom $\exists n_{oo} \forall p_o [n_{oo} p_o \equiv \sim p_o]$, and which is denoted by $[\lambda p_o \sim p_o]$, is clearly just the negation function, so we see that the negation function has type (oo). This is natural, since it maps truth values to truth values. Similarly, disjunction and conjunction (etc.) are binary functions from truth values to truth values, so they have type (ooo).

The statement $\forall \mathbf{x}_\alpha \mathbf{A}_o$ is true iff the set $[\lambda \mathbf{x}_\alpha \mathbf{A}_o]$ contains all elements of type α. The comprehension axiom $\exists p_{o(o\alpha)} \forall s_{(o\alpha)} [p_{o(o\alpha)} s_{o\alpha} \equiv \forall x_\alpha \centerdot s_{o\alpha} x_\alpha]$ asserts the existence of a property $\Pi_{o(o\alpha)}$ of sets such that $\forall s_{o\alpha} [\Pi_{o(o\alpha)} s_{o\alpha} \equiv \forall x_\alpha \centerdot s_{o\alpha} x_\alpha]$, i.e., a set $s_{o\alpha}$ has the property $\Pi_{o(o\alpha)}$ iff $s_{o\alpha}$ contains all elements of type α. In particular, $\Pi_{o(o\alpha)}[\lambda \mathbf{x}_\alpha \mathbf{A}_o] \equiv \forall \mathbf{x}_\alpha [[\lambda \mathbf{x}_\alpha \mathbf{A}_o] \mathbf{x}_\alpha]$. Since $[\lambda \mathbf{x}_\alpha \mathbf{A}_o] \mathbf{x}_\alpha \equiv \mathbf{A}_o$ by λ-conversion, we have $\Pi_{o(o\alpha)}[\lambda \mathbf{x}_\alpha \mathbf{A}_o] \equiv \forall \mathbf{x}_\alpha \mathbf{A}_o$, so if we introduce $\Pi_{o(o\alpha)}$ as a constant, we can *define* $\forall \mathbf{x}_\alpha$ in terms of it, and λ is the only variable-binder that we need.

A formulation of type theory based on these ideas was introduced by Church in [Church, 1940], and proved complete by Henkin in [Henkin 1950].[3] In this system equality can be defined using connectives and quantifiers. However, it is also possible to define connectives and quantifiers in terms of equality. Equality is a very basic and simple notion, so instead of using Church's formulation of type theory, we shall use a slight variant of it (first introduced in [Henkin, 1963b] and simplified in [Andrews, 1963]) in which equality is taken as the basic primitive notion. (The last paragraph of [Andrews, 1972a] gives an additional reason for this choice of primitive symbols.)

We shall denote the binary relation of equality between elements of type α by the constant $\mathsf{Q}_{o\alpha\alpha}$. (Naturally, elements which are the same have the same type.) We can easily find an expression T_o to denote truth; any expression of the form $\mathbf{A} = \mathbf{A}$ will do. Then $[\lambda x_\alpha T_o]$ denotes the set of all elements of type α, so we may regard $\forall \mathbf{x}_\alpha \mathbf{A}_o$ as an abbreviation for $[\lambda x_\alpha T_o] = [\lambda \mathbf{x}_\alpha \mathbf{A}_o]$. If anything is false, the assertion $\forall x_o x_o$ that everything is true must be false, so we let this wff, which we also write as F_o, denote falsehood. The expression $[\lambda g_{ooo} \centerdot g_{ooo} x_o y_o]$ can be used to represent the ordered pair $\langle x_0, y_o \rangle$, and the conjunction $x_o \wedge y_o$ is true iff x_o and y_o are both true, i.e., iff $\langle T_o, T_o \rangle = \langle x_o, y_o \rangle$. Hence $x_o \wedge y_o$ can be expressed by the formula $[\lambda g_{ooo} \centerdot g_{ooo} T_o T_o] = [\lambda g_{ooo} \centerdot g_{ooo} x_o y_o]$. From falsehood, conjunction, and equivalence (the equality relation Q_{ooo} between truth values), all other propositional connectives can be defined. Of course, primitive constants could be introduced to denote all these concepts, but the resulting system would be redundant.

There is one additional notation which we wish to have available in our language. If $\mathbf{A}_o$ is a wff of type o in which the variable $\mathbf{y}_\imath$ occurs free, and there is a unique element $\mathbf{y}_\imath$ of which $\mathbf{A}_o$ is true, we denote that element by

[3] See exercise X8030.

$[℩\,\mathbf{y}_\iota \mathbf{A}_o]$, and call it "the $\mathbf{y}_\iota$ such that $\mathbf{A}_o$".[4] What we have here, in essence, is a function which maps the set $[\lambda \mathbf{y}_\iota \mathbf{A}_o]$ to its unique element. We shall denote such a function by $\iota_{\iota(o\iota)}$ and define $[℩\,\mathbf{y}_\iota \mathbf{A}_o]$ to be an abbreviation for $\iota_{\iota(o\iota)}[\lambda \mathbf{y}_\iota \mathbf{A}_o]$. $\iota_{\iota(o\iota)}$ is called a *description operator* (for individuals), since $\mathbf{A}_o$ describes $[℩\,\mathbf{y}_\iota \mathbf{A}_o]$. (We shall see in §53 how description operators for other types can be introduced by definition.) Actually, if there is a unique individual $\mathbf{y}_\iota$ in the set $[\lambda \mathbf{y}_\iota \mathbf{A}_o]$, then this set is simply the set whose only member is $\mathbf{y}_\iota$, which can be denoted by $\mathsf{Q}_{o\iota\iota} \mathbf{y}_\iota$ since for any individual $\mathbf{x}_\iota$, $\mathbf{x}_\iota$ is in $\mathsf{Q}_{o\iota\iota} \mathbf{y}_\iota$ iff $\mathbf{y}_\iota = \mathbf{x}_\iota$, i.e., iff $\mathsf{Q}_{o\iota\iota} \mathbf{y}_\iota \mathbf{x}_\iota$. Thus the assertion that $\iota_{\iota(o\iota)}$ maps one-element sets to their unique members can be expressed by the simple proposition that $\iota_{\iota(o\iota)}[\mathsf{Q}_{o\iota\iota} y_\iota] = y_\iota$ (for each individual y_ι). (Thus $\iota_{\iota(o\iota)}$ is an inverse for $\mathsf{Q}_{o\iota\iota}$. $\mathsf{Q}_{o\iota\iota}$ maps individuals to the unit sets containing them, and $\iota_{\iota(o\iota)}$ maps these sets back to their unique members.)

In §51 we shall present a system $\mathcal{Q}_0$ of type theory based on these ideas.

EXERCISES

Give proofs for the theorems $\mathcal{F}^\omega$ stated below. Use appropriate analogues of the derived rules of inference of §21.

X5000. $\vdash x_\tau = x_\tau$

X5001. $\vdash \mathbf{x}_\tau = \mathbf{y}_\tau \supset \centerdot\, \mathsf{S}^{\mathbf{z}_\tau}_{\mathbf{x}_\tau} \mathbf{A} \supset \mathsf{S}^{\mathbf{z}_\tau}_{\mathbf{y}_\tau} \mathbf{A}$ provided $\mathbf{x}_\tau$ and $\mathbf{y}_\tau$ are free for $\mathbf{z}_\tau$ in $\mathbf{A}$.

X5002. $\vdash x_\tau = y_\tau \supset y_\tau = x_\tau$

X5003. $\vdash x_\tau = y_\tau \supset \centerdot\, y_\tau = z_\tau \supset x_\tau = z_\tau$

X5004. $\vdash x_{(\tau)} = y_{(\tau)} \supset \forall z_\tau \centerdot\, xz \supset yz$

X5005. $\vdash x_{(\tau)} = y_{(\tau)} \supset \forall z_\tau \centerdot\, yz \supset xz$

X5006. $\vdash x_{(\tau)} = y_{(\tau)} \equiv \forall z_\tau \centerdot\, xz \equiv yz$

X5007. $\vdash x_\tau = y_\tau \equiv \forall p_{(\tau\tau)}[\forall z_\tau p_{(\tau\tau)} z_\tau z_\tau \supset p_{(\tau\tau)} x_\tau y_\tau]$

One formulation of the Axiom of Choice says that for each collection u of nonempty pairwise disjoint sets (whose elements have type τ), there is a set r containing exactly one element from each set in u. This can be expressed in $\mathcal{F}^\omega$ by

[4] A strong tradition of using the inverted iota for this purpose was established by its use in [Whitehead and Russell, 1913].

$$
\begin{aligned}
AC_1^\tau \quad : \quad & \forall u_{((\tau))} \centerdot \forall p_{(\tau)}[u_{((\tau))}p_{(\tau)} \supset \exists x_\tau p_{(\tau)}x_\tau] \\
\wedge \quad & \forall p_{(\tau)}\forall q_{(\tau)}[u_{((\tau))}p_{(\tau)} \wedge u_{((\tau))}q_{(\tau)} \wedge p_{(\tau)} \neq q_{(\tau)} \\
\supset \quad & \forall x_\tau \sim \centerdot p_{(\tau)}x_\tau \wedge q_{(\tau)}x_\tau] \\
\supset \quad & \exists r_{(\tau)}\forall p_{(\tau)} \centerdot u_{((\tau))}p_{(\tau)} \supset \exists_1 x_\tau \centerdot p_{(\tau)}x_\tau \wedge r_{(\tau)}x_\tau.
\end{aligned}
$$

A second formulation of the Axiom of Choice says that there is a function (a special sort of binary relation) f which chooses from each nonempty set p (whose elements have type τ) a member of that set. This can be expressed in $\mathcal{F}^\omega$ by

$$
\begin{aligned}
AC_2^\tau \quad : \quad & \exists f_{((\tau)\tau)}\forall p_{(\tau)} \centerdot [\exists x_\tau p_{(\tau)}x_\tau \supset \exists_1 x_\tau f_{((\tau)\tau)}p_{(\tau)}x_\tau] \\
\wedge \quad & \forall x_\tau \centerdot f_{((\tau)\tau)}p_{(\tau)}x_\tau \supset p_{(\tau)}x_\tau
\end{aligned}
$$

X5008. $\vdash AC_2^\tau \supset AC_1^\tau$

X5009. $\vdash AC_1^{((\tau)\tau)} \supset AC_2^\tau$

§51. The Primitive Basis of $\mathcal{Q}_0$

We now describe a system $\mathcal{Q}_0$ of type theory which is based on the ideas discussed in §50. We shall intersperse with our formal description of the language $\mathcal{Q}_0$ occasional informal comments concerning the intended interpretations of the language.

We use α, β, γ, etc., as syntactical variables ranging over *type symbols*, which are defined inductively as follows:

(a) $\imath$ is a type symbol (denoting the type of individuals.)
(b) o is a type symbol (denoting the type of truth values).
(c) If α and β are type symbols, then $(\alpha\beta)$ is a type symbol (denoting the type of functions from elements of type β to elements of type α).

The *primitive symbols* of $\mathcal{Q}_0$ are the following:

(a) Improper symbols: [ ] λ
(b) For each type symbol α, a denumerable list of *variables* of type α :
$f_\alpha\ g_\alpha\ h_\alpha \ldots x_\alpha\ y_\alpha\ z_\alpha\ f_\alpha^1\ g_\alpha^1 \ldots z_\alpha^1\ f_\alpha^2 \ldots$

We shall use $\mathbf{f}_\alpha$, $\mathbf{g}_\alpha, \ldots$, $\mathbf{x}_\alpha$, $\mathbf{y}_\alpha$, $\mathbf{z}_\alpha$, etc., as syntactical variables for variables of type α.

(c) Logical constants: $\mathsf{Q}_{((o\alpha)\alpha)}$ $\iota_{(\imath(o\imath))}$

The types of these constants are indicated by their subscripts. $\mathsf{Q}_{((o\alpha)\alpha)}$ denotes the identity relation between elements of type α. $\iota_{\imath(o\imath))}$ denotes a description or selection operator for individuals.

(d) In addition, there may be other constants of various types, which will be called *nonlogical constants* or *parameters.* Each choice of parameters determines a particular formulation of $\mathcal{Q}_0$.

We write *wff*$_\alpha$ as an abbreviation for *wff of type* α, and use $\mathbf{A}_\alpha$, $\mathbf{B}_\alpha$, $\mathbf{C}_\alpha$, etc., as syntactical variables ranging over wffs$_\alpha$, which are defined inductively as follows:

(a) A primitive variable or constant of type α is a wff$_\alpha$.
(b) $[\mathbf{A}_{\alpha\beta}\mathbf{B}_\beta]$ is a wff$_\alpha$.
(c) $[\lambda\mathbf{x}_\beta\mathbf{A}_\alpha]$ is a wff$_{(\alpha\beta)}$.

In the intended interpretations of $\mathcal{Q}_0$, $[\mathbf{A}_{\alpha\beta}\mathbf{B}_\beta]$ denotes the value of the function denoted by $\mathbf{A}_{\alpha\beta}$ at the argument denoted by $\mathbf{B}_\beta$. Also, $[\lambda\mathbf{x}_\beta\mathbf{A}_\alpha]$ denotes the function whose value at any element $\mathbf{x}_\beta$ of type β is (the denotation of) $\mathbf{A}_\alpha$ (which depends on $\mathbf{x}_\beta$, in general). More precise discussion of interpretations of $\mathcal{Q}_0$ will occur in §54.

5100 Principle of Induction on the Construction of a Wff. Let $\mathcal{R}$ be a property of formulas, and let $\mathcal{R}(\mathbf{A})$ mean that $\mathbf{A}$ has property $\mathcal{R}$. Suppose

(1) Every formula consisting of a single variable or constant standing alone has property $\mathcal{R}$.
(2) Whenever $\mathbf{A}_{\alpha\beta}$ is a wff$_{(\alpha\beta)}$ and $\mathbf{B}_\beta$ is a wff$_\beta$ and $\mathcal{R}(\mathbf{A}_{\alpha\beta})$ and $\mathcal{R}(\mathbf{B}_\beta)$, then $\mathcal{R}([\mathbf{A}_{\alpha\beta}\mathbf{B}_\beta])$.
(3) Whenever $\mathbf{x}_\beta$ is a variable and $\mathcal{R}(\mathbf{A}_\alpha)$, then $\mathcal{R}([\lambda\mathbf{x}_\beta\mathbf{A}_\alpha])$.

Then every wff of $\mathcal{Q}_0$ has property $\mathcal{R}$.

Proof: Left to the reader. ■

Brackets, parentheses in type symbols, and type symbols may be omitted when no ambiguity is thereby introduced . The type symbol o shall usually be omitted, and we often use **A, B, C**, etc., as syntactical variables for wffs$_o$ unless otherwise stated. A dot in an abbreviation for a wff stands for a left bracket whose mate is as far to the right as is consistent with the pairing of brackets already present and with the formula being well-formed. Otherwise brackets and parentheses are to be restored using the convention of association to the left.

We introduce the following definitions and abbreviations:

$[\mathbf{A}_\alpha = \mathbf{B}_\alpha]$	stands for $[\mathsf{Q}_{o\alpha\alpha}\mathbf{A}_\alpha\mathbf{B}_\alpha]$.
$[\mathbf{A} \equiv \mathbf{B}]$	stands for $[\mathsf{Q}_{ooo}\mathbf{A}\ \mathbf{B}]$.
T_o	stands for $[\mathsf{Q}_{ooo} = \mathsf{Q}_{ooo}]$.
F_o	stands for $[\lambda x_o T] = [\lambda x_o x_o]$.
$\Pi_{o(o\alpha)}$	stands for $[\mathsf{Q}_{o(o\alpha)(o\alpha)}[\lambda x_\alpha T]]$.
$[\forall \mathbf{x}_\alpha \mathbf{A}]$	stands for $[\Pi_{o(o\alpha)}[\lambda \mathbf{x}_\alpha \mathbf{A}]]$.
$\wedge_{ooo}$	stands for $[\lambda x_o \lambda y_o \centerdot [\lambda g_{ooo} \centerdot g_{ooo}TT] = [\lambda g_{ooo} \centerdot g_{ooo}x_o y_o]]$.
$[\mathbf{A} \wedge \mathbf{B}]$	stands for $[\wedge_{ooo}\mathbf{A}\ \mathbf{B}]$.
$\supset_{ooo}$	stands for $[\lambda x_o \lambda y_o \centerdot x_o = \ \centerdot x_o \wedge y_o]$.
$[\mathbf{A} \supset \mathbf{B}]$	stands for $[\supset_{ooo}\ \mathbf{A}\ \mathbf{B}]$.
$\sim_{oo}$	stands for $[\mathsf{Q}_{ooo}F]$.
$\vee_{ooo}$	stands for $[\lambda x_o \lambda y_o \centerdot \sim\ \centerdot [\sim x_o] \wedge [\sim y_o]]$.
$[\mathbf{A} \vee \mathbf{B}]$	stands for $[\vee_{ooo}\mathbf{A}\ \mathbf{B}]$.
$[\exists \mathbf{x}_\alpha \mathbf{A}]$	stands for $[\sim \forall \mathbf{x}_\alpha \sim \mathbf{A}]$.
$[\mathbf{A}_\alpha \neq \mathbf{B}_\alpha]$	stands for $[\sim\ \centerdot \mathbf{A}_\alpha = \mathbf{B}_\alpha]$.

We supplement our conventions for restoring brackets in abbreviations of wffs$_o$ so as to incorporate the conventions used previously for propositional calculus and first-order logic. Naturally, brackets must be restored at each stage so that the pairing of brackets already present is not altered, and so that the entire formula is well-formed. After restoring left brackets indicated by dots (and the mates of these brackets), enclose parts of the forms $[\sim \mathbf{A}]$, $[\forall \mathbf{x}_\alpha \mathbf{A}]$ and $[\exists \mathbf{x}_\alpha \mathbf{A}]$ in such a way that $\sim$, $\forall$, and $\exists$ have the smallest possible scope. Then brackets enclosing parts of the form $[\mathbf{A} \wedge \mathbf{B}]$ are to be restored in such a way that each $\wedge$ has the smallest possible scope, starting from the left, so that the law of association to the left governs the relationship of brackets between several such parts. Then brackets enclosing parts of the form $[\mathbf{A} \vee \mathbf{B}]$, followed by parts of the form $[\mathbf{A} \supset \mathbf{B}]$, and finally $[\mathbf{A} \equiv \mathbf{B}]$ (or $[\mathbf{A} = \mathbf{B}]$, where **A** and **B** are wffs$_o$), are to be similarly restored.

We next state the axioms for $\mathcal{Q}_0$. A remark about Axiom Schemata 4_1

$-\ 4_5$ will be made after some necessary terminology has been introduced.

Axioms for $\mathcal{Q}_0$

(1) $g_{oo}T_o \wedge g_{oo}F_o = \forall x_o \,\centerdot\, g_{oo}x_o$

(2^α) $[x_\alpha = y_\alpha] \supset \,\centerdot\, h_{o\alpha}x_\alpha = h_{o\alpha}y_\alpha$

($3^{\alpha\beta}$) $f_{\alpha\beta} = g_{\alpha\beta} = \forall x_\beta \,\centerdot\, f_{\alpha\beta}x_\beta = g_{\alpha\beta}x_\beta$

(4_1) $[\lambda \mathbf{x}_\alpha \mathbf{B}_\beta]\mathbf{A}_\alpha = \mathbf{B}_\beta$ where $\mathbf{B}_\beta$ is a primitive constant or variable distinct from $\mathbf{x}_\alpha$

(4_2) $[\lambda \mathbf{x}_\alpha \mathbf{x}_\alpha]\mathbf{A}_\alpha = \mathbf{A}_\alpha$

(4_3) $[\lambda \mathbf{x}_\alpha \,\centerdot\, \mathbf{B}_{\beta\gamma}\mathbf{C}_\gamma]\mathbf{A}_\alpha = [[\lambda \mathbf{x}_\alpha \mathbf{B}_{\beta\gamma}]\mathbf{A}_\alpha][[\lambda \mathbf{x}_\alpha \mathbf{C}_\gamma]\mathbf{A}_\alpha]$

(4_4) $[\lambda \mathbf{x}_\alpha \,\centerdot\, \lambda \mathbf{y}_\gamma \mathbf{B}_\delta]\mathbf{A}_\alpha = [\lambda \mathbf{y}_\gamma \,\centerdot\, [\lambda \mathbf{x}_\alpha \mathbf{B}_\delta]\mathbf{A}_\alpha]$ where $\mathbf{y}_\gamma$ is distinct from $\mathbf{x}_\alpha$ and from all variables in $\mathbf{A}_\alpha$.

(4_5) $[\lambda \mathbf{x}_\alpha \,\centerdot\, \lambda \mathbf{x}_\alpha \mathbf{B}_\delta]\mathbf{A}_\alpha = [\lambda \mathbf{x}_\alpha \mathbf{B}_\delta]$

(5) $\iota_{\imath(o\imath)}[\mathrm{Q}_{o\imath\imath} y_\imath] = y_\imath$

$\mathcal{Q}_0$ has a single rule of inference, which is the following:

Rule R. From $\mathbf{C}$ and $\mathbf{A}_\alpha = \mathbf{B}_\alpha$ to infer the result of replacing one occurrence of $\mathbf{A}_\alpha$ in $\mathbf{C}$ by an occurrence of $\mathbf{B}_\alpha$, provided that the occurrence of $\mathbf{A}_\alpha$ in $\mathbf{C}$ is not (an occurrence of a variable) immediately preceded by λ.

As in other logistic systems, a *proof* of a wff $\mathbf{A}$ in $\mathcal{Q}_0$ is a finite sequence of wffs, the last of which is $\mathbf{A}$, such that every wff in the sequence is an axiom or is inferred from previous wffs by the rule of inference. A *theorem* of $\mathcal{Q}_0$ is a wff which has a proof in $\mathcal{Q}_0$.

An occurrence of $\mathbf{x}_\alpha$ is *bound* [*free*] in $\mathbf{B}_\beta$ iff it is [is not] in a wf part of $\mathbf{B}_\beta$ of the form $[\lambda \mathbf{x}_\alpha \mathbf{C}_\delta]$. $\mathbf{A}_\alpha$ is *free for* $\mathbf{x}_\alpha$ in $\mathbf{B}_\beta$ iff no free occurrence of $\mathbf{x}_\alpha$ in $\mathbf{B}_\beta$ is in a wf part of $\mathbf{B}_\beta$ of the form $[\lambda \mathbf{y}_\gamma \mathbf{C}_\delta]$ such that $\mathbf{y}_\gamma$ is a free variable of $\mathbf{A}_\alpha$.

A *closed wff (cwff)* is a wff without free variables. A *sentence* is a cwff_o.

If $\mathbf{x}^1_{\alpha_1}, \ldots, \mathbf{x}^n_{\alpha_n}$ are distinct variables, $\mathsf{S}^{\mathbf{x}^1_{\alpha_1} \ldots \mathbf{x}^n_{\alpha_n}}_{\mathbf{A}^1_{\alpha_1} \ldots \mathbf{A}^n_{\alpha_n}} \mathbf{B}_\beta$ denotes the result of simultaneously substituting $\mathbf{A}^i_{\alpha_i}$ for all free occurrences of $\mathbf{x}^i_{\alpha_i}$ in $\mathbf{B}_\beta$ for $1 \leq i \leq n$.

REMARK. Axiom Schemata 4_1 - 4_5 express the basic properties of λ-notation, and will be used in conjunction with Axiom Schema 3 in §52 to derive Theorem Schema 5207, which says that $[\lambda \mathbf{x}_\alpha \mathbf{B}_\beta]\mathbf{A}_\alpha = \mathsf{S}^{\mathbf{x}_\alpha}_{\mathbf{A}_\alpha} \mathbf{B}_\beta$, provided that

$\mathbf{A}_\alpha$ is free for $\mathbf{x}_\alpha$ in $\mathbf{B}_\beta$. 5207 could be taken as an axiom schema in place of 4_1 - 4_5, and for some purposes this would be desirable, since 5207 has a conceptual simplicity and unity which is not apparent in 4_1 - 4_5. However, there is a sense in which it is simpler to state 4_1 - 4_5, which do not require prior understanding of substitution and related notions. This simplicity will facilitate our work in §70, where we shall arithmetize the syntax of $\mathcal{Q}_0$.

When one considers the primitive symbols of $\mathcal{Q}_0$, it is apparent that from a conceptual point of view, the axioms of $\mathcal{Q}_0$ are really quite simple and natural. Axiom 1 expresses the idea that Truth and Falsehood are the only truth values, Axioms 2^α express one of the basic properties of equality, Axioms $3^{\alpha\beta}$ are Axioms of Extensionality, Axioms 4_1 - 4_5 express the properties of λ-notation, and Axiom 5 is the Axiom of Descriptions.

If $\mathcal{H}$ is a finite set of wffs$_o$, we shall write $\mathcal{H} \vdash \mathbf{A}$ as an abbreviation for the phrase "**A** is derivable from the set $\mathcal{H}$ of hypotheses", and we shall define this concept so that:

(a) If $\mathbf{A} \in \mathcal{H}$, then $\mathcal{H} \vdash \mathbf{A}$

(b) If **A** is a theorem of $\mathcal{Q}_0$, then $\mathcal{H} \vdash \mathbf{A}$.

(c) (Rule R$'$) If $\mathcal{H} \vdash \mathbf{A}_\alpha = \mathbf{B}_\alpha$, and $\mathcal{H} \vdash \mathbf{C}$, and **D** is obtained from **C** by replacing one occurrence of $\mathbf{A}_\alpha$ in **C** by an occurrence of $\mathbf{B}_\alpha$, then $\mathcal{H} \vdash \mathbf{D}$, provided that the occurrence of $\mathbf{A}_\alpha$ in **C** is not an occurrence of a variable immediately preceded by λ, and the occurrence of $\mathbf{A}_\alpha$ in **C** is not in a wf part $[\lambda\mathbf{x}_\beta\mathbf{E}_\gamma]$ of **C**, where $\mathbf{x}_\beta$ is free in a member of $\mathcal{H}$ and free in $[\mathbf{A}_\alpha = \mathbf{B}_\alpha]$.

DEFINITION. A *proof of* **A** *from the hypotheses* $\mathcal{H}$ consists of two finite sequences $\mathcal{S}_1$ and $\mathcal{S}_2$ of wffs$_o$, such that $\mathcal{S}_1$ is a proof in $\mathcal{Q}_0$, and **A** is the last member of the sequence $\mathcal{S}_2$, and each wff $\mathbf{D}^i$ in $\mathcal{S}_2$ satisfies at least one of the following conditions:

(a) $\mathbf{D}^i$ is in $\mathcal{H}$.

(b) $\mathbf{D}^i$ is in $\mathcal{S}_1$.

(c) $\mathbf{D}^i$ is obtained by Rule R$'$ from wffs $\mathbf{D}^j$ and $\mathbf{D}^k$ which precede $\mathbf{D}^i$ in the sequence $\mathcal{S}_2$.

We cannot replace condition (b) in this definition by the requirement that $\mathbf{D}^i$ is a theorem of $\mathcal{Q}_0$, since (as we shall see in Corollary 7302 below) one cannot effectively test whether an arbitrary wff is a theorem of $\mathcal{Q}_0$, and it is essential that we be able to effectively test whether an alleged proof from hypotheses is in fact a proof from hypotheses. Therefore one must supply a proof for any theorem one uses in a proof from hypotheses; this is the purpose of the sequence $\mathcal{S}_1$.

We now define $\mathcal{H} \vdash \mathbf{A}$ to mean that there is a proof of $\mathbf{A}$ from the hypotheses $\mathcal{H}$. Note that $\emptyset \vdash \mathbf{A}$ iff $\mathbf{A}$ is a theorem of $\mathcal{Q}_0$. Hence we write $\vdash \mathbf{A}$ to mean that $\mathbf{A}$ is a theorem of $\mathcal{Q}_0$.

EXERCISES

X5100. Prove that $\mathsf{S}^{\mathbf{y}_\alpha}_{.\mathbf{A}_\alpha}\mathsf{S}^{\mathbf{x}_\alpha}_{.\mathbf{y}_\alpha}\mathbf{B}_\beta = \mathsf{S}^{\mathbf{x}_\alpha}_{.\mathbf{A}_\alpha}\mathbf{B}_\beta$ provided that $\mathbf{y}_\alpha$ does not occur in $\mathbf{B}_\beta$.

X5101. Prove that $\mathsf{S}^{\mathbf{z}_\beta}_{.\mathbf{y}_\beta}\mathsf{S}^{\mathbf{x}_\beta}_{.\mathbf{z}_\beta}\mathbf{A}_\alpha = \mathsf{S}^{\mathbf{x}_\beta}_{.\mathbf{y}_\beta}\mathbf{A}_\alpha$, provided that $\mathbf{z}_\beta$ does not occur free in $\mathbf{A}_\alpha$ and $\mathbf{z}_\beta$ is free for $\mathbf{x}_\beta$ in $\mathbf{A}_\alpha$. Note explicitly where the conditions on the variables are used in your proof.

§52. Elementary Logic in $\mathcal{Q}_0$

We shall now prove certain theorem schemata and derived rules of inference of $\mathcal{Q}_0$. Among other things, we shall show that appropriate analogues of theorems and rules of first-order logic can be derived in $\mathcal{Q}_0$.

5200. $\vdash \mathbf{A}_\alpha = \mathbf{A}_\alpha$

Proof:

(.1) $\vdash [\lambda x_\alpha x_\alpha]\mathbf{A}_\alpha = \mathbf{A}_\alpha$ Axiom 4_2
(.2) $\vdash \mathbf{A}_\alpha = \mathbf{A}_\alpha$ R: .1, .1

■

Note the special case: $\vdash T_o$

5201 Equality Rules (= **Rules**).

If $\mathcal{H} \vdash \mathbf{A}$ and $\mathcal{H} \vdash \mathbf{A} \equiv \mathbf{B}$ then $\mathcal{H} \vdash \mathbf{B}$.
If $\mathcal{H} \vdash \mathbf{A}_\alpha = \mathbf{B}_\alpha$ then $\mathcal{H} \vdash \mathbf{B}_\alpha = \mathbf{A}_\alpha$.
If $\mathcal{H} \vdash \mathbf{A}_\alpha = \mathbf{B}_\alpha$ and $\mathcal{H} \vdash \mathbf{B}_\alpha = \mathbf{C}_\alpha$ then $\mathcal{H} \vdash \mathbf{A}_\alpha = \mathbf{C}_\alpha$.
If $\mathcal{H} \vdash \mathbf{A}_{\alpha\beta} = \mathbf{B}_{\alpha\beta}$ and $\mathcal{H} \vdash \mathbf{C}_\beta = \mathbf{D}_\beta$ then $\mathcal{H} \vdash \mathbf{A}_{\alpha\beta}\mathbf{C}_\beta = \mathbf{B}_{\alpha\beta}\mathbf{D}_\beta$.
If $\mathcal{H} \vdash \mathbf{A}_{\alpha\beta} = \mathbf{B}_{\alpha\beta}$ then $\mathcal{H} \vdash \mathbf{A}_{\alpha\beta}\mathbf{C}_\beta = \mathbf{B}_{\alpha\beta}\mathbf{C}_\beta$.
If $\mathcal{H} \vdash \mathbf{C}_\beta = \mathbf{D}_\beta$ then $\mathcal{H} \vdash \mathbf{A}_{\alpha\beta}\mathbf{C}_\beta = \mathbf{A}_{\alpha\beta}\mathbf{D}_\beta$.

Proof: By 5200 and Rule R′. ∎

5202 Rule RR. If $\vdash \mathbf{A}_\alpha = \mathbf{B}_\alpha$ or $\vdash \mathbf{B}_\alpha = \mathbf{A}_\alpha$, and if $\mathcal{H} \vdash \mathbf{C}$, then $\mathcal{H} \vdash \mathbf{D}$, where $\mathbf{D}$ is obtained from $\mathbf{C}$ by replacing one occurrence of $\mathbf{A}_\alpha$ in $\mathbf{C}$ by an occurrence of $\mathbf{B}_\alpha$, and the occurrence of $\mathbf{A}_\alpha$ is not an occurrence of a variable immediately preceded by λ. (Note that since hypotheses were not needed to derive $A_\alpha = B_\alpha$, the restriction on Rule R′ involving variables free in $\mathcal{H}$ does not apply.)

Proof:

(.1) $\vdash \mathbf{A}_\alpha = \mathbf{B}_\alpha$	given, and 5201 if necessary
(.2) $\vdash \mathbf{C} = \mathbf{C}$	5200
(.3) $\vdash \mathbf{C} = \mathbf{D}$	R
(.4) $\mathcal{H} \vdash \mathbf{C}$	given
(.5) $\mathcal{H} \vdash \mathbf{C} = \mathbf{D}$	theorem
(.6) $\mathcal{H} \vdash \mathbf{D}$	R′

∎

5203. $\vdash [\lambda \mathbf{x}_\alpha \mathbf{B}_\beta]\mathbf{A}_\alpha = \mathsf{S}^{\mathbf{x}_\alpha}_{\mathbf{A}_\alpha} \mathbf{B}_\beta$ provided no variable in $\mathbf{A}_\alpha$ is bound in $\mathbf{B}_\beta$.

Proof: By induction on the number of occurrences of [in $\mathbf{B}_\beta$.

Case: $\mathbf{B}_\beta$ is a primitive constant or variable. Use 4_1 or 4_2.
Case: $\mathbf{B}_\beta$ has the form $[\mathbf{D}_{\beta\gamma}\mathbf{C}_\gamma]$.

$\vdash [\lambda \mathbf{x}_\alpha \mathbf{D}_{\beta\gamma}]\mathbf{A}_\alpha = \mathsf{S}^{\mathbf{x}_\alpha}_{\mathbf{A}_\alpha} \mathbf{D}_{\beta\gamma}$	by inductive hypothesis
$\vdash [\lambda \mathbf{x}_\alpha \mathbf{C}_\gamma]\mathbf{A}_\alpha = \mathsf{S}^{\mathbf{x}_\alpha}_{\mathbf{A}_\alpha} \mathbf{C}_\gamma$	by inductive hypothesis
$\vdash [\lambda \mathbf{x}_\alpha \centerdot \mathbf{D}_{\beta\gamma}\mathbf{C}_\gamma]\mathbf{A}_\alpha = [[\lambda \mathbf{x}_\alpha \mathbf{D}_{\beta\gamma}]\mathbf{A}_\alpha][[\lambda \mathbf{x}_\alpha \mathbf{C}_\gamma]\mathbf{A}_\alpha]$	4_3

Hence the desired theorem follows by Rule R.

Case: $\mathbf{B}_\beta$ has the form $[\lambda \mathbf{y}_\gamma \mathbf{D}_\delta]$. If $\mathbf{y}_\gamma$ is $\mathbf{x}_\alpha$, the result follows by 4_5. If $\mathbf{y}_\gamma$ is not $\mathbf{x}_\alpha$, then since $\mathbf{y}_\gamma$ is bound in $\mathbf{B}_\beta$, it does not occur in $\mathbf{A}_\alpha$, so

$\vdash [\lambda \mathbf{x}_\alpha \centerdot \lambda \mathbf{y}_\gamma \mathbf{D}_\delta]\mathbf{A}_\alpha = [\lambda \mathbf{y}_\gamma \centerdot [\lambda \mathbf{x}_\alpha \mathbf{D}_\delta]\mathbf{A}_\alpha]$	4_4
$\vdash [\lambda \mathbf{x}_\alpha \mathbf{D}_\delta]\mathbf{A}_\alpha = \mathsf{S}^{\mathbf{x}_\alpha}_{\mathbf{A}_\alpha} \mathbf{D}_\delta$	by inductive hypothesis.

Hence the desired theorem follows by Rule R. ∎

REMARK. Theorem Schema 5203 is a preliminary result, which will occur in more general form in 5207. We next prove a restricted substitution rule, which will occur in more general form in 5209, and in full generality in 5221.

5204. If $\vdash \mathbf{B}_\beta = \mathbf{C}_\beta$, then $\vdash \mathsf{S}^{\mathbf{x}_\alpha}_{\mathbf{A}_\alpha}[\mathbf{B}_\beta = \mathbf{C}_\beta]$, provided no variable in $\mathbf{A}_\alpha$ is bound in $\mathbf{B}_\beta$ or $\mathbf{C}_\beta$.

Proof:

(.1) $\vdash [\lambda \mathbf{x}_\alpha \mathbf{B}_\beta]\mathbf{A}_\alpha = [\lambda \mathbf{x}_\alpha \mathbf{B}_\beta]\mathbf{A}_\alpha$	5200
(.2) $\vdash [\lambda \mathbf{x}_\alpha \mathbf{B}_\beta]\mathbf{A}_\alpha = [\lambda \mathbf{x}_\alpha \mathbf{C}_\beta]\mathbf{A}_\alpha$	R: .1
(3) $\vdash [\lambda \mathbf{x}_\alpha \mathbf{B}_\beta]\mathbf{A}_\alpha = \mathsf{S}^{\mathbf{x}_\alpha}_{\mathbf{A}_\alpha}\mathbf{B}_\beta$	5203
(.4) $\vdash [\lambda \mathbf{x}_\alpha \mathbf{C}_\beta]\mathbf{A}_\alpha = \mathsf{S}^{\mathbf{x}_\alpha}_{\mathbf{A}_\alpha}\mathbf{C}_\beta$	5203
(.5) $\vdash \mathsf{S}^{\mathbf{x}_\alpha}_{\mathbf{A}_\alpha}[\mathbf{B}_\beta = \mathbf{C}_\beta]$	R: .2, .3, 4.

■

5205. $\vdash f_{\alpha\beta} = [\lambda \mathbf{y}_\beta \centerdot f_{\alpha\beta}\mathbf{y}_\beta]$

Proof: First we handle the case where $\mathbf{y}_\beta$ is distinct from x_β (the particular variable which is bound in Axiom 3^α).

(.1) $\vdash f_{\alpha\beta} = [\lambda \mathbf{y}_\beta \centerdot f_{\alpha\beta}\mathbf{y}_\beta] = \forall x_\beta \centerdot f_{\alpha\beta}x_\beta = [\lambda \mathbf{y}_\beta \centerdot f_{\alpha\beta}\mathbf{y}_\beta]x_\beta$	5204: 3^α
(.2) $\vdash [\lambda \mathbf{y}_\beta \centerdot f_{\alpha\beta}\mathbf{y}_\beta]x_\beta = f_{\alpha\beta}x_\beta$	5203
(.3) $\vdash f_{\alpha\beta} = [\lambda \mathbf{y}_\beta \centerdot f_{\alpha\beta}\mathbf{y}_\beta] = \forall x_\beta \centerdot f_{\alpha\beta}x_\beta = f_{\alpha\beta}x_\beta$	R: .1, .2
(.4) $\vdash f_{\alpha\beta} = f_{\alpha\beta} = \forall x_\beta \centerdot f_{\alpha\beta}x_\beta = f_{\alpha\beta}x_\beta$	5204: 3^α
(.5) $\vdash f_{\alpha\beta} = [\lambda \mathbf{y}_\beta \centerdot f_{\alpha\beta}\mathbf{y}_\beta]$	where $\mathbf{y}_\beta$ is distinct from x_β. = Rules: .3, .4, 5200

Next we handle the case where $\mathbf{y}_\beta$ is x_β.

(.6) $\vdash f_{\alpha\beta} = [\lambda y_\beta \centerdot f_{\alpha\beta}y_\beta]$	instance of .5
(.7) $\vdash [\lambda x_\beta \centerdot f_{\alpha\beta}x_\beta] = [\lambda y_\beta \centerdot [\lambda x_\beta \centerdot f_{\alpha\beta}x_\beta]y_\beta]$	5204: .6
(.8) $\vdash [\lambda x_\beta \centerdot f_{\alpha\beta}x_\beta] = [\lambda y_\beta \centerdot f_{\alpha\beta}y_\beta]$	R: .7, 5203
(.9) $\vdash f_{\alpha\beta} = [\lambda x_\beta \centerdot f_{\alpha\beta}x_\beta]$	= Rules: .6, .8

■

5206 $\vdash [\lambda \mathbf{x}_\beta \mathbf{A}_\alpha] = [\lambda \mathbf{z}_\beta \mathsf{S}^{\mathbf{x}_\beta}_{\mathbf{z}_\beta}\mathbf{A}_\alpha]$, provided that $\mathbf{z}_\beta$ does not occur free in $\mathbf{A}_\alpha$ and $\mathbf{z}_\beta$ is free for $\mathbf{x}_\beta$ in $\mathbf{A}_\alpha$.

Remark. A natural way to try to prove this is to start with

$$[\lambda \mathbf{x}_\beta \mathbf{A}_\alpha] = [\lambda \mathbf{z}_\beta \centerdot [\lambda \mathbf{x}_\beta \mathbf{A}_\alpha]\mathbf{z}_\beta] \qquad 5204\text{: } 5205$$

and then use 5203. However, this application of 5204 will not work if $\mathbf{z}_\beta$ occurs (as a bound variable) in $\mathbf{A}_\alpha$, so we use the method below.

Proof: Let $\mathbf{y}_\beta$ be distinct from $\mathbf{x}_\beta$, $\mathbf{z}_\beta$, and all variables in $\mathbf{A}_\alpha$.

(.1) $\vdash [\lambda \mathbf{x}_\beta \mathbf{A}_\alpha] = [\lambda \mathbf{y}_\beta \centerdot [\lambda \mathbf{x}_\beta \mathbf{A}_\alpha]\mathbf{y}_\beta]$ 5204: 5205
(.2) $\vdash [\lambda \mathbf{x}_\beta \mathbf{A}_\alpha] = [\lambda \mathbf{y}_\beta \centerdot \mathsf{S}^{\mathbf{x}_\beta}_{\mathbf{y}_\beta} \mathbf{A}_\alpha]$ R: .1, 5203
(.3) $\vdash [\lambda \mathbf{z}_\beta \mathsf{S}^{\mathbf{x}_\beta}_{\mathbf{z}_\beta} \mathbf{A}_\alpha] = [\lambda \mathbf{y}_\beta \centerdot [\lambda \mathbf{z}_\beta \mathsf{S}^{\mathbf{x}_\beta}_{\mathbf{z}_\beta} \mathbf{A}_\alpha]\mathbf{y}_\beta]$ 5204: 5205
(.4) $\vdash [\lambda \mathbf{z}_\beta \mathsf{S}^{\mathbf{x}_\beta}_{\mathbf{z}_\beta} \mathbf{A}_\alpha] = [\lambda \mathbf{y}_\beta \centerdot \mathsf{S}^{\mathbf{x}_\beta}_{\mathbf{y}_\beta} \mathbf{A}_\alpha]$ R: .3, 5203

(.4) is obtained as specified because $\mathsf{S}^{\mathbf{z}}_{\mathbf{y}} \mathsf{S}^{\mathbf{x}}_{\mathbf{z}} \mathbf{A}_\alpha = \mathsf{S}^{\mathbf{x}}_{\mathbf{y}} \mathbf{A}_\alpha$, as shown in Exercise X5101. The theorem then follows from (.2) and (.4) by = Rules. ■

5207 $\vdash [\lambda \mathbf{x}_\alpha \mathbf{B}_\beta]\mathbf{A}_\alpha = \mathsf{S}^{\mathbf{x}_\alpha}_{\mathbf{A}_\alpha} \mathbf{B}_\beta$, provided that $\mathbf{A}_\alpha$ is free for $\mathbf{x}_\alpha$ in $\mathbf{B}_\beta$.

Proof: This theorem follows from 5203 and 5206. For a careful proof, we proceed as in the proof of 5203 by induction on the number of left brackets in $\mathbf{B}_\beta$. The cases where $\mathbf{B}_\beta$ is a primitive constant or variable, has the form $[\mathbf{D}_{\beta\gamma} \mathbf{C}_\gamma]$, or has the form $[\lambda \mathbf{y}_\gamma \mathbf{D}_\delta]$ where $\mathbf{y}_\gamma$ is $\mathbf{x}_\alpha$, follow just as in the proof of 5203.

Hence we need deal only with the case where $\mathbf{B}_\beta$ has the form $[\lambda \mathbf{y}_\gamma \mathbf{D}_\delta]$, and $\mathbf{y}_\gamma$ is not $\mathbf{x}_\alpha$. Thus we must prove $\vdash [\lambda \mathbf{x}_\alpha \centerdot \lambda \mathbf{y}_\gamma \mathbf{D}_\delta]\mathbf{A}_\alpha = [\lambda \mathbf{y}_\gamma \mathsf{S}^{\mathbf{x}_\alpha}_{\mathbf{A}_\alpha} \mathbf{D}_\delta]$. If $\mathbf{y}_\gamma$ does not occur in $\mathbf{A}_\alpha$, we can use Axiom 4_4 and the inductive hypothesis. Of course, $\mathbf{y}_\gamma$ may actually occur in $\mathbf{A}_\alpha$. However, bound occurrences of $\mathbf{y}_\gamma$ can be changed to occurrences of some new variable $\mathbf{z}_\gamma$, and this is done in both cases below.

Let $\mathbf{z}_\gamma$ be a variable distinct from $\mathbf{x}_\alpha$, $\mathbf{y}_\gamma$, and all variables in $\mathbf{A}_\alpha$ or $\mathbf{D}_\delta$.

Subcase. $\mathbf{x}_\alpha$ is not free in $\mathbf{D}_\delta$. Let $\mathbf{D}'_\delta$ be $\mathsf{S}^{\mathbf{y}_\gamma}_{\mathbf{z}_\gamma} \mathbf{D}_\delta$. (We shall replace $[\lambda \mathbf{y}_\gamma \mathbf{D}_\delta]$ by $[\lambda \mathbf{z}_\gamma \mathbf{D}'_\delta]$ so that the restriction on Axiom 4_4 will be satisfied.)

(.1) $\vdash [\lambda \mathbf{x}_\alpha [\lambda \mathbf{z}_\gamma \mathbf{D}'_\delta]]\mathbf{A}_\alpha = [\lambda \mathbf{z}_\gamma \centerdot [\lambda \mathbf{x}_\alpha \mathbf{D}'_\delta]\mathbf{A}_\alpha]$ 4_4
(.2) $\vdash [\lambda \mathbf{x}_\alpha \mathbf{D}'_\delta]\mathbf{A}_\alpha = \mathbf{D}'_\delta$ by inductive hypothesis; the right side is D'_δ since $\mathbf{x}_\alpha$ is not free in $\mathbf{D}'_\delta$.
(.3) $\vdash [\lambda \mathbf{x}_\alpha [\lambda \mathbf{z}_\gamma \mathbf{D}'_\delta]]\mathbf{A}_\alpha = [\lambda \mathbf{z}_\gamma \mathbf{D}'_\delta]$ R: .1, .2
(.4) $\vdash [\lambda \mathbf{y}_\gamma \mathbf{D}_\delta] = [\lambda \mathbf{z}_\gamma \mathbf{D}'_\delta]$ 5206
(.5) $\vdash [\lambda \mathbf{x}_\alpha [\lambda \mathbf{y}_\gamma \mathbf{D}_\delta]]\mathbf{A}_\alpha = [\lambda \mathbf{y}_\gamma \mathbf{D}_\delta]$ RR: .3, .4

This is the desired theorem for this subcase.

Subcase. $\mathbf{x}_\alpha$ is free in $\mathbf{D}_\delta$. (Here we will change $\mathbf{y}_\gamma$ in $\mathbf{A}_\alpha$.) Since $\mathbf{A}_\alpha$ is free for $\mathbf{x}_\alpha$ in $\mathbf{B}_\beta$, $\mathbf{y}_\gamma$ does not occur free in $\mathbf{A}_\alpha$. Let $\mathbf{A}'_\alpha$ be the result of replacing all (bound) occurrences of $\mathbf{y}_\gamma$ in $\mathbf{A}_\alpha$ by occurrences of $\mathbf{z}_\gamma$.

(.6) $\vdash \mathbf{A}_\alpha = \mathbf{A}'_\alpha$ R: 5200, 5206

(Several different instances of 5206 may be needed.)

(.7) $\vdash [\lambda \mathbf{x}_\alpha \centerdot \lambda \mathbf{y}_\gamma \mathbf{D}_\delta]\mathbf{A}'_\alpha = [\lambda \mathbf{y}_\gamma \centerdot [\lambda \mathbf{x}_\alpha \mathbf{D}_\delta]\mathbf{A}'_\alpha]$ 4_4

(.8) $\vdash [\lambda \mathbf{x}_\alpha \centerdot \lambda \mathbf{y}_\gamma \mathbf{D}_\delta]\mathbf{A}_\alpha = [\lambda \mathbf{y}_\gamma \centerdot [\lambda \mathbf{x}_\alpha \mathbf{D}_\delta]\mathbf{A}_\alpha]$ RR: .6, .7

(.9) $\vdash [\lambda \mathbf{x}_\alpha \mathbf{D}_\delta]\mathbf{A}_\alpha = \mathsf{S}^{\mathbf{x}_\alpha}_{\mathbf{A}_\alpha} \mathbf{D}_\delta$ by inductive hypothesis

(.10) $\vdash [\lambda \mathbf{x}_\alpha \centerdot \lambda \mathbf{y}_\gamma \mathbf{D}_\delta]\mathbf{A}_\alpha = [\lambda \mathbf{y}_\gamma \mathsf{S}^{\mathbf{x}_\alpha}_{\mathbf{A}_\alpha} \mathbf{D}_\delta]$ R: .8, .9

This is the desired theorem. ■

We now introduce some traditional terminology by defining the derived rules of inference listed below, which are obtained directly from 5206, 5207, and 5205 by using Rule RR and 5204.

- **α-Conversion (Alphabetic Change of Bound Variables) (α).** Replace a wf part $[\lambda \mathbf{x}_\beta \mathbf{A}_\alpha]$ of a wff by $[\lambda \mathbf{z}_\beta \mathsf{S}^{\mathbf{x}_\beta}_{\mathbf{z}_\beta} \mathbf{A}_\alpha]$, provided that $\mathbf{z}_\beta$ does not occur free in $\mathbf{A}_\alpha$ and $\mathbf{z}_\beta$ is free for $\mathbf{x}_\beta$ in $\mathbf{A}_\alpha$.

- **β-Contraction** or **β-Reduction**. Replace a wf part $[\lambda \mathbf{x}_\alpha \mathbf{B}_\beta]\mathbf{A}_\alpha$ of a wff by $\mathsf{S}^{\mathbf{x}_\alpha}_{\mathbf{A}_\alpha} \mathbf{B}_\beta$, provided $\mathbf{A}_\alpha$ is free for $\mathbf{x}_\alpha$ in $\mathbf{B}_\beta$.

- **β-Expansion**. Apply the inverse of β-Contraction.

- **β-Conversion (β)**. Apply β-Contraction or β-Expansion.

- **η-Contraction** or **η-Reduction**. Replace a wf part $[\lambda \mathbf{y}_\beta \centerdot \mathbf{B}_{\alpha\beta} \mathbf{y}_\beta]$ of a wff by $\mathbf{B}_{\alpha\beta}$, provided $\mathbf{y}_\beta$ does not occur free in $\mathbf{B}_{\alpha\beta}$.

- **η-Expansion**. Apply the inverse of η-Contraction.

- **η-Conversion (η)**. Apply η-Contraction or η-Expansion.

- **λ-Conversion (λ)**. Apply any combination of α-Conversions, β-Conversions, and η-Conversions.

5208 $\vdash [\lambda \mathbf{x}^1_{\alpha_1} \ldots \lambda \mathbf{x}^n_{\alpha_n} \mathbf{B}_\beta]\mathbf{x}^1_{\alpha_1} \ldots \mathbf{x}^n_{\alpha_n} = \mathbf{B}_\beta$ where $n \geq 1$.

Proof: By induction on n. For $n = 1$, the theorem is an instance of 5207.

Suppose all instances of 5208 can be proved for $n = m$; we must prove it for $n = m+1$. (Note that we do not require the variables $\mathbf{x}^i_{\alpha_i}$ to be distinct.)

(.1) $\vdash [\lambda \mathbf{x}^1 \lambda \mathbf{x}^2 \ldots \lambda \mathbf{x}^{m+1} \mathbf{B}]\mathbf{x}^1 \mathbf{x}^2 \ldots \mathbf{x}^{m+1} = [\lambda \mathbf{x}^2 \ldots \lambda \mathbf{x}^{m+1} \mathbf{B}]\mathbf{x}^2 \ldots \mathbf{x}^{m+1}$

by 5200, 5207

(.2) $\vdash [\lambda \mathbf{x}^2 \ldots \lambda \mathbf{x}^{m-1}\mathbf{B}]\mathbf{x}^2 \ldots \mathbf{x}^{m-1} = \mathbf{B}$ by inductive hypothesis

The theorem follows by Rule R. ■

5209. If $\vdash \mathbf{B}_\beta = \mathbf{C}_\beta$, then $\vdash \mathsf{S}^{\mathbf{x}_\alpha}_{\mathbf{A}_\alpha}[\mathbf{B}_\beta = \mathbf{C}_\beta]$, provided $\mathbf{A}_\alpha$ is free for $\mathbf{x}_\alpha$ in $[\mathbf{B}_\beta = \mathbf{C}_\beta]$.

The proof follows as for 5204, with 5207 used in place of 5203.

5210 $\vdash T = [\mathbf{B}_\beta = \mathbf{B}_\beta]$

Proof:

(.1) $\vdash [\lambda y_\beta y_\beta] = [\lambda y_\beta y_\beta] = \forall x_\beta \,\centerdot [\lambda y_\beta y_\beta] x_\beta = [\lambda y_\beta y_\beta] x_\beta$ 5209: $3^{\beta\beta}$
(.2) $\vdash [\lambda x_\beta T] = [\lambda x_\beta \,\centerdot\, x_\beta = x_\beta]$ R: 5200, .1; λ, def of $\forall$
(.3) $\vdash [\lambda x_\beta T]\mathbf{B}_\beta = [\lambda x_\beta \,\centerdot\, x_\beta = x_\beta]\mathbf{B}_\beta$ =: .2
(.4) $T = \,\centerdot\, \mathbf{B}_\beta = \mathbf{B}_\beta$ λ: .3

■

5211. $\vdash [T \wedge T] = T$

Proof:

(.1) $\vdash [\lambda y_o T]T \wedge [\lambda y_o T]F = \forall x_o [\lambda y_o T] x_o$ 5209: Axiom 1
(.2) $\vdash T \wedge T = \forall x_o T$ λ: .1
(.3) $\vdash T = \forall x_o T$ 5210, def of $\forall$
(.4) $\vdash T \wedge T = T$ =: .2, .3

■

REMARK. An alternative proof can be given which does not use Axiom 1, but does use the definition of $\wedge$.

5212. $\vdash T \wedge T$

Proof: by = Rules (5201) from 5211, 5200, and the definition of T. Alternatively, by 5200, the definition of $\wedge$, and λ.

5213. If $\vdash \mathbf{A}_\alpha = \mathbf{B}_\alpha$ and $\vdash \mathbf{C}_\beta = \mathbf{D}_\beta$ then $\vdash [\mathbf{A}_\alpha = \mathbf{B}_\alpha] \wedge [\mathbf{C}_\beta = \mathbf{D}_\beta]$.

Proof:

(.1) $\vdash \mathbf{A}_\alpha = \mathbf{B}_\alpha$ given
(.2) $\vdash T = [\mathbf{A}_\alpha = \mathbf{B}_\alpha]$ R: 5210, .1
(.3) $\vdash \mathbf{C}_\beta = \mathbf{D}_\beta$ given

(.4) $\vdash T = [\mathbf{C}_\beta = \mathbf{D}_\beta]$ R: 5210, .3
(.5) $\vdash [\mathbf{A}_\alpha = \mathbf{B}_\alpha] \wedge [\mathbf{C}_\beta = \mathbf{D}_\beta]$ R: 5212, .2, .4
■

5214. $\vdash T \wedge F = F$

Proof:

(.1) $\vdash [\lambda x_o x_o]T \wedge [\lambda x_o x_o]F = \forall x_o[\lambda x_o x_o]x_o$ 5209: Axiom 1
(.2) $\vdash T \wedge F = \forall x_o x_o$ λ: .1

By the definitions of $\forall$ and F, we see that this is the desired theorem. ■

5215 Universal Instantiation ($\forall$I). If $\mathcal{H} \vdash \forall \mathbf{x}_\alpha \mathbf{B}$, then $\mathcal{H} \vdash \mathsf{S}^{\mathbf{x}_\alpha}_{\mathbf{A}_\alpha} \mathbf{B}$, provided that $\mathbf{A}_\alpha$ is free for $\mathbf{x}_\alpha$ in $\mathbf{B}$.

Proof:

(.1) $\mathcal{H} \vdash [\lambda x_\alpha T] = [\lambda \mathbf{x}_\alpha \mathbf{B}]$ given, def of $\forall$
(.2) $\mathcal{H} \vdash [\lambda x_\alpha T]\mathbf{A}_\alpha = [\lambda \mathbf{x}_\alpha \mathbf{B}]\mathbf{A}_\alpha$ =: .1
(.3) $\mathcal{H} \vdash T = \mathsf{S}^{\mathbf{x}_\alpha}_{\mathbf{A}_\alpha} \mathbf{B}$ λ: .2
(.4) $\mathcal{H} \vdash \mathsf{S}^{\mathbf{x}_\alpha}_{\mathbf{A}_\alpha} \mathbf{B}$ R: 5200, .3
■

5216. $\vdash [T \wedge \mathbf{A}] = \mathbf{A}$

Proof:

(.1) $\vdash [\lambda x_o \centerdot T \wedge x_o = x_o]T \wedge [\lambda x_o \centerdot T \wedge x_o = x_o]F$
$= \forall x_o \centerdot [\lambda x_o \centerdot T \wedge x_o = x_o]x_o$ 5209: Axiom 1
(.2) $\vdash [T \wedge T = T] \wedge [T \wedge F = F] = \forall x_o \centerdot T \wedge x_o = x_o$ λ: .1
(.3) $\vdash [T \wedge T = T] \wedge [T \wedge F = F]$ 5213: 5211, 5214
(.4) $\vdash \forall x_o \centerdot T \wedge x_o = x_o$ R: .2, .3
(.5) $\vdash T \wedge \mathbf{A} = \mathbf{A}$ $\forall$I: .4
■

5217. $\vdash [T = F] = F$

Proof:

(.1) $\vdash [\lambda x_o \centerdot T = x_o]T \wedge [\lambda x_o \centerdot T = x_o]F = \forall x_o \centerdot [\lambda x_o \centerdot T = x_o]x_o$
5209: Axiom 1
(.2) $\vdash [T = T] \wedge [T = F] = \forall x_o \centerdot T = x_o$ λ: .1

(.3) $\vdash T \wedge [T = F] = \forall x_o \centerdot T = x_o$ RR: 5210, .2
(.4) $\vdash T = F = \forall x_o \centerdot T = x_o$ R: 5216, .3
(.5) $\vdash [\lambda x_o T] = [\lambda x_o x_o] = \forall x_o \centerdot [\lambda x_o T]x_o = [\lambda x_o x_o]x_o$ 5209: Axiom 3^{oo}
(.6) $\vdash F = \forall x_o \centerdot T = x_o$ λ: .5; def of F
(.7) $\vdash [T = F] = F$ =: .4, .6 ∎

5218 $\vdash [T = \mathbf{A}] = \mathbf{A}$

Proof:

(.1) $\vdash [[T = T] = T] \wedge [[T = F] = F] = \forall x_o \centerdot [T = x_o] = x_o$
5209: Axiom 1; λ
(.2) $\vdash [T = T] = T$ =: 5210
(.3) $\vdash [T = T = T] \wedge [T = F = F]$ 5213: .2, 5217
(.4) $\vdash \forall x_o \centerdot T = x_o = x_o$ R: .3, .1
(.5) $\vdash T = \mathbf{A} = \mathbf{A}$ $\forall$I: .3 ∎

5219 Rule T. $\mathcal{H} \vdash \mathbf{A}$ if and only if $\mathcal{H} \vdash T = \mathbf{A}$; $\mathcal{H} \vdash \mathbf{A}$ if and only if $\mathcal{H} \vdash \mathbf{A} = T$.

Proof: By RR(5202), 5218, and =(5201). ∎

5220 Rule of Universal Generalization (Gen). If $\mathcal{H} \vdash \mathbf{A}$ then $\mathcal{H} \vdash \forall \mathbf{x}_\alpha \mathbf{A}$, provided that $\mathbf{x}_\alpha$ is not free in any wff in $\mathcal{H}$.

Proof:

(.1) $\mathcal{H} \vdash \mathbf{A}$ given
(.2) $\mathcal{H} \vdash T = \mathbf{A}$ Rule T: .1
(.3) $\mathcal{H} \vdash [\lambda x_\alpha T] = [\lambda \mathbf{x}_\alpha T]$ 5206
(.4) $\mathcal{H} \vdash \forall \mathbf{x}_\alpha \mathbf{A}$ R$'$: .2, .3; def of $\forall$
Note that the restrictions on Rule R$'$ are satisfied. ∎

5221$_n$ Rule of Substitution (Sub). If $\mathcal{H} \vdash \mathbf{B}$, then $\mathcal{H} \vdash \mathsf{S}^{\mathbf{x}^1_{\alpha_1} \cdots \mathbf{x}^n_{\alpha_n}}_{\mathbf{A}^1_{\alpha_1} \cdots \mathbf{A}^n_{\alpha_n}} \mathbf{B}$, provided that the variables $\mathbf{x}^1, \ldots, \mathbf{x}^n$ are distinct from one another and from all free variables of wffs in $\mathcal{H}$, and that $\mathbf{A}^i_{\alpha_i}$ is free for $\mathbf{x}^i_{\alpha_i}$ in $\mathbf{B}$ for all i $(1 \leq i \leq n)$.

Proof:

First we consider the case $n = 1$.

(.1) $\mathcal{H} \vdash \mathbf{B}$ given

(.2) $\mathcal{H} \vdash \forall \mathbf{x}_\alpha \mathbf{B}$ Gen: .1

(.3) $\mathcal{H} \vdash \mathsf{S}^{\mathbf{x}_\alpha}_{\mathbf{A}_\alpha} \mathbf{B}$. $\forall$I(5215): .2

Next consider the case of an arbitrary number n. Let $\mathbf{y}^1_{\alpha_1}, \ldots, \mathbf{y}^n_{\alpha_n}$ be distinct from one another, from $\mathbf{x}^1, \ldots, \mathbf{x}^n$, and from all variables in $\mathbf{B}$, $\mathbf{A}^1_{\alpha_1}, \ldots, \mathbf{A}^n_{\alpha_n}$, and wffs in $\mathcal{H}$.

(.4) $\mathcal{H} \vdash \mathsf{S}^{\mathbf{x}^1 \ldots \mathbf{x}^n}_{\mathbf{y}^1 \ldots \mathbf{y}^n} \mathbf{B}$ 5221_1 (applied n times)

(.5) $\mathcal{H} \vdash \mathsf{S}^{\mathbf{y}^1 \ldots \mathbf{y}^n}_{\mathbf{A}^1 \ldots \mathbf{A}^n} [\mathsf{S}^{\mathbf{x}^1 \ldots \mathbf{x}^n}_{\mathbf{y}^1 \ldots \mathbf{y}^n} \mathbf{B}]$ 5221_1 (applied n times): .4

This is the desired wff. ■

Since we have now established that we can substitute freely for variables of $\mathcal{Q}_0$, we can henceforth indicate that all wffs of $\mathcal{Q}_0$ having a certain form are theorems in either of two ways: by stating a theorem schema (such as $T = \mathbf{A} = \mathbf{A}$), or by stating one instance of it (such as $T = x_o = x_o$) from which all others can be obtained by substitution. In the former case we use syntactical variables from our meta-language, and in the latter we use variables from the object language $\mathcal{Q}_0$. We shall generally prefer to use the variables of $\mathcal{Q}_0$.

When no confusion seems likely, we shall often use the number of a theorem to designate any theorem which can be derived from it by the α and Substitution Rules, and omit explicit reference to these rules in justifying the steps of a proof.

5222 Rule of Cases. If $\mathcal{H} \vdash \mathsf{S}^{\mathbf{x}_o}_{T} \mathbf{A}$ and $\mathcal{H} \vdash \mathsf{S}^{\mathbf{x}_o}_{F} \mathbf{A}$, then $\mathcal{H} \vdash \mathbf{A}$.

Proof:

(.1) $\mathcal{H} \vdash T = [\lambda \mathbf{x}_o \mathbf{A}] T$ Rule T, λ: hypothesis

(.2) $\mathcal{H} \vdash T = [\lambda \mathbf{x}_o \mathbf{A}] F$ Rule T, λ: hypothesis

(.3) $\mathcal{H} \vdash T \wedge T$ 5212

(.4) $\mathcal{H} \vdash [\lambda \mathbf{x}_o \mathbf{A}] T \wedge [\lambda \mathbf{x}_o \mathbf{A}] F$ R′: .1, .2, .3

(.5) $\vdash [\lambda \mathbf{x}_o \mathbf{A}] T \wedge [\lambda \mathbf{x}_o \mathbf{A}] F = \forall \mathbf{x}_o \mathbf{A}$ α, Sub, λ: Axiom 1

(.6) $\mathcal{H} \vdash \forall \mathbf{x}_o \mathbf{A}$ RR: .4, .5

(.7) $\mathcal{H} \vdash \mathbf{A}$ $\forall$I(5215): .6

■

5223. $\vdash [T \supset y_o] = y_o$

Proof:

(.1) $\vdash [T \supset y] = \centerdot T = [T \wedge y]$ $\quad$ λ, def of $\supset$
(.2) $\vdash [T \supset y] = \centerdot T \wedge y$ $\quad$ R: .1, 5218
(.3) $\vdash [T \supset y] = y$ $\quad$ R: .2, 5216

■

5224 Modus Ponens (MP). If $\mathcal{H} \vdash \mathbf{A}$ and $\mathcal{H} \vdash \mathbf{A} \supset \mathbf{B}$, then $\mathcal{H} \vdash \mathbf{B}$.

Proof:

(.1) $\mathcal{H} \vdash \mathbf{A} \supset \mathbf{B}$ $\quad$ given
(.2) $\mathcal{H} \vdash \mathbf{A} = T$ $\quad$ Rule T: hypothesis
(.3) $\mathcal{H} \vdash T \supset \mathbf{B}$ $\quad$ R′: .1, .2
(.4) $\mathcal{H} \vdash \mathbf{B}$ $\quad$ RR: .3, 5223

■

5225. $\vdash \Pi_{o(o\alpha)} f_{o\alpha} \supset f_{o\alpha} x_\alpha$

Proof:

(.1) $\vdash \Pi_{o(o\alpha)} f_{o\alpha} \supset \centerdot [\lambda f_{o\alpha} \centerdot f_{o\alpha} x_\alpha][\lambda x_\alpha T] = [\lambda f_{o\alpha} \centerdot f_{o\alpha} x_\alpha] f_{o\alpha}$
Sub: Axiom $2^{o\alpha}$; def of $\Pi_{o(o\alpha)}$
(Note that this is our first use of Axiom 2.)
(.2) $\vdash \Pi_{o(o\alpha)} f_{o\alpha} \supset \centerdot T = f_{o\alpha} x_\alpha$ $\quad$ λ: .1
(.3) $\vdash \Pi_{o(o\alpha)} f_{o\alpha} \supset f_{o\alpha} x_\alpha$ $\quad$ R: 5218, .2

■

5226. $\vdash \forall \mathbf{x}_\alpha \mathbf{B} \supset \mathsf{S}^{\mathbf{x}_\alpha}_{\mathbf{A}_\alpha} \mathbf{B}$ provided that $\mathbf{A}_\alpha$ is free for $\mathbf{x}_\alpha$ in $\mathbf{B}$.

Proof:

(.1) $\vdash \Pi_{o(o\alpha)}[\lambda \mathbf{x}_\alpha \mathbf{B}] \supset [\lambda \mathbf{x}_\alpha \mathbf{B}] \mathbf{A}_\alpha$ $\quad$ Sub: 5225
(.2) $\vdash \forall \mathbf{x}_\alpha \mathbf{B} \supset \mathsf{S}^{\mathbf{x}_\alpha}_{\mathbf{A}_\alpha} \mathbf{B}$ $\quad$ λ: .1

■

5227. $\vdash F \supset x_o$

This is a special case of 5226, since F_o is $\forall x_o x_o$.

5228. $\vdash [T \supset T] = T$; $\vdash [T \supset F] = F$; $\vdash [F \supset T] = T$; $\vdash [F \supset F] = T$

Proof: by 5223, 5227, and Rule T. ■

5229. $\vdash [T \wedge T] = T$; $\vdash [T \wedge F] = F$; $\vdash [F \wedge T] = F$; $\vdash [F \wedge F] = F$

Proof: $\vdash F \wedge \mathbf{A} = F$ by 5227, the definition of $\supset$, and $=$ Rules. Also use 5216. ■

5230. $\vdash [T = T] = T$; $\vdash [T = F] = F$; $\vdash [F = T] = F$; $\vdash [F = F] = T$

Proof: From 5210 and 5218 one obtains all of these except $[F = T] = F$; to prove this we proceed as follows:

(.1) $\vdash [F = T] \supset \centerdot [\lambda x_o \centerdot x_o = F]F = [\lambda x_o \centerdot x_o = F]T$	Sub: Axiom 2^0
(.2) $\vdash [F = T] \supset \centerdot [F = F] = [T = F]$	λ: .1
(.3) $\vdash [F = T] \supset F$	RR: .2, 5210, 5218
(.4) $\vdash [F = T] = \centerdot [F = T] \wedge F$	λ: .3, def of $\supset$
(.5) $\vdash x_o \wedge F = F$	Rule of Cases (5222): 5229
(.6) $\vdash [F = T] \wedge F = F$	Sub: .5
(.7) $\vdash [F = T] = F$	=: .4, .6

■

5231. $\vdash \sim T = F$; $\vdash \sim F = T$

Proof: By 5230 and the definition of $\sim$. ■

5232. $\vdash T \vee T = T$; $\vdash T \vee F = T$; $\vdash F \vee T = T$; $\vdash F \vee F = F$

Proof: By 5229, 5231, λ, and the definition of $\vee_{ooo}$. ■

DEFINITION. The class of *propositional wffs* is the smallest class of wffs$_o$ which contains T_o, F_o, all variables of type o, and which, if it contains $\mathbf{A}$ and $\mathbf{B}$, also contains $[\sim \mathbf{A}]$, $[\mathbf{A} \wedge \mathbf{B}]$, $[\mathbf{A} \vee \mathbf{B}]$, $[\mathbf{A} \supset \mathbf{B}]$, and $[\mathbf{A} = \mathbf{B}]$. Given any assignment φ of truth values to variables of type o and a propositional wff $\mathbf{A}$, we define $\mathcal{V}_\varphi \mathbf{A}$ (*the truth value of* $\mathbf{A}$ *with respect to* φ) as follows by induction on the number of occurrences of [in $\mathbf{A}$:

(1) $\mathcal{V}_\varphi T_o = \mathsf{T}$ (truth)
(2) $\mathcal{V}_\varphi F_0 = \mathsf{F}$ (falsehood)
(3) $\mathcal{V}_\varphi \mathbf{p}_o = \varphi \mathbf{p}_o$
(4) $\mathcal{V}_\varphi \sim \mathbf{A} = \mathsf{T}$ if $\mathcal{V}_\varphi \mathbf{A} = \mathsf{F}$
$\quad = \mathsf{F}$ if $\mathcal{V}_\varphi \mathbf{A} = \mathsf{T}$
(5) $\mathcal{V}_\varphi [\mathbf{A} \wedge \mathbf{B}] = \mathsf{T}$ if $\mathcal{V}_\varphi \mathbf{A} = \mathcal{V}_\varphi \mathbf{B} = \mathsf{T}$
$\quad = \mathsf{F}$ otherwise

$$\begin{aligned}
(6)\quad \mathcal{V}_\varphi[\mathbf{A} \lor \mathbf{B}] &= \mathsf{T} \text{ if } \mathcal{V}_\varphi\mathbf{A} = \mathsf{T} \text{ or } \mathcal{V}_\varphi\mathbf{B} = \mathsf{T} \\
&= \mathsf{F} \text{ otherwise}
\end{aligned}$$

$$\begin{aligned}
(7)\quad \mathcal{V}_\varphi[\mathbf{A} \supset \mathbf{B}] &= \mathsf{T} \text{ if } \mathcal{V}_\varphi\mathbf{A} = \mathsf{F} \text{ or } \mathcal{V}_\varphi\mathbf{B} = \mathsf{T} \\
&= \mathsf{F} \text{ otherwise}
\end{aligned}$$

$$\begin{aligned}
(8)\quad \mathcal{V}_\varphi[\mathbf{A} = \mathbf{B}] &= \mathsf{T} \text{ if } \mathcal{V}_\varphi\mathbf{A} = \mathcal{V}_\varphi\mathbf{B} \\
&= \mathsf{F} \text{ otherwise}
\end{aligned}$$

REMARK. If $\mathbf{A}$ is a propositional wff without free variables, then $\mathcal{V}_\varphi\mathbf{A} = \mathcal{V}_\psi\mathbf{A}$ for all assignments φ and ψ, so we let $\mathcal{V}\mathbf{A} = \mathcal{V}_\varphi\mathbf{A}$ for any assignment φ.

DEFINITION. A wff $\mathbf{A}$ is a *tautology* iff $\mathbf{A}$ is a propositional wff and $\mathcal{V}_\varphi\mathbf{A} = \mathsf{T}$ for all truth-value assignments φ. A wff $\mathbf{B}$ is a *substitution instance of a tautology*, or is *tautologous*, iff there exists a tautology $\mathbf{A}$, wffs$_o$ $\mathbf{C}^1, \ldots, \mathbf{C}^n$, and distinct variables $\mathbf{p}_o^1, \ldots, \mathbf{p}_o^n$ such that $\mathbf{B} = \mathsf{S}^{\mathbf{p}_o^1 \ldots \mathbf{p}_o^n}_{\mathbf{C}^1 \ldots \mathbf{C}^n}\mathbf{A}$.

5233. If $\mathbf{A}$ is a tautology, then $\vdash \mathbf{A}$.

Proof: By induction on the number n of free variables in $\mathbf{A}$.

Case: $n = 0$

We prove by induction on the number of occurrences of [in $\mathbf{B}$ that if $\mathbf{B}$ is a propositional wff without free variables, then $\vdash \mathbf{B} = T$ if $\mathcal{V}\mathbf{B} = \mathsf{T}$, and $\vdash \mathbf{B} = F$ if $\mathcal{V}\mathbf{B} = \mathsf{F}$.

Subcase. B is T or F. Use 5200.

Subcase. B has the form $[\sim \mathbf{C}]$.

If $\mathcal{V}\mathbf{C} = \mathsf{F}$ then $\vdash \mathbf{C} = F$ by inductive hypothesis, so $\vdash \sim \mathbf{C} = T$ by 5231 and = Rules. Also $\mathcal{V}\mathbf{B} = \mathsf{T}$.

If $\mathcal{V}\mathbf{C} = \mathsf{T}$ then $\vdash \mathbf{B} = F$ and $\mathcal{V}\mathbf{B} = \mathsf{F}$ by a similar argument.

The cases where $\mathbf{B}$ is a conjunction, disjunction, implication, or equivalence are handled similarly using 5229, 5232, 5228, and 5230, respectively.

Induction step. If $\mathbf{A}$ is a tautology with at least one free variable $\mathbf{p}_o$, then $\mathsf{S}^{\mathbf{p}_o}_{T}\mathbf{A}$ and $\mathsf{S}^{\mathbf{p}_o}_{F}\mathbf{A}$ are tautologies with fewer free variables than $\mathbf{A}$, so by the inductive hypothesis (and Rule T when $n = 1$) $\vdash \mathsf{S}^{\mathbf{p}_o}_{T}\mathbf{A}$ and $\vdash \mathsf{S}^{\mathbf{p}_o}_{F}\mathbf{A}$. Therefore $\vdash \mathbf{A}$ by the Rule of Cases (5222). ■

5234 Rule P. If $\mathcal{H} \vdash \mathbf{A}^1, \ldots, \mathcal{H} \vdash \mathbf{A}^n$, and if $[\mathbf{A}^1 \wedge \ldots \wedge \mathbf{A}^n] \supset \mathbf{B}$ is tautologous, then $\mathcal{H} \vdash \mathbf{B}$. Also, if $\mathbf{B}$ is tautologous, then $\mathcal{H} \vdash \mathbf{B}$.

Proof: Clearly $\mathbf{A}^1 \supset \centerdot \ldots \supset \centerdot \mathbf{A}^n \supset \mathbf{B}$ is also tautologous, and therefore a theorem by Rule Sub (5221) and 5233. Hence $\mathcal{H} \vdash \mathbf{A}^1 \supset \centerdot \ldots \supset \centerdot \mathbf{A}^n \supset \mathbf{B}$. Then $\mathcal{H} \vdash \mathbf{B}$ by n applications of Modus Ponens (5224). ■

5235. $\vdash \forall \mathbf{x}_\alpha[\mathbf{A} \vee \mathbf{B}] \supset \centerdot \mathbf{A} \vee \forall \mathbf{x}_\alpha \mathbf{B}$ provided $\mathbf{x}_\alpha$ is not free in $\mathbf{A}$.

Proof: Let $\mathbf{p}_o$ be a variable which does not occur in the wff to be proved.

(.1) $\vdash \forall \mathbf{x}_\alpha[T \vee \mathbf{B}] \supset \centerdot T \vee \forall \mathbf{x}_\alpha \mathbf{B}$	Rule P
(.2) $\vdash \forall \mathbf{x}_\alpha \mathbf{B} \supset \centerdot F \vee \forall \mathbf{x}_\alpha \mathbf{B}$	Rule P
(.3) $\vdash \mathbf{B} = \centerdot F \vee \mathbf{B}$	Rule P
(.4) $\vdash \forall \mathbf{x}_\alpha[F \vee \mathbf{B}] \supset \centerdot F \vee \forall \mathbf{x}_\alpha \mathbf{B}$	R: .2, .3
(.5) $\vdash \forall \mathbf{x}_\alpha[\mathbf{p}_o \vee \mathbf{B}] \supset \centerdot \mathbf{p}_o \vee \forall \mathbf{x}_\alpha \mathbf{B}$	Rule of Cases (5222): .1, .4
(.6) $\vdash \forall \mathbf{x}_\alpha[\mathbf{A} \vee \mathbf{B}] \supset \centerdot \mathbf{A} \vee \forall \mathbf{x}_\alpha \mathbf{B}$	Sub: .5

■

REMARK. In view of 5220 (Gen), 5234 (Rule P), 5226, and 5235, it is now clear that all substitution instances of theorems of quantification theory (first-order logic) are theorems of $\mathcal{Q}_0$. We may sometimes make use of this fact by citing the following metatheorem to indicate informally how certain inferences in an abbreviated proof can be justified.

5236 "Rule Q". If $\mathcal{H} \vdash \mathbf{A}^1, \ldots$, and $\mathcal{H} \vdash \mathbf{A}^n$ (where $n \geq 0$), and if $\bar{\mathbf{A}}^i$ can be obtained from $\mathbf{A}^i$ by generalizing on zero or more free variables of $\mathbf{A}^i$ which do not occur free in $\mathcal{H}$, and if $[\bar{\mathbf{A}}^1 \wedge \ldots \wedge \bar{\mathbf{A}}^n \supset \mathbf{B}]$ is a substitution instance of a theorem of quantification theory, then $\mathcal{H} \vdash \mathbf{B}$.

REMARK. Rule Q is not an effective rule of inference, since there is no effective procedure for determining whether an alleged application of Rule Q is in fact an application of the rule. Of course one can always cite the particular theorem of quantification theory which one is using in applying Rule Q, and supply a proof of that theorem. However, we will normally use Rule Q only when it will be rather obvious which well-known theorem of quantification theory is being used.

5237 $\supset \forall$ Rule. If $\mathcal{H} \vdash \mathbf{A} \supset \mathbf{B}$, then $\mathcal{H} \vdash \mathbf{A} \supset \forall \mathbf{x} \mathbf{B}$, provided that $\mathbf{x}_\alpha$ is not free in $\mathbf{A}$ or in $\mathcal{H}$.

Proof: by Rule Q, using Axiom 5 of $\mathcal{F}$. ■

5238. $\vdash [\lambda \mathbf{x}^1_{\beta_1} \ldots \lambda \mathbf{x}^n_{\beta_n} \mathbf{A}_\alpha] = [\lambda \mathbf{x}^1_{\beta_1} \ldots \lambda \mathbf{x}^n_{\beta_n} \mathbf{B}_\alpha] = \forall \mathbf{x}^1_{\beta_1} \ldots \forall \mathbf{x}^n_{\beta_n} [\mathbf{A}_\alpha = \mathbf{B}_\alpha]$

Proof: By induction on n.

(.1) $\vdash [\lambda \mathbf{x}_\beta \mathbf{A}_\alpha] = [\lambda \mathbf{x}_\beta \mathbf{B}_\alpha] = \forall \mathbf{x}_\beta \centerdot \mathbf{A}_\alpha = \mathbf{B}_\alpha$ $\quad$ α, Sub, λ: Axiom $3^{\alpha\beta}$

This takes care of the case $n = 1$. For the induction step we proceed as follows:

(.2) $\vdash [\lambda \mathbf{x}^1 \ldots \lambda \mathbf{x}^{n-1} [\lambda \mathbf{x}^n \mathbf{A}_\alpha]] = [\lambda \mathbf{x}^1 \ldots \lambda \mathbf{x}^{n-1} [\lambda \mathbf{x}^n \mathbf{B}_\alpha]]$
$= \forall \mathbf{x}^1 \ldots \forall \mathbf{x}^{n-1} \centerdot [\lambda \mathbf{x}^n \mathbf{A}_\alpha] = [\lambda \mathbf{x}^n \mathbf{B}_\alpha]$ $\quad$ by inductive hypothesis

The desired result follows from .2 and an appropriate instance of .1 by Rule R. ∎

5239. $\vdash \forall \mathbf{x}^1_{\beta_1} \ldots \forall \mathbf{x}^n_{\beta_n} [\mathbf{A}_\alpha = \mathbf{B}_\alpha] \supset \centerdot \mathbf{C} = \mathbf{D}$, where $n \geq 0$ and

(a) $\mathbf{D}$ is obtained from $\mathbf{C}$ and $[\mathbf{A}_\alpha = \mathbf{B}_\alpha]$ as in Rule R;
(b) $\mathbf{x}^1_{\beta_1}, \ldots, \mathbf{x}^n_{\beta_n}$ is a complete list of those variables $\mathbf{x}_\beta$ such that $\mathbf{x}_\beta$ occurs free in $[\mathbf{A}_\alpha = \mathbf{B}_\alpha]$, and the occurrence of $\mathbf{A}_\alpha$ in $\mathbf{C}$ (which is replaced by $\mathbf{B}_\alpha$ in $\mathbf{D}$) is in a wf part of $\mathbf{C}$ of the form $[\lambda \mathbf{x}_\beta \mathbf{E}_\gamma]$.

Proof: Let $\mathbf{t}_{\alpha\beta_n \ldots \beta_1}$ be a variable which does not occur in the wff to be proved, and let $\mathbf{G}_o$ be the result of replacing the occurrence $\mathbf{A}_\alpha$ in $\mathbf{C}$ by an occurrence of $[\mathbf{t}_{\alpha\beta_n \ldots \beta_1} \mathbf{x}^1_{\beta_1} \ldots \mathbf{x}^n_{\beta_n}]$.

(.1) $\vdash [[\lambda \mathbf{x}^1 \ldots \lambda \mathbf{x}^n \mathbf{A}_\alpha] = [\lambda \mathbf{x}^1 \ldots \lambda \mathbf{x}^n \mathbf{B}_\alpha]]$
$\supset \centerdot [\lambda \mathbf{t}_{\alpha\beta_n \ldots \beta_1} \mathbf{G}][\lambda \mathbf{x}^1 \ldots \lambda \mathbf{x}^n \mathbf{A}_\alpha] = [\lambda \mathbf{t}_{\alpha\beta_n \ldots \beta_1} \mathbf{G}][\lambda \mathbf{x}^1 \ldots \lambda \mathbf{x}^n \mathbf{B}_\alpha]$
Sub: Axiom $2^{(\alpha\beta_n \ldots \beta_1)}$

The desired theorem follows by 5238 and λ (5207, 5208). Note that the case where $n = 0$ presents no difficulties. ∎

NOTATION. $\mathcal{H}$, $\mathbf{H} \vdash \mathbf{P}$ means $\mathcal{H} \cup \{\mathbf{H}\} \vdash \mathbf{P}$.

5240 Deduction Theorem. If $\mathcal{H}$, $\mathbf{H} \vdash \mathbf{P}$ then $\mathcal{H} \vdash \mathbf{H} \supset \mathbf{P}$.

Proof: Let $\mathcal{S}_1$; $\mathcal{S}_2$ be a proof of $\mathbf{P}$ from the hypotheses $\mathcal{H}$, $\mathbf{H}$. We prove by induction on i that $\mathcal{H} \vdash \mathbf{H} \supset \mathbf{R}^i$ for each wff $\mathbf{R}^i$ in the sequence $\mathcal{S}_2$.

Case: $\mathbf{R}^i$ is a theorem or $\mathbf{H}$.

(.1) $\vdash \mathbf{H} \supset \mathbf{R}^i$ Rule P
(.2) $\mathcal{H} \vdash \mathbf{H} \supset \mathbf{R}^i$ theorem

Case: $\mathbf{R}^i \in \mathcal{H}$.

(.3) $\mathcal{H} \vdash \mathbf{R}^i$ hypothesis
(.4) $\mathcal{H} \vdash \mathbf{H} \supset \mathbf{R}^i$ Rule P: .3

Case: $\mathbf{R}^i$ is obtained from $\mathbf{C}$ and $[\mathbf{A}_\alpha = \mathbf{B}_\alpha]$ by Rule $\mathbf{R}'$.

(.5) $\mathcal{H} \vdash \mathbf{H} \supset \mathbf{C}$ by inductive hypothesis
(.6) $\mathcal{H} \vdash \mathbf{H} \supset \centerdot\, \mathbf{A}_\alpha = \mathbf{B}_\alpha$ by inductive hypothesis

Let $\mathbf{x}^1_{\beta_1}, \ldots, \mathbf{x}^n_{\beta_n}$ be a complete list of those variables $\mathbf{x}_\beta$ such that $\mathbf{x}_\beta$ occurs free in $[\mathbf{A}_\alpha = \mathbf{B}_\alpha]$ but the occurrence of $\mathbf{A}_\alpha$ in $\mathbf{C}$ is in a wf part $[\lambda \mathbf{x}_\beta \mathbf{E}_\gamma]$ of $\mathbf{C}$. Note that the condition on Rule $\mathbf{R}'$ assures that none of these variables is free in $\mathcal{H}$ or $\mathbf{H}$.

(.7) $\mathcal{H} \vdash \mathbf{H} \supset \forall \mathbf{x}^1_{\beta_1} \ldots \forall \mathbf{x}^n_{\beta_n} \centerdot\, \mathbf{A}_\alpha = \mathbf{B}_\alpha$ 5237: .6
(.8) $\mathcal{H} \vdash \mathbf{H} \supset \mathbf{R}^i$ Rule P: .5, .7, 5239

■

5241. If $\mathcal{H} \vdash \mathbf{A}$ and $\mathcal{H} \subseteq \mathcal{G}$ then $\mathcal{G} \vdash \mathbf{A}$.

Proof: This is clear when $\mathcal{H} = \emptyset$; if $\mathcal{H} \neq \emptyset$, let $\mathcal{H} = \{\mathbf{H}^1, \ldots, \mathbf{H}^n\}$.

(.1) $\vdash \mathbf{H}^1 \supset \centerdot \ldots \supset \centerdot\, \mathbf{H}^n \supset \mathbf{A}$ 5240
(.2^i) $\mathcal{G} \vdash \mathbf{H}^i$ for $1 \leq i \leq n$ hypothesis
(.3) $\mathcal{G} \vdash \mathbf{A}$ Rule P

■

DEFINITION. If $\mathcal{G}$ is an infinite set of wffs$_o$, we define $\mathcal{G} \vdash A$ to mean there is a finite subset $\mathcal{H}$ of $\mathcal{G}$ such that $\mathcal{H} \vdash A$. Note that 5240 and 5241 still apply when $\mathcal{H}$ is infinite.

5242 Rule of Existential Generalization (∃Gen). If $\mathcal{H} \vdash \mathsf{S}^{\mathbf{x}_\alpha}_{\mathbf{A}_\alpha} \mathbf{B}$, and $\mathbf{A}_\alpha$ is free for $\mathbf{x}_\alpha$ in $\mathbf{B}$, then $\mathcal{H} \vdash \exists \mathbf{x}_\alpha \mathbf{B}$.

Proof: by Rule P, 5226, and the definition of $\exists$. ■

5243 Comprehension Theorem.
$\vdash \exists \mathbf{u}_{\beta\alpha_n\ldots\alpha_1} \forall \mathbf{x}^1_{\alpha_1} \ldots \forall \mathbf{x}^n_{\alpha_n} \centerdot\, \mathbf{u}_{\beta\alpha_n\ldots\alpha_1} \mathbf{x}^1_{\alpha_1} \ldots \mathbf{x}^n_{\alpha_n} = \mathbf{B}_\beta$
when $n \geq 0$ and $\mathbf{u}_{\beta\alpha_n\ldots\alpha_1}$ is not free in $\mathbf{B}_\beta$.

Proof:

(.1) $\vdash \forall \mathbf{x}^1_{\alpha_1} \ldots \forall \mathbf{x}^n_{\alpha_n} \blacksquare [\lambda \mathbf{x}^1_{\alpha_1} \ldots \lambda \mathbf{x}^n_{\alpha_n} \mathbf{B}_\beta] \mathbf{x}^1_{\alpha_1} \ldots \mathbf{x}^n_{\alpha_n} = \mathbf{B}_\beta$
Gen: 5208. (Use 5200 when $n = 0$.)

Then use Existential Generalization. ■

5244 Existential Rule ($\exists$ Rule). If $\mathcal{H}, \mathbf{B} \vdash \mathbf{A}$, and $\mathbf{x}_\alpha$ is not free in $\mathcal{H}$ or in $\mathbf{A}$, then $\mathcal{H},\ \exists \mathbf{x}_\alpha \mathbf{B} \vdash \mathbf{A}$.

Proof:

(.1) $\mathcal{H} \vdash \mathbf{B} \supset \mathbf{A}$	Deduction Theorem
(.2) $\mathcal{H} \vdash \exists \mathbf{x}_\alpha \mathbf{B} \supset \mathbf{A}$	Rule Q: .1
(.3) $\mathcal{H},\ \exists \mathbf{x}_\alpha \mathbf{B} \vdash \exists \mathbf{x}_\alpha \mathbf{B} \supset \mathbf{A}$	5241: .2
(.4) $\mathcal{H},\ \exists \mathbf{x}_\alpha \mathbf{B} \vdash \mathbf{A}$	MP: .3, hypothesis

■

5245 Rule C. If $\mathcal{H} \vdash \exists \mathbf{x}_\alpha \mathbf{B}$ and $\mathcal{H},\ \mathsf{S}^{\mathbf{x}_\alpha}_{\mathbf{y}_\alpha} \mathbf{B} \vdash \mathbf{A}$, where $\mathbf{y}_\alpha$ is free for $\mathbf{x}_\alpha$ in $\mathbf{B}$ and $\mathbf{y}_\alpha$ is not free in $\mathcal{H}$, $\exists \mathbf{x}_\alpha \mathbf{B}$, or $\mathbf{A}$, then $\mathcal{H} \vdash \mathbf{A}$.

Proof:

(.1) $\mathcal{H}, \mathsf{S}^{\mathbf{x}_\alpha}_{\mathbf{y}_\alpha} \mathbf{B} \vdash \mathbf{A}$	given
(.2) $\mathcal{H},\ \exists \mathbf{y}_\alpha [\mathsf{S}^{\mathbf{x}_\alpha}_{\mathbf{y}_\alpha} \mathbf{B}] \vdash \mathbf{A}$	$\exists$ Rule: .1
(.3) $\mathcal{H} \vdash \exists \mathbf{y}_\alpha [\mathsf{S}^{\mathbf{x}_\alpha}_{\mathbf{y}_\alpha} \mathbf{B}] \supset \mathbf{A}$	Deduction Theorem: .2
(.4) $\mathcal{H} \vdash \exists \mathbf{x}_\alpha \mathbf{B} \supset \mathbf{A}$	α: .3
(.5) $\mathcal{H} \vdash \exists \mathbf{x}_\alpha \mathbf{B}$	given
(.6) $\mathcal{H} \vdash \mathbf{A}$	MP: .4, .5

■

DEFINITIONS.

$\subseteq_{o(o\alpha)(o\alpha)}$ stands for $[\lambda x_{o\alpha}\lambda y_{o\alpha}\forall z_\alpha \,\blacksquare\, xz \supset yz]$
(subset)

$\mathbf{A}_{o\alpha} \subseteq \mathbf{B}_{o\alpha}$ stands for $\subseteq_{o(o\alpha)(o\alpha)} \mathbf{A}_{o\alpha}\mathbf{B}_{o\alpha}$

$\mathcal{P}_{(o(o\alpha))(o\alpha)}$ stands for $[\lambda y_{o\alpha}\lambda x_{o\alpha} \,\blacksquare\, x_{o\alpha} \subseteq y_{o\alpha}]$
(power set)

$\cup_{o\alpha(o\alpha)(o\alpha)}$ stands for $[\lambda x_{o\alpha}\lambda y_{o\alpha}\lambda z_\alpha \,\blacksquare\, xz \vee yz]$
(union)

$\mathbf{A}_{o\alpha} \cup \mathbf{B}_{o\alpha}$ stands for $\cup_{o\alpha(o\alpha)(o\alpha)} \mathbf{A}_{o\alpha}\mathbf{B}_{o\alpha}$

$\bigcup_{o\alpha(o(o\alpha))}$ stands for $[\lambda w_{o(o\alpha)}\lambda z_a \,\blacksquare\, \exists x_{o\alpha} \,\blacksquare\, wx \wedge xz]$
(union of a collection)

$\cap_{o\alpha(o\alpha)(o\alpha)}$ stands for $[\lambda x_{o\alpha}\lambda y_{o\alpha}\lambda z_\alpha \,\blacksquare\, xz \wedge yz]$
(intersection)

$\mathbf{A}_{o\alpha} \cap \mathbf{B}_{o\alpha}$ stands for $\cap_{o\alpha(o\alpha)(o\alpha)} \mathbf{A}_{o\alpha}\mathbf{B}_{o\alpha}$

$\bigcap_{o\alpha(o(o\alpha))}$ stands for $[\lambda w_{o(o\alpha)}\lambda z_\alpha \,\blacksquare\, \forall x_{o\alpha} \,\blacksquare\, wx \supset xz]$
(intersection of a collection)

$\#_{o\alpha(o\beta)(\alpha\beta)}$ stands for $[\lambda f_{\alpha\beta}\lambda x_{o\beta}\lambda z_\alpha \,\blacksquare\, \exists t_\beta \,\blacksquare\, xt \wedge z = ft]$
(image of a set under a function)

$\{\mathbf{F}_\alpha \mid \mathbf{A}_o\}$ stands for $[\lambda \mathbf{z}_\alpha \exists \mathbf{x}^1 \ldots \exists \mathbf{x}^n \,\blacksquare\, \mathbf{A}_o \wedge \mathbf{z}_\alpha = \mathbf{F}_\alpha]$,
where $\mathbf{x}^1, \ldots, \mathbf{x}^n$ are the variables which occur free in both $\mathbf{A}_o$ and $\mathbf{F}_\alpha$, and $\mathbf{z}_\alpha$ is the first variable$_\alpha$ which does not occur free in $\mathbf{F}_\alpha$ or in $\mathbf{A}_o$.

EXERCISES

Prove the following wffs.

X5200. $x_{o\alpha} \cup y_{o\alpha} = \bigcup_{o\alpha(o(o\alpha))}[\lambda v_{o\alpha} \,\blacksquare\, v = x \vee v = y]$

X5201. $x_{o\alpha} \cap y_{o\alpha} = \bigcap_{o\alpha(o(o\alpha))}[\lambda v_{o\alpha} \,\blacksquare\, v = x \vee v = y]$

X5202. $\# f_{\alpha\beta}[x_{o\beta} \cup y_{o\beta}] = \,\blacksquare\, [\# f_{\alpha\beta} x_{o\beta}] \cup [\# f_{\alpha\beta} y_{o\beta}]$

X5203. $\# f_{\alpha\beta}[x_{o\beta} \cap y_{o\beta}] \subseteq \,\blacksquare\, [\# f_{\alpha\beta} x_{o\beta}] \cap [\# f_{\alpha\beta} y_{o\beta}]$

X5204. $\# f_{\alpha\beta}[\bigcup_{o\beta(o(o\beta))} w_{o(o\beta)}]$
$= \bigcup_{o\alpha(o(o\alpha))} \,\blacksquare\, [\#_{o(o\alpha)(o(o\beta))((o\alpha)(o\beta))} \,\blacksquare\, \#_{o\alpha(o\beta)(\alpha\beta)} f_{\alpha\beta}] w_{o(o\beta)}$

X5205. $\#f_{\alpha\beta}[\bigcap_{o\beta(o(o\beta))} w_{o(o\beta))}]$
$$\subseteq \bigcap_{o\alpha(o(o\alpha))} \mathbin{\blacksquare} [\#_{o(o\alpha)(o(o\beta))((o\alpha)(o\beta))} \mathbin{\blacksquare} \#_{o\alpha(o\beta)(\alpha\beta)} f_{\alpha\beta}] w_{o(o\beta)}$$

X5206. Derive X5202 from X5204 and X5200.

X5207. Derive X5203 from X5205 and X5201.

X5208. $\exists s_{o\iota} \forall x_\iota[[sx \vee \mathbin{\blacksquare} p_{o\iota}x] \wedge \mathbin{\blacksquare} \sim sx \vee \mathbin{\blacksquare} q_{o\iota}x] \equiv \forall y_\iota \mathbin{\blacksquare} py \vee qy$

X5209. $\mathcal{P}_{(o(o\alpha))(o\alpha)}[u_{o\alpha} \cap v_{o\alpha}] = \mathbin{\blacksquare} [\mathcal{P}u] \cap [\mathcal{P}v]$

X5210. $\mathrm{Q}_{o\alpha\alpha} x_\alpha = \{y_\alpha \mid y_\alpha = x_\alpha\}$

X5211. $y_{o\alpha} = \bigcup_{o\alpha(o(o\alpha))} \{\mathrm{Q}_{o\alpha\alpha} x_\alpha \mid y_{o\alpha} x_\alpha\}$

X5212. $\{f_{\alpha\beta} x_\beta \mid g_{o\beta} x_\beta\} = \# f_{\alpha\beta} g_{o\beta}$

X5213. A wff **D** is in *β-normal form* iff it has no wf parts of the form $[\lambda \mathbf{x}_\alpha \mathbf{B}_\beta]\mathbf{A}_\alpha$. Show that every wff of $\mathcal{Q}_0$ can be transformed to β-normal form by a sequence of β-contractions and α-conversions.

§53. Equality and Descriptions

5300. $\vdash [x_\alpha = y_\alpha] \supset \mathbin{\blacksquare} h_{\beta\alpha} x_\alpha = h_{\beta\alpha} y_\alpha$

Proof:

(.1) $\vdash [x_\alpha = y_\alpha] \supset \mathbin{\blacksquare} [h_{\beta\alpha} x_\alpha = h_{\beta\alpha} x_\alpha] = [h_{\beta\alpha} x_\alpha = h_{\beta\alpha} y_\alpha]$ Sub: Axiom 2; λ

(.2) $\vdash [x_\alpha = y_\alpha] \supset \mathbin{\blacksquare} h_{\beta\alpha} x_\alpha = h_{\beta\alpha} y_\alpha$ Rule P: .1, 5200

■

5301. $\vdash [x_\alpha = y_\alpha] \wedge [f_{\beta\alpha} = g_{\beta\alpha}] \supset \mathbin{\blacksquare} f_{\beta\alpha} x_\alpha = g_{\beta\alpha} y_\alpha$

Proof:

(.1) $x_\alpha = y_\alpha,\ f_{\beta\alpha} = g_{\beta\alpha} \vdash f_{\beta\alpha} x_\alpha = g_{\beta\alpha} y_\alpha$ 5201: hypotheses

Then use the Deduction Theorem and Rule P. ■

5302. $\vdash [x_\alpha = y_\alpha] = [y_\alpha = x_\alpha]$

Proof:

(.1) $x_\alpha = y_\alpha \vdash y_\alpha = x_\alpha$	5201: hypothesis
(.2) $\vdash [x_\alpha = y_\alpha] \supset \centerdot\, y_\alpha = x_\alpha$	Deduction Theorem: .1
(.3) $\vdash [y_\alpha = x_\alpha] \supset \centerdot\, x_\alpha = y_\alpha$	Sub: .2
(.4) $\vdash [x_\alpha = y_\alpha] = [y_\alpha = x_\alpha]$	Rule P: .2, .3

■

5303. $\vdash [x_\alpha = y_\alpha] \supset \centerdot\, [x_\alpha = z_\alpha] = [y_\alpha = z_\alpha]$

Proof: Substitute $[\lambda t_\alpha \centerdot\, t_\alpha = z_\alpha]$ for $h_{o\alpha}$ in Axiom 2^α, and apply λ-conversion. ■

Note that the transitive law of equality follows from 5303 by Rule P.

DEFINITIONS:

$\sum^1_{o(o\alpha)}$ stands for $[\lambda p_{o\alpha} \exists y_\alpha \centerdot\, p_{o\alpha} = \mathsf{Q}_{o\alpha\alpha} y_\alpha]$

$\exists_1 \mathbf{x}_\alpha \mathbf{A}$ stands for $\sum^1_{o(o\alpha)} [\lambda \mathbf{x}_\alpha \mathbf{A}]$

("$\exists_1 \mathbf{x}_\alpha \mathbf{A}$" means "there is exactly one $\mathbf{x}_\alpha$ such that $\mathbf{A}$".)

We shall leave the proofs of the next four theorems to the reader.

5304. $\vdash \exists_1 y_\alpha p_{o\alpha} y_\alpha = \exists y_\alpha \centerdot\, p_{o\alpha} = \mathsf{Q}_{o\alpha\alpha} y_\alpha$

5305. $\vdash \exists_1 y_\alpha p_{o\alpha} y_\alpha = \exists y_\alpha \forall z_\alpha \centerdot\, p_{o\alpha} z_\alpha = \centerdot\, y_\alpha = z_\alpha$

5306. $\vdash \exists_1 y_\alpha p_{o\alpha} y_\alpha = \exists y_\alpha \centerdot\, p_{o\alpha} y_\alpha \wedge \forall z_\alpha \centerdot\, p_{o\alpha} z_\alpha \supset y_\alpha = z$

5307. $\vdash \exists_1 y_\alpha p_{o\alpha} y_\alpha = \centerdot\, \exists y_\alpha p_{o\alpha} y_\alpha \wedge \forall y_\alpha \forall z_\alpha \centerdot\, p_{o\alpha} y_\alpha \wedge p_{o\alpha} z_\alpha \supset y_\alpha = z_\alpha$

We next introduce description operators for all types.

DEFINITION. $\iota_{o(oo)}$ stands for $\mathsf{Q}_{o(oo)(oo)}[\lambda x_o x_o]$

5308. $\vdash \iota_{o(oo)}[\mathsf{Q}_{ooo} y_o] = y_o$

Proof:

(.1) $\vdash \iota_{o(oo)}[\mathsf{Q}_{ooo} y_o] = \forall x_o \centerdot\, [\lambda x_o x_o] x_o = [\mathsf{Q}_{ooo} y_o] x_o$

Sub: Axiom 3, def of ι

(.2) $\vdash \iota_{o(oo)}[Q_{ooo}y_o] = \forall x_o \,.\, x_o = \,.\, y_o = x_o$ — λ
(.3) $\vdash [x_o = \,.\, y_o = x_o] = y_o$ — Rule P
(.4) $\vdash \iota_{o(oo)}[Q_{ooo}y_o] = \forall x_o y_o$ — R: .2, .3
(.5) $\vdash [\forall x_o y_o] = y_o$ — Rule Q
(.6) $\vdash \iota_{o(oo)}[Q_{ooo}y_o] = y_o$ — R: .4, .5

∎

REMARK. We could have defined $\iota_{o(oo)}$ as $[\lambda g_{oo} \,.\, g_{oo}T_o]$. Then $\iota_{o(oo)}[Q_{ooo}y_o] = y_o$ would have followed immediately from the tautology $[y_o = T] = y_o$. We also could have defined $\iota_{o(oo)}$ as $Q_{o(oo)(oo)}[Q_{ooo}T_o]$.

DEFINITION. We define $\iota_{\gamma(o\gamma)}$ for all type symbols γ by induction on the length of γ and let $[\imath\, \mathbf{z}_\gamma \mathbf{A}]$ stand for $\iota_{\gamma(o\gamma)}[\lambda \mathbf{z}_\gamma \mathbf{A}]$. Then "$[\imath\, \mathbf{z}_\gamma \mathbf{A}]$" means "that $\mathbf{z}_\gamma$ such that $\mathbf{A}$", when there is a unique such $\mathbf{z}_\gamma$. When $\gamma = (\alpha\beta)$ and $\iota_{\alpha(o\alpha)}$ is already defined, $\iota_{(\alpha\beta)(o(\alpha\beta))}$ stands for

$$[\lambda h_{o(\alpha\beta)} \lambda x_\beta \,.\, \imath z_\alpha \exists f_{\alpha\beta} \,.\, h_{o(\alpha\beta)} f_{\alpha\beta} \wedge f_{\alpha\beta}x_\beta = z_\alpha].$$

5309$^\gamma$. $\vdash \iota_{\gamma(o\gamma)}[Q_{o\gamma\gamma}y_\gamma] = y_\gamma$ for each type symbol γ.

Proof: We prove this by induction on the length of γ. 5309$^\iota$ is Axiom 5, and 5309° is 5308. Suppose $\gamma = (\alpha\beta)$.

(.1) $\vdash \iota_{(\alpha\beta)(o(\alpha\beta))}[Q_{o(\alpha\beta)(\alpha\beta)}y_{\alpha\beta}] = [\lambda x_\beta\, \imath z_\alpha \exists f_{\alpha\beta} \,.\, y_{\alpha\beta} = f_{\alpha\beta} \wedge f_{\alpha\beta}x_\beta = z_\alpha]$
by λ and the definitions of $=$ and $\iota_{\gamma(o\gamma)}$
(.2) $y_{\alpha\beta} = f_{\alpha\beta} \wedge f_{\alpha\beta}x_\beta = z_\alpha \vdash y_{\alpha\beta}x_\beta = z_\alpha$ — Rule P, =, Rule R′: hyp
(.3) $\vdash \exists f_{\alpha\beta}[y_{\alpha\beta} = f_{\alpha\beta} \wedge f_{\alpha\beta}x_\beta = z_{\alpha\beta}] \supset y_{\alpha\beta}x_\beta = z_\alpha$
$\exists$ Rule (5244), Deduction Theorem: .2
(.4) $y_{\alpha\beta}x_\beta = z_\alpha \vdash y_{\alpha\beta} = y_{\alpha\beta} \wedge y_{\alpha\beta}x_\beta = z_\alpha$ — Rule P: 5200, hyp
(.5) $y_{\alpha\beta}x_\beta = z_\alpha \vdash \exists f_{\alpha\beta} \,.\, y_{\alpha\beta} = f_{\alpha\beta} \wedge f_{\alpha\beta}x_\beta = z_\alpha$ — $\exists$ Gen: .4
(.6) $\vdash \exists f_{\alpha\beta}[y_{\alpha\beta} = f_{\alpha\beta} \wedge f_{\alpha\beta}x_\beta = z_\alpha] = \,.\, y_{\alpha\beta}x_\beta = z_\alpha$
Deduction Theorem, Rule P: 3, .5
(.7) $\vdash \iota_{(\alpha\beta)(o(\alpha\beta))}[Q_{o(\alpha\beta)(\alpha\beta)}y_{\alpha\beta}] = [\lambda x_\beta \,.\, \iota_{\alpha(o\alpha)}\lambda z_\alpha \,.\, Q_{o\alpha\alpha}[y_{\alpha\beta}x_\beta]z_\alpha]$
R: .1, .6, def of =
(.8) $\vdash \iota_{(\alpha\beta)(o(\alpha\beta))}[Q_{o(\alpha\beta)(\alpha\beta)}y_{\alpha\beta}] = [\lambda x_\beta \,.\, \iota_{\alpha(o\alpha)}Q_{o\alpha\alpha}[y_{\alpha\beta}x_\beta]]$
Sub: 5205; Rule RR: .7
(.9) $\vdash \iota_{(\alpha\beta)(o(\alpha\beta))}[Q_{o(\alpha\beta)(\alpha\beta)}y_{\alpha\beta}] = [\lambda x_\beta \,.\, y_{\alpha\beta}x_\beta]$
Sub: 5309$^\alpha$ (induction hypothesis); Rule R: .8
(.10) $\vdash \iota_{(\alpha\beta)(o(\alpha\beta))}[Q_{o(\alpha\beta)(\alpha\beta)}y_{\alpha\beta}] = y_{\alpha\beta}$ — RR: 5205, .9

∎

5310. $\vdash \forall z_\alpha[p_{o\alpha}z_\alpha = \blacksquare\, y_\alpha = z_\alpha] \supset \blacksquare\, \iota_{\alpha(o\alpha)}p_{o\alpha} = y_\alpha$

Proof:

(.1) $.1 \vdash \forall z_\alpha \blacksquare\, p_{o\alpha}z_\alpha = [\mathsf{Q}_{o\alpha\alpha}y_\alpha]z_\alpha$	hyp
(.2) $.1 \vdash p_{o\alpha} = \mathsf{Q}_{o\alpha\alpha}y_\alpha$	Sub: Axiom 3; RR: .1
(.3) $.1 \vdash \iota_{\alpha(o\alpha)}p_{o\alpha} = y_\alpha$	=: .2, R′: 5309

Then use the Deduction Theorem. ■

5311. $\vdash \exists_1 y_\alpha p_{o\alpha}y_\alpha \supset p_{o\alpha}[\iota_{\alpha(o\alpha)}p_{o\alpha}]$

Proof:

(.1) $.1 \vdash p_{o\alpha} = \blacksquare\, \mathsf{Q}_{o\alpha\alpha}y_\alpha$	hyp
(.2) $.1 \vdash p_{o\alpha}y_\alpha = \blacksquare\, \mathsf{Q}_{o\alpha\alpha}y_\alpha y_\alpha$	=: .1
(.3) $.1 \vdash p_{o\alpha}y_\alpha$	=, R′: .2, 5200
(.4) $.1 \vdash p_{o\alpha}[\iota_{\alpha(o\alpha)} \blacksquare\, \mathsf{Q}_{o\alpha\alpha}y_\alpha]$	RR: 5309, .3
(.5) $.1 \vdash p_{o\alpha}[\iota_{\alpha(o\alpha)}p_{o\alpha}]$	=, R′: .1, .4
(.6) $\vdash \exists y_\alpha[p_{o\alpha} = \blacksquare\, \mathsf{Q}_{o\alpha\alpha}y_\alpha] \supset p_{o\alpha}[\iota_{\alpha(o\alpha)}p_{o\alpha}]$	∃ Rule, Deduction Theorem: .5

Then use 5304 and .6 to obtain the desired theorem. ■

5312. $\vdash \exists_1 y_\alpha p_{o\alpha}y_\alpha \supset \forall z_\alpha \blacksquare\, p_{o\alpha}z_\alpha = \blacksquare\, \iota_{\alpha(o\alpha)}p_{o\alpha} = z_\alpha$

Proof:

(.1) $.1 \vdash p_{o\alpha} = \mathsf{Q}_{o\alpha\alpha}y_\alpha$	hyp
(.2) $.1 \vdash \iota_{\alpha(o\alpha)}p_{o\alpha} = y_\alpha$	=: .1, 5309
(.3) $.1 \vdash p_{o\alpha} = \mathsf{Q}_{o\alpha\alpha}[\iota_{(o\alpha)}p_{o\alpha}]$	=, R′: .1, .2
(.4) $.1 \vdash \forall z_\alpha \blacksquare\, p_{o\alpha}z_\alpha = \blacksquare\, \iota_{\alpha(o\alpha)}p_{o\alpha} = z_\alpha$	Sub: Axiom 3; R′: .3
(.5) $\vdash \exists y_\alpha[p_{o\alpha} = \mathsf{Q}_{o\alpha\alpha}y_\alpha] \supset \forall z_\alpha \blacksquare\, p_{o\alpha}z_\alpha = \blacksquare\, \iota_{\alpha(o\alpha)}p_{o\alpha} = z_\alpha$	∃ Rule, Deduction Theorem: .4

The Theorem follows from .5 and 5304. ■

DEFINITION: Let $C_{\gamma o\gamma\gamma}$ be

$$[\lambda x_\gamma \lambda y_\gamma \lambda p_o\, \imath q_\gamma \blacksquare\, [p_o \wedge \blacksquare\, x_\gamma = q_\gamma] \vee [\sim p_o \wedge \blacksquare\, y_\gamma = q_\gamma]]$$

$C_{\gamma o\gamma\gamma}x_\gamma y_\gamma p_o$ can be read "if p_o then x_γ, else y_γ".

5313. $\vdash [C_{\gamma o\gamma\gamma}x_\gamma y_\gamma T_o = x_\gamma] \wedge \blacksquare\, C_{\gamma o\gamma\gamma}x_\gamma y_\gamma F_o = y_\gamma$

Proof:

(.1)	$\vdash C_{\gamma o\gamma\gamma}x_\gamma y_\gamma T = \imath q_\gamma \centerdot [T_o \wedge \centerdot x_\gamma = q_\gamma] \vee \centerdot \sim T \wedge \centerdot y_\gamma = q_\gamma$	λ
(.2)	$\vdash [[T \wedge \centerdot x_\gamma = q_\gamma] \vee \centerdot \sim T \wedge \centerdot y_\gamma = q_\gamma] = \centerdot x_\gamma = q_\gamma$	Rule P
(.3)	$\vdash C_{\gamma o\gamma\gamma}x_\gamma y_\gamma T = \iota_{\gamma(o\gamma)}\lambda q_\gamma[x_\gamma = q_\gamma]$	R: .1, .2
(.4)	$\vdash C_{\gamma o\gamma\gamma}x_\gamma y_\gamma T = \iota_{\gamma(o\gamma)}[\mathsf{Q}_{o\gamma\gamma}x_\gamma]$	Sub: 5205; RR: .3; def of =
(.5)	$\vdash C_{\gamma o\gamma\gamma}x_\gamma y_\gamma T = x_\gamma$	Sub: 5309; R: .4
(.6)	$\vdash C_{\gamma o\gamma\gamma}x_\gamma y_\gamma F = \iota_{\gamma(o\gamma)}\lambda q_\gamma[y_\gamma = q_\gamma]$	λ, Rule P, R
(.7)	$\vdash C_{\gamma o\gamma\gamma}x_\gamma y_\gamma F = y_\gamma$	Sub: 5205; RR: .6, 5309
	Then use Rule P with .5 and .7.	■

DEFINITION: Let AC^α be $\exists j_{\alpha(o\alpha)}\forall p_{o\alpha} \centerdot \exists x_\alpha p_{o\alpha}x_\alpha \supset p_{o\alpha} \centerdot j_{\alpha(o\alpha)}p_{o\alpha}$.
This is a formulation of the Axiom of Choice. AC^α says that there is a choice function j which chooses from every nonempty set p (of type $(o\alpha)$) an element (which is designated as jp) of that set. Normally when one assumes the Axiom of Choice in type theory, one assumes it as an axiom schema, and asserts AC^α for each type symbol α. For more information about the Axiom of Choice see [Moore, 1982] and [Rubin and Rubin, 1985].

EXERCISES

X5300. Prove $\sum^1_{o(o\alpha)} p_{o\alpha} = \centerdot p_{o\alpha} = \mathsf{Q}_{o\alpha\alpha} \centerdot \iota_{\alpha(o\alpha)}p_{o\alpha}$

X5301. Prove that for each type symbol α,

$$\forall x_\iota \forall y_\iota [x_\iota = y_\iota] \supset AC^\alpha.$$

X5302. Prove that if β occurs in α, then $\vdash AC^\alpha \supset AC^\beta$. (*Hint:* X5301 may be useful.)

X5303. Prove $\mathsf{Q}_{o\alpha\alpha} = [\lambda x_\alpha \lambda y_\alpha \forall p_{o\alpha\alpha} \centerdot \forall z_\alpha p_{o\alpha\alpha} z_\alpha z_\alpha \supset p_{o\alpha\alpha} x_\alpha y_\alpha]$

X5304. A famous theorem due to Georg Cantor says that the power set (set of all subsets) of a set always has greater cardinality than the set. This simply means that every set has more subsets than members, and this can be expressed simply by saying that there is no function which maps a set onto its power set. The latter statement can be expressed in type theory by the following wff: $\forall s_{o\alpha} \sim \exists g_{o\alpha\alpha} \forall f_{o\alpha} \centerdot [f_{o\alpha} \subseteq s_{o\alpha}] \supset \exists j_\alpha \centerdot s_{o\alpha} j_\alpha \wedge g_{o\alpha\alpha} j_\alpha = f_{o\alpha}$. (Here α is an arbitrary type symbol.) If we take $s_{o\alpha}$ to be $[\lambda x_\alpha T_o]$, the wff above reduces to

$$\sim \exists g_{o\alpha\alpha} \forall f_{o\alpha} \exists j_\alpha \centerdot g_{o\alpha\alpha} j_\alpha = f_{o\alpha}$$

Prove this wff in $\mathcal{Q}_0$.

X5305. Prove $\forall s_{o\alpha} \sim \exists g_{o\alpha\alpha} \forall f_{o\alpha} \centerdot [f_{o\alpha} \subseteq s_{o\alpha}] \supset \exists j_\alpha \centerdot s_{o\alpha} j_\alpha \wedge g_{o\alpha\alpha} j_\alpha = f_{o\alpha}$

X5306. A variant of the second wff in exercise X5304 is

$$\exists s_{\imath\imath} \forall m_\imath \sim [s_{\imath\imath} m_\imath = m_\imath] \supset \sim \exists g_{\imath\imath\imath} \forall f_{\imath\imath} \exists j_\imath \centerdot g_{\imath\imath\imath} j_\imath = f_{\imath\imath}$$

Explain what this says intuitively, and prove it.

X5307. Let α and β be arbitrary type symbols, and let $P^{\alpha\beta}$ (pairing function) be the wff

$$[\lambda x_\alpha \lambda y_\beta \lambda g_{o\beta\alpha} \centerdot g_{o\beta\alpha} x_\alpha y_\beta].$$

(a) What is the type of the wff $P^{\alpha\beta}$? Let $L^{\alpha\beta}$ (left projection) be the wff

$$[\lambda p_\delta \centerdot \imath x_\alpha \exists y_\beta \centerdot p_\delta = P^{\alpha} x_\alpha y_\beta],$$

where δ is chosen so that $L^{\alpha\beta}$ is a wff.

(b) What type symbol must δ be?

(c) Prove the following theorem of $\mathcal{Q}_0$:

$$L^{\alpha\beta}[P^{\alpha\beta} x_\alpha y_\beta] = x_\alpha.$$

(d) Define a wff $R^{\alpha\beta}$ (right projection) such that $[R^{\alpha\beta}[P^{\alpha\beta} x_\alpha y_\beta] = y_\beta]$ is a theorem $\mathcal{Q}_0$. What is the type of $R^{\alpha\beta}$?

(e) To define the left projection operator another way, define $M^{\alpha\beta}$ as $[\lambda p_{o(o\beta\alpha)} \centerdot \imath x_\alpha \centerdot \forall g_{o\beta\alpha} \centerdot p_{o(o\beta\alpha)} g_{o\beta\alpha} \supset \exists y_\beta \centerdot g_{o\beta\alpha} x_\alpha y_\beta]$.
Prove $M^{\alpha\beta}[P^{\alpha\beta} x_\alpha y_\beta] = x_\alpha$.

X5308. Prove $AC^\beta \supset \centerdot \forall x_\alpha \exists y_\beta p_{o\beta\alpha} x_\alpha y_\beta \equiv \exists f_{\beta\alpha} \forall x_\alpha \centerdot p_{o\beta\alpha} x_\alpha \centerdot f_{\beta\alpha} x_\alpha$

X5309. Prove $\sim \exists h_{\iota(o\iota)} \forall p_{o\iota} \forall q_{o\iota} . h p = h q \supset p = q$
This is called the Injective Cantor Theorem, since it asserts that there is no injection from the power set of the domain of individuals into that domain.

X5310. Prove $\forall r_{o\beta(o\beta)} [\forall x_{o\beta} \exists y_\beta r x y \supset \exists f_{\beta(o\beta)} \forall x r x . f x] \supset AC^\beta$
(Compare this with X5308.)

X5311. Prove $\sim [x_\iota = y_\iota] \supset \exists f_{\iota\iota} \centerdot f x = a_\iota \wedge f y = b_\iota$

§54. Semantics of $\mathcal{Q}_0$

A *frame* is a collection $\{\mathcal{D}_\alpha\}_\alpha$ of nonempty domains (sets) $\mathcal{D}_\alpha$, one for each type symbol α, such that $\mathcal{D}_o = \{\mathsf{T}, \mathsf{F}\}$ and $\mathcal{D}_{\alpha\beta}$ is some collection of functions mapping $\mathcal{D}_\beta$ into $\mathcal{D}_\alpha$. The members of $\mathcal{D}_o$ are called *truth values* and the members of $\mathcal{D}_\iota$ are called *individuals*.

If $x \in \mathcal{D}_\alpha$, the one-element set whose sole member is x (sometimes denoted $\{x\}$) is that function h from $\mathcal{D}_\alpha$ into $\mathcal{D}_o$ such that $hx = \mathsf{T}$ and $hy = \mathsf{F}$ for each element y of $\mathcal{D}_\alpha$ distinct from x. The *identity relation* on $\mathcal{D}_\alpha$ is that function q with domain $\mathcal{D}_\alpha$ such that for each $x \in \mathcal{D}_\alpha$, $qx = \{x\}$. Thus if x and y are in $\mathcal{D}_\alpha$, $(qx)y = \mathsf{T}$ iff $x = y$.

An *interpretation* $\langle\{\mathcal{D}_\alpha\}_\alpha, \mathcal{J}\rangle$ of $\mathcal{Q}_0$ consists of a frame and a function $\mathcal{J}$ which maps each constant of type α of $\mathcal{Q}_0$ to an element of $\mathcal{D}_\alpha$, so that $\mathcal{J}\mathsf{Q}_{o\alpha\alpha}$ is the identity relation on $\mathcal{D}_\alpha$, and $\mathcal{J}\iota_{\iota(o\iota)}$ is some function from $\mathcal{D}_{o\iota}$ into $\mathcal{D}_\iota$ which maps each one-element set in $\mathcal{D}_{o\iota}$ to its unique member in $\mathcal{D}_\iota$. For any constant C, $\mathcal{J}C$ is called the *denotation of* C.

An *assignment* into a frame $\{\mathcal{D}_\alpha\}_\alpha$ (sometimes called an assignment of values in the frame $\{\mathcal{D}_\alpha\}_\alpha$ to the variables of $\mathcal{Q}_0$) is a function φ with domain the set of variables of $\mathcal{Q}_0$ such that for each variable $\mathbf{x}_\alpha$, $\varphi\mathbf{x}_\alpha \in \mathcal{D}_\alpha$. An assignment into an interpretation is an assignment into the frame of the interpretation. In contexts where an interpretation is under discussion, it will be assumed that all assignments discussed are into that interpretation unless otherwise indicated. Given an assignment φ, a variable $\mathbf{x}_\alpha$, and $z \in \mathcal{D}_\alpha$, let $(\varphi : \mathbf{x}_\alpha/z)$ be that assignment ψ such that $\psi\mathbf{x}_\alpha = z$ and $\psi\mathbf{y}_\beta = \varphi\mathbf{y}_\beta$ if $\mathbf{y}_\beta \neq \mathbf{x}_\alpha$.

An interpretation $\mathcal{M} = \langle\{\mathcal{D}_\alpha\}_\alpha, \mathcal{J}\rangle$ is a *general model* for $\mathcal{Q}_0$ iff there is a binary function $\mathcal{V}^\mathcal{M}$ such that for each assignment φ and wff $\mathbf{A}_\alpha$, $\mathcal{V}^\mathcal{M}_\varphi\mathbf{A}_\alpha \in \mathcal{D}_\alpha$, and the following conditions are satisfied for all assignments φ and all wffs:

(a) $\mathcal{V}^\mathcal{M}_\varphi\mathbf{x}_\alpha = \varphi\mathbf{x}_\alpha$.
(b) $\mathcal{V}^\mathcal{M}_\varphi A_\alpha = \mathcal{J}A_\alpha$ if A_α is a primitive constant.
(c) $\mathcal{V}^\mathcal{M}_\varphi[\mathbf{A}_{\alpha\beta}\mathbf{B}_\beta] = (\mathcal{V}^\mathcal{M}_\varphi\mathbf{A}_{\alpha\beta})(\mathcal{V}^\mathcal{M}_\varphi\mathbf{B}_\beta)$ (the value of a function $\mathcal{V}^\mathcal{M}_\varphi\mathbf{A}_{\alpha\beta}$ at the argument $\mathcal{V}^\mathcal{M}_\varphi\mathbf{B}_\beta$).
(d) $\mathcal{V}^\mathcal{M}_\varphi[\lambda\mathbf{x}_\alpha\mathbf{B}_\beta]$ = that function from $\mathcal{D}_\alpha$ into $\mathcal{D}_\beta$ whose value for each argument $z \in \mathcal{D}_\alpha$ is $\mathcal{V}^\mathcal{M}_{(\varphi:\mathbf{x}_\alpha/z)}\mathbf{B}_\beta$.

If an interpretation $\mathcal{M}$ is a general model, the function $\mathcal{V}^\mathcal{M}$ is uniquely determined. $\mathcal{V}^\mathcal{M}_\varphi\mathbf{A}_\alpha$ is called the *value* of $\mathbf{A}_\alpha$ in $\mathcal{M}$ with respect to φ. We sometimes write $\mathcal{V}_\varphi^\mathcal{M}$ as $\mathcal{V}_\varphi$, as $\mathcal{V}^\mathcal{M}$, or as $\mathcal{V}$ when the interpretation or assignment is clear from the context, or irrelevant.

REMARK. Not all frames belong to interpretations, and not all interpretations are general models. In order to be a general model, an interpretation must have a frame satisfying certain closure conditions which are discussed further in [Andrews, 1972b]. Basically, in a general model every wff must have a value with respect to each assignment. Thus, the identity function mapping $\mathcal{D}_\iota$ onto itself must be a member of $\mathcal{D}_{\iota\iota}$ so that $\mathcal{V}_\varphi^\mathcal{M}[\lambda x_\iota x_\iota]$ will be defined. Similarly, if $\mathcal{M}$ is a model where $\mathcal{D}_\iota$ is the set of natural numbers and $\mathcal{D}_{\iota\iota\iota}$ contains the addition function j such that $jxy = x + y$ for all numbers x and y, then $\mathcal{D}_{\iota\iota}$ must contain the function k such that $kx = 2x + 5$ for all numbers x, since $k = \mathcal{V}_\varphi^\mathcal{M}[\lambda x_\iota \centerdot f_{\iota\iota\iota}[f_{\iota\iota\iota}x_\iota x_\iota]y_\iota]$ when $\varphi f_{\iota\iota\iota} = j$ and $\varphi y_\iota = 5$.

An interpretation $\langle\{\mathcal{D}_\alpha\}_\alpha, \mathcal{J}\rangle$ is a *standard model* iff for all α and β, $\mathcal{D}_{\alpha\beta}$ is the set of all functions from $\mathcal{D}_\beta$ into $\mathcal{D}_\alpha$. Clearly a standard model is a general model.

5400 Proposition. Let $\mathcal{M}$ be a general model, $\mathbf{A}_\alpha$ a wff, and φ and ψ assignments which agree on all free variables of $\mathbf{A}_\alpha$. Then $\mathcal{V}_\varphi^\mathcal{M}\mathbf{A}_\alpha = \mathcal{V}_\psi^\mathcal{M}\mathbf{A}_\alpha$.

Proof: By induction on the construction of $\mathbf{A}_\alpha$. The details are left to the reader. ■

It is clear that if $\mathbf{A}_\alpha$ is a closed wff, then $\mathcal{V}^\mathcal{M}\mathbf{A}_\alpha$ may be considered meaningful without regard to any assignment. In this case, $\mathcal{V}^\mathcal{M}\mathbf{A}_\alpha$ is called the *denotation* of $\mathbf{A}_\alpha$ in $\mathcal{M}$.

DEFINITIONS: Let $\mathbf{A}$ be a wff$_o$, $\mathcal{M}$ a general model, and φ an assignment into $\mathcal{M}$.

(1) φ *satisfies* $\mathbf{A}$ in $\mathcal{M}$ iff $\mathcal{V}_\varphi^\mathcal{M}\mathbf{A} = \mathsf{T}$.
(2) $\mathbf{A}$ is *satisfiable* in $\mathcal{M}$ iff there is an assignment which satisfies $\mathbf{A}$ in $\mathcal{M}$.
(3) $\mathbf{A}$ is *valid* in $\mathcal{M}$ iff for every assignment φ into $\mathcal{M}$, $\mathcal{V}_\varphi^\mathcal{M}\mathbf{A} = \mathsf{T}$.
(4) A sentence $\mathbf{A}$ is *true* in $\mathcal{M}$ iff $\mathcal{V}^\mathcal{M}\mathbf{A} = \mathsf{T}$, and *false* in $\mathcal{M}$ iff $\mathcal{V}^\mathcal{M}\mathbf{A} = \mathsf{F}$.
(5) $\mathbf{A}$ is *valid in the general* [*standard*] *sense* iff $\mathbf{A}$ is valid in every general [standard] model of $\mathcal{Q}_0$.
(6) A *model for a set* $\mathcal{G}$ *of wffs*$_o$ is a model for $\mathcal{Q}_0$ in which each wff of $\mathcal{G}$ is valid.

We write $\mathcal{M} \models_\varphi \mathbf{A}$, $\mathcal{M} \models \mathbf{A}$, and $\models \mathbf{A}$ to indicate that φ satisfies $\mathbf{A}$ in $\mathcal{M}$, $\mathbf{A}$ is valid in $\mathcal{M}$, and $\mathbf{A}$ is valid in the general sense, respectively.

Clearly a wff which is valid in the general sense is valid in the standard sense, though we shall see in Chapter 7 that the converse of this statement is false. We shall show that a wff$_o$ of $\mathcal{Q}_0$ is a theorem iff it is valid in the general sense.

5401 Lemma. Let $\mathcal{M}$ be a general model, and φ be an assignment into $\mathcal{M}$.

(a) $\mathcal{V}^{\mathcal{M}}_{\varphi}[[\lambda \mathbf{x}_\alpha \mathbf{B}_\beta]\mathbf{A}_\alpha] = \mathcal{V}^{\mathcal{M}}_{(\varphi:\mathbf{x}_\alpha/\mathcal{V}^{\mathcal{M}}_{\varphi}\mathbf{A}_\alpha)}\mathbf{B}_\beta$.
(b) $\mathcal{V}^{\mathcal{M}}_{\varphi}[\mathbf{A}_\alpha = \mathbf{B}_\alpha] = \mathsf{T}$ iff $\mathcal{V}^{\mathcal{M}}_{\varphi}\mathbf{A}_\alpha = \mathcal{V}^{\mathcal{M}}_{\varphi}\mathbf{B}_\alpha$.
(c) $\mathcal{V}^{\mathcal{M}} T_o = \mathsf{T}$.
(d) $\mathcal{V}^{\mathcal{M}} F_o = \mathsf{F}$.
(e) If $x, y \in \mathcal{D}_o$, then $(\mathcal{V}^{\mathcal{M}} \wedge_{ooo})xy = \mathsf{T}$ if $x = \mathsf{T}$ and $y = \mathsf{T}$; $= \mathsf{F}$ otherwise.
(f) If $x,\ y \in \mathcal{D}_o$, then $(\mathcal{V}^{\mathcal{M}} \supset_{ooo})xy = \mathsf{T}$ if $x = \mathsf{F}$ or $y = \mathsf{T}$; $= \mathsf{F}$ if $x = \mathsf{T}$ and $y = \mathsf{F}$.
(g) $\mathcal{M} \models_\varphi \forall \mathbf{x}\mathbf{A}$ iff $\mathcal{M} \models_\psi \mathbf{A}$ for all assignments ψ which agree with φ off $\mathbf{x}_\alpha$.

Proof: These results follow rather directly from the relevant definitions, so we leave most of the details to the reader. In the case of (e), one shows the aid of (a) and (b) that if $\varphi x_o = x$ and $\varphi y_o = y$, then $(\mathcal{V}^{\mathcal{M}} \wedge_{ooo})xy = \mathsf{T}$ iff $\mathcal{V}[\lambda g_{ooo} \centerdot g_{ooo}TT] = \mathcal{V}_\varphi[\lambda g_{ooo} \centerdot g_{ooo}x_o y_o]$, which is true iff for every function $g \in \mathcal{D}_{ooo}$, $g\mathsf{TT} = (\mathcal{V}[\lambda g_{ooo} \centerdot g_{ooo}TT])g = (\mathcal{V}_\varphi[\lambda g_{ooo} \centerdot g_{ooo}x_o y_o])g = gxy$. Clearly $g\mathsf{TT} = gxy$ for every $g \in \mathcal{D}_{ooo}$ if $x = \mathsf{T} = y$. On the other hand, if $x \neq \mathsf{T}$ (for example), then $(\mathcal{V}[\lambda x_o \lambda y_o x_o])\mathsf{TT} = \mathsf{T} \neq x = \mathcal{V}_\varphi x_o = (\mathcal{V}_\varphi[\lambda y_o x_o])y = (\mathcal{V}[\lambda x_o \lambda y_o x_o])xy$, so $(\mathcal{V}^{\mathcal{M}} \wedge_{ooo})xy = \mathsf{F}$.

In the case of (g), one sees that $\models_\varphi \forall \mathbf{x}_\alpha \mathbf{A}$ iff $\mathcal{V}_\varphi[\lambda x_\alpha T] = \mathcal{V}_\varphi[\lambda \mathbf{x}_\alpha \mathbf{A}]$, which is true iff for every $z \in \mathcal{D}_\alpha$, $\mathsf{T} = (\mathcal{V}_\varphi[\lambda x_\alpha T])z = (\mathcal{V}_\varphi[\lambda \mathbf{x}_\alpha \mathbf{A}])z = \mathcal{V}_{(\varphi:\mathbf{x}_\alpha/z)}\mathbf{A}$. This is equivalent to the stated condition. ∎

REMARK. Speaking loosely, we might expect that a sentence $\forall \mathbf{x}_{o\beta}\mathbf{A}$ is true in a model $\mathcal{M} = \langle \{\mathcal{D}_\alpha\}_\alpha, \mathcal{J} \rangle$ iff $\mathbf{A}$ is true for all subsets x of $\mathcal{D}_\beta$. However, we see from 5401(g) that it would be more accurate to say that $\forall \mathbf{x}_{o\beta}\mathbf{A}$ is true in $\mathcal{M}$ iff $\mathbf{A}$ is true for all of those subsets x of $\mathcal{D}_\beta$ which are in $\mathcal{D}_{o\beta}$. When $\mathcal{M}$ is nonstandard, these are quite different statements.

5402 Soundness Theorem.

(a) Every theorem of $\mathcal{Q}_0$ is valid in the general sense, and hence in the standard sense.

(b) If $\mathcal{G}$ is a set of wffs$_o$ and $\mathcal{M}$ is a model for $\mathcal{G}$ and $\mathcal{G} \vdash \mathbf{A}$, then $\mathcal{M} \models \mathbf{A}$.

Proof: We show that Rule R preserves validity in every general model for $\mathcal{Q}_0$, and that the axioms of $\mathcal{Q}_0$ are valid in the general sense.

Rule R. Suppose that $\mathcal{M}$ is a general model such that $\mathcal{V}_\varphi^{\mathcal{M}}\mathbf{A}_\alpha = \mathcal{V}_\varphi^{\mathcal{M}}\mathbf{B}_\alpha$ for all assignments φ into $\mathcal{M}$, and $\mathbf{C}_\beta$ and $\mathbf{C}'_\beta$ are wffs such that $\mathbf{C}'_\beta$ is obtained from $\mathbf{C}_\beta$ by replacing at most one occurrence of $\mathbf{A}_\alpha$ (which is not a variable immediately preceded by λ) in $\mathbf{C}_\beta$ by an occurrence of $\mathbf{B}_\alpha$. We prove that $\mathcal{V}_\varphi^{\mathcal{M}}\mathbf{C}_\beta = \mathcal{V}_\varphi^{\mathcal{M}}\mathbf{C}'_\beta$ for all assignments φ, by induction on the construction of $\mathbf{C}_\beta$.

Case: $\mathbf{C}'_\beta$ is $\mathbf{C}_\beta$. This case is trivial.

Case: $\mathbf{C}_\beta$ is $\mathbf{A}_\alpha$. Then $\mathbf{C}'_\beta$ is $\mathbf{B}_\alpha$ or $\mathbf{A}_\alpha$, so this case is trivial.

In the remaining cases we assume that $\mathbf{A}_\alpha$ is a proper part of $\mathbf{C}_\beta$.

Case: $\mathbf{C}_\beta$ has the form $[\mathbf{G}_{\beta\gamma}\mathbf{H}_\gamma]$.
$\mathbf{C}'_\beta$ has the form $[\mathbf{G}'_{\beta\gamma}\mathbf{H}'_\gamma]$, so by inductive hypothesis we have $\mathcal{V}_\varphi\mathbf{G}_{\beta\gamma} = \mathcal{V}_\varphi\mathbf{G}'_{\beta\gamma}$ and $\mathcal{V}_\varphi\mathbf{H}_\gamma = \mathcal{V}_\varphi\mathbf{H}'_\gamma$, so $\mathcal{V}_\varphi\mathbf{C}_\beta = (\mathcal{V}_\varphi\mathbf{G}_{\beta\gamma})(\mathcal{V}_\varphi\mathbf{H}_\gamma) = (\mathcal{V}_\varphi\mathbf{G}'_{\beta\gamma})(\mathcal{V}_\varphi\mathbf{H}'_\gamma)$ $= \mathcal{V}_\varphi\mathbf{C}'_\beta$.

Case: $\mathbf{C}_\beta$ has the form $[\lambda\mathbf{x}_\gamma\mathbf{E}_\delta]$.
$\mathbf{C}'_\beta$ has the form $[\lambda\mathbf{x}_\gamma\mathbf{E}'_\delta]$ and by induction hypothesis $\mathcal{V}_\psi\mathbf{E}_\delta = \mathcal{V}_\psi\mathbf{E}'_\delta$ for all assignments ψ, so by the definition of $\mathcal{V}_\varphi$ we see that $\mathcal{V}_\varphi\mathbf{C}_\beta = \mathcal{V}_\varphi\mathbf{C}'_\beta$.

Now suppose that $\mathcal{M} \models [\mathbf{A}_\alpha = \mathbf{B}_\alpha]$ and $\mathcal{M} \models \mathbf{C}$, and $\mathbf{C}'$ is obtained from $\mathbf{C}$ by Rule R. Then $\mathcal{V}_\varphi^{\mathcal{M}}\mathbf{A}_\alpha = \mathcal{V}_\varphi^{\mathcal{M}}\mathbf{B}_\alpha$ for all assignments φ, so for any assignment φ we have $\mathcal{V}_\varphi^{\mathcal{M}}\mathbf{C}' = \mathcal{V}_\varphi^{\mathcal{M}}\mathbf{C} = \mathsf{T}$, so $\mathcal{M} \models \mathbf{C}'$.

To show that the axioms are valid, we let $\mathcal{M}$ be any general model and φ be any assignment, and show for each axiom $\mathbf{A}$ that $\mathcal{M} \models_\varphi \mathbf{A}$.

Axiom 1. Suppose first that $(\varphi g_{oo})\mathsf{T} = \mathsf{T}$ and $(\varphi g_{oo})\mathsf{F} = \mathsf{T}$. Then by Lemma 5401 $\mathcal{V}_\varphi[g_{oo}T \wedge g_{oo}F] = \mathsf{T}$, and for any assignment ψ which agrees with φ off x_o, $\mathcal{V}_\psi[g_{oo}x_o] = (\psi g_{oo})(\psi x_o) = (\varphi g_{oo})(\psi x_o) = \mathsf{T}$, so $\mathcal{V}_\varphi\forall x_o[g_{oo}x_o] = \mathsf{T}$ by 5401(g). Hence $\models_\varphi$[Axiom 1] in this case.
On the other hand suppose there exists $z \in \mathcal{D}_o$ such that $(\varphi g_{oo})z = \mathsf{F}$. Define ψ to agree with φ off x_o, with $\psi x_o = z$. Then $\mathcal{V}_\psi[g_{oo}x_o] = (\psi g_{oo})(\psi x_o) = (\varphi g_{oo})z = \mathsf{F}$, so $\mathcal{V}_\varphi\forall x_o[g_{oo}x_o] = \mathsf{F}$ by 5401(g). But $\mathcal{V}_\varphi[g_{oo}T \wedge g_{oo}F] = \mathsf{F}$ too, so $\models_\varphi$[Axiom 1] in this case also.

Axiom 2. If $\varphi x_\alpha \neq \varphi y_\alpha$, then $\models_\varphi$ [Axiom 2] by 5401(b) and (f). If $\varphi x_\alpha = \varphi y_\alpha$, then $\mathcal{V}_\varphi[h_{o\alpha}x_\alpha] = (\varphi h_{o\alpha})(\varphi x_\alpha) = (\varphi h_{o\alpha})(\varphi y_\alpha) = \mathcal{V}_\varphi[h_{o\alpha}y_\alpha]$, so $\models_\varphi$[Axiom 2] by 5401(b) and (f).

Axiom 3. Suppose $\varphi f_{\alpha\beta} = \varphi g_{\alpha\beta}$, and let ψ be any assignment which agrees with φ off x_β. Then $\mathcal{V}_\psi[f_{\alpha\beta}x_\beta] = (\psi f_{\alpha\beta})(\psi x_\beta) = (\varphi f_{\alpha\beta})(\psi x_\beta) = (\varphi g_{\alpha\beta})(\psi x_\beta) = (\psi g_{\alpha\beta})(\psi x_\beta) = \mathcal{V}_\psi[g_{\alpha\beta}x_\beta]$, so $\mathcal{V}_\psi[f_{\alpha\beta}x_\beta = g_{\alpha\beta}x_\beta] = \mathsf{T}$, so $\mathcal{V}_\varphi \forall x_\beta\, [f_{\alpha\beta}x_\beta = g_{\alpha\beta}x_\beta] = \mathsf{T}$. Also, $\mathcal{V}_\varphi[f_{\alpha\beta} = g_{\alpha\beta}] = \mathsf{T}$, so $\models_\varphi$[Axiom 3].

Suppose $(\varphi f_{\alpha\beta}) \neq (\varphi g_{\alpha\beta})$. Then there is an element $z \in \mathcal{D}_\beta$ such that $(\varphi f_{\alpha\beta})z \neq (\varphi g_{\alpha\beta})z$. Let $\psi = (\varphi : x_\beta/z)$. Then $\mathcal{V}_\psi[f_{\alpha\beta}x_\beta] = (\psi f_{\alpha\beta})(\psi x_\beta) = (\varphi f_{\alpha\beta})z \neq (\varphi g_{\alpha\beta})z = (\psi g_{\alpha\beta})(\psi x_\beta) = \mathcal{V}_\psi[g_{\alpha\beta}x_\beta]$, so $\mathcal{V}_\psi[f_{\alpha\beta}x_\beta = g_{\alpha\beta}x_\beta] = \mathsf{F}$, and $\mathcal{V}_\varphi \forall x_\beta[f_{\alpha\beta}x_\beta = g_{\alpha\beta}x_\beta] = \mathsf{F}$ by 5401(g). Also $\mathcal{V}_\varphi[f_{\alpha\beta} = g_{\alpha\beta}] = \mathsf{F}$, so $\models_\varphi$[Axiom 3].

For each of the Axioms $4_1 - 4_5$, the left side of the equality has the form $[[\lambda \mathbf{x}_\alpha \mathbf{D}_\beta]\mathbf{A}_\alpha]$. Let $\psi = (\varphi : \mathbf{x}_\alpha/\mathcal{V}_\varphi \mathbf{A}_\alpha)$. Then by 5401(a) $\mathcal{V}_\varphi[[\lambda \mathbf{x}_\alpha \mathbf{D}_\beta]\mathbf{A}_\alpha] = \mathcal{V}_\psi \mathbf{D}_\beta$.

Axiom 4_1. $[\lambda \mathbf{x}_\alpha \mathbf{D}_\beta]\mathbf{A}_\alpha = \mathbf{D}_\beta$, where $\mathbf{D}_\beta$ is a primitive constant or variable distinct from $\mathbf{x}_\alpha$. $\mathcal{V}_\psi \mathbf{D}_\beta = \mathcal{V}_\varphi \mathbf{D}_\beta$ by 5400, so $\models_\varphi$[Axiom 4_1].

Axiom 4_2. $[\lambda \mathbf{x}_\alpha \mathbf{x}_\alpha]\mathbf{A}_\alpha = \mathbf{A}_\alpha$. Here $\mathbf{D}_\beta$ is $\mathbf{x}_\alpha$ and $\mathcal{V}_\psi \mathbf{D}_\beta = \psi \mathbf{x}_\alpha = \mathcal{V}_\varphi \mathbf{A}_\alpha$, so $\models$[Axiom 4_2].

Axiom 4_3. $[\lambda \mathbf{x}_\alpha \,\blacksquare\, \mathbf{B}_{\beta\gamma}\mathbf{C}_\gamma]\mathbf{A}_\alpha = [[\lambda \mathbf{x}_\alpha \mathbf{B}_{\beta\gamma}]\mathbf{A}_\alpha][[\lambda \mathbf{x}_\alpha \mathbf{C}_\gamma]\mathbf{A}_\alpha]$. $\mathbf{D}_\beta$ is $[\mathbf{B}_{\beta\gamma}\mathbf{C}_\gamma]$.

$$\begin{aligned}\mathcal{V}_\psi \mathbf{D}_\beta &= (\mathcal{V}_\psi \mathbf{B}_{\beta\gamma})(\mathcal{V}_\psi \mathbf{C}_\gamma) = (\mathcal{V}_\varphi[[\lambda \mathbf{x}_\alpha \mathbf{B}_{\beta\gamma}]\mathbf{A}_\alpha])(\mathcal{V}_\varphi[[\lambda \mathbf{x}_\alpha \mathbf{C}_\gamma]\mathbf{A}_\alpha]) \\ &= \mathcal{V}_\varphi[[[\lambda \mathbf{x}_\alpha \mathbf{B}_{\beta\gamma}]\mathbf{A}_\alpha][[\lambda \mathbf{x}_\alpha \mathbf{C}_\gamma]\mathbf{A}_\alpha]],\end{aligned}$$

so $\models_\varphi$[Axiom 4_3].

Axiom 4_4. $[\lambda \mathbf{x}_\alpha \,\blacksquare\, \lambda \mathbf{y}_\gamma \mathbf{B}_\delta]\mathbf{A}_\alpha = [\lambda \mathbf{y}_\gamma \,\blacksquare\, [\lambda \mathbf{x}_\alpha \mathbf{B}_\delta]\mathbf{A}_\alpha]$, where $\mathbf{y}_\gamma$ is distinct from $\mathbf{x}_\alpha$ and from all variables in $\mathbf{A}_\alpha$. Here $\mathbf{D}_\beta$ is $[\lambda \mathbf{y}_\gamma \mathbf{B}_\delta]$. Let y be an arbitrary member of $\mathcal{D}_\gamma$. Let $\varphi' = (\varphi : \mathbf{y}_\gamma/y)$. Then $\mathcal{V}_\varphi \mathbf{A}_\alpha = \mathcal{V}_{\varphi'} \mathbf{A}_\alpha$ by 5400, and $(\varphi' : \mathbf{x}_\alpha/\mathcal{V}_{\varphi'} \mathbf{A}_\alpha) = (\psi : \mathbf{y}_\gamma/y)$ since $\mathbf{x}_\alpha$ and $\mathbf{y}_\gamma$ are distinct. Hence $(\mathcal{V}_\psi \mathbf{D}_\beta)y = \mathcal{V}_{(\psi:\mathbf{y}_\gamma/y)}\mathbf{B}_\delta = \mathcal{V}_{\varphi'}[[\lambda \mathbf{x}_\alpha \mathbf{B}_\delta]\mathbf{A}_\alpha] = (\mathcal{V}_\varphi[\lambda \mathbf{y}_\gamma \,\blacksquare\, [\lambda \mathbf{x}_\alpha \mathbf{B}_\delta]\mathbf{A}_\alpha])y$, so the functions $\mathcal{V}_\psi \mathbf{D}_\beta$ and $\mathcal{V}_\varphi[\lambda \mathbf{y}_\gamma \,\blacksquare\, [\lambda \mathbf{x}_\alpha \mathbf{B}_\beta]\mathbf{A}_\alpha]$ are the same. Hence $\models_\varphi$[Axiom 4_4].

Axiom 4_5. $[\lambda \mathbf{x}_\alpha \,\blacksquare\, \lambda \mathbf{x}_\alpha \mathbf{B}_\delta]\mathbf{A}_\alpha = [\lambda \mathbf{x}_\alpha \mathbf{B}_\delta]$. Here $\mathbf{D}_\beta$ is $[\lambda \mathbf{x}_\alpha \mathbf{B}_\delta]$ and $\mathcal{V}_\psi \mathbf{D}_\beta = \mathcal{V}_\varphi \mathbf{D}_\beta$ by 5400 since $\mathbf{x}_\alpha$ is not free in $\mathbf{D}_\beta$, so $\models_\varphi$[Axiom 4_5].

Axiom 5. $\iota_{\imath(o\imath)}[\mathrm{Q}_{o\imath\imath}y] = y_\imath$. Clearly $\mathcal{V}_\varphi[\iota_{\imath(o\imath)} \centerdot \mathrm{Q}_{o\imath\imath}y_\imath] = (\mathcal{V}\iota_{\imath(o\imath)})((\mathcal{V}\mathrm{Q}_{o\imath\imath})\mathcal{V}_\varphi y_\imath) = \mathcal{V}_\varphi y_\imath$ by the definitions of $\mathcal{V}\iota_{\imath(o\imath)}$ and $\mathcal{V}\mathrm{Q}_{o\imath\imath}$, so $\models_\varphi$ [Axiom 5].

It is now clear that every wff which occurs in a proof is valid in the general sense, so (a) is proved.

To prove (b), suppose $\mathcal{G} \vdash \mathbf{A}$. Then there is a finite subset $\{\mathbf{H}^1, \ldots, \mathbf{H}^n\}$ of $\mathcal{G}$ such that $\mathbf{H}^1, \ldots, \mathbf{H}^n \vdash \mathbf{A}$, so by the Deduction Theorem $\vdash \mathbf{H}^1 \supset \centerdot \ldots \supset \centerdot \mathbf{H}^n \supset \mathbf{A}$, so by (a) $\mathcal{M} \models \mathbf{H}^1 \supset \centerdot \ldots \supset \centerdot \mathbf{H}^n \supset \mathbf{A}$. Also $\mathcal{M} \models \mathbf{H}^i$ for $1 \leq i \leq n$ since $\mathcal{M}$ is a model for $\mathcal{G}$, so $\mathcal{M} \models \mathbf{A}$.

REMARK. We can now see that $\mathcal{Q}_0$ is a conservative extension of the system $\mathcal{F}$ of first-order logic, in the sense that if $\mathbf{C}$ is any wff of $\mathcal{F}$ and $\mathbf{C}'$ is the result of translating $\mathbf{C}$ into the notation of $\mathcal{Q}_0$, then $\vdash_\mathcal{F} \mathbf{C}$ if and only if $\vdash_{\mathcal{Q}_0} \mathbf{C}'$. The proof in one direction follows directly from Rule Q (5236). For the other direction, suppose $\vdash_{\mathcal{Q}_0} \mathbf{C}'$. Then $\models \mathbf{C}'$ by 5402. It is not hard to see that every model of $\mathcal{F}$ can be extended to a model of $\mathcal{Q}_0$, and hence that $\mathbf{C}$ is valid in the sense of first-order logic. Therefore $\vdash_\mathcal{F} \mathbf{C}$ by Gödel's Completeness Theorem (2510). Similar considerations show that $\mathcal{F}^\omega$ is a conservative extension of $\mathcal{F}$. In summary, every first-order theorem of type theory has a first-order proof.

However, some theorems of first-order logic can be proved most efficiently by using concepts which can be expressed only in higher-order logic. Examples may be found in [Andrews and Bishop, 1996] and [Boolos, 1998, Chapter 25]. Statman proved [Statman, 1978, Proposition 6.3.5] that the minimal length of a proof in first-order logic of a wff of first-order logic may be extraordinarily longer than that the minimal length of a proof of the same wff in second-order logic. A related result by Gödel [Gödel, 1936] is that in general "passing to the logic of the next higher order has the effect, not only of making provable certain propositions that were not provable before, but also of making it possible to shorten, by an extraordinary amount, infinitely many of the proofs already available". A complete proof of this may be found in [Buss, 1994].

DEFINITION. A set $\mathcal{G}$ of wffs$_o$ is *inconsistent* iff $\mathcal{G} \vdash F_o$. $\mathcal{Q}_0$ is inconsistent iff $\vdash F_o$. As usual, "consistent" means "not inconsistent". A wff$_o$ $\mathbf{B}$ is *consistent with* $\mathcal{G}$ iff $\mathcal{G} \cup \{\mathbf{B}\}$ is consistent. Note that by $\forall$I (5215) and the definition of F_o, $\mathcal{G}$ is inconsistent iff for every wff$_o$ $\mathbf{A}$, $\mathcal{G} \vdash \mathbf{A}$.

5403 Consistency Theorem $\mathcal{Q}_0$ is consistent, and every set of wffs$_o$ which has a model is consistent.

Proof: If $\mathcal{G}$ is a set of wffs$_o$ which has a model $\mathcal{M}$ but $\mathcal{G}$ is inconsistent, then $\mathcal{G} \vdash F_o$ so $\mathcal{M} \models F_o$ by 5402, contradicting 5401(d). Hence a set of wffs which has a model must be consistent.

To show that $\mathcal{Q}_0$ is consistent, i.e., that not $\vdash F_o$, we must simply note that there is a model $\mathcal{M}$ for the empty set of wffs, i.e., models exist. Let d be any object, and let $\mathcal{M}$ be the standard model with $\mathcal{D}_\imath = \{d\}$. (Each of the domains $\mathcal{D}_\alpha$ is finite, and can be described very explicitly.) Note that $\mathcal{V}^{\mathcal{M}}\iota_{\imath(o\imath)}$ must be the constant function whose value is d for all arguments. ■

DEFINITION. A *countable model* is a model $\langle\{\mathcal{D}_\alpha\}_\alpha, \mathcal{J}\rangle$ in which each of the domains $\mathcal{D}_\alpha$ is countable (finite or denumerable). A *finite model* is a model in which each of the domains $\mathcal{D}_\alpha$ is finite.

Note that a model is finite iff $\mathcal{D}_\imath$ is finite.

5404 Theorem. Every finite model for $\mathcal{Q}_0$ is standard.

Proof:[5] Let $\mathcal{M} = \langle\{\mathcal{D}_\alpha\}_\alpha, \mathcal{J}\rangle$ be any finite model for $\mathcal{Q}_0$. For arbitrary type symbols α and β we must show that $\mathcal{D}_{\alpha\beta}$ contains all functions from $\mathcal{D}_\beta$ to $\mathcal{D}_\alpha$. So let g be any such function. Let $\mathcal{D}_\beta = \{m^1, \ldots, m^k\}$. Let φ be an assignment with values on the variables $w^1_\beta, \ldots, w^k_\beta$, $z^1_\alpha, \ldots, z^k_\alpha$ as follows: $\varphi w^1_\beta = m^1, \ldots, \varphi w^k_\beta = m^k$, $\varphi z^1_\alpha = gm^1, \ldots, \varphi z^k_\alpha = gm^k$. Let A be the wff

$$[[x_\beta = w^1_\beta \;\wedge\; y_\alpha = z^1_\alpha] \vee \ldots \vee [x_\beta = w^k_\beta \;\wedge\; y_\alpha = z^k_\alpha]].$$

We shall show that $g = \mathcal{V}_\varphi[\lambda x_\beta \,\blacksquare\, \imath y_\alpha A]$, which is in $\mathcal{D}_{\alpha\beta}$ since $\mathcal{M}$ is a general model.

We first prove a theorem of $\mathcal{Q}_0$ which will be useful. Let $1 \le j \le k$, and let P be the wff

$$[[x_\beta = w^j_\beta] \;\wedge \bigwedge_{i \in \{1,\ldots,k\}-\{j\}} [x_\beta \neq w^i_\beta]]$$

(.1)	$.1 \vdash P$	hyp
(.2)	$.1 \vdash A = \blacksquare\, z^j_\alpha = y_\alpha$	Rule P: .1, 5302
(.3)	$.1 \vdash \imath y_\alpha A = \iota_{\alpha(o\alpha)} \lambda y_\alpha A$	5200, def of $\imath y_\alpha$
(.4)	$.1 \vdash \imath y_\alpha A = \iota_{\alpha(o\alpha)} \lambda y_\alpha \,\blacksquare\, \mathsf{Q}_{o\alpha\alpha} z^j_\alpha y_\alpha$ (Note that y_α is not free in the hypothesis P.)	R′ : .2, .3, def of =
(.5)	$.1 \vdash \imath y_\alpha A = \iota_{\alpha(o\alpha)} \,\blacksquare\, \mathsf{Q}_{o\alpha\alpha} z^j_\alpha$	R′: 5205, .4

[5]The idea for this proof comes from [Henkin, 1963b].

(.6) $.1 \vdash \imath y_\alpha A = z^j_\alpha$ R′: 5309, .5
(.7) $\vdash P \supset \centerdot \imath y_\alpha A = z^j_\alpha$ 5240: .6

Now let m^j be an arbitrary element of $\mathcal{D}_\beta$, and let $\psi = (\varphi : x_\beta / m^j)$. Note that $\mathcal{M} \models_\psi P$, so $\mathcal{V}_\psi[\imath y_\alpha A] = \mathcal{V}_\psi z^j_\alpha$ by .7, 5402, and 5401. Therefore

$$(\mathcal{V}_\varphi[\lambda x_\beta \centerdot \imath y_\alpha A])m^j = \mathcal{V}_\psi[\imath y_\alpha A] = \mathcal{V}_\psi z^j_\alpha = \varphi z^j_\alpha = gm^j.$$

Thus $\mathcal{V}_\varphi[\lambda x_\beta \centerdot \imath y_\alpha A]$ and g have the same values for all arguments in $\mathcal{D}_\beta$, so they are the same function. ∎

In 5504-5506 we shall see that many infinite models are nonstandard.

REMARK. It is easy to see that all instances of the Axiom Schema of Choice are true in every finite model, since the elements of each type can be enumerated, and we can define (in the meta-language) a function which chooses from each nonempty set the first member under this enumeration. Since (by 5404) the model is standard, this function will actually be in the model, so the Axiom of Choice will be true in the model. Note that the proof of the consistency of $\mathcal{Q}_0$ (5403) given above would still work if we added to $\mathcal{Q}_0$ the Axiom Schema of Choice. Of course, this result is rather trivial. Significant consistency questions arise only when an Axiom of Infinity is added to $\mathcal{Q}_0$. Such an axiom will be introduced in Chapter 6.

DEFINITION. Let $\mathcal{N} = \langle \{\mathcal{E}_\alpha\}_\alpha, \mathcal{J} \rangle$ and $\mathcal{M} = \langle \{\mathcal{D}_\alpha\})_\alpha, \mathcal{I} \rangle$ be general models of $\mathcal{Q}_0$. A mapping τ is an *isomorphism* from $\mathcal{N}$ to $\mathcal{M}$ iff for all type symbols α and β,

(1) τ maps $\mathcal{E}_\alpha$ onto $\mathcal{D}_\alpha$;
(2) $\tau\mathsf{T} = \mathsf{T}$ and $\tau\mathsf{F} = \mathsf{F}$;
(3) For all $f \in \mathcal{E}_{\alpha\beta}$ and $x \in \mathcal{E}_\beta$, $(\tau f)(\tau x) = \tau(fx)$ (see Figure 5.1);
(4) For each primitive constant $\mathbf{c}_\alpha$ of $\mathcal{Q}_0$, $\tau\mathcal{J}\mathbf{c}_\alpha = \mathcal{I}\mathbf{c}_\alpha$.

$\mathcal{N}$ and $\mathcal{M}$ are *isomorphic* iff there is an isomorphism from $\mathcal{N}$ to $\mathcal{M}$.

5405 Proposition. An isomorphism of general models is a one-to-one mapping.

Proof: We use the notation in the definition above. Suppose τ is an isomorphism, and x and y are distinct elements of $\mathcal{E}_\alpha$. We shall show that $\tau x \neq \tau y$.

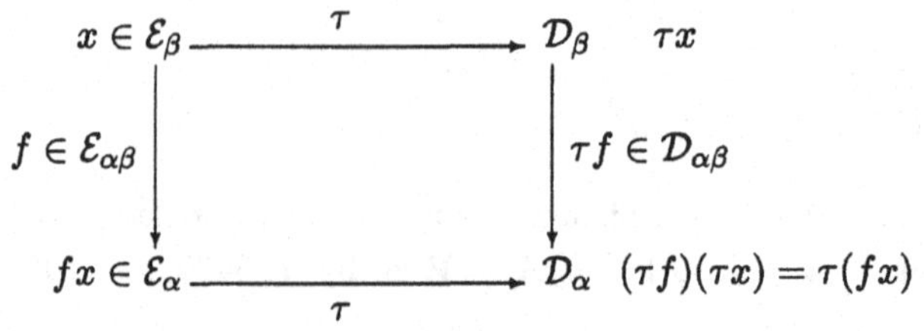

Figure 5.1: Isomorphism condition (3)

Let q be the identity relation on $\mathcal{E}_\alpha$. $\tau q = \tau(\mathcal{J}\mathsf{Q}_{o\alpha\alpha}) = \mathcal{I}\mathsf{Q}_{o\alpha\alpha}$, which is the identity relation on $\mathcal{D}_\alpha$, and

$$
\begin{aligned}
((\tau q)(\tau x))(\tau y) &= (\tau(qx))(\tau y) && \text{(by 3)} \\
&= \tau(qxy) && \text{(by 3)} \\
&= \tau\mathsf{F} && \text{(since } x \neq y\text{)} \\
&= \mathsf{F} && \text{(by 2)}
\end{aligned}
$$

so $\tau x \neq \tau y$. ■

It is readily verified that the relation of being isomorphic is an equivalence relation (X5404).

5406 Proposition. Let $\mathcal{N}$ and $\mathcal{M}$ be general models for $\mathcal{Q}_0$, and let τ be an isomorphism from $\mathcal{N}$ to $\mathcal{M}$. For each assignment φ into $\mathcal{N}$, let $\tau \circ \varphi$ be the assignment into $\mathcal{M}$ such that for each variable $\mathbf{x}_\gamma$, $(\tau \circ \varphi)\mathbf{x}_\gamma = \tau(\varphi \mathbf{x}_\gamma)$. Then for every wff $\mathbf{C}_\gamma$ of $\mathcal{Q}_0$, $\tau(\mathcal{V}_\varphi^{\mathcal{N}}\mathbf{C}_\gamma) = \mathcal{V}_{\tau\circ\varphi}^{\mathcal{M}}\mathbf{C}_\gamma$. (See Figure 5.2.)

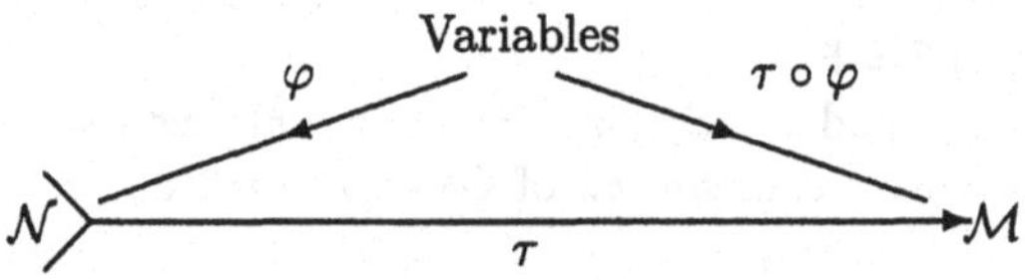

Figure 5.2: Assignments into $\mathcal{N}$ and $\mathcal{M}$

Proof: Suppose $\mathcal{N} = \langle\{\mathcal{E}_\alpha\}_\alpha, \mathcal{J}\rangle$ and $\mathcal{M} = \langle\{\mathcal{D}_\alpha\}_\alpha, \mathcal{I}\rangle$. We prove the proposition for all assignments φ by induction on the construction of $\mathbf{C}_\gamma$.

(a) If $\mathbf{C}_\gamma$ is a variable $\mathbf{x}_\gamma$, then $\tau(\mathcal{V}_\varphi^{\mathcal{N}}\mathbf{C}_\gamma) = \tau(\varphi\mathbf{x}_\gamma) = (\tau\circ\varphi)\mathbf{x}_\gamma = \mathcal{V}_{\tau\circ\varphi}^{\mathcal{M}}\mathbf{C}_\gamma$.

(b) If $\mathbf{C}_\gamma$ is a primitive constant of $\mathcal{Q}_0$, then $\tau(\mathcal{V}_\varphi^{\mathcal{N}}\mathbf{C}_\gamma) = \tau(\mathcal{J}\mathbf{C}_\gamma) = \mathcal{I}\mathbf{C}_\gamma = \mathcal{V}_{\tau\circ\varphi}^{\mathcal{M}}\mathbf{C}_\gamma$.

(c) If $\mathbf{C}_\gamma = [\mathbf{A}_{\gamma\beta}\mathbf{B}_\beta]$, then $\tau(\mathcal{V}^{\mathcal{N}}_\varphi \mathbf{C}_\gamma) = \tau((\mathcal{V}^{\mathcal{N}}_\varphi \mathbf{A}_{\gamma\beta})(\mathcal{V}^{\mathcal{N}}_\varphi \mathbf{B}_\beta))$
$= (\tau\mathcal{V}^{\mathcal{N}}_\varphi \mathbf{A}_{\gamma\beta})\ (\tau\mathcal{V}^{\mathcal{N}}_\varphi \mathbf{B}_\beta)$
$= (\mathcal{V}^{\mathcal{M}}_{\tau\circ\varphi} \mathbf{A}_{\gamma\beta})(\mathcal{V}^{\mathcal{M}}_{\tau\circ\varphi} \mathbf{B}_\beta)$ (by inductive hypothesis)
$= \mathcal{V}^{\mathcal{M}}_{\tau\circ\varphi}[\mathbf{A}_{\gamma\beta}\mathbf{B}_\beta]$.

(d) Suppose $\mathbf{C}_\gamma = [\lambda \mathbf{x}_\beta \mathbf{A}_\alpha]$. Let z be an arbitrary member of $\mathcal{D}_\beta$. Let $w = \tau^{-1}z$, so $w \in \mathcal{E}_\beta$ and $\tau w = z$. It is readily checked that $\tau \circ (\varphi : \mathbf{x}_\beta / w) = (\tau \circ \varphi : \mathbf{x}_\beta / z)$.

$$\begin{aligned}(\tau\mathcal{V}^{\mathcal{N}}_\varphi[\lambda \mathbf{x}_\beta \mathbf{A}_\alpha])z &= \tau((\mathcal{V}^{\mathcal{N}}_\varphi[\lambda \mathbf{x}_\beta \mathbf{A}_\alpha])w) = \tau(\mathcal{V}^{\mathcal{N}}_{(\varphi:\mathbf{x}_\beta/w)}\mathbf{A}_\alpha) \\ &= \mathcal{V}^{\mathcal{M}}_{\tau\circ(\varphi:\mathbf{x}_\beta/w)}\mathbf{A}_\alpha \text{ (by inductive hypothesis)} \\ &= \mathcal{V}^{\mathcal{M}}_{(\tau\circ\varphi:\mathbf{x}_\beta/z)}\mathbf{A}_\alpha = (\mathcal{V}^{\mathcal{M}}_{\tau\circ\varphi}[\lambda \mathbf{x}_\beta \mathbf{A}_\alpha])z.\end{aligned}$$

Since this is true for each $z \in \mathcal{D}_\beta$, $\tau\mathcal{V}^{\mathcal{N}}_\varphi[\lambda \mathbf{x}_\beta \mathbf{A}_\alpha] = \mathcal{V}^{\mathcal{M}}_{\tau\circ\varphi}[\lambda \mathbf{x}_\beta \mathbf{A}_\alpha]$. ■

5407 Corollary. If $\mathcal{N}$ and $\mathcal{M}$ are isomorphic general models of $\mathcal{Q}_0$, then $\mathcal{M} \models \mathbf{C}$ iff $\mathcal{N} \models \mathbf{C}$.

Proof: Suppose $\mathcal{M} \models$ C, and τ is an isomorphism from $\mathcal{N}$ to $\mathcal{M}$. Then for any assignment φ into $\mathcal{N}$, since τ is the identity function on $\mathcal{D}_o$, $\mathcal{V}^{\mathcal{N}}_\varphi \mathbf{C} = \tau(\mathcal{V}^{\mathcal{N}}_\varphi \mathbf{C}) = \mathcal{V}^{\mathcal{M}}_{\tau\circ\varphi}\mathbf{C} = \mathsf{T}$, so $\mathcal{N} \models \mathbf{C}$.

Similarly, if $\mathcal{N} \models \mathbf{C}$, then $\mathcal{M} \models \mathbf{C}$. ■

EXERCISES

X5400. Prove that $\models$ 5207 without using 5402.

X5401. Show that not every wff of the form $[\mathbf{A}_\alpha = \mathbf{B}_\alpha] \supset \centerdot\, [\lambda \mathbf{z}_\beta \mathbf{A}_\alpha] = [\lambda \mathbf{z}_\beta \mathbf{B}_\alpha]$ is valid.

X5402. Show that every wff of the form $[\mathbf{A}_\alpha = \mathbf{B}_\alpha] \supset \centerdot\, [\lambda \mathbf{z}_\beta \mathbf{A}_\alpha] = [\lambda \mathbf{z}_\beta \mathbf{B}_\alpha]$ is satisfiable in every general model.

X5403. Show that if $\mathcal{M}$ is any general model, φ is any assignment into $\mathcal{M}$, and $\mathbf{A}$ is any wff$_o$, then $\mathcal{M} \models_\varphi \exists \mathbf{x}_\alpha \mathbf{A}$ iff there is an assignment ψ which agrees with φ off $\mathbf{x}_\alpha$ such that $\mathcal{M} \models_\psi \mathbf{A}$.

X5404. Show that the relation of being isomorphic is an equivalence relation between models of $\mathcal{Q}_0$.

X5405. Prove that any theorem of $\mathcal{Q}_0$ in β-normal form has the form $[\mathbf{A}_\alpha = \mathbf{B}_\alpha]$. (Therefore, by Exercise X5213, every theorem of $\mathcal{Q}_0$ can be transformed to the form $[\mathbf{A}_\alpha = \mathbf{B}_\alpha]$ by a sequence of β-contractions and α-conversions.)

X5406. Let $\mathcal{B}$ be the system which is obtained from $\mathcal{Q}_0$ when Axiom Schema 2^α is replaced by the schema $[x_\alpha = y_\alpha] = \forall h_{o\alpha}[h_{o\alpha}x_\alpha = h_{o\alpha}y_\alpha]$. Is 2^α a theorem schema of $\mathcal{B}$?

§55. Completeness of $\mathcal{Q}_0$

Formulations of $\mathcal{Q}_0$ differ from one another in having different primitive constants. We denote by $\mathcal{L}(\mathcal{Q}_0)$ the *language* of $\mathcal{Q}_0$, i.e., the set of wffs of $\mathcal{Q}_0$. If $\mathcal{Q}_0^1$ and $\mathcal{Q}_0^2$ are formulations of $\mathcal{Q}_0$ such that every primitive symbol of $\mathcal{Q}_0^1$ is a primitive symbol of $\mathcal{Q}_0^2$, we say that $\mathcal{Q}_0^2$ is an *expansion* of $\mathcal{Q}_0^1$. Clearly, $\mathcal{Q}_0^2$ is an expansion of $\mathcal{Q}_0^1$ iff $\mathcal{L}(\mathcal{Q}_0^1) \subseteq \mathcal{L}(\mathcal{Q}_0^2)$.

We denote the cardinality of any set $\mathcal{S}$ by card$(\mathcal{S})$. Since every formulation of $\mathcal{Q}_0$ contains the denumerable set of logical constants, and every wff is a finite sequence of primitive symbols, it is easy to see that card$(\mathcal{L}(\mathcal{Q}_0))$ is the cardinality of the set of primitive constants of $\mathcal{Q}_0$. For most purposes we shall deal with formulations of $\mathcal{Q}_0$ in which the set of nonlogical primitive constants is countable (denumerably infinite or finite, and possibly empty), so that card$(\mathcal{L}(\mathcal{Q}_0)) = \aleph_o$, and we shall call such formulations of $\mathcal{Q}_0$ *countable*. Indeed, unless otherwise indicated, the reader may usually assume that we are dealing with a formulation of $\mathcal{Q}_0$ which has no nonlogical constants. However, we shall impose no actual restrictions on the cardinality of the set of nonlogical constants.

DEFINITIONS: Let $\mathcal{H}$ be a set of sentences of $\mathcal{Q}_0$.

(1) $\mathcal{H}$ is *complete* in $\mathcal{Q}_0$ iff for every sentence $\mathbf{A}$ of $\mathcal{Q}_0$, either $\mathcal{H} \vdash \mathbf{A}$ or $\mathcal{H} \vdash \sim \mathbf{A}$.

(2) $\mathcal{H}$ is *extensionally complete* in $\mathcal{Q}_0$ iff for every sentence of $\mathcal{Q}_0$ of the form $[\mathbf{A}_{\alpha\beta} = \mathbf{B}_{\alpha\beta}]$, there is a cwff $\mathbf{C}_\beta$ of $\mathcal{Q}_0$ such that

$$\mathcal{H} \vdash [\mathbf{A}_{\alpha\beta}\mathbf{C}_\beta = \mathbf{B}_{\alpha\beta}\mathbf{C}_\beta] \supset \centerdot\, \mathbf{A}_{\alpha\beta} = \mathbf{B}_{\alpha\beta}. \qquad (*)$$

REMARK. The wff (*) is equivalent to $\mathbf{A}_{\alpha\beta} \neq \mathbf{B}_{\alpha\beta} \supset \centerdot\, \mathbf{A}_{\alpha\beta}\mathbf{C}_\beta \neq \mathbf{B}_{\alpha\beta}\mathbf{C}_\beta$, which says that if $\mathbf{A}_{\alpha\beta}$ and $\mathbf{B}_{\alpha\beta}$ are different functions, then $\mathbf{C}_\beta$ is an argument on which they differ. Note that a special case of the wff schema (*) is $[\lambda x_\beta T]\mathbf{C}_\beta = [\lambda \mathbf{x}_\beta \mathbf{D}]\mathbf{C}_\beta \supset \centerdot\, [\lambda x_\beta T] = [\lambda \mathbf{x}_\beta \mathbf{D}]$, which is equivalent to $[\lambda \mathbf{x}_\beta \mathbf{D}]\mathbf{C}_\beta \supset \forall \mathbf{x}_\beta \mathbf{D}$. If $\mathbf{D}$ is $\sim \mathbf{E}$. This is equivalent to $\exists \mathbf{x}_\beta \mathbf{E} \supset [\lambda \mathbf{x}_\beta \mathbf{E}]\mathbf{C}_\beta$.

REMARK. If from $\mathcal{H}$ one can derive the Axiom of Choice (for each type symbol β) in a form such as $\forall p_{o\beta} \centerdot\, \exists x_\beta p_{o\beta} x_\beta \supset p_{o\beta} \centerdot\, \mathbf{C}_{\beta(o\beta)} p_{o\beta}$ (for some cwff $\mathbf{C}_{\beta(o\beta)}$), then $\mathcal{H}$ is extensionally complete, since

$$\mathcal{H} \vdash \forall f_{\alpha\beta} \forall g_{\alpha\beta} \centerdot f_{\alpha\beta}[\mathbf{C}_{\beta(o\beta)} \centerdot \lambda \mathbf{x}_\beta \centerdot f_{\alpha\beta} x_\beta \neq g_{\alpha\beta} x_\beta]$$

$$= g_{\alpha\beta}[\mathbf{C}_{\beta(o\beta)} \centerdot \lambda x_\beta \centerdot f_{\alpha\beta} x_\beta \neq g_{\alpha\beta} x_\beta] \supset f_{\alpha\beta} = g_{\alpha\beta}.$$

(See Exercise X5500 below.)

REMARK. Some examples of complete sets of sentences are $\{\mathbf{A} \mid \mathcal{M} \models \mathbf{A}$ and $\mathbf{A}$ is a sentence$\}$ for any particular model $\mathcal{M}$, $\{\forall x_\imath \forall y_\imath \centerdot x_\imath = y_\imath\}$ (see Exercise X8032), and $\{F_o\}$.

5500 Extension Lemma. Let $\mathcal{G}$ be any consistent set of sentences of $\mathcal{Q}_0$. Then there is an expansion $\mathcal{Q}_0^+$ of $\mathcal{Q}_0$ and a set of $\mathcal{H}$ of sentences of $\mathcal{Q}_0^+$ such that

(1) $\mathcal{G} \subseteq \mathcal{H}$.
(2) $\mathcal{H}$ is consistent.
(3) $\mathcal{H}$ is complete in $\mathcal{Q}_0^+$.
(4) $\mathcal{H}$ is extensionally complete in $\mathcal{Q}_0^+$.
(5) $\text{card}(\mathcal{L}(\mathcal{Q}_0^+)) = \text{card}(\mathcal{L}(\mathcal{Q}))$.

Proof: Let $K = \text{card}(\mathcal{L}(\mathcal{Q}_0))$ and for each type symbol α let $\mathcal{C}_\alpha$ be a well-ordered set of cardinality K of new primitive constants of type α, and let $\mathcal{C} = \bigcup_\alpha \mathcal{C}_\alpha$ (the union being taken over all type symbols α). Let $\mathcal{Q}_0^+$ be the expansion of $\mathcal{Q}_0$ obtained by adding the constants in $\mathcal{C}$ to the primitive symbols of $\mathcal{Q}_0$. Clearly $\text{card}(\mathcal{L}(\mathcal{Q}_0^+)) = K$.

Let κ be the initial ordinal of cardinality K. Well-order the sentences of $\mathcal{Q}_0^+$, and for each ordinal $\tau < \kappa$, let $\mathbf{S}^\tau$ be the τth sentence of $\mathcal{Q}_0^+$ under this well-ordering.[6]

By transfinite induction, for each ordinal $\tau \leq \kappa$ we define a set $\mathcal{G}_\tau$ of sentences of $\mathcal{Q}_0^+$, so that $\mathcal{G}_\sigma \subseteq \mathcal{G}_\tau$ whenever $\sigma \leq \tau$, and there is a finite cardinal n such that at most $n + \text{card}(\tau)$ constants from $\mathcal{C}$ occur in the sentences of $\mathcal{G}_\tau$. $\mathcal{G}_0 = \mathcal{G}$. If τ is a limit ordinal, $\mathcal{G}_\tau = \bigcup_{\sigma<\tau} \mathcal{G}_\sigma$. The definition for successor ordinals $\tau + 1$ is by cases:

[6]Very explicit well-orderings of sentences can easily be specified for countable formulations of $\mathcal{Q}_0$. The method of assigning numbers to wffs given in §70 can easily be modified to apply to any formulation of $\mathcal{Q}_0$ with countably many parameters. The natural ordering of the numbers assigned to the sentences induces an ordering (which is a well-ordering) of the sentences.

(a) If $\mathbf{S}^\tau$ is consistent with $\mathcal{G}_\tau$, then $\mathcal{G}_{\tau+1} = \mathcal{G}_\tau \cup \{\mathbf{S}^\tau\}$.
(b) If $\mathbf{S}^\tau$ is not consistent with $\mathcal{G}_\tau$, and $\mathbf{S}^\tau$ is not of the form $[\mathbf{A}_{\alpha\beta} = \mathbf{B}_{\alpha\beta}]$, then $\mathcal{G}_{\tau+1} = \mathcal{G}_\tau$.
(c) If $\mathbf{S}^\tau$ is not consistent with $\mathcal{G}_\tau$ and $\mathbf{S}^\tau$ is of the form $[\mathbf{A}_{\alpha\beta} = \mathbf{B}_{\alpha\beta}]$, then $\mathcal{G}_{\tau+1} = \mathcal{G}_\tau \cup \{\sim \mathbin{\text{■}} \mathbf{A}_{\alpha\beta}\mathbf{c}_\beta = \mathbf{B}_{\alpha\beta}\mathbf{c}_\beta\}$, where $\mathbf{c}_\beta$ is the first constant in $\mathcal{C}_\beta$ which does not occur in $\mathcal{G}_\tau$ or $\mathbf{S}^\tau$. (Note that there must be such a constant.)

We next prove by induction on τ that each $\mathcal{G}_\tau$ is consistent in $\mathcal{Q}_0^+$.

To prove that $\mathcal{G}_0$ is consistent in $\mathcal{Q}_0^+$, suppose $\mathcal{P}$ is a proof of F_o in $\mathcal{Q}_0^+$ from some finite subset $\mathcal{S}$ of $\mathcal{G}_0$. Let θ be a substitution which replaces the constants from $\mathcal{C}$ which occur in $\mathcal{P}$ by distinct variables of the same type which do not occur in $\mathcal{P}$. It is easily seen that $\theta\mathcal{P}$ is a proof in $\mathcal{Q}_0$ of F_o from $\mathcal{S}$, since no constant of $\mathcal{C}$ occurs in $\mathcal{G}_0$. This contradicts the consistency of $\mathcal{G}_0$ in $\mathcal{Q}_0$.

Next, we suppose $\mathcal{G}_\tau$ is consistent and prove $\mathcal{G}_{\tau+1}$ consistent. In case (a) or (b) this is trivial. Suppose $\mathcal{G}_{\tau+1}$ is obtained by case (c) and is inconsistent. Thus $\mathcal{G}_{\tau+1} = \mathcal{G}_\tau \cup \{\sim \mathbin{\text{■}} \mathbf{A}_{\alpha\beta}\mathbf{c}_\beta = \mathbf{B}_{\alpha\beta}\mathbf{c}_\beta\}$ so by the Deduction Theorem and Rule P, $\mathcal{G}_\tau \vdash \mathbf{A}_{\alpha\beta}\mathbf{c}_\beta = \mathbf{B}_{\alpha\beta}\mathbf{c}_\beta$. Let $\mathcal{P}$ be a proof of $\mathbf{A}_{\alpha\beta}\mathbf{c}_\beta = \mathbf{B}_{\alpha\beta}\mathbf{c}_\beta$ from a finite subset $\mathcal{S}$ of $\mathcal{G}_\tau$, and let $\mathbf{x}_\beta$ be a variable of type β which does not occur in $\mathcal{P}$ or $\mathcal{S}$. Since $\mathbf{c}_\beta$ does not occur in $\mathcal{G}_\tau$, $\mathbf{A}_{\alpha\beta}$, or $\mathbf{B}_{\alpha\beta}$, it is easily seen that $\mathsf{S}^{\mathbf{c}_\beta}_{\mathbf{x}_\beta}\mathcal{P}$ (the result of substituting $\mathbf{x}_\beta$ for $\mathbf{c}_\beta$ throughout $\mathcal{P}$) is a proof from $\mathcal{S}$ of $\mathbf{A}_{\alpha\beta}\mathbf{x}_\beta = \mathbf{B}_{\alpha\beta}\mathbf{x}_\beta$. Therefore

$\mathcal{S} \vdash \mathbf{A}_{\alpha\beta}\mathbf{x}_\beta = \mathbf{B}_{\alpha\beta}\mathbf{x}_\beta$
$\mathcal{S} \vdash \forall \mathbf{x}_\beta \mathbin{\text{■}} \mathbf{A}_{\alpha\beta}\mathbf{x}_\beta = \mathbf{B}_{\alpha\beta}\mathbf{x}_\beta$ by Gen, since $\mathbf{x}_\beta$ does not occur in $\mathcal{S}$
$\mathcal{S} \vdash \mathbf{A}_{\alpha\beta} = \mathbf{B}_{\alpha\beta}$ by RR and Axiom 3

Thus $\mathcal{G}_\tau \vdash \mathbf{A}_{\alpha\beta} = \mathbf{B}_{\alpha\beta}$. But by the definition of $\mathcal{G}_{\tau+1}$ in case (c), the wff $[\mathbf{A}_{\alpha\beta} = \mathbf{B}_{\alpha\beta}]$ is inconsistent with $\mathcal{G}_\tau$, so $\mathcal{G}_\tau$, $[\mathbf{A}_{\alpha\beta} = \mathbf{B}_{\alpha\beta}] \vdash F$, so $\mathcal{G}_\tau$ is inconsistent. This contradiction shows that $\mathcal{G}_{\tau+1}$ must be consistent when $\mathcal{G}_\tau$ is.

Finally, if τ is a limit ordinal and $\mathcal{G}_\sigma$ is consistent for each $\sigma < \tau$, then $\mathcal{G}_\tau$ must be consistent, since each finite subset of $\mathcal{G}_\tau$ is a subset of some $\mathcal{G}_\sigma$ with $\sigma < \tau$.

Let $\mathcal{H} = \mathcal{G}_\kappa$. Clearly $\mathcal{H}$ is a consistent set of sentences of $\mathcal{Q}_0^+$, and $\mathcal{G} \subseteq \mathcal{H}$.

To see that $\mathcal{H}$ is complete and extensionally complete, let $\mathbf{S}$ be any sentence of $\mathcal{Q}_0^+$. Then there exists $\tau < \kappa$ such that $\mathbf{S}$ is $\mathbf{S}^\tau$. If $\mathbf{S}$ is consistent with $\mathcal{G}_\tau$ then by case (a) $\mathbf{S} \in \mathcal{G}_{\tau+1} \subseteq \mathcal{H}$ so $\mathcal{H} \vdash \mathbf{S}$. Otherwise $\mathcal{G}_\tau, \mathbf{S} \vdash F$ so $\mathcal{G}_\tau \vdash \sim \mathbf{S}$ and $\mathcal{H} \vdash \sim \mathbf{S}$. Thus $\mathcal{H}$ is complete. Moreover, suppose $\mathbf{S}$ has the form $\mathbf{A}_{\alpha\beta} = \mathbf{B}_{\alpha\beta}$; in the case where $\mathbf{S}$ is consistent with $\mathcal{G}_\tau$, by Rule P $\mathcal{H} \vdash [\mathbf{A}_{\alpha\beta}\mathbf{C}_\beta = \mathbf{B}_{\alpha\beta}\mathbf{C}_\beta] \supset \mathbf{S}$ for every wff $\mathbf{C}_\beta$; in the case where $\mathbf{S}$ is inconsistent with $\mathcal{G}_\tau$, by (c) there is a $\mathbf{c}_\beta \in \mathcal{C}$ such that $\mathcal{H} \vdash \sim \centerdot\, \mathbf{A}_{\alpha\beta}\mathbf{c}_\beta = \mathbf{B}_{\alpha\beta}\mathbf{c}_\beta$, so by Rule P, $\mathcal{H} \vdash \mathbf{A}_{\alpha\beta}\mathbf{c}_\beta = \mathbf{B}_{\alpha\beta}\mathbf{c}_\beta \supset \centerdot\, \mathbf{A}_{\alpha\beta} = \mathbf{B}_{\alpha\beta}$. ■

We shall say that a model $\langle\{\mathcal{D}_\alpha\}_\alpha,\ \mathcal{J}\rangle$ for $\mathcal{Q}_0$ is *frugal* iff for each type symbol α, $\text{card}(\mathcal{D}_\alpha) \leq \text{card}(\mathcal{L}(\mathcal{Q}_0))$. Thus, in the usual case where $\mathcal{Q}_0$ is countable, a *frugal* model is a *countable* model. Note that by the same reasoning as was used in §25 when the concept of frugality was introduced, if one has an uncountable formulation of $\mathcal{Q}_0$, one cannot expect that every consistent set of sentences will have a countable model.

5501 Henkin's Theorem. Every consistent set of sentences of $\mathcal{Q}_0$ has a frugal general model.

Proof: Let $\mathcal{G}$ be a consistent set of sentences of $\mathcal{Q}_0$, and let $\mathcal{H}$ and $\mathcal{Q}_0^+$ be as described in Lemma 5500.

REMARK. If $\mathbf{A}_\alpha$ and $\mathbf{B}_\alpha$ are cwffs of $\mathcal{Q}_0^+$, we shall say that $\mathbf{A}_\alpha$ is *equivalent* to $\mathbf{B}_\alpha$ iff $\mathcal{H} \vdash \mathbf{A}_\alpha = \mathbf{B}_\alpha$. Clearly this is an equivalence relation. In essence, we shall take the equivalence classes of this relation to be the elements in our general model. However, an equivalence class of cwffs is not really a function, so we define a map $\mathcal{V}$ such that $\mathcal{V}\mathbf{A}_{\alpha\beta}$ is a function which corresponds, in an appropriate sense, to the equivalence class containing $\mathbf{A}_{\alpha\beta}$.

We define, by induction on γ, the domains $\mathcal{D}_\gamma$ and a function $\mathcal{V}$ whose domain is the set of cwffs of $\mathcal{Q}_0^+$, so that

(1^γ) $\mathcal{D}_\gamma = \{\mathcal{V}\mathbf{A}_\gamma \mid \mathbf{A}_\gamma \text{ is a cwff}_\gamma\}$ and
(2^γ) for all cwffs $\mathbf{A}_\gamma$ and $\mathbf{B}_\gamma$, $\mathcal{V}\mathbf{A}_\gamma = \mathcal{V}\mathbf{B}_\gamma$ iff $\mathcal{H} \vdash \mathbf{A}_\gamma = \mathbf{B}_\gamma$.

Let $\mathcal{D}_o = \{\mathsf{T},\ \mathsf{F}\}$, and for each sentence $\mathbf{A}$, let $\mathcal{V}\mathbf{A} = \mathsf{T}$ if $\mathcal{H} \vdash \mathbf{A}$; otherwise let $\mathcal{V}\mathbf{A} = \mathsf{F}$. Since $\mathcal{H}$ is complete and consistent, exactly one of the conditions $\mathcal{H} \vdash \mathbf{A}$ and $\mathcal{H} \vdash \sim \mathbf{A}$ holds for each sentence $\mathbf{A}$, so (1°) and (2°) are satisfied.

For each cwff$_\imath$ $\mathbf{A}_\imath$, let $\mathcal{V}\mathbf{A}_\imath = \{\mathbf{B}_\imath \mid \mathbf{B}_\imath \text{ is a cwff}_\imath \text{ and } \mathcal{H} \vdash \mathbf{A}_\imath = \mathbf{B}_\imath\}$, and let $\mathcal{D}_\imath = \{\mathcal{V}\mathbf{A}_\imath \mid \mathbf{A}_\imath \text{ is a cwff}_\imath\}$. Clearly $\mathcal{V}\mathbf{A}_\imath = \mathcal{V}\mathbf{B}_\imath$ iff $\mathcal{H} \vdash \mathbf{A}_\imath = \mathbf{B}_\imath$.

To define $\mathcal{D}_{\alpha\beta}$, suppose $\mathcal{D}_\alpha$ and $\mathcal{D}_\beta$ are defined, and $\mathbf{A}_{\alpha\beta}$ is a cwff$_{\alpha\beta}$. Let $\mathcal{V}\mathbf{A}_{\alpha\beta}$ be the function from $\mathcal{D}_\beta$ to $\mathcal{D}_\alpha$ whose value, for any argument $\mathcal{V}\mathbf{C}_\beta \in \mathcal{D}_\beta$, is $\mathcal{V}[\mathbf{A}_{\alpha\beta}\mathbf{C}_\beta]$. To show that this definition is unambiguous, we must show that it is independent of the particular cwff $\mathbf{C}_\beta$ used to represent the argument. So suppose $\mathcal{V}\mathbf{B}_\beta = \mathcal{V}\mathbf{C}_\beta$; then $\mathcal{H} \vdash \mathbf{B}_\beta = \mathbf{C}_\beta$, so $\mathcal{H} \vdash$

$\mathbf{A}_{\alpha\beta}\mathbf{B}_\beta = \mathbf{A}_{\alpha\beta}\mathbf{C}_\beta$, so $\mathcal{V}[\mathbf{A}_{\alpha\beta}\mathbf{B}_\beta] = \mathcal{V}[\mathbf{A}_{\alpha\beta}\mathbf{C}_\beta]$. Finally, let $\mathcal{D}_{\alpha\beta} = \{\mathcal{V}\mathbf{A}_{\alpha\beta} \mid \mathbf{A}_{\alpha\beta}$ is a cwff$_{\alpha\beta}\}$.

We must show that $(2^{\alpha\beta})$ is satisfied. If $\mathcal{H} \vdash \mathbf{A}_{\alpha\beta} = \mathbf{B}_{\alpha\beta}$, then for all cwffs $\mathbf{C}_\beta$, $\mathcal{H} \vdash \mathbf{A}_{\alpha\beta}\mathbf{C}_\beta = \mathbf{B}_{\alpha\beta}\mathbf{C}_\beta$, so for all arguments $\mathcal{V}\mathbf{C}_\beta$, $(\mathcal{V}\mathbf{A}_{\alpha\beta})(\mathcal{V}\mathbf{C}_\beta) = \mathcal{V}[\mathbf{A}_{\alpha\beta}\mathbf{C}_\beta] = \mathcal{V}[\mathbf{B}_{\alpha\beta}\mathbf{C}_\beta] = (\mathcal{V}\mathbf{B}_{\alpha\beta})(\mathcal{V}\mathbf{C}_\beta)$, so $\mathcal{V}\mathbf{A}_{\alpha\beta} = \mathcal{V}\mathbf{B}_{\alpha\beta}$. On the other hand, suppose $\mathcal{V}\mathbf{A}_{\alpha\beta} = \mathcal{V}\mathbf{B}_{\alpha\beta}$, and let $\mathbf{C}_\beta$ be a cwff such that $\mathcal{H} \vdash [\mathbf{A}_{\alpha\beta}\mathbf{C}_\beta = \mathbf{B}_{\alpha\beta}\mathbf{C}_\beta \supset \centerdot\, \mathbf{A}_{\alpha\beta} = \mathbf{B}_{\alpha\beta}]$; then $\mathcal{V}[\mathbf{A}_{\alpha\beta}\mathbf{C}_\beta] = (\mathcal{V}\mathbf{A}_{\alpha\beta})(\mathcal{V}\mathbf{C}_\beta) = (\mathcal{V}\mathbf{B}_{\alpha\beta})(\mathcal{V}\mathbf{C}_\beta) = \mathcal{V}[\mathbf{B}_{\alpha\beta}\mathbf{C}_\beta]$, so $\mathcal{H} \vdash \mathbf{A}_{\alpha\beta}\mathbf{C}_\beta = \mathbf{B}_{\alpha\beta}\mathbf{C}_\beta$, so $\mathcal{H} \vdash \mathbf{A}_{\alpha\beta} = \mathbf{B}_{\alpha\beta}$.

Next we shall show that $\mathcal{M} = \langle\{\mathcal{D}_\alpha\}_\alpha, \mathcal{V}\rangle$ constitutes an interpretation. Clearly $\mathcal{V}$ maps each primitive constant of $\mathcal{Q}_0^+$ of type γ into $\mathcal{D}_\gamma$. We must show that $\mathcal{V}\mathsf{Q}_{o\alpha\alpha}$ is the identity relation on $\mathcal{D}_\alpha$. Let $\mathcal{V}\mathbf{A}_\alpha$ and $\mathcal{V}\mathbf{B}_\alpha$ be arbitrary members of $\mathcal{D}_\alpha$. Then $\mathcal{V}\mathbf{A}_\alpha = \mathcal{V}\mathbf{B}_\alpha$ iff $\mathcal{H} \vdash \mathbf{A}_\alpha = \mathbf{B}_\alpha$ iff $\mathcal{H} \vdash \mathsf{Q}_{o\alpha\alpha}\mathbf{A}_\alpha\mathbf{B}_\alpha$ iff $\mathsf{T} = \mathcal{V}[\mathsf{Q}_{o\alpha\alpha}\mathbf{A}_\alpha\mathbf{B}_\alpha] = (\mathcal{V}\mathsf{Q}_{o\alpha\alpha})(\mathcal{V}\mathbf{A}_\alpha)(\mathcal{V}\mathbf{B}_\alpha)$, so $\mathcal{V}\mathsf{Q}_{o\alpha\alpha}$ is indeed the identity relation on $\mathcal{D}_\alpha$.

We must show that $\mathcal{V}\iota_{\imath(o\imath)}$ maps one-element sets in $\mathcal{D}_{o\imath}$ to their unique members. Let $\mathcal{V}\mathbf{A}_\imath$ be an arbitrary member of $\mathcal{D}_\imath$. Since $\mathcal{V}\mathsf{Q}_{o\imath\imath}$ is the identity relation on $\mathcal{D}_\imath$, $((\mathcal{V}\mathsf{Q}_{o\imath\imath})\mathcal{V}\mathbf{A}_\imath)$ is the one-element set whose sole member is $(\mathcal{V}\mathbf{A}_\imath)$. $\mathcal{H} \vdash \mathbf{A}_\imath = \iota_{\imath(o\imath)}[\mathsf{Q}_{o\imath\imath}\mathbf{A}_\imath]$ by Axiom 5, so $\mathcal{V}\mathbf{A}_\imath = \mathcal{V}(\iota_{\imath(o\imath)}[\mathsf{Q}_{o\imath\imath}\mathbf{A}_\imath]) = (\mathcal{V}\iota_{\imath(o\imath)})((\mathcal{V}\mathsf{Q}_{o\imath\imath})\mathcal{V}\mathbf{A}_\imath)$, as required.

Thus $\mathcal{M}$ is indeed an interpretation, and we proceed to show that it is a general model for $\mathcal{Q}_0^+$. For each assignment φ into $\mathcal{M}$ and wff $\mathbf{C}_\gamma$, let $\mathbf{C}_\gamma^\varphi = \mathsf{S}^{\mathbf{x}^1_{\delta_1}\ldots\mathbf{x}^n_{\delta_n}}_{\mathbf{E}^1_{\delta_1}\ldots\mathbf{E}^n_{\delta_n}}\,\mathbf{C}_\gamma$, where $\mathbf{x}^1_{\delta_1}\ldots\mathbf{x}^n_{\delta_n}$ are the free variables of $\mathbf{C}_\gamma$, and (for $1 \le i \le n$) $\mathbf{E}^i_{\delta_i}$ is the first cwff (in some fixed enumeration) of $\mathcal{Q}_0^+$ such that $\varphi\mathbf{x}^i_{\delta_i} = \mathcal{V}\mathbf{E}^i_{\delta_i}$. Let $\mathcal{V}_\varphi\mathbf{C}_\gamma = \mathcal{V}\mathbf{C}_\gamma^\varphi$. Clearly $\mathbf{C}_\gamma^\varphi$ is a cwff$_\gamma$ so $\mathcal{V}_\varphi\mathbf{C}_\gamma \in \mathcal{D}_\gamma$.

(a) For any variable $\mathbf{x}_\delta$, choose $\mathbf{E}_\delta$ so that $\varphi\mathbf{x}_\delta = \mathcal{V}\mathbf{E}_\delta$ as above; then $\mathcal{V}_\varphi\mathbf{x}_\delta = \mathcal{V}\mathbf{x}_\delta^\varphi = \mathcal{V}\mathbf{E}_\delta = \varphi\mathbf{x}_\delta$.

(b) For each cwff (and hence for each primitive constant) $\mathbf{C}_\gamma$, $\mathcal{V}_\varphi\mathbf{C}_\gamma = \mathcal{V}\mathbf{C}_\gamma^\varphi = \mathcal{V}\mathbf{C}_\gamma$.

(c) $\mathcal{V}_\varphi[\mathbf{A}_{\alpha\beta}\mathbf{B}_\beta] = \mathcal{V}[\mathbf{A}^\varphi_{\alpha\beta}\mathbf{B}^\varphi_\beta] = (\mathcal{V}\mathbf{A}^\varphi_{\alpha\beta})(\mathcal{V}\mathbf{B}^\varphi_\beta) = (\mathcal{V}_\varphi\mathbf{A}_{\alpha\beta})(\mathcal{V}_\varphi\mathbf{B}_\beta)$.

(d) Let $\mathcal{V}\mathbf{E}_\delta$ be an arbitrary member of $\mathcal{D}_\delta$; we may assume $\mathbf{E}_\delta$ is the first cwff which represents this member of $\mathcal{D}_\delta$. Given an assignment φ and a variable $\mathbf{x}_\delta$, let $\psi = (\varphi : \mathbf{x}_\delta/\mathcal{V}\mathbf{E}_\delta)$. It can be seen with the aid of 5207 that $\vdash [\lambda\mathbf{x}_\delta\mathbf{B}_\beta]^\varphi\mathbf{E}_\delta = \mathbf{B}^\psi_\beta$, so $(\mathcal{V}_\varphi[\lambda\mathbf{x}_\delta\mathbf{B}_\beta])(\mathcal{V}\mathbf{E}_\delta) = (\mathcal{V}[\lambda\mathbf{x}_\delta\mathbf{B}_\beta]^\varphi)(\mathcal{V}\mathbf{E}_\delta) = \mathcal{V}([\lambda\mathbf{x}_\delta\mathbf{B}_\beta]^\varphi\mathbf{E}_\delta) = \mathcal{V}\mathbf{B}^\psi_\beta = \mathcal{V}_\psi\mathbf{B}_\beta$. Thus $\mathcal{V}_\varphi[\lambda\mathbf{x}_\delta\mathbf{B}_\beta]$ satisfies condition (d) in the definition of a general model.

Thus $\mathcal{M}$ is a general model for $\mathcal{Q}_0^+$ (and hence for $\mathcal{Q}_0$). Also, if $\mathbf{A} \in \mathcal{G}$, then $\mathbf{A} \in \mathcal{H}$, so $\mathcal{H} \vdash \mathbf{A}$, so $\mathcal{V}\mathbf{A} = \mathsf{T}$ and $\mathcal{M} \models \mathbf{A}$, so $\mathcal{M}$ is a general model for $\mathcal{G}$. Clearly each $\mathcal{D}_\alpha$ has cardinality $\leq$ card($\mathcal{L}(\mathcal{Q}_0)$), since $\mathcal{V}$ maps the set of cwffs$_\alpha$ of $\mathcal{Q}_0^+$ onto $\mathcal{D}_\alpha$, and card($\mathcal{L}(\mathcal{Q}_0^+)$) = card($\mathcal{L}(\mathcal{Q}_0)$). ∎

5502 Henkin's Completeness and Soundness Theorem. Let $\mathbf{A}$ be a wff$_o$ and let $\mathcal{G}$ be a set of sentences.

(a) $\vdash \mathbf{A}$ iff $\models \mathbf{A}$.
(b) $\mathcal{G} \vdash \mathbf{A}$ iff $\mathbf{A}$ is valid in every general model for $\mathcal{G}$.
(c) $\mathcal{G} \vdash \mathbf{A}$ iff $\mathbf{A}$ is valid in every frugal general model for $\mathcal{G}$.

Proof: (a) is a special case of (b) with $\mathcal{G} = \emptyset$, and (b) follows from (c) and the Soundness Theorem (5402), so we prove only (c).

Suppose $\mathbf{A}$ is valid in every frugal general model for $\mathcal{G}$. Let $\mathbf{B}$ be a universal closure of $\mathbf{A}$. Then $\mathbf{B}$ is valid in every frugal general model of $\mathcal{G}$ by 5401(g). Suppose $\mathcal{G} \cup \{\sim \mathbf{B}\}$ is consistent. Then by Henkin's Theorem, this set of sentences has a frugal general model $\mathcal{M}$, and $\mathcal{M} \models \sim \mathbf{B}$. But $\mathcal{M}$ is a frugal model for $\mathcal{G}$, so $\mathcal{M} \models \mathbf{B}$. This contradiction shows that $\mathcal{G} \cup \{\sim \mathbf{B}\}$ must be inconsistent, so $\mathcal{G} \vdash \mathbf{B}$ by the Deduction Theorem and Rule P. Hence $\mathcal{G} \vdash \mathbf{A}$ by $\forall$I(5215). This, with the Soundness Theorem, completes the Proof. ∎

The reader may have noticed a similarity between the basic approaches used to prove 5502 (Henkin's Completeness Theorem for $\mathcal{Q}_0$) and 2509 (the Generalized Completeness Theorem for first order logic). However, the approach to 2509 was carried out in a more abstract setting, since we have not presented anything analogous to Smullyan's Unifying Principle for $\mathcal{Q}_0$. Actually, an extension of Smullyan's Unifying Principle to a formulation of type theory which differs slightly from $\mathcal{Q}_0$ is presented in [Andrews, 1971].

5503 Weak Compactness Theorem. If every finite subset of the set $\mathcal{S}$ of sentences has a general model, then $\mathcal{S}$ has a general model.

Proof: By 5403, every finite subset of $\mathcal{S}$ is consistent. Therefore $\mathcal{S}$ is consistent (since every proof is finite), so $\mathcal{S}$ has a general model by 5501. ∎

REMARK. It is natural to consider the Strong Compactness conjecture that if every finite subset of a set $\mathcal{S}$ of sentences has a standard model, then $\mathcal{S}$ has a standard model. We shall show that this is false at the end of §72.

An important application of the Weak Compactness Theorem is the construction of nonstandard models of the real number system, which leads to the subject known as nonstandard analysis. The basic techniques are the same as those used to apply the Compactness Theorem for first-order logic (2514) in §25.

REMARK. Our results about frugal models are somewhat disturbing. As we shall see in §60, we can add to $\mathcal{Q}_0$ an axiom of infinity (which we shall call Axiom 6) which is consistent with $\mathcal{Q}_0$ (Corollary 7204) and which is true only in models whose domains of individuals are infinite. If $\mathcal{M} = \langle\{\mathcal{D}_\alpha\}_\alpha, \mathcal{J}\rangle$ is any such model, Cantor's Theorem (X5304) must be valid in it, so it would appear that $\mathcal{D}_{o\iota}$ must be uncountably infinite, and hence that any model of Axiom 6 is uncountable. However, Henkin's Theorem implies that {Axiom 6} must have a countable model. This seemingly paradoxical situation is called Skolem's Paradox. (Recall the discussion following 2513.) It can be resolved by looking carefully at Cantor's Theorem, i.e., $\sim \exists g_{o\iota\iota} \forall f_{o\iota} \exists j_\iota \centerdot g_{o\iota\iota} j_\iota = f_{o\iota}$, and considering what it means in a model. The theorem says that there is no function $g \in \mathcal{D}_{o\iota\iota}$ from $\mathcal{D}_\iota$ into $\mathcal{D}_{o\iota}$ which has every set $f_{o\iota} \in \mathcal{D}_{o\iota}$ in its range. The usual interpretation of the statement is that $\mathcal{D}_{o\iota}$ is bigger (in cardinality) than $\mathcal{D}_\iota$. However, this obviously cannot be true in an infinite countable model, since $\mathcal{D}_\iota$ and $\mathcal{D}_{o\iota}$ are both countably infinite, so there must be a bijection g between them. Of course, g cannot be in $\mathcal{D}_{o\iota\iota}$, or Cantor's Theorem would be contradicted. Thus, the model must be nonstandard.

We shall explore these ideas more carefully in 5504–5506 below. However, let us first consider what Henkin's Theorem tells us about models of the theory of real numbers. One can introduce finitely many constants to denote the basic operations and relations in the field of real numbers, and postulates which state that the domain of individuals is the set of real numbers (for example), and if one really believes in the real numbers, and expresses the postulates carefully, one would expect that these postulates would constitute a consistent set of sentences. Therefore, by Henkin's Theorem, there must be a countable model for these postulates, in spite of the fact that the set of real numbers is uncountable. Of course, the model will turn out to be nonstandard. One can express in $\mathcal{Q}_0$, and prove with the aid of postulates, the statement that there are uncountably many real numbers, but in a nonstandard model this statement turns out not to mean quite what one expected it to. Even if one could take as postulates the set of all sentences true in the standard model whose domain of individuals is the set of real numbers, the resulting system would still have a countable model, as shown by Henkin's Theorem. Note that there is no sentence of $\mathcal{Q}_0$ by which one can distinguish between this countable model and the standard model of the real numbers (i.e., there is no sentence true in one model, but false in the other). Thus one who speaks the language $\mathcal{Q}_0$ cannot tell whether he lives in a standard or nonstandard world, even if he can answer all the questions he can ask.

Of course, one can abstractly consider a system with uncountably many individual constants, say one to denote each real number. However, even this system would have a general model in which the set $\mathcal{D}_{o\iota}$ of sets of real numbers would have only the cardinality of the continuum.

5504 Lemma. Let $\mathcal{M} = \langle\{\mathcal{D}_\alpha\}_\alpha, \mathcal{J}\rangle$ be a frugal model for $\mathcal{Q}_0$ such that $\text{card}(\mathcal{D}_\iota) = \text{card}(\mathcal{L}(\mathcal{Q}_0))$. Then $\mathcal{M}$ is nonstandard.

Proof: Suppose $\mathcal{M}$ were standard. Then all subsets of $\mathcal{D}_\iota$ would be in $\mathcal{D}_{o\iota}$. By Cantor's Theorem (see X5304), $\mathcal{D}_\iota$ has more subsets than members, so $\text{card}(\mathcal{D}_{o\iota}) > \text{card}(\mathcal{D}_\iota) = \text{card}(\mathcal{L}(\mathcal{Q}_0)) \geq \text{card}(\mathcal{D}_{o\iota})$ (by frugality). This contradiction shows that $\mathcal{M}$ must be nonstandard. ∎

5505 Corollary. Every infinite frugal model for a countable formulation of $\mathcal{Q}_0$ is nonstandard.

5506 Theorem. Let $\mathcal{H}$ be any set of sentences of $\mathcal{Q}_0$ which has infinite models. Then $\mathcal{H}$ has nonstandard models.

Proof: Let $\kappa = \text{card}(\mathcal{L}(\mathcal{Q}_0))$ and let $\mathcal{Q}_0^+$ be an expansion of $\mathcal{Q}_0$ obtained by adding to $\mathcal{Q}_0$ a set $\mathcal{C}$ of new individual constants, where $\text{card}(\mathcal{C}) = \kappa$. Let $\mathcal{S}$ be the set of all sentences of the form $\mathbf{c}_\iota \neq \mathbf{d}_\iota$, where $\mathbf{c}_\iota$ and $\mathbf{d}_\iota$ are distinct constants in $\mathcal{C}$. It is easy to see that any finite subset of $\mathcal{S} \cup \mathcal{H}$ has a model, since the finitely many sentences from $\mathcal{S}$ can all be made true in any infinite model of $\mathcal{H}$ by giving the constants from $\mathcal{C}$ which occur in those sentences distinct interpretations. Therefore, by 5503, 5403, and 5501 $\mathcal{S} \cup \mathcal{H}$ has a frugal general model $\mathcal{M} = \langle\{\mathcal{D}_\alpha\}_\alpha, \mathcal{J}\rangle$. Since all sentences of $\mathcal{S}$ are true in $\mathcal{M}$, $\mathcal{D}_\iota$ has κ distinct members, so $\mathcal{M}$ is nonstandard by 5504. ∎

Thus, there are always nonstandard models for infinite mathematical systems, but one has the option of ignoring them, and concentrating on standard models. From the point of view of standard models, however, the system $\mathcal{Q}_0$ is not complete, since there are sentences true in all standard models which are not theorems. For example, the Axiom of Choice is presumably true in all standard models (though it seems difficult to prove this without making assumptions so strong as to rob the proof of any significance), but is false in certain nonstandard models. (Such a nonstandard model is used in [Andrews, 1972b] to prove the independence of the Axiom of Choice.)

It would be desirable, therefore, to add new axioms to $\mathcal{Q}_0$ to obtain a strengthened system in which the theorems are precisely the wffs$_o$ valid in all standard models. Unfortunately, this is impossible, unless we abandon the rather fundamental principle that there must be an effective procedure for deciding, of an arbitrary wff, whether or not it is an axiom. We shall see this in Theorem 7206 of Chapter 7.

EXERCISES

X5500. Prove:

$$\forall p_{o\beta}[\exists x_\beta p_{o\beta} x_\beta \supset p_{o\beta} \,\blacksquare\, j_{\beta(o\beta)} p_{o\beta}]$$

$$\supset \forall f_{\alpha\beta} \forall g_{\alpha\beta} \,\blacksquare\, f_{\alpha\beta}[j_{\beta(o\beta)} \,\blacksquare\, \lambda x_\beta \,\blacksquare\, f_{\alpha\beta} x_\beta \neq g_{\alpha\beta} x_\beta]$$

$$= g_{\alpha\beta}[j_{\beta(o\beta)} \,\blacksquare\, \lambda x_\beta \,\blacksquare\, f_{\alpha\beta} x_\beta \neq g_{\alpha\beta} x_\beta] \supset \,\blacksquare\, f_{\alpha\beta} = g_{\alpha\beta}.$$

X5501. Justify the Remark following the definition of extensional completeness.

X5502. A set of sentences is *categorical* iff all its models are isomorphic to each other. Prove that every categorical set of sentences is complete.

X5503. Is there a sentence **B** such that for every general model $\mathcal{M}$, it is true that $\mathcal{M} \models \mathbf{B}$ if and only if $\mathcal{M}$ is a standard model?

Chapter 6

Formalized Number Theory

§60. Cardinal Numbers and the Axiom of Infinity

In the preface to *Principia Mathematica*, it is stated that "what were formerly taken, tacitly or explicitly, as axioms, are either unnecessary or demonstrable".[1] In this spirit, we shall in this chapter define the natural numbers as the finite cardinals, and derive Peano's Postulates from an Axiom of Infinity. We then establish those results about natural numbers and recursive functions which we need in order to establish the incompleteness results in Chapter 7.

We shall not develop the general theory of cardinal numbers, but it may be helpful to introduce and discuss a few relevant definitions.

DEFINITIONS. $E_{o(o\alpha)(o\beta)}$ stands for

$$[\lambda p_{o\beta}\lambda q_{o\alpha}\exists s_{\alpha\beta} \centerdot \forall x_{\beta}[p_{o\beta}x_{\beta} \supset q_{o\alpha} \centerdot s_{\alpha\beta}x_{\beta}]$$

$$\wedge\ \forall y_{\alpha} \centerdot q_{o\alpha}y_{\alpha} \supset \exists_1 x_{\beta} \centerdot p_{o\beta}x_{\beta} \wedge y_{\alpha} = s_{\alpha\beta}x_{\beta}].$$

$$[NC]_{o(o(o\beta))} \text{ stands for } \left[\lambda u_{o(o\beta)}\exists p_{o\beta} \centerdot u_{o(o\beta)} = E_{o(o\beta)(o\beta)}p_{o\beta}\right].$$

$E_{o(o\alpha)(o\beta)}p_{o\beta}q_{o\alpha}$ means that the sets $p_{o\beta}$ and $q_{o\alpha}$ are equipollent, i.e., $p_{o\beta}$ and $q_{o\alpha}$ have the same cardinality (number of members). This is expressed by saying that there is a function $s_{\alpha\beta}$ which is one-to-one on $p_{o\beta}$ and maps $p_{o\beta}$ onto $q_{o\alpha}$. It is easy to establish that $E_{o(o\beta)(o\beta)}$ is an equivalence relation.

[1][Whitehead and Russell, 1913, Second Edition, Volume 1, p. v.]

(See Exercises X6000-X6002.) The *cardinality* of a set $p_{o\beta}$ (sometimes denoted $\overline{\overline{p_{o\beta}}}$) is $E_{o(o\beta)(o\beta)}p_{o\beta}$, the equivalence class of sets (of the same type as $p_{o\beta}$) which are equipollent to $p_{o\beta}$. $[NC]_{o(o(o\beta))}$ is the set of *cardinal numbers* (for type β), i.e., the set of cardinalities (of sets of type $(o\beta)$).

In order to assure that all of the finite cardinals exist and are distinct, we must assume an axiom of infinity (i.e., a sentence which is satisfiable, but has no finite models.) The assertion that there are infinitely many individuals can be expressed naturally in a variety of ways, and in the absence of the Axiom of Choice it is not always possible to prove the equivalence of various natural formulations of this assertion. For our Axiom of Infinity we shall take the assertion that there is an irreflexive partial ordering of the individuals with respect to which there is no maximal element.[2]

Axiom 6:

$$\exists r_{o\iota\iota}\forall x_\iota\forall y_\iota\forall z_\iota[\exists w_\iota r_{o\iota\iota}x_\iota w_\iota \wedge \sim r_{o\iota\iota}x_\iota x_\iota$$

$$\wedge \,\blacksquare\, r_{o\iota\iota}x_\iota y_\iota \supset \,\blacksquare\, r_{o\iota\iota}y_\iota z_\iota \supset r_{o\iota\iota}x_\iota z_\iota].$$

Without going into the subject very far, let us consider some alternative axioms of infinity which might naturally occur to the reader. One way to say that a set is infinite is to say that it is equipollent with a proper subset of itself. (For example, the set of integers is equipollent with the set of even integers.) This can be expressed by the wff

$$Inf^I: \quad \exists n_{o\iota}\exists m_{o\iota}[\forall x_\iota[m_{o\iota}x_\iota \supset n_{o\iota}x_\iota]$$

$$\wedge \exists o_\iota[n_{o\iota}o_\iota \wedge \sim m_{o\iota}o_\iota] \wedge E_{o(o\iota)(o\iota)}n_{o\iota}m_{o\iota}].$$

Another way to say that a set is infinite is to say that it has a subset which is isomorphic to the set of natural numbers. The natural numbers can be characterized by Peano's Postulates, which can be stated informally as follows:

(P1) There is an entity called 0 which is a natural number.
(P2) Every natural number n has a successor Sn which is also a natural number.
(P3) 0 is not the successor of any natural number.

[2] This axiom of infinity was first suggested (in essence) in [Bernays and Schönfinkel, 1928].

(P4) If n and m are natural numbers with the same successors, then n and m are the same.

(P5) Principle of Mathematical Induction (see §10A).

Actually, any set satisfying postulates P1–P4 must be infinite. The Induction Principle simply limits the size of the set. Postulates P1–P4 can be expressed by the wff

$$Inf^{II}: \ \exists n_{o\iota} \exists o_\iota \exists s_{\iota\iota} [n_{o\iota} o_\iota \ \wedge \ \forall x_\iota [n_{o\iota} x_\iota \supset n_{o\iota} \, \centerdot \, s_{\iota\iota} x_\iota]$$

$$\wedge \ \forall x_\iota [n_{o\iota} x_\iota \supset s_{\iota\iota} x_\iota \neq o_\iota]$$

$$\wedge \ \forall x_\iota \forall y_\iota \, \centerdot \, n_{o\iota} x_\iota \wedge n_{o\iota} y_\iota \wedge s_{\iota\iota} x_\iota = s_{\iota\iota} y_\iota \supset x_\iota = y_\iota].$$

It is not hard to see that $\vdash_{\mathcal{Q}_0} Inf^I \equiv Inf^{II}$. Indeed, if we assume that there are sets $n_{o\iota}$ and $m_{o\iota}$ of individuals and there is an individual o_ι as described in Inf^I, and let $s_{\iota\iota}$ be a function demonstrating that $En_{o\iota}m_{o\iota}$, then $n_{o\iota}$, o_ι, and $s_{\iota\iota}$ verify Inf^{II}, so $\vdash Inf^I \supset Inf^{II}$. On the other hand, if there exist $n_{o\iota}$, o_ι, and $s_{\iota\iota}$ as described in Inf^{II}, and $m_{o\iota}$ is the image of $n_{o\iota}$ under $s_{\iota\iota}$, then $n_{o\iota}$ and $m_{o\iota}$ verify Inf^I, so $\vdash Inf^{II} \supset Inf^I$.

It can also be seen that $\vdash_{\mathcal{Q}_0} Inf^{II} \supset$ Axiom 6. The details of this argument are best postponed to §62, but we may give a preview of it here. If there exist $n_{o\iota}$, o_ι, and $s_{\iota\iota}$ as described in Inf^{II}, we may let $n'_{o\iota}$ be the intersection of all sets containing o_ι and closed under $s_{\iota\iota}$ (as in the definition of $\mathbb{N}_{o\sigma}$ below). Then $n'_{o\iota}$ is isomorphic to the natural numbers, and the natural ordering $<$ is definable on $n'_{o\iota}$ (as in §62). The relation $[\lambda x_\iota \lambda y_\iota \, \centerdot \, n'_{o\iota} y_\iota \wedge \, \centerdot \, \sim n'_{o\iota} x_\iota \vee x_\iota < y_\iota]$ satisfies the requirements on $r_{o\iota\iota}$ in Axiom 6.

Thus, all the consequences of Axiom 6 which we shall derive below are also consequences of Inf^I and Inf^{II}, so any of these three sentences could serve as our axiom of infinity. Of course, there are many additional possibilities.

We shall use $\mathcal{Q}_0^\infty$ as a name for the system obtained by adding Axiom 6 to the list of axioms of $\mathcal{Q}_0$. Henceforth in this chapter, $\vdash$ **A** shall mean that **A** is a theorem of $\mathcal{Q}_0^\infty$ (i.e., $\vdash_{\mathcal{Q}_0^\infty}$ **A**) unless otherwise indicated. Note that $\vdash_{\mathcal{Q}_0^\infty}$ **A** iff [Axiom 6] $\vdash_{\mathcal{Q}_0}$ **A**, and that Axiom 6 is a sentence, so we can use the derived rules of inference established in §52 for $\mathcal{Q}_0^\infty$ as well as for $\mathcal{Q}_0$. In presenting (condensed) proofs, we shall often omit explicit reference to the rules of inference used to infer a line of the proof, and simply indicate from what lines of the proof and previously established theorems the given line is obtained. When a line of a proof is inferred from the immediately preceding line, we may also omit explicit mention of the preceding line. From time to time the reader may find it a useful exercise to fill in all intermediate steps of a proof, and indicate the rule of inference used for each step. We shall often omit type symbols from occurrences of a variable after the first occurrence in a wff, or when the context in some other way makes it clear what the type must be.

We next define the natural numbers. These are equivalence classes of sets of individuals, and so have type $(o(o\imath))$.

DEFINITIONS. Let σ be the type symbol $(o(o\imath))$.

0_σ stands for $\mathsf{Q}_{o(o\imath)(o\imath)}[\lambda x_\imath F_o]$

$S_{\sigma\sigma}$ stands for $[\lambda n_{o(o\imath)}\lambda p_{o\imath}\exists x_\imath \,.\, p_{o\imath}x_\imath \wedge n_{o(o\imath)}[\lambda t_\imath \,.\, t_\imath \neq x_\imath \wedge p_{o\imath}t_\imath]]$

$\mathbb{N}_{o\sigma}$ stands for $[\lambda n_\sigma \forall p_{o\sigma} \,.\, [p_{o\sigma}0_\sigma \wedge \forall x_\sigma \,.\, p_{o\sigma}x_\sigma \supset p_{o\sigma} \,.\, S_{\sigma\sigma}x_\sigma] \supset p_{o\sigma}n_\sigma]$

$\mathit{Finite}_{o(o\imath)}$ stands for $[\lambda p_{o\imath}\exists n_\sigma \,.\, \mathbb{N}_{o\sigma}n_\sigma \wedge n_\sigma p_{o\imath}]$

Thus zero is the collection of all sets with zero members, i.e., the collection containing just the empty set $[\lambda x_\imath F_o]$. The wff S represents the successor function. If $n_{(o(o\imath))}$ is a finite cardinal (say 2), then a set $p_{o\imath}$ (say $\{a,b,c\}$) is in Sn iff there is an individual (say c) which is in $p_{o\imath}$ and whose deletion from $p_{o\imath}$ leaves a set ($\{a,b\}$) which is in n. $\mathbb{N}_{o\sigma}$ represents the set of natural numbers, i.e., the intersection of all sets which contain 0 and are closed under S. (See Exercise X6007.) It can be shown that every natural number is a cardinal number (Exercise X6102).

DEFINITIONS. Let $\mathbf{M}_{o\alpha}$ be any wff in which $\mathbf{x}_\alpha$ does not occur free.

$(\forall \mathbf{x}_\alpha \in \mathbf{M}_{o\alpha})\mathbf{A}$	stands for $\forall \mathbf{x}_\alpha[\mathbf{M}_{o\alpha}\mathbf{x}_\alpha \supset \mathbf{A}]$.
$(\exists \mathbf{x}_\alpha \in \mathbf{M}_{o\alpha})\mathbf{A}$	stands for $\exists \mathbf{x}_\alpha[\mathbf{M}_{o\alpha}\mathbf{x}_\alpha \wedge \mathbf{A}]$.
$\dot{\forall} \mathbf{x}_\sigma \mathbf{A}$	stands for $(\forall \mathbf{x}_\sigma \in \mathbb{N}_{o\sigma})\mathbf{A}$.
$\dot{\exists} \mathbf{x}_\sigma \mathbf{A}$	stands for $(\exists \mathbf{x}_\sigma \in \mathbb{N}_{o\sigma})\mathbf{A}$.

We leave to the reader the proofs of the following simple but useful propositions of $\mathcal{Q}_0$:

6000. $\vdash \sim (\forall \mathbf{x}_\alpha \in \mathbf{M}_{o\alpha})\mathbf{A} = (\exists \mathbf{x}_\alpha \in \mathbf{M}_{o\alpha}) \sim \mathbf{A}$ provided $\mathbf{x}_\alpha$ is not free in $\mathbf{M}_{o\alpha}$.

6001. $\vdash \sim (\exists \mathbf{x}_\alpha \in \mathbf{M}_{o\alpha})\mathbf{A} = (\forall \mathbf{x}_\alpha \in \mathbf{M}_{o\alpha}) \sim \mathbf{A}$ provided $\mathbf{x}_\alpha$ is not free in $\mathbf{M}_{o\alpha}$.

EXERCISES

Prove the following theorems:

X6000. $\forall p_{o\beta} \centerdot E_{o(o\beta)(o\beta)} p_{o\beta} p_{o\beta}$

X6001. $\forall p_{o\beta} \forall q_{o\alpha} \centerdot E_{o(o\alpha)(o\beta)} p_{o\beta} q_{o\alpha} \supset E_{o(o\beta)(o\alpha)} q_{o\alpha} p_{o\beta}$

X6002. $\forall p_{o\beta} \forall q_{o\alpha} \forall r_{o\gamma} \centerdot E_{o(o\alpha)(o\beta)} p_{o\beta} q_{o\alpha} \wedge E_{o(o\gamma)(o\alpha)} q_{o\alpha} r_{o\gamma} \supset E_{o(o\gamma)(o\beta)} p_{o\beta} r_{o\gamma}$

X6003. $\forall p_{o\beta} \forall q_{o\beta} \centerdot E_{o(o\beta)(o\beta)} p_{o\beta} q_{o\beta} = \centerdot \overline{\overline{p_{o\beta}}} = \overline{\overline{q_{o\beta}}}$

X6004. $E_{o(o\alpha)(o\beta)} [\mathsf{Q}_{o\beta\beta} x_\beta][\mathsf{Q}_{o\alpha\alpha} y_\alpha]$

X6005. $[NC]_{o(o(o\beta))} \overline{\overline{p_{o\beta}}}$

X6006. $Inf^I = Inf^{II}$

X6007. $\mathbb{N} = \bigcap_{o(o\sigma)} [\lambda p_{o\sigma} \centerdot p_{o\sigma} 0_\sigma \wedge \forall x_\sigma \centerdot p_{o\sigma} x_\sigma \supset p_{o\alpha} \centerdot S_{\sigma\sigma} x_\sigma]$

X6008. Is there a sentence S such that for every general model $\mathcal{M}$, it is true that $\mathcal{M} \models S$ if and only if $\mathcal{M}$ has an infinite domain of individuals? (*Hint*: use 5503.)

§61. Peano's Postulates

We now prove Peano's Postulates (Theorems 6100, 6101, 6102, 6106, and 6109), plus a few related results.

6100. $\vdash \mathbb{N}_{o\sigma} 0_\sigma$ by the def of $\mathbb{N}$

6101. $\vdash \forall x_\sigma \centerdot \mathbb{N}_{o\sigma} x_\sigma \supset \mathbb{N}_{o\sigma} \centerdot S_{o\sigma} x_\sigma$

Proof:

(.1)	$\mathbb{N} x_\sigma, .1 \vdash p_{o\alpha} 0 \wedge \forall x_\sigma \centerdot px \supset p \centerdot Sx$	hyp
(.2)	$\mathbb{N} x_\sigma, .1 \vdash p_{o\sigma} x_\sigma$	.1, hyp., def of $\mathbb{N}$
(.3)	$\mathbb{N} x_\sigma, .1 \vdash p_{o\sigma} \centerdot S x_\sigma$	.1, .2
(.4)	$\mathbb{N} x_\sigma \vdash \mathbb{N} \centerdot S x_\sigma$	.3, def of $\mathbb{N}$

■

6102. The Induction Theorem.

$$\vdash \forall p_{o\sigma} \centerdot [p_{o\sigma} 0_\sigma \ \wedge \ \dot{\forall} x_\sigma \centerdot p_{o\sigma} x_\sigma \supset p_{o\sigma} \centerdot S_{\sigma\sigma} x_\sigma] \supset \dot{\forall} x_\sigma p_{o\sigma} x_\sigma$$

Proof: Let $P_{o\sigma}$ be $[\lambda t_\sigma \centerdot \mathbb{N} t \ \wedge \ p_{o\sigma} t]$.

(.1) $.1 \vdash p_{o\sigma} 0 \ \wedge \ \forall x_\sigma \centerdot \mathbb{N} x \supset \centerdot px \supset p \centerdot Sx$	hyp
(.2) $\mathbb{N} y_\sigma \vdash [P0 \ \wedge \ \forall x_\sigma \centerdot Px \supset P \centerdot Sx] \supset Py_\sigma$	hyp, def of $\mathbb{N}$
(.3) $.1 \vdash P0$	def of P, .1, 6100
(.4) $.1 \vdash \forall x_\sigma \centerdot Px \supset P \centerdot Sx$	def of P, .1, 6101
(.5) $.1, \ \mathbb{N} y_\sigma \vdash Py_\sigma$	.2, .3, .4
(.6) $.1 \vdash \dot{\forall} y_\sigma p y_\sigma$	.5, def of $\dot{\forall}$, P

■

6103. $\vdash p_{o\sigma} 0_\sigma \ \wedge \ \dot{\forall} x_\sigma p_{o\sigma} S_{\sigma\sigma} x_\sigma \supset \dot{\forall} n_\sigma p_{o\sigma} n_\sigma$

Proof:

(.1) $\vdash \dot{\forall} x_\sigma p_{o\sigma} S_{\sigma\sigma} x_\sigma \supset \dot{\forall} x_\sigma \centerdot p_{o\sigma} x_\sigma \supset p_{o\sigma} S_{\sigma\sigma} x_\sigma$	Rule Q
The theorem follows by 6102.	■

6104. $\vdash \dot{\forall} n_\sigma \centerdot n_\sigma = 0 \ \vee \ \dot{\exists} m_\sigma \centerdot n_\sigma = S_{\sigma\sigma} m_\sigma$

Proof by 6103:

(.1) $\vdash 0_\sigma = 0_\sigma \ \vee \ \dot{\exists} m_\sigma \centerdot 0_\sigma = S_{\sigma\sigma} m_\sigma$	5200
(.2) $\mathbb{N} x_\sigma \vdash \mathbb{N} x_\sigma \ \wedge \ \centerdot Sx_\sigma = Sx_\sigma$	5200
(.3) $\mathbb{N} x_\sigma \vdash \exists m_\sigma \centerdot \mathbb{N} m_\sigma \ \wedge \ Sx_\sigma = Sm_\sigma$	$\exists$ Gen: .2
(.4) $\vdash \dot{\forall} x_\sigma \centerdot Sx_\sigma = 0 \ \vee \ \dot{\exists} m_\sigma \centerdot Sx_\sigma = Sm_\sigma$	.3
Then use 6103 with .1 and .4.	■

6105. $\vdash \forall n_\sigma \centerdot S_{\sigma\sigma} n_\sigma \neq 0_\sigma$

Proof by contradiction:

(.1) $.1 \vdash Sn_\sigma = 0$	hyp
(.2) $\vdash 0_\sigma [\lambda x_\iota F]$	def of 0
(.3) $.1 \vdash Sn_\sigma [\lambda x_\iota F]$	=: .1, .2
(.4) $.1 \vdash \exists x_\iota F$	.3, def of S
(.5) $\vdash Sn_\sigma \neq 0$	.4

■

6106. $\vdash \dot\forall n_\sigma \centerdot S_{\sigma\sigma} n_\sigma \neq 0_\sigma$ by 6105 and the def of $\dot\forall$

We have now proven all of Peano's Postulates except one, which will be proved below as Theorem 6109. Our first step in proving it is to show that if we remove any element from a set of cardinality Sn we obtain a set of cardinality n.

6107. $\vdash \dot\forall n_\sigma \forall p_{o\iota} \centerdot Sn_\sigma p_{o\iota} \wedge p_{o\iota} w_\iota \supset n_\sigma[\lambda t_\iota \centerdot t_\iota \neq w_\iota \wedge p_{o\iota} t_\iota]$

Proof: The proof is by induction on n. First we treat the case $n = 0$.

(.1) $.1 \vdash S0p_{o\iota} \wedge pw_\iota$ hyp
(.2) $.1 \vdash \exists x_\iota .3$ .1, def of S
(.3) $.1,\ .3 \vdash p_{o\iota} x_\iota \wedge 0[\lambda t_\iota \centerdot t \neq x \wedge pt]$ choose x
(.4) $.1,\ .3 \vdash \sim [\lambda t_\iota \centerdot t \neq x_\iota \wedge p_{o\iota} t] w_\iota$ .3, def of 0
(.5) $.1,\ .3 \vdash w_\iota = x_\iota$ .1, .4
(.6) $.1,\ .3 \vdash 0[\lambda t_\iota \centerdot t \neq w_\iota \wedge p_{o\iota} t]$ .3, .5
(.7) $.1 \vdash .6$ .2, .6
(.8) $\vdash \forall p_{o\iota} \centerdot S0p \wedge pw_\iota \supset 0[\lambda t_\iota \centerdot t \neq w \wedge pt]$ .7

Next we treat the induction step.

(.9) $.9 \vdash \mathbb{N} n_\sigma \wedge \forall p_{o\iota} \centerdot Snp \wedge pw_\iota \supset n[\lambda t_\iota \centerdot t \neq w \wedge pt]$ (inductive) hyp
(.10) $.10 \vdash [SSn_\sigma] p_{o\iota} \wedge pw_\iota$ hyp
(.11) $.10 \vdash \exists x_\iota .12$ .10, def of S
(.12) $.10, .12 \vdash p_{o\iota} x_\iota \wedge Sn_\sigma[\lambda t_\iota \centerdot t \neq x \wedge pt]$ choose x

From .10 we must prove $Sn[\lambda t \centerdot t \neq w \wedge pt]$. We consider two cases in .13 and .15.

(.13) $.13 \vdash x_\iota = w_\iota$ hyp (case 1)
(.14) $.10,\ .12,\ .13 \vdash Sn_\sigma[\lambda t_\iota \centerdot t \neq w_\iota \wedge p_{o\iota} t]$ .12, .13

In case 2 we shall use the inductive hypothesis.

(.15) $.15 \vdash x_\iota \neq w_\iota$ hyp (case 2)
(.16) $.9 \vdash Sn_\sigma[\lambda t_\iota \centerdot t \neq x_\iota \wedge p_{o\iota} t]$
$\wedge [\lambda t_\iota \centerdot t \neq x \wedge pt] w_\iota \supset n[\lambda t_\iota \centerdot t \neq w \wedge \centerdot t \neq x \wedge pt]$ .9
(.17) $.9,\ .10,\ .12,\ .15 \vdash n_\sigma[\lambda t_\iota \centerdot t \neq w_\iota \wedge t \neq x_\iota \wedge p_{o\iota} t]$ .10, .12, .15, .16
(.18) $.9,\ .10,\ .12,\ .15 \vdash x_\iota \neq w_\iota \wedge p_{o\iota} x \wedge n_\sigma[\lambda t_\iota \centerdot t \neq x \wedge \centerdot t \neq w \wedge pt]$
.12, .15, .17
(.19) $.9,\ .10,\ .12,\ ,15 \vdash Sn_\sigma[\lambda t_\iota \centerdot t \neq w_\iota \wedge p_{o\iota} t]$ .18, def of S

This completes case 2.

(.20) .9, .10 $\vdash Sn_\sigma[\lambda t_\iota \,\centerdot\, t \neq w_\iota \wedge p_{o\iota}t]$ .14, .19, .11
(.21) .9 $\vdash \forall p_{o\iota} \,\centerdot\, [SSn_\sigma]p \wedge pw_\iota \supset Sn[\lambda t_\iota \,\centerdot\, t \neq w \wedge pt]$ .20

This completes the induction step. The theorem now follows from .8 and .21 by the Induction Theorem. ■

REMARK. It can now be seen (Exercise X6107) that

$$\vdash S_{\sigma\sigma}n_\sigma p_{o\iota} = \,\centerdot\, \sim 0_\sigma p \wedge \forall x_\iota \,\centerdot\, px \supset n_\sigma[\lambda t_\iota \,\centerdot\, t \neq x \wedge pt].$$

This provides an alternative possible definition for $S_{\sigma\sigma}$.

It will be observed so far in this section we have not used the Axiom of Infinity. We shall use it in proving the next theorem, which will be used to prove Theorem 6109.

6108. $\vdash \dot{\forall} n_\sigma \,\centerdot\, n_\sigma p_{o\iota} \supset \exists w_\iota \sim p_{o\iota}w_\iota$

Proof:

(.1) .1 $\vdash \forall x_\iota \forall y_\iota \forall z_\iota \,\centerdot\, \exists w_\iota r_{o\iota\iota}xw \wedge \sim rxx \wedge \,\centerdot\, rxy \supset \,\centerdot\, ryz \supset rxz$
choose r (Axiom 6)

Let $P_{o\sigma}$ be $[\lambda n_\sigma \forall p_{o\iota} \,\centerdot\, np \supset \exists z_\iota \forall w_\iota \,\centerdot\, pw \supset \sim r_{o\iota\iota}zw]$. We may informally interpret rzw as meaning that z is below w. Thus Pn means that if p is any set of cardinality n, then there is an element z which is below no member of p. We shall prove $\dot{\forall} n_\sigma Pn$ by induction on n.

(.2) $0p_{o\iota} \vdash \sim p_{o\iota}w_\iota$ def of 0
(.3) $\vdash P0$ .2, def of P

Next we treat the induction step.

(.4) .4 $\vdash \mathbb{N}n_\sigma \wedge Pn$ (inductive) hyp
(.5) .5 $\vdash Sn_\sigma p_{o\iota}$ hyp
(.6) .5 $\vdash \exists x_\iota$ .7 .5, def of S
(.7) .5, .7 $\vdash p_{o\iota}x_\iota \wedge n_\sigma[\lambda t_\iota \,\centerdot\, t \neq x \wedge pt]$ choose x (.6)
(.8) .4, .5, .7 $\vdash \exists z_\iota$.9 .4, def of P, .7
(instantiate $\forall p_{o\iota}$ in Pn with $[\lambda t_\iota \,\centerdot\, t \neq x_\iota \wedge p_{o\iota}t]$)
(.9) .4, .5, .7 .9 $\vdash \forall w_\iota \,\centerdot\, w \neq x_\iota \wedge p_{o\iota}w \supset \sim r_{o\iota\iota}zw$ choose z (.8)

Thus from the inductive hypothesis we see that there is an element z which is below nothing in $p - \{x\}$. We must show that there is an element which is below nothing in p. We consider two cases, .10 and .14.

(.10) $.10 \vdash \sim r_{o\iota\iota} z x$	hyp (case 1)

In this case we show that z is under nothing in p.

(.11) $.4,\ .5,\ .7,\ .10 \vdash w_\iota = x_\iota \supset \sim r_{o\iota\iota} z w$	.10, Axiom 2
(.12) $.4,\ .5,\ .7,\ .9,\ .10 \vdash \forall w_\iota \centerdot p_{o\iota} w \supset \sim r_{o\iota\iota} z w$	.9, .11
(.13) $.4,\ .5,\ .7,\ .9,\ .10 \vdash \exists z_\iota\ .12$	.12

Next we consider case 2, and show by the indirect method that x is below nothing in p.

(.14) $.14 \vdash r_{o\iota\iota} z x$	hyp (case 2)
(.15) $.1,\ .14,\ .15 \vdash p_{o\iota} w_\iota \ \wedge\ r_{o\iota\iota} x_\iota w$	hyp
(.16) $.1,\ .14,\ .15 \vdash r_{o\iota\iota} z_\iota w_\iota$	.1, .14, .15
(.17) $.1,\ .9,\ .14,\ .15 \vdash w_\iota = x_\iota$	.9, .15, .16
(.18) $.1,\ .9,\ .14,\ .15 \vdash r_{o\iota\iota} x_\iota x_\iota$	.15, .17
(.19) $.1,\ .4,\ .5,\ .7,\ .9,\ .14,\ .15 \vdash F_o$	.1, .18
(.20) $.1,\ .4,\ .5,\ .7,\ .9,\ .14 \vdash \forall w_\iota \centerdot p_{o\iota} w \supset \sim r_{o\iota\iota} x_\iota w$	.19
(.21) $.1,\ .4,\ .5,\ .7,\ .9,\ .14 \vdash \exists z_\iota \forall w_\iota \centerdot p_{o\iota} w \supset \sim r_{o\iota\iota} z w$	.20
(.22) $.1,\ .4,\ .5 \vdash .21$	Cases: .13, .21; Rule C: .8, .6
(.23) $.1 \vdash \mathbb{N} n_\sigma \ \wedge\ Pn \supset P \centerdot Sn$	.22, def of P
(.24) $.1 \vdash \dot\forall n_\sigma P n_\sigma$	.3, .23, Induction Theorem

Having finished the inductive proof, we proceed to prove the main theorem.

(.25) $.25 \vdash \mathbb{N} n_\sigma \ \wedge\ n p_{o\iota}$	hyp
(.26) $.1,\ .25 \vdash \exists z_\iota \forall w_\iota \centerdot p_{o\iota} w \supset \sim r_{o\iota\iota} z w$	.24, .25, def of P
(.27) $.1 \vdash \forall z_\iota \exists w_\iota r_{o\iota\iota} z w$	.1
(.28) $.1,\ .25 \vdash \exists w_\iota \sim p_{o\iota} w$	.26, .27
(.29) $.1 \vdash \dot\forall n_\sigma \centerdot n p_{o\iota} \supset \exists w_\iota \sim p_{o\iota} w_\iota$	.28
(.30) $\vdash .29$	Rule C: Axiom 6, .29

■

6109. $\vdash \dot{\forall} n_\sigma \dot{\forall} m_\sigma \,\blacksquare\, S_{\sigma\sigma} n_\sigma = S_{\sigma\sigma} m_\sigma \supset n_\sigma = m_\sigma$

Proof:

(.1)	$.1 \vdash \mathbb{N} n_\sigma \wedge \mathbb{N} m_\sigma \wedge Sn = Sm$	hyp
(.2)	$.2 \vdash n_\sigma p_{o\imath}$	hyp
(.3)	$.1,\ .2 \vdash \exists w_\imath \sim p_{o\imath} w$	.1, .2, 6108
(.4)	$.1,\ .2,\ .4 \vdash \sim p_{o\imath} w_\imath$	choose w (.3)
(.5)	$.1,\ .2,\ .4 \vdash p_{o\imath} = [\lambda t_\imath \,\blacksquare\, t \neq w_\imath \wedge \,\blacksquare\, t = w \vee pt]$	.4, Axiom 2, Axiom 3
(.6)	$.1,\ .2,\ .4 \vdash n_\sigma [\lambda t_\imath \,\blacksquare\, t \neq w_\imath \wedge \,\blacksquare\, t = w \vee p_{o\imath} t]$	.2, .5
(.7)	$.1,\ .2,\ .4 \vdash S n_\sigma [\lambda t_\imath \,\blacksquare\, t = w_\imath \vee p_{o\imath} t]$	.6, def of S
(.8)	$.1,\ .2,\ .4 \vdash S m_\sigma [\lambda t_\imath \,\blacksquare\, t = w_\imath \vee p_{o\imath} t]$	.1, .7
(.9)	$.1,\ .2,\ .4 \vdash S m_\sigma [\lambda t_\imath \,\blacksquare\, t = w_\imath \vee p_{o\imath} t] \wedge [\lambda t_\imath \,\blacksquare\, t = w \vee p_{o\imath} t] w$ $\supset m[\lambda t_\imath \,\blacksquare\, t \neq w \wedge \,\blacksquare\, t = w \vee p_{o\imath} t]$	.1, 6107
(.10)	$.1,\ .2,\ .4 \vdash m_\sigma [\lambda t_\imath \,\blacksquare\, t \neq w_\imath \wedge \,\blacksquare\, t = w \vee p_{o\imath} t]$	.8, .9, 5200
(.11)	$.1,\ .2,\ .4 \vdash m_\sigma p_{o\imath}$	.5, .10
(.12)	$.1 \vdash n_\sigma p_{o\imath} \supset m_\sigma p$	.3, .11
(.13)	$.1 \vdash m_\sigma p_{o\imath} \supset n_\sigma p$	Deduction Theorem, Sub: .12
(.14)	$.1 \vdash \forall p_{o\imath} \,\blacksquare\, n_\sigma p = m_\sigma p$	.12, .13
(.15)	$.1 \vdash n_\sigma = m_\sigma$	.14, Axiom 3

■

This completes the derivation of Peano's Postulates for the natural numbers. Note that the Axiom of Infinity was needed to prove only the last of these. Henceforth our development of number theory will be based entirely on Peano's Postulates without further reference to the Axiom of Infinity or to the definitions of the natural numbers. Indeed, the reader may henceforth imagine that we are developing a language $\mathcal{Q}^\sigma$ which is like $\mathcal{Q}_0^\infty$ except that the type of individuals is called σ rather than $\imath$. Also, $\mathcal{Q}^\sigma$ has primitive constants 0_σ and $S_{\sigma\sigma}$, and has 6105 and 6109 as axioms in place of Axiom 6. $\mathbb{N}_{o\sigma}$ is defined as above, and 6100–6104 are theorems of $\mathcal{Q}^\sigma$.

DEFINITIONS. Given wffs $\mathbf{F}_{\alpha\alpha}$ and $\mathbf{A}_\alpha$, we define the wff$_\alpha$ $\mathbf{F}^n_{\alpha\alpha}\mathbf{A}_\alpha$ for each natural number $n \geq 0$ as follows by induction on n:

(a) $\mathbf{F}^0_{\alpha\alpha}\mathbf{A}_\alpha$ is $\mathbf{A}_\alpha$.
(b) $\mathbf{F}^{n+1}_{\alpha\alpha}\mathbf{A}_\alpha$ is $[\mathbf{F}_{\alpha\alpha} \,\blacksquare\, \mathbf{F}^n_{\alpha\alpha}\mathbf{A}_\alpha]$.

If n is any natural number, $\bar{n}$ is $[S^n_{\sigma\sigma}0_\sigma]$; this wff is called the *representation of* n in $\mathcal{Q}_0^\infty$. The wffs $\bar{n}$ represent natural numbers and are called *numerals.*

We also wish to define what it means for a wff to represent a function of natural numbers. This could be expressed by saying that the wff denotes the function in some specified model(s), but we wish to have a purely syntactic definition of representability, so we use the following:

DEFINITION. Let $k \geq 1$, and let g be a k-ary function from natural numbers to natural numbers. The cwff $\mathbf{G}_{\sigma\sigma\ldots\sigma}$ *represents* g in any extension of $\mathcal{Q}_0^\infty$ iff for all k-tuples $(n_1, \ldots, n_k)$ of natural numbers,

$$\vdash \mathbf{G}_{\sigma\sigma\ldots\sigma}\overline{n_1}\ldots\overline{n_k} = \overline{g(n_1, \ldots, n_k)}.$$

Let $k \geq 1$, and let g be a k-ary relation between natural numbers. (If $k = 1$, let g be a set of natural numbers.) The cwff $\mathbf{G}_{o\sigma\ldots\sigma}$ *represents* g iff for all k-tuples $(n_1, \ldots, n_k)$ of natural numbers,

$$\vdash \mathbf{G}_{o\sigma\ldots\sigma}\overline{n_1}\ldots\overline{n_k} = T_o \quad \text{if } g(n_1, \ldots, n_k), \text{ and}$$
$$\vdash \mathbf{G}_{o\sigma\ldots\sigma}\overline{n_1}\ldots\overline{n_k} = F_o \quad \text{if not } g(n_1, \ldots, n_k).$$

A numerical function or relation or set is *representable* in $\mathcal{Q}_0^\infty$ iff there is a wff which represents it.

6110 $\vdash \mathbb{N}_{o\sigma}\bar{n}$ where n is any natural number.

The proof is by induction (in the meta-language) on n, using 6100 and 6101.

6111. If k is a natural number greater than 0, $\vdash \dot{\forall} x_\sigma \,\blacksquare\, S^k x_\sigma \neq x_\sigma$.

Proof: By induction on x.

(.1) $\vdash S_{\sigma\sigma}[S^{k-1}0_\sigma] \neq 0_\sigma$	6105
(.2) $\mathbb{N}x_\sigma \vdash \mathbb{N}S^k x_\sigma$	by 6101 (used k times)
(.3) $\mathbb{N}x_\sigma \vdash S_{\sigma\sigma}[S^k x_\sigma] = S_{\sigma\sigma}x_\sigma \supset \,\blacksquare\, S^k x_\sigma = x_\sigma$	6109, .2
(.4) $\vdash \dot{\forall} x_\sigma \,\blacksquare\, S^k x_\sigma \neq x_\sigma \supset \,\blacksquare\, S^k[S_{\sigma\sigma}x_\sigma] \neq S_{\sigma\sigma}x_\sigma$	.3, def of $S^k\mathbf{A}_\sigma$

Then use 6102 with .1 and .4. ■

6112 Theorem. $\mathsf{Q}_{o\sigma\sigma}$ represents the equality relation. Thus, if m and n are distinct natural numbers, $\vdash \bar{m} \neq \bar{n}$.

Proof: If $m = n$, then $\vdash \bar{m} = \bar{n}$ by 5200. If $m \neq n$, we may assume $m > n$, and let $k = m - n$.

(.1) $\vdash \mathbb{N}\bar{n} \supset \,\blacksquare\, S^k\bar{n} \neq \bar{n}$ 6111

(.2) $\vdash \bar{m} \neq \bar{n}$ 6110, .1, def of $\bar{m}$

■

DEFINITION. The *characteristic function* χ_r of a k-ary relation r is defined as follows:

$$\begin{aligned} \chi_r(n_1, \ldots, n_k) &= 1 \quad \text{if } r(n_1, \ldots, n_k) \\ &= 0 \quad \text{if not } r(n_1, \ldots, n_k). \end{aligned}$$

6113 Theorem. Let r be any relation between natural numbers. Then r is representable iff χ_r is representable.

Proof: Let k be the number of arguments of r. If r is represented by $G_{o\sigma\ldots\sigma}$, let $C_{\sigma o\sigma\sigma}$ be the wff of theorem 5313 and let $H_{\sigma\sigma\ldots\sigma}$ be $[\lambda x_\sigma^1 \ldots \lambda x_\sigma^k \centerdot C_{\sigma o\sigma\sigma}\bar{1}\,\bar{0} \centerdot G_{o\sigma\ldots\sigma} x_\sigma^1 \ldots x_\sigma^k]$. We show that $H_{\sigma\sigma\ldots\sigma}$ represents χ_r. Let $n_1, \ldots,$ n_k be natural numbers.

(.1) $\vdash H_{\sigma\sigma\ldots\sigma}\ \overline{n_1} \ldots \overline{n_k} = C_{\sigma o\sigma\sigma}\bar{1}\,\bar{0} \centerdot G_{o\sigma\ldots\sigma}\overline{n_1} \ldots \overline{n_k}$ def of H

If $r(n_1, \ldots, n_k)$, then

(.2) $\vdash G_{o\sigma\ldots\sigma}\overline{n_1} \ldots \overline{n_k} = T$ since G represents r

(.3) $\vdash H_{\sigma\sigma\ldots\sigma}\overline{n_1} \ldots \overline{n_k} = \bar{1}$ .1, .2, 5313

If not $r(n_1, \ldots, n_k)$, then

(.4) $\vdash G_{o\sigma\ldots\sigma}\overline{n_1} \ldots \overline{n_k} = F$ since G represents r

(.5) $\vdash H_{\sigma\sigma\ldots\sigma}\overline{n_1} \ldots \overline{n_k} = \bar{0}$ .1, .4, 5313

Thus $H_{\sigma\sigma\ldots\sigma}$ does indeed represent χ_r.

On the other hand, if a wff $H_{\sigma\sigma\ldots\sigma}$ represents χ_r, it is easy to see with the aid of 5200 and 6112 that $[\lambda x_\sigma^1 \ldots \lambda x_\sigma^k \centerdot Q_{o\sigma\sigma}\bar{1} \centerdot H_{\sigma\sigma\ldots\sigma} x_\sigma^1 \ldots x_\sigma^k]$ represents r. ■

EXERCISES

X6100. Let f be the function from natural numbers to natural numbers such that $f(0) = 3$ and $f(n+1) = f(n)+1$ for all natural numbers n. Show that f is representable in $\mathcal{Q}_0^\infty$ by giving an explicit, detailed, and complete demonstration that it satisfies the definition of representability.

Prove the theorems stated in exercises X6101-X6108.

X6101. $\bar{1} = \sum_{o(o\iota)}^{1}$

X6102. $\forall n_\sigma \,\blacksquare\, \mathrm{N}_{o\sigma} n_\sigma \supset [NC]_{o\sigma} n_\sigma$ (Every natural number is a cardinal number).

X6103. $\forall n_\sigma \sim n_\sigma[\lambda x_\imath T_o]$

X6104. The *composition* of functions f and g is the function $f \circ g$ defined as follows: $(f \circ g)(x) = f(g(x))$. An *iterate* of a function g is a function of the form $g \circ \ldots \circ g$ (the composition of k copies of g, where k is any natural number). In particular, we regard the identity function (here $k = 0$), g itself ($k = 1$), and $g \circ g$ ($k = 2$) as iterates of g.

(a) Define a closed wff $I_{o(aa)(aa)}$ such that $I_{o(aa)(aa)} g_{aa}$ is a name in type theory for the set of iterates of g_{aa}.
(b) Prove $I_{o(aa)(aa)} g_{aa}[\lambda x_a x_a]$.
(c) Prove $I_{o(aa)(aa)} g_{aa}[\lambda x_a \,\blacksquare\, g_{aa} \,\blacksquare\, g_{aa} x_a]$.
(d) Prove $I_{o(aa)(aa)}[\lambda x_a y_a] f_{aa} \supset f_{aa} y_a = y_a$.

X6105. $\forall n_{o(o\iota)} . \mathrm{N} n \supset \forall q_{o\iota} . n\, q \supset \exists j_{\iota(o\iota)} \forall r_{o\iota} . r \subseteq q \wedge \exists x_\iota r\, x \supset r . j\, r$

X6106. $Finite[\lambda x_\iota T_o] \supset AC^\imath$

X6107. $S_{\sigma\sigma} n_\sigma p_{o\imath} = \,\blacksquare\, \sim 0_\sigma p \;\wedge\; \forall x_\imath \,\blacksquare\, px \supset n_\sigma[\lambda t_\imath \,\blacksquare\, t \neq x \;\wedge\; pt]$

X6108. $\bar{n} p_{o\imath} \supset \exists x_\imath^1 \ldots \exists x_\imath^n \,\blacksquare\, \bigwedge_{i=1}^{n} p_{o\imath} x_\imath^i \;\wedge\; \bigwedge_{1 \leq i < j \leq n} x_\imath^i \neq x_\imath^j$ for each positive integer n. (*Hint*: use induction.)

X6109. Show that if $\mathcal{M}$ is a model for $\mathcal{Q}_0$ with exactly n individuals (members of $\mathcal{D}_\imath$), $\mathcal{M} \models \bar{n} = \mathrm{Q}_{o(o\imath)(o\imath)}[\lambda x_\imath T_o]$ and $\mathcal{M} \models \bar{m} = [\lambda p_{o\imath} F_o]$ for each integer $m > n$.

X6110. Show that if n is any positive integer and $\mathcal{M}$ is any model of $\mathcal{Q}_0$, then $\mathcal{M} \models \exists p_{o\imath} \bar{n}_\sigma p_{o\imath}$ if and only if $\mathcal{M}$ has at least n individuals.

§62. Order

Intuitively, if x and y are natural numbers, $x \leq y$ iff one can get from x to y by zero or more applications of the successor operation, i.e., iff $y \in \{x, \mathrm{S}x, \mathrm{SS}x, \mathrm{SSS}x, \ldots\}$. We formalize this idea in the following definition:

DEFINITION. $\leq_{o\sigma\sigma}$ stands for

$$[\lambda x_\sigma \lambda y_\sigma \forall p_{o\sigma} \,\blacksquare\, [p_{o\sigma} x_\sigma \;\wedge\; \forall z_\sigma \,\blacksquare\, p_{o\sigma} z_\sigma \supset p_{o\sigma} \,\blacksquare\, Sz_\sigma] \supset p_{o\sigma} y_\sigma].$$

$\mathbf{A}_\sigma \leq \mathbf{B}_\sigma$ stands for $\leq_{o\sigma\sigma} \mathbf{A}_\sigma \mathbf{B}_\sigma$.

6200. $\vdash \dot{\forall} y_\sigma \centerdot 0 \leq y_\sigma$

Proof:

(.1)	$\vdash \mathbb{N} y_\sigma = \centerdot 0 \leq y_\sigma$	defs of $\mathbb{N}$ and $\leq$
(.2)	$\vdash \dot{\forall} y_\sigma \centerdot 0 \leq y_\sigma$	.1

■

REMARK. Reflection on 6200.1 shows that we could have simply defined $\mathbb{N}_{o\sigma}$ as $\leq_{o\sigma\sigma} 0_\sigma$.

6201. $\vdash x_\sigma \leq x_\sigma$

Proof:

(.1)	$\vdash p_{o\sigma} x_\sigma \ \wedge \ \forall z_\sigma [pz \supset p \centerdot Sz] \supset px$	
(.2)	$\vdash x_\sigma \leq x$	.1, def of $\leq$

■

6202. $\vdash x_\sigma = y_\sigma \supset x_\sigma \leq y_\sigma$

Proof: By 6201. ■

6203. $\vdash x_\sigma \leq Sx_\sigma$

Proof:

(.1)	$.1 \vdash p_{o\sigma} x_\sigma \ \wedge \ \forall z_\sigma \centerdot pz \supset p \centerdot Sz$	hyp
(.2)	$.1 \vdash p_{o\sigma} \centerdot Sx_\sigma$	.1
(.3)	$\vdash x_\sigma \leq Sx$	.2, def of $\leq$

■

6204. $\vdash x_\sigma \leq y_\sigma \ \wedge \ y_\sigma \leq z_\sigma \supset x_\sigma \leq z_\sigma$

Proof:

(.1)	$.1 \vdash x_\sigma \leq y_\sigma$	hyp
(.2)	$.2 \vdash y_\sigma \leq z_\sigma$	hyp
(.3)	$.3 \vdash p_{o\sigma} x_\sigma \ \wedge \ \forall w_\sigma \centerdot pw \supset p \centerdot Sw$	hyp
(.4)	$.1, .2, .3 \vdash p_{o\sigma} y_\sigma$	.1, .3, def of $\leq$
(.5)	$.1, .2, .3 \vdash p_{o\sigma} z_\sigma$	.2, .3, .4, def of $\leq$
(.6)	$.1, .2 \vdash x_\sigma \leq z_\sigma$	.5, def of $\leq$

■

6205. $\vdash \dot{\forall} x_\sigma \centerdot x_\sigma \leq y_\sigma \supset \centerdot x_\sigma = y_\sigma \lor \dot{\exists} z_\sigma \centerdot x_\sigma \leq z_\sigma \land Sz = y$

Proof: Let $P_{o\sigma}$ be $[\lambda w_\sigma \centerdot x_\sigma = w \lor \dot{\exists} z_\sigma \centerdot x \leq z \land Sz = w]$.

(.1)	$\vdash x_\sigma \leq y_\sigma \supset \centerdot [Px \land \forall w_\sigma \centerdot Pw \supset P \centerdot Sw] \supset Py$	def of $\leq$
(.2)	$\vdash Px_\sigma$	5200, def of P
(.3)	$Pw_\sigma \vdash x_\sigma = w \lor \dot{\exists} z_\sigma \centerdot x \leq z \land Sz = w$	hyp, def of P
(.4)	$\mathbb{N} x_\sigma, x = w_\sigma \vdash \mathbb{N} x \land x \leq x \land Sx = Sw$	hyp, 6201
(.5)	$\mathbb{N} x_\sigma, x = w_\sigma \vdash \dot{\exists} z_\sigma \centerdot x \leq z \land Sz = Sw$	$\exists$ Gen: .4
(.6)	$.6 \vdash \mathbb{N} z_\sigma \land x_\sigma \leq z \land Sz = w_\sigma$	hyp
(.7)	$.6 \vdash \mathbb{N} Sz_\sigma$	.6, 6101
(.8)	$.6 \vdash x_\sigma \leq Sz_\sigma$	.6, 6203, 6204
(.9)	$.6 \vdash SSz = Sw$	.6
(.10)	$.6 \vdash \mathbb{N} Sz_\sigma \land x_\sigma \leq Sz \land SSz = Sw_\sigma$	.7, .8, .9
(.11)	$.6 \vdash \dot{\exists} z_\sigma \centerdot x_\sigma \leq z \land Sz = Sw_\sigma$	$\exists$ Gen: .10
(.12)	$\exists z_\sigma [.6] \vdash .11$	.11
(.13)	$\mathbb{N} x_\sigma, Pw_\sigma \vdash .11$	.3, .5, .12
(.14)	$\mathbb{N} x_\sigma \vdash \forall w_\sigma \centerdot Pw \supset P \centerdot Sw$	.13, def of P
(.15)	$\mathbb{N} x_\sigma \vdash x \leq y_\sigma \supset Py$	.1, .2, 14

■

6206. $\vdash \dot{\forall} x_\sigma \dot{\forall} y_\sigma \centerdot x_\sigma \leq Sy_\sigma \equiv \centerdot x_\sigma \leq y_\sigma \lor x_\sigma = Sy_\sigma$

Proof:

(.1)	$.1 \vdash \mathbb{N} x_\sigma \land \mathbb{N} y_\sigma \land x_\sigma \leq Sy_\sigma$	hyp
(.2)	$.1 \vdash x_\sigma = Sy_\sigma \lor \dot{\exists} z_\sigma \centerdot x \leq z \land Sz = Sy$	.1, 6205
(.3)	$.3 \vdash \mathbb{N} z_\sigma \land x_\sigma \leq z \land Sz = Sy$	hyp (choose z)
(.4)	$.1, .3 \vdash z_\sigma = y_\sigma$	.1, .3, 6109
(.5)	$.1, .3 \vdash x_\sigma \leq y_\sigma$	.3, .4
(.6)	$.1, \exists z_\sigma [.3] \vdash x_\sigma \leq y_\sigma$	.5
(.7)	$.1 \vdash x_\sigma \leq y_\sigma \lor x = Sy$	.2, .6
(.8)	$\vdash x_\sigma \leq y_\sigma \supset x \leq Sy$	6203, 6204
(.9)	$\vdash x_\sigma = Sy_\sigma \supset x \leq Sy$	6202
(.10)	$\vdash \mathbb{N} x_\sigma \land \mathbb{N} y_\sigma \supset \centerdot x \leq Sy \equiv \centerdot x \leq y \lor x = Sy$	.7, .8, .9

■

6207⁰. $\vdash \dot{\forall} x_\sigma \centerdot x \leq 0 \equiv \centerdot x = 0$

6207¹. $\vdash \dot{\forall} x_\sigma \centerdot x \leq \overline{1} \equiv \centerdot x = 0 \lor x = \overline{1}$

6207ᵏ. $\vdash \dot{\forall} x_\sigma \centerdot x \leq \overline{k} \equiv \centerdot x = 0 \lor x = \overline{1} \lor \ldots \lor x = \overline{k}$

We prove these theorems by induction (in the meta-language) on k.

(.1) $\vdash \mathbb{N}x_\sigma \supset \centerdot x \leq 0 \supset \centerdot x = 0 \lor \dot{\exists} z_\sigma \centerdot x \leq z \land Sz = 0$ 6205
(.2) $\vdash \sim \dot{\exists} z_\sigma \centerdot x_\sigma \leq z \land Sz = 0$ 6106
(.3) $\vdash \dot{\forall} x_\sigma \centerdot x \leq 0 \equiv \centerdot x = 0$ .1, .2, 6202

This proves 6207^0. Next we assume we have proved 6207^k (where $k \geq 0$), and prove 6207^{k+1}.

(.4) $\mathbb{N}x_\sigma \vdash x \leq \overline{k} \equiv \centerdot x = 0 \lor \ldots \lor x = \overline{k}$ by inductive hypothesis
(.5) $\mathbb{N}x_\sigma \vdash x \leq S\overline{k} \equiv \centerdot x \leq \overline{k} \lor x = S\overline{k}$ 6206, 6110
(.6) $\mathbb{N}x_\sigma \vdash x \leq S\overline{k} \equiv \centerdot x = 0 \lor \ldots \lor x = \overline{k} \lor x = S\overline{k}$ .4, .5

This gives the desired theorem, since $\overline{k+1}$ is $S\overline{k}$. ■

6208 Theorem. $\leq_{o\sigma\sigma}$ represents the relation $\leq$ between natural numbers.

Proof: $\vdash \overline{n} \leq \overline{k} = \centerdot \overline{n} = 0 \lor \ldots \lor \overline{n} = \overline{k}$, and $\mathsf{Q}_{o\sigma\sigma}$ represents equality (6112). ■

6209^k. $\vdash \dot{\forall} y_\sigma [y_\sigma \leq \overline{k} \supset p_{o\sigma} y_\sigma] = \centerdot p_{o\sigma} 0 \land \ldots \land p_{o\sigma} \overline{k}$

Proof: Let $j \leq k$ in each line below where j occurs.

($.1^j$) $\vdash \dot{\forall} y_\sigma [y \leq \overline{k} \supset p_{o\sigma} y] \supset \centerdot \overline{j} \leq \overline{k} \supset p\overline{j}$ 6110
($.2^j$) $\vdash \overline{j} \leq \overline{k}$ 6208
(.3) $\vdash \dot{\forall} y_\sigma [y \leq \overline{k} \supset p_{o\sigma} y] \supset \centerdot p0 \land \ldots \land p\overline{k}$ $.1^0 - .1^k$, $.2^0 - .2^k$
(.4) .4 $\vdash p_{o\sigma} 0 \land \ldots \land p\overline{k} \land \mathbb{N} y_\sigma \land y \leq \overline{k}$ hyp
(.5) .4 $\vdash y_\sigma = 0 \lor \ldots \lor y = \overline{k}$ .4, 6207^k
($.6^j$) $\vdash y_\sigma = \overline{j} \supset \centerdot p_{o\sigma} y = p\overline{j}$ Axiom 2
(.7) .4 $\vdash p_{o\sigma} y_\sigma$ .4, .5, $.6^0 - .6^k$
(.8) $\vdash p_{o\sigma} 0 \land \ldots \land p\overline{k} \supset \dot{\forall} y_\sigma \centerdot y \leq \overline{k} \supset py$ .7

The theorem follows from .3 and .8. ■

6210^k. $\vdash \dot{\exists} y_\sigma [y_\sigma \leq \overline{k} \land p_{o\sigma} y_\sigma] = \centerdot p_{o\sigma} 0 \lor \ldots \lor p_{o\sigma} \overline{k}$

Proof:

(.1) $\vdash \dot{\forall} y_\sigma [y \leq \overline{k} \supset \sim p_{o\sigma} y] = \centerdot \sim p0 \land \ldots \land \sim p\overline{k}$ Sub: 6209^k
(.2) $\vdash \sim \dot{\forall} y_\sigma [y \leq \overline{k} \supset \sim p_{o\sigma} y] = \dot{\exists} y_\sigma \centerdot y \leq \overline{k} \land py$ 6000, defs of $\dot{\forall}$ and $\dot{\exists}$

The theorem follows from .1 and .2 by Rule P. ■

6211. $\vdash \dot{\forall} y_\sigma \dot{\forall} x_\sigma \leq y_\sigma \supset Sx_\sigma \leq Sy_\sigma$

Proof by induction on y:

(.1)	$.1 \vdash \mathbb{N}x_\sigma \ \wedge \ x \leq 0$	hyp
(.2)	$.1 \vdash x_\sigma = 0$	.1, 6207^0
(.3)	$.1 \vdash Sx_\sigma = S0$	.2
(.4)	$.1 \vdash Sx_\sigma \leq S0$	.3, 6202
(.5)	$\vdash \dot{\forall} x_\sigma \,\blacksquare\, x \leq 0 \supset Sx \leq S0$	.4
(.6)	$.6 \vdash \mathbb{N}y_\sigma \ \wedge \ \dot{\forall} x_\sigma \,\blacksquare\, x \leq y_\sigma \supset Sx \leq Sy$	(inductive) hyp
(.7)	$.7 \vdash \mathbb{N}x_\sigma \ \wedge \ x \leq Sy$	hyp
(.8)	$.6,\ .7 \vdash x_\sigma \leq y_\sigma \ \vee \ x = Sy$	.6, .7, 6206
(.9)	$.6,\ .7,\ x_\sigma \leq y_\sigma \vdash Sx \leq Sy$	hyp, .6, .7
(.10)	$.6,\ .7,\ x_\sigma \leq y_\sigma \vdash Sx \leq SSy$	.9, 6203, 6204
(.11)	$.6,\ .7,\ x_\sigma = Sy_\sigma \vdash Sx = SSy$	hyp
(.12)	$.6,\ .7,\ x_\sigma = Sy_\sigma \vdash Sx \leq SSy$	.11, 6202
(.13)	$.6,\ .7 \vdash Sx_\sigma \leq SSy_\sigma$	.8, .10, .12
(.14)	$.6 \vdash \dot{\forall} x_\sigma \,\blacksquare\, x \leq Sy_\sigma \supset Sx \leq SSy$	.13
(.15)	$\vdash \dot{\forall} y_\sigma \dot{\forall} x_\sigma \,\blacksquare\, x \leq y \supset Sx \leq Sy$	6102, .5, .14

■

6212. $\vdash \dot{\forall} x_\sigma \dot{\forall} y_\sigma \,\blacksquare\, x \leq y \ \vee \ y \leq x$

Proof by induction on x:

(.1)	$\vdash \dot{\forall} y_\sigma \,\blacksquare\, 0 \leq y \ \vee \ y \leq 0$	6200
(.2)	$.2 \vdash \mathbb{N}x_\sigma \ \wedge \ \dot{\forall} y_\sigma \,\blacksquare\, x \leq y \ \vee \ y \leq x$	(inductive) hyp
(.3)	$.2,\ \mathbb{N}y_\sigma \vdash x_\sigma \leq y \ \vee \ y \leq x$	.2, hyp
(.4)	$\vdash y_\sigma \leq x_\sigma \supset y \leq Sx$	6203, 6204
(.5)	$.2 \vdash x_\sigma \leq y_\sigma \supset \,\blacksquare\, x = y \ \vee \ \dot{\exists} z_\sigma \,\blacksquare\, x \leq z \ \wedge \ Sz = y$	.2, 6205
(.6)	$\vdash x_\sigma = y_\sigma \supset y \leq Sx$	5302, 6202, .4
(.7)	$.7 \vdash \mathbb{N}z_\sigma \ \wedge \ x_\sigma \leq z \ \wedge \ Sz = y$	hyp (choose *z*)
(.8)	$.2,\ .7 \vdash Sx_\sigma \leq Sz_\sigma$	.2, .7, 6211
(.9)	$.2,\ .7 \vdash Sx_\sigma \leq y_\sigma$	.7, .8
(.10)	$.2,\ \exists z_\sigma[.7] \vdash Sx_\sigma \leq y_\sigma$	.9
(.11)	$.2,\ \mathbb{N}y_\sigma \vdash Sx_\sigma \leq y \vee \ y \leq Sx$	.3, .4, .5, .6, .10
(.12)	$\vdash \dot{\forall} x_\sigma \,\blacksquare\, \dot{\forall} y_\sigma[x \leq y \ \vee \ y \leq x] \supset \dot{\forall} y_\sigma \,\blacksquare\, Sx \leq y \ \vee \ y \leq Sx$	.11

The theorem follows from .1 and .12 by the Induction Theorem. ■

6213. $\vdash \dot\forall y_\sigma \dot\forall x_\sigma \centerdot x \le y \wedge y \le x \supset x = y$

Proof by induction on y:

(.1)	$\vdash \dot\forall x_\sigma \centerdot x \le 0 \wedge 0 \le x \supset x = 0$	6207^0
(.2)	$.2 \vdash \mathbb{N} y_\sigma \wedge \dot\forall x_\sigma \centerdot x \le y \wedge y \le x \supset x = y$	(inductive) hyp
(.3)	$.3 \vdash \mathbb{N} x_\sigma \wedge x \le S y_\sigma \wedge S y_\sigma \le x$	hyp
(.4)	$.2, .3 \vdash x_\sigma \le y_\sigma \vee x = Sy$	.3, 6206
(.5)	$.5 \vdash x_\sigma \le y_\sigma$	hyp
(.6)	$.2, .3, .5 \vdash y_\sigma \le x_\sigma$	.3, 6203, 6204
(.7)	$.2, .3, .5 \vdash x_\sigma = y_\sigma$	.2, .3, .5, .6

Intuitively, .3 and .7 are contradictory, so under the assumptions .2 and .3, we have found that .5 is impossible, so $x = Sy$ by .4. Actually, it proves easier to derive $x = Sy$ than to derive an actual contradiction.

(.8)	$.2, .3, .5 \vdash y_\sigma \le S x_\sigma \wedge Sx \le y$	R′: .7, .3
(.9)	$.2, .3, .5 \vdash \mathbb{N} \centerdot S x_\sigma$	.3, 6101
(.10)	$.2, .3, .5 \vdash S x_\sigma = y_\sigma$	.2, .8, .9
(.11)	$.2, .3, .5 \vdash S y_\sigma = x_\sigma$	R′: .7, .10
(.12)	$.2, .3 \vdash x_\sigma = S y_\sigma$	.4, .11, 5302
(.13)	$\vdash \dot\forall y_\sigma \centerdot \dot\forall x_\sigma [x \le y \wedge y \le x \supset x = y]$ $\supset \dot\forall x_\sigma \centerdot x \le Sy \wedge Sy \le x \supset x = Sy$	.12

The theorem follows from .1 and .13 by the Induction Theorem. ■

6214. $\vdash \dot\forall x_\sigma \dot\forall y_\sigma \centerdot \sim x \le y \supset Sy \le x$

Proof:

(.1)	$.1 \vdash \mathbb{N} x_\sigma \wedge \mathbb{N} y_\sigma \wedge \sim x \le y$	hyp
(.2)	$.1 \vdash x_\sigma \le S y_\sigma \vee Sy \le x$	.1, 6101, 6212
(.3)	$.1 \vdash x_\sigma \le S y_\sigma \equiv \centerdot x \le y \vee x = Sy$	.1, 6206
(.4)	$.1 \vdash x_\sigma = S y_\sigma \supset Sy \le x$	5302, 6202
(.5)	$.1 \vdash S y_\sigma \le x_\sigma$	.1, .2, .3, .4

■

We may now supply more details of the discussion about axioms of infinity begun in §60. Let $Infin^{\alpha}$ be the wff

$$\exists r_{o\alpha\alpha}\forall x_{\alpha}\forall y_{\alpha}\forall z_{\alpha}[\exists w_{\alpha}rxw \wedge \sim rxx \wedge \centerdot rxy \supset \centerdot ryz \supset rxz].$$

Note that $Infin^{\imath}$ is Axiom 6.

6215. $\vdash Infin^{\sigma}$

Proof: Let $R_{o\sigma\sigma}$ be the wff $[\lambda n_{\sigma}\lambda m_{\sigma}\centerdot \mathbb{N}m \wedge \centerdot \sim \mathbb{N}n \vee Sn \leq m]$.
Note that $Sn \leq m$ means that $n < m$.

(.1) $\vdash \sim \mathbb{N}x_{\sigma} \supset Rx0$	6100, def of R
(.2) $\vdash \sim \mathbb{N}x_{\sigma} \supset \exists w_{\sigma}Rxw$	.1
(.3) $\vdash \mathbb{N}x_{\sigma} \supset Rx\centerdot Sx$	6101, 6201, def of R
(.4) $\vdash \mathbb{N}x_{\sigma} \supset \exists w_{\sigma}Rxw$	.3
(.5) $\vdash \exists w_{\sigma}Rx_{\sigma}w$	.2, .4
(.6) $.6 \vdash Rx_{\sigma}x$	hyp
(.7) $.6 \vdash \mathbb{N}x_{\sigma} \wedge Sx \leq x$	.6, def of R
(.8) $.6 \vdash x_{\sigma} = Sx$	.7, 6203, 6101, 6213
(.9) $.6 \vdash F_{o}$	.7, .8, 6111
(.10) $\vdash \sim Rx_{\sigma}x$	.9
(.11) $.11 \vdash Rx_{\sigma}y_{\sigma} \wedge Ryz_{\sigma}$	hyp
(.12) $.11 \vdash \mathbb{N}y_{\sigma} \wedge \mathbb{N}z_{\sigma}$	.11, def of R
(.13) $.11 \vdash \sim \mathbb{N}x_{\sigma} \supset Rxz_{\sigma}$	.12, def of R
(.14) $.11,\ .14 \vdash \mathbb{N}x_{\sigma}$	hyp
(.15) $.11,\ .14 \vdash Sx_{\sigma} \leq y_{\sigma} \wedge Sy \leq z_{\sigma}$	.11, .14, def of R
(.16) $.11,\ .14 \vdash Sx_{\sigma} \leq z_{\sigma}$	.15, 6203, 6204
(.17) $.11 \vdash \mathbb{N}x_{\sigma} \supset Rx_{\sigma}z_{\sigma}$	.12, .16, def of R
(.18) $\vdash Rx_{\sigma}y_{\sigma} \wedge Ryz_{\sigma} \supset Rxz$	.13, .17
(.19) $\vdash Infin^{\sigma}$	.5, .10, .18

■

Note that we have now shown that $\vdash_{\mathcal{Q}_0} Infin^{\imath} \supset Infin^{\sigma}$. It is clear that $\vdash_{\mathcal{Q}_0} Infin^{\alpha} \supset Infin^{(o(o\alpha))}$. Also, we have shown that without using the definitions of $\sigma, 0$, or S, one can prove $\vdash_{\mathcal{Q}_0}$ [6106] $\wedge$ [6109] $\supset Infin^{\sigma}$. (See the discussion following the proof of 6109. For the sake of economy we have used 6105 in several places where we could have used 6106.) The reader will now find it easy to fill in the details of the argument in §60 that $\vdash_{\mathcal{Q}_0} Inf^{II} \supset$ Axiom 6.

EXERCISES

X6200. Note that we have used the definition of $\leq$ only in the proofs of theorems 6200, 6201, 6203, 6204, and 6205. Of course, there are many theorems about $\leq$ which remain to be proved. The question naturally arises whether the definition of $\leq$ must be used (directly) again to derive some of these, or whether 6200–6205 characterize $\leq$ on the natural numbers. (In the latter case, of course, first-order formulations of these theorems could be taken as axioms in a first-order theory of natural numbers in which $\leq$ is a primitive predicate constant.) Investigate this question by considering whether the following wff is a theorem:

$$\forall r_{o\sigma\sigma} \mathbin{\blacksquare} \dot{\forall} x_\sigma \dot{\forall} y_\sigma \dot{\forall} z_\sigma [r0y \wedge rxx \wedge rx[Sx] \wedge [rxy \wedge ryz \supset rxz] \\ \wedge \mathbin{\blacksquare} rxy \supset \mathbin{\blacksquare} x = y \vee \dot{\exists} w_\sigma \mathbin{\blacksquare} rxw \wedge Sw = y] \supset \dot{\forall} x_\sigma \dot{\forall} y_\sigma \mathbin{\blacksquare} rxy \equiv x \leq y.$$

X6201. Show that $\vdash_{\mathcal{Q}_0} Infin^\alpha \supset Infin^{(o\alpha)}$.

X6202. Let max be the numerical function such that

$$\begin{aligned} \max(x,y) &= x \text{ if } x \geq y; \\ &= y \text{ otherwise.} \end{aligned}$$

Show that max is a representable function.

§63. Minimization

The theorems of §62 show that $\leq$ orders the set $\mathbb{N}$ of natural numbers. We shall now prove that this ordering is a well-ordering, i.e., that every nonempty subset p of $\mathbb{N}$ has a least member μp. Indeed, we shall define a mapping $\mu_{\sigma(o\sigma)}$ which maps each set $p_{o\sigma}$ to the least natural number in $p_{o\sigma}$ (if there is one). Thus, $\mu_{\sigma(o\sigma)}$ will be a choice function for $\mathbb{N}$.

DEFINITIONS.

$\mu_{\sigma(o\sigma)}$ stands for $[\lambda p_{o\sigma} \imath x_\sigma \mathbin{\blacksquare} \mathrm{N}_{o\sigma} x_\sigma \wedge p_{o\sigma} x_\sigma \wedge \dot{\forall} y_\sigma \mathbin{\blacksquare} p_{o\sigma} y_\sigma \supset x_\sigma \leq y_\sigma]$.
$[\mu \mathbf{x}_\sigma \mathbf{A}]$ stands for $\mu_{\sigma(o\sigma)}[\lambda \mathbf{x}_\sigma \mathbf{A}]$.

$[\mu \mathbf{x}_\sigma \mathbf{A}]$ denotes "the least natural number $\mathbf{x}_\sigma$ such that $\mathbf{A}$" (when one exists). For this reason, μ is sometimes called "the least-number operator".

6300. $\vdash \mathbb{N}_{o\sigma} t_\sigma \wedge p_{o\sigma} t_\sigma \supset \bullet \mathbb{N}_{o\sigma}[\mu_{\sigma(o\sigma)} p_{o\sigma}] \wedge p_{o\sigma}[\mu_{\sigma(o\sigma)} p_{o\sigma}] \wedge \mu_{\sigma(o\sigma)} p_{o\sigma} \leq t_\sigma$

Proof: Let $G_{o\sigma}$ be $[\lambda x_\sigma \bullet \mathbb{N} x \wedge px \wedge \dot{\forall} y_\sigma \bullet py \supset x \leq y]$. Note that $p_{o\sigma}$ is the sole free variable in $G_{o\sigma}$, which denotes the set of smallest natural numbers in the set denoted by $p_{o\sigma}$.

(.1)	$\vdash \mu_{\sigma(o\sigma)} p_{o\sigma} = \iota_{\sigma(o\sigma)} G_{o\sigma}$	defs of μ and G

We wish to prove $\exists_1 x_\sigma G_{o\sigma} x_\sigma$ from the hypothesis $\mathbb{N}_{o\sigma} t_\sigma \wedge p_{o\sigma} t_\sigma$. Intuitively, it is clear that if t_σ is a natural number in $p_{o\sigma}$, then there is a least natural number in $p_{o\sigma}$ and it will be in $G_{o\sigma}$. How do we "find" this number? We can imagine examining natural numbers $\leq t_\sigma$ in order until we first come upon one in $p_{o\sigma}$. This suggests that some kind of an inductive process may be useful. Of course, a proof by induction will work only if we seek to prove a statement which is actually true for all natural numbers. Therefore, let us prove $\dot{\forall} z_\sigma \bullet \dot{\exists} x_\sigma [p_{o\sigma} x_\sigma \wedge x_\sigma \leq z_\sigma] \supset \exists x_\sigma G_{o\sigma} x_\sigma$ by induction on z.

(.2)	$.2 \vdash \mathbb{N} x_\sigma \wedge p_{o\sigma} x \wedge x \leq 0$	hyp
(.3)	$.2 \vdash x_\sigma = 0$	.2, 6207^0
(.4)	$.2 \vdash \dot{\forall} y_\sigma \bullet p_{o\sigma} y \supset x_\sigma \leq y$	.3, 6200
(.5)	$.2 \vdash G x_\sigma$	.2, .4, def of G
(.6)	$\vdash \dot{\exists} x_\sigma [p_{o\sigma} x \wedge x \leq 0] \supset \exists x_\sigma G x$	.5

Next we treat the induction step.

(.7)	$.7 \vdash \mathbb{N} z_\sigma \wedge \bullet \dot{\exists} x_\sigma [p_{o\sigma} x \wedge x \leq z] \supset \exists x_\sigma G x$	(inductive) hyp
(.8)	$.8 \vdash \mathbb{N} x_\sigma \wedge p_{o\sigma} x \wedge x \leq S z_\sigma$	hyp

From .7 and .8 we must prove $\exists x_\sigma G x$. From .7 it can be seen that we really need be concerned only with the case in line .9.

(.9)	$.7, .8, .9 \vdash \sim \dot{\exists} x_\sigma \bullet p_{o\sigma} x \wedge x \leq z_\sigma$	hyp
(.10)	$.7, .8, .9 \vdash \dot{\forall} y_\sigma \bullet p_{o\sigma} y_\sigma \supset \sim y \leq z_\sigma$	.9
(.11)	$.7, .8, .9, .11 \vdash \mathbb{N} y_\sigma \wedge p_{o\sigma} y_\sigma$	hyp
(.12)	$.7, .8, .9, .11 \vdash \sim y \leq z_\sigma$	.10, .11
(.13)	$.7, .8, .9, .11 \vdash S z_\sigma \leq y_\sigma$	.7, .11, .12, 6214
(.14)	$.7, .8, .9, .11 \vdash x_\sigma \leq y_\sigma$	.8, .13, 6204
(.15)	$.7, .8, .9 \vdash \dot{\forall} y_\sigma \bullet p_{o\sigma} y \supset x \leq y$	.14
(.16)	$.7, .8, .9 \vdash G x_\sigma$	.8, .15, def of G
(.17)	$.7, .8, .9 \vdash \exists x_\sigma G x$	.16
(.18)	$.7, .8, \vdash \exists x_\sigma G x$	.7, .17

(.19) $.7, \exists x_\sigma[.8] \vdash \exists x_\sigma Gx$ .18

(.20) $\vdash \dot{\forall} z_\sigma \centerdot \dot{\exists} x_\sigma[p_{o\sigma}x \land x \leq z] \supset \exists x_\sigma Gx \supset \centerdot \dot{\exists} x_\sigma[p_{o\sigma}x \land x \leq Sz] \supset \exists x_\sigma Gx$ .19

This completes the induction step.

(.21) $\vdash \dot{\forall} z_\sigma \centerdot \dot{\exists} x_\sigma[p_{o\sigma}x \land x \leq z] \supset \exists x_\sigma Gx$ 6102, .6, .20

(.22) $.22 \vdash \mathbb{N} t_\sigma \land p_{o\sigma}t$ hyp

(.23) $.22 \vdash t_\sigma \leq t$ 6201

(.24) $.22 \vdash \dot{\exists} x_\sigma \centerdot p_{o\sigma}x \land x \leq t_\sigma$ .22, .23

(.25) $.22 \vdash \exists x_\sigma Gx$ .21, .22, .24

Next we prove the uniqueness of membership in $G_{o\sigma}$.

(.26) $.26 \vdash Gx_\sigma \land Gz_\sigma$ hyp

(.27) $.26 \vdash \mathbb{N} x_\sigma \land p_{o\sigma}x \land \mathbb{N} z_\sigma \land pz$ .26

(.28) $.26 \vdash \dot{\forall} y_\sigma \centerdot p_{o\sigma}y \supset x_\sigma \leq y$ .26

(.29) $.26 \vdash \dot{\forall} y_\sigma \centerdot p_{o\sigma}y \supset z_\sigma \leq y$ .26

(.30) $.26 \vdash x_\sigma \leq z_\sigma \land z \leq x$ .27, .28, .29

(.31) $.26 \vdash x_\sigma = z_\sigma$ 6213, .30, .27

(.32) $\vdash \forall x_\sigma \forall z_\sigma \centerdot Gx \land Gz \supset x = z$ .31

(.33) $.22 \vdash \exists_1 x_\sigma Gx$ 5307, .25, .32

(.34) $.22 \vdash G_{o\sigma} \centerdot \iota_{\sigma(o\sigma)} G_{o\sigma}$ 5311, .33

(.35) $.22 \vdash G_{o\sigma} \centerdot \mu_{\sigma(o\sigma)} p_{o\sigma}$ .1, .34

(.36) $.22 \vdash \mathbb{N} \mu p_{o\sigma} \land p\mu p \land \dot{\forall} y_\sigma \centerdot py \supset \mu p \leq y$ .35, def of G

(.37) $.22 \vdash \mu p_{o\sigma} \leq t_\sigma$ .22, .36

The theorem follows from .36 and .37. ■

6301. $\vdash \mathbb{N}_{o\sigma} t_\sigma \land p_{o\sigma} t_\sigma \land \dot{\forall} y_\sigma[y_\sigma \leq t_\sigma \supset \centerdot p_{o\sigma} y_\sigma \supset y_\sigma = t_\sigma] \supset \mu_{\sigma(o\sigma)} p_{o\sigma} = t_\sigma$

Proof:

(.1) $.1 \vdash \mathbb{N} t_\sigma \land p_{o\sigma}t \land \dot{\forall} y_\sigma \centerdot y \leq t \supset \centerdot py \supset y = t$ hyp

(.2) $.1 \vdash \mathbb{N} \mu p_{o\sigma} \land p\mu p \land \mu p \leq t_\sigma$ 6300, .1

(.3) $.1 \vdash \mu p_{o\sigma} = t_\sigma$ Rule P, $\forall$I: .1, .2

■

We now need to introduce some terminology. We shall refer to functions from natural numbers to natural numbers as *numerical functions.* For the sake of uniformity of terminology, we shall sometimes refer to a natural number as a numerical function of zero arguments. Relations between natural numbers shall be referred to as *numerical relations*.

It will be convenient to use λ-notation in our meta-language to define numerical functions and relations. Thus, if G is a numerical expression (such as $2 \times n_1 \times (n_2 + n_3)$) in our meta-language, possibly containing numerical parameters $n_1, \ldots, n_k$, we let $((\lambda n_1, \ldots, n_k)G)$ denote that k-ary numerical function f such that $f(n_1, \ldots, n_k) = G$ for all natural numbers $n_1, \ldots, n_k$. Similarly, if B is a statement (such as $n_1 + n_2 = n_3$) in our meta-language about natural numbers $n_1, \ldots, n_k$, we let $((\lambda n_1, \ldots, n_k)B)$ denote that k-ary numerical relation r such that $r(n_1, \ldots, n_k)$ iff B, for all natural numbers $n_1, \ldots, n_k$. (Of course, in other contexts one might use λ-notation to denote functions and relations defined over domains other than the set of natural numbers.)

We shall also use the least-number operator μ in our meta-language. If B is a statement about a number n, then $(\mu n)B$ shall denote the least natural number n for which B is true, if one exists; if none exists, then $(\mu n)B$ will be undefined.

As an example of these notations, $((\lambda n)(\mu m)n \leq m^2)$ denotes that numerical function f whose value at n is the least natural number $\geq \sqrt{n}$; in particular if $n = k^2$, $f(n) = k = \sqrt{n}$.

In principle, we could also introduce the description operator $\imath$ into our meta-language. However, this is unnecessary as long as we confine our attention to natural numbers, since if there is a unique natural number n such that B, then "that n such that B" is "the least n such that B".

A $(k+1)$-ary numerical function g is said to be *regular* iff for all natural numbers $n_1, \ldots, n_k$ there is a natural number m such that $g(n_1, \ldots, n_k, m) = 1$. A $(k+1)$-ary numerical relation r is said to be *regular* iff its characteristic function χ_r is regular.

If g is a regular $(k+1)$-ary numerical function and

$$h = ((\lambda n_1, \ldots, n_k)(\mu m)g(n_1, \ldots, n_k, m) = 1),$$

we say that h is obtained from g by *minimization*. Similarly, if r is a regular $(k+1)$-ary numerical relation and

$$h = ((\lambda n_1, \ldots, n_k)(\mu m)r(n_1, \ldots, n_k, m)),$$

we say that h is obtained from r by *minimization*. Note that the regularity of g (or of r) assures that h is defined everywhere. Also note that h is obtained from a relation r by minimization iff h is obtained from χ_r by minimization.

6302 Theorem. If r is a representable $(k+1)$-ary regular numerical relation, then $((\lambda n_1, \ldots, n_k)(\mu m) r(n_1, \ldots, n_k, m))$ is a representable function.

Proof: Let $h = ((\lambda n_1, \ldots, n_k)(\mu m) r(n_1, \ldots, n_k, m))$. Suppose $R_{o\sigma\sigma\ldots\sigma}$ represents r, and let $H_{\sigma\sigma\ldots\sigma}$ be $[\lambda x^1_\sigma \ldots \lambda x^k_\sigma \mu_{\sigma(o\sigma)} \centerdot R_{o\sigma\sigma\ldots\sigma} x^1_\sigma \ldots x^k_\sigma]$. We show that H represents h. Suppose $h(n_1, \ldots, n_k) = m$. Thus

(.1) $\vdash R\overline{n_1} \ldots \overline{n_k}\overline{m} = T_o$ and
(.2) $\vdash R\overline{n_1} \ldots \overline{n_k}\overline{j} = F_o$ for each $j < m$.
(.3) $\vdash \mathbb{N}\overline{m} \wedge R\overline{n_1} \ldots \overline{n_k}\overline{m} \wedge \dot{\forall} y_\sigma[y_\sigma \leq \overline{m} \supset \centerdot R\overline{n_1} \ldots \overline{n_k} y \supset \centerdot y = \overline{m}] \supset$
$\centerdot \mu R\overline{n}_1 \ldots \overline{n}_k = \overline{m}$ 6301

Let $P_{o\sigma}$ be $[\lambda y_\sigma \centerdot R\overline{n_1} \ldots \overline{n_k} y_\sigma \supset y_\sigma = \overline{m}]$.

(.4) $\vdash \dot{\forall} y_\sigma[y \leq \overline{m} \supset P_{o\sigma} y_\sigma] = \centerdot P_{o\sigma} 0 \wedge \ldots \wedge P_{o\sigma}\overline{m}$ 6209^m
(.5) $\vdash P_{o\sigma}\overline{j}$ for each $j < m$ .2, def of P
(.6) $\vdash P_{o\sigma}\overline{m}$ 5200, def of P
(.7) $\vdash \mu R\overline{n_1} \ldots \overline{n_k} = \overline{m}$ .3, 6110, .1, .4, .5, .6, def of P
(.8) $\vdash H_{\sigma\sigma\ldots\sigma}\overline{n_1} \ldots \overline{n_k} = \overline{m}$ def of H, .7

■

6303 Theorem. If g is a representable $(k+1)$-ary regular numerical function, and h is obtained from g by minimization, then h is representable.

Proof: Let the relation r be $(\lambda n_1, \ldots, n_k, m)(g(n_1, \ldots, n_k, m) = 1)$. Clearly r is regular, so by 6302 it suffices to show that r is representable. Let $G_{\sigma\sigma\sigma\ldots\sigma}$ represent g, and let $R_{o\sigma\sigma\ldots\sigma}$ be $[\lambda x^1_\sigma \ldots \lambda x^k_\sigma \lambda y_\sigma \centerdot G x^1 \ldots x^k y = \overline{1}]$. We show that R represents r. Let $n_1, \ldots, n_k,\ m$ be any natural numbers.

(.1) $\vdash R\overline{n_1} \ldots \overline{n_k}\overline{m} \equiv \centerdot G\overline{n_1} \ldots \overline{n_k}\overline{m} = \overline{1}$ def of R
(.2) $\vdash G\overline{n_1} \ldots \overline{n_k}\overline{m} = \overline{g(n_1, \ldots, n_k, m)}$ since G represents g

Suppose $r(n_1, \ldots, n_k, m)$. Then $g(n_1, \ldots, n_k, m) = 1$.

(.3) $\vdash R\overline{n_1} \ldots \overline{n_k}\overline{m} = T_o$ .1, .2

Suppose not $r(n_1, \ldots, n_k, m)$. Then $g(n_1, \ldots, n_k, m) \neq 1$.

(.4) $\vdash \overline{g(n_1, \ldots, n_k, m)} \neq \overline{1}$ 6112
(.5) $\vdash R\overline{n_1} \ldots \overline{n_k}\overline{m} = F_o$ .1, .2, .4

■

§64. Recursive Functions

DEFINITION. Let g be a k-ary numerical function (where $k \geq 0$), and h be a $(k+2)$-ary numerical function. A $(k+1)$-ary function f is said to be defined by *primitive recursion* from g and h iff the following equations hold for all natural numbers $m_1, \ldots, m_k, n$:

$$f(m_1, \ldots, m_k, 0) = g(m_1, \ldots, m_k)$$

$$f(m_1, \ldots, m_k, \mathrm{S}n) = h(m_1, \ldots, m_k, n, f(m_1, \ldots, m_k, n))$$

EXAMPLE: $m + 0 = m$ and $m+\mathrm{S}n = \mathrm{S}(m + n)$ for all natural numbers m and n, where S is the successor function. Thus, if we let $g(m) = m$ and $h(m, n, p) = \mathrm{S}p$ for all natural numbers m, n, and p, and define

$$f(m, 0) = g(m)$$

$$f(m, \mathrm{S}n) = h(m, n, f(m, n)),$$

then $f(m, n) = m+n$ for all natural numbers m and n, so addition is defined by primitive recursion from g and h.

It can be seen that for given numerical functions g and h, there is a unique numerical function f satisfying these equations. Indeed, if f and f' both satisfy the equations, then it is easy to prove by induction on n that $f(m_1, \ldots, m_k, n) = f'(m_1, \ldots, m_k, n)$ for all natural numbers $m_1, \ldots, m_k$, and n, so $f = f'$. The existence of a function f satisfying these equations may seem more obvious at first glance than is really justified. (For an excellent discussion of this matter, see [Henkin, 1960].) In any case, we shall give a proof within $\mathcal{Q}_0^\infty$ of the existence of such a function. Indeed, since f is uniquely determined by h and g, we can define a recursion operator R such that $f = Rhg$. Since $m_1, \ldots, m_k$ can be regarded as parameters, it suffices to define R for the case where $k = 0$ (so g is simply a natural number).

Before giving the formal definition of R, let us provide an informal explanation of the method to be used. R yields f as a function of h and g, so it suffices to define f in terms of h and g. To do this it suffices to define the set C of ordered pairs $\langle n, m\rangle$ such that $f(n) = m$. Since $f(0) = g$, the pair $\langle 0, g\rangle$ is in C. Also, whenever $\langle n, m\rangle \in C$, we know that $f(n) = m$, so $f(\mathrm{S}n) = h(n, f(n)) = h(n, m)$, so $\langle \mathrm{S}n, h(n, m)\rangle \in C$. Thus, C can be defined inductively by the conditions that $\langle 0, g\rangle \in C$ and that $\forall n \forall m[\langle n, m\rangle \in C \Rightarrow \langle \mathrm{S}n,\ h(n, m)\rangle \in C]$. This leads to the following definition:

DEFINITION. $R_{\sigma\sigma\sigma(\sigma\sigma\sigma)}$ stands for $[\lambda h_{\sigma\sigma\sigma} \lambda g_\sigma \lambda n_\sigma \imath m_\sigma \forall w_{o\sigma\sigma} \centerdot [w_{o\sigma\sigma} 0_\sigma g_\sigma \wedge \forall x_\sigma \forall y_\sigma \centerdot w_{o\sigma\sigma} x_\sigma y_\sigma \supset w_{o\sigma\sigma}[S_{\sigma\sigma} x_\sigma] h_{\sigma\sigma\sigma} x_\sigma y_\sigma] \supset w_{o\sigma\sigma} n_\sigma m_\sigma]$.

6400. $\vdash [R_{\sigma\sigma\sigma(\sigma\sigma\sigma)} h_{\sigma\sigma\sigma} g_\sigma] 0_\sigma = g_\sigma \ \wedge\ \dot\forall n_\sigma \centerdot [R h_{\sigma\sigma\sigma} g_\sigma][S_{\sigma\sigma} n] = h_{\sigma\sigma\sigma} n_\sigma \centerdot [R h_{\sigma\sigma\sigma} g_\sigma] n_\sigma$

Proof: Let $B_{o(o\sigma\sigma)}$ be $[\lambda w_{o\sigma\sigma} \centerdot \forall x_\sigma \forall y_\sigma \centerdot w_{o\sigma\sigma} x_\sigma y_\sigma \supset w_{o\sigma\sigma}[S_{\sigma\sigma} x_\sigma] h_{\sigma\sigma\sigma} x_\sigma y_\sigma]$ and let $C_{o\sigma\sigma}$ be $[\lambda n_\sigma \lambda m_\sigma \centerdot \forall w_{o\sigma\sigma} \centerdot w_{o\sigma\sigma} 0_\sigma g_\sigma \ \wedge\ B w_{o\sigma\sigma} \supset w_{o\sigma\sigma} n_\sigma m_\sigma]$. Note that the free variables of $C_{o\sigma\sigma}$ are g_σ and $h_{\sigma\sigma\sigma}$. Also note that if $W_{o\sigma\sigma}$ is any wff such that $W 0 g_\sigma \ \wedge\ BW$ is derivable from certain hypotheses, then $Cxy \supset Wxy$ is also derivable from those hypotheses. We shall use this fact twice below.

(.1) $\vdash R_{\sigma\sigma\sigma(\sigma\sigma\sigma)} h_{\sigma\sigma\sigma} g_\sigma n_\sigma = \iota_{\sigma(o\sigma)}[C_{o\sigma\sigma} n_\sigma]$ λ, defs of R, C

Intuitively, $C n_\sigma m_\sigma$ means that m_σ is a possible value for $Rhgn$. We first prove that $\dot\forall n_\sigma \exists m_\sigma [Cnm \ \wedge\ \forall k_\sigma \centerdot Cnk \supset m = k]$ by induction on n.

(.2) $\vdash \exists m_\sigma [C0m \ \wedge \forall k_\sigma \centerdot C0k \supset m = k] \ \wedge$
$\dot\forall n_\sigma [\exists m_\sigma [Cnm \ \wedge\ \forall k_\sigma \centerdot Cnk \supset m = k]$
$\supset \exists m_\sigma [C[Sn]m \ \wedge\ \forall k_\sigma \centerdot C[Sn]k \supset m = k]]$
$\supset \dot\forall n_\sigma \exists m_\sigma \centerdot Cnm \ \wedge\ \forall k_\sigma \centerdot Cnk \supset m = k$ 6102

Essentially we are seeking to prove that $\dot\forall n \exists_1 m\ Cnm$. First we establish theorems that will enable us to prove $\exists m\ Cnm$.

(.3) $\vdash C 0 g_\sigma$ def of C
(.4) $C n_\sigma m_\sigma,\ w_{o\sigma\sigma} 0 g_\sigma \ \wedge\ Bw \vdash wnm$ def of C
(.5) $C n_\sigma m_\sigma,\ w_{o\sigma\sigma} 0 g_\sigma \ \wedge\ Bw \vdash w[Sn][h_{\sigma\sigma\sigma} nm]$ def of B, hyp, .4
(.6) $\vdash C n_\sigma m_\sigma \supset C[Sn][h_{\sigma\sigma\sigma} nm]$ def of C, .5

We shall need .8 as a lemma.

(.7) $\vdash C n_\sigma m_\sigma \supset \centerdot [\lambda x_\sigma \lambda y_\sigma \centerdot \mathbb{N} x] 0 g_\sigma \ \wedge\ B[\lambda x_\sigma \lambda y_\sigma \centerdot \mathbb{N} x] \supset$
$[\lambda x_\sigma \lambda y_\sigma \centerdot \mathbb{N} x] nm$ def of C
(.8) $\vdash C n_\sigma m_\sigma \supset \mathbb{N}_{o\sigma} n$ .7, def of B, 6100, 6101

Now we start the inductive proof with the case $n = 0$. Let $W_{o\sigma\sigma}$ be $[\lambda x_\sigma \lambda y_\sigma \centerdot x = 0 \supset g_\sigma = y]$.

(.9)	$C0k_\sigma \vdash W0g_\sigma \;\wedge\; Bw \supset W0k$	def of C
(.10)	$\vdash W0g_\sigma$	5200, def of W
(.11)	$\vdash W[Sx_\sigma][h_{\sigma\sigma\sigma}xy_\sigma] = \centerdot\, Sx = 0 \supset \centerdot\, g_\sigma = hxy$	def of W
(.12)	$\vdash W[Sx_\sigma][h_{\sigma\sigma\sigma}xy_\sigma]$	6105, .11
(.13)	$\vdash BW$	.12, def of B
(.14)	$C0k_\sigma \vdash W0k$	.9, .10, .13
(.15)	$\vdash \forall k_\sigma \centerdot\, C0k \supset g_\sigma = k$	.14, def of W, 5200
(.16)	$\vdash \exists m_\sigma \centerdot\, C0m \;\wedge\; \forall k_\sigma \centerdot\, C0k \supset m = k$	.3, .15

Next we prove the induction part of .2.

(.17)	$.17 \vdash Cn_\sigma m_\sigma \;\wedge\; \forall k_\sigma \centerdot\, Cnk \supset \centerdot\, m = k$	(inductive) hyp
(.18)	$.17 \vdash C[Sn_\sigma][h_{\sigma\sigma\sigma}nm_\sigma]$	.17, .6

We shall show that from .17 and $C[Sn_\sigma]k_\sigma$ we can prove $h_{\sigma\sigma\sigma}n_\sigma m_\sigma = k_\sigma$. So let $W^2_{o\sigma\sigma}$ be $[\lambda x_\sigma \lambda y_\sigma \centerdot\, Cxy \;\wedge\; \centerdot\, x = Sn_\sigma \supset \centerdot\, hnm_\sigma = y]$.

(.19)	$C[Sn_\sigma]k_\sigma \vdash W^2 0g_\sigma \;\wedge\; BW^2 \supset W^2[Sn]k$	def of C
(.20)	$C[Sn_\sigma]k_\sigma \vdash W^2 0g_\sigma \;\wedge\; BW^2 \supset \centerdot\, hnm_\sigma = k$	.19, 5200, def of W^2

Thus we wish to prove $W^2 0g$ and BW^2.

(.21)	$\vdash W^2 0g_\sigma$	.3, 6105, def of W^2
(.22)	$W^2 x_\sigma y_\sigma \vdash Cxy$	hyp, def of W^2
(.23)	$W^2 x_\sigma y_\sigma \vdash C[Sx]h_{\sigma\sigma\sigma}xy$	.6, .22
(.24)	$W^2 x_\sigma y_\sigma \vdash \mathbb{N}x$	.8, .22
(.25)	$.17,\; W^2 x_\sigma y_\sigma,\; Sx = Sn_\sigma \vdash x = n$	hyp, 6109, .24, .17, .8
(.26)	$.17,\; W^2 x_\sigma y_\sigma,\; Sx = Sn_\sigma \vdash Cny$	.25, .22
(.27)	$.17,\; W^2 x_\sigma y_\sigma,\; Sx = Sn_\sigma \vdash m_\sigma = y$	.17, .26
(.28)	$.17,\; W^2 x_\sigma y_\sigma,\; Sx = Sn_\sigma \vdash h_{\sigma\sigma\sigma}nm_\sigma = hxy$	.25, .27
(.29)	$.17,\; W^2 x_\sigma y_\sigma \vdash W^2[Sx][h_{\sigma\sigma\sigma}xy]$	.23, .28, def of W^2
(.30)	$.17 \vdash BW^2$	.29, def of B
(.31)	$.17 \vdash \forall k_\sigma \centerdot\, C[Sn_\sigma]k \supset \centerdot\, h_{\sigma\sigma\sigma}nm_\sigma = k$	.20, .21, .30
(.32)	$.17 \vdash \exists m_\sigma \centerdot\, C[Sn_\sigma]m \;\wedge\; \forall k_\sigma \centerdot\, C[Sn]k \supset \centerdot\, m = k$	.18, .31
(.33)	$\exists \dot{m}_\sigma[.17] \vdash [.32]$	$\exists$ Rule: .32
(.34)	$\vdash \dot{\forall} n_\sigma \exists m_\sigma \centerdot\, Cnm \;\wedge\; \forall k_\sigma \centerdot\, Cnk \supset m = k$	.2, .16, .33

This completes the real work in this proof.

(.35)	$\vdash \exists_1 m_\sigma Cn_\sigma m = \exists m_\sigma \centerdot\, Cnm \;\wedge\; \forall k_\sigma \centerdot\, Cnk \supset m = k$	5306
(.36)	$\mathbb{N}n_\sigma \vdash \exists_1 m_\sigma Cnm$	.34, .35
(.37)	$\vdash \exists_1 m_\sigma Cn_\sigma m \supset Cn[\iota_{\sigma(o\sigma)} \centerdot\, C_{o\sigma\sigma}n_\sigma]$	5311

(.38) $\mathbb{N}n_\sigma \vdash Cn[\iota_{\sigma(o\sigma)} \centerdot Cn]$ .36, .37
(.39) $\mathbb{N}n_\sigma \vdash Cn[R_{\sigma\sigma\sigma(\sigma\sigma\sigma)}h_{\sigma\sigma\sigma}g_\sigma n]$ .1, .38
(.40) $\mathbb{N}n_\sigma \vdash Cnm_\sigma \supset \centerdot Rh_{\sigma\sigma\sigma}g_\sigma n = m$ 5307, .36, .39

Now we can derive the results we wish.

(.41) $\vdash Rh_{\sigma\sigma\sigma}g_\sigma 0 = g$ .40, .3, 6100
(.42) $\vdash Cn_\sigma[Rh_{\sigma\sigma\sigma}g_\sigma n] \supset \centerdot C[Sn][hn \centerdot Rhgn]$ .6
(.43) $\mathbb{N}n_\sigma \vdash C[Sn][h_{\sigma\sigma\sigma}n \centerdot Rhg_\sigma n]$ .39, .42
(.44) $\mathbb{N}n_\sigma \vdash \mathbb{N}[Sn]$ 6101
(.45) $\mathbb{N}n_\sigma \vdash Rh_{\sigma\sigma\sigma}g_\sigma[Sn] = \centerdot hn \centerdot Rhgn$ .40, .44, .43

Our theorem follows from .41 and .45. ■

To illustrate the usefulness of R, define $+_{\sigma\sigma\sigma}$ as $R_{\sigma\sigma\sigma(\sigma\sigma\sigma)}[\lambda x_\sigma S_{\sigma\sigma}]$, and $\mathbf{A}_\sigma + \mathbf{B}_\sigma$ as $+_{\sigma\sigma\sigma}\mathbf{A}_\sigma\mathbf{B}_\sigma$. Then as an immediate consequence of 6400 we obtain $\vdash g_\sigma + 0 = g_\sigma \wedge \dot\forall n_\sigma \centerdot g_\sigma + [Sn_\sigma] = S[g_\sigma + n_\sigma]$. From this one can derive the usual properties of addition. (See the exercises below.) Similarly, $\times_{\sigma\sigma\sigma}$ can be defined as $[\lambda m_\sigma \centerdot R_{\sigma\sigma\sigma(\sigma\sigma\sigma)}[\lambda x_\sigma \centerdot +_{\sigma\sigma\sigma}m_\sigma]0_\sigma]$, and $\mathbf{A}_\sigma \times \mathbf{B}_\sigma$ as $\times_{\sigma\sigma\sigma}\mathbf{A}_\sigma\mathbf{B}_\sigma$. Then from 6400 one obtains $\vdash m_\sigma \times 0 = 0 \wedge \dot\forall n_\sigma \centerdot m_\sigma \times [Sn_\sigma] = m_\sigma + [m_\sigma \times n_\sigma]$.

Recall that a numerical function of 0 arguments is simply a number.

DEFINITION. The set of *primitive recursive* numerical functions (of zero or more arguments) is the intersection of all sets $\mathcal{C}$ of numerical functions which have the following five properties:

(a) $0 \in \mathcal{C}$.
(b) $\mathrm{S} \in \mathcal{C}$ (where S is the successor function).
(c) For each k and i such that $1 \leq i \leq k$, $((\lambda n_1, \cdots, n_k)n_i) \in \mathcal{C}$.
(d) If h is an m-ary function (where $m \geq 1$) in $\mathcal{C}$ and $g_1, \ldots, g_m$ are k-ary functions (where $k \geq 0$) in $\mathcal{C}$, then

$$((\lambda n_1, \ldots, n_k)h(g_1(n_1, \ldots, n_k), \ldots, g_m(n_1, \ldots, n_k))) \in \mathcal{C}.$$

(When $k = 0$, this means that $h(g_1, \ldots, g_m) \in \mathcal{C}$.)
(e) If g is a k-ary function in $\mathcal{C}$ (where $k \geq 0$) and h is a $(k+2)$-ary function in $\mathcal{C}$, and f is obtained from g and h by primitive recursion, then $f \in \mathcal{C}$.

The functions $((\lambda n_1, \ldots, n_k)n_i)$ of (c) are called *projection* functions; these, together with the successor function and the number zero are called *initial functions*. The function of (d) is obtained by *composition* from $g_1, \ldots, g_m$, and h. Thus a function is primitive recursive iff it can be obtained from the initial functions by a finite number of applications of composition and primitive recursion. Most of the numerical functions which are actually used in mathematics (outside the fields of logic and recursive function theory) are primitive recursive. Further discussion of primitive recursive functions will occur in §65.

DEFINITION. The set of *general recursive* numerical functions is the intersection of all sets $\mathcal{C}$ of numerical functions which have properties (a) – (e) (in the definition of the primitive recursive functions), and also the following property:

(f) If g is a regular $(k+1)$-ary numerical function (where $k \geq 0$) in $\mathcal{C}$, and h is obtained from g by minimization, then $h \in \mathcal{C}$.

Clearly, every primitive recursive function is general recursive. There are general recursive functions which are not primitive recursive. Indeed, it can be shown (see [Mendelson, 1964, p. 250]) that the following equations define such a function:

$$\begin{aligned} f(n, 0) &= Sn \\ f(0, Sm) &= f(1, m) \\ f(Sn, Sm) &= f(f(n, Sm), m). \end{aligned}$$

The *partial recursive* functions are defined in the same way as the general recursive functions, except that minimization may be applied to functions which are not regular. Hence partial recursive functions may not be defined for certain arguments. General recursive functions are often referred to simply as recursive functions.

A numerical relation (or a set of natural numbers) is said to be *primitive recursive, general recursive, or partial recursive* iff its characteristic function is primitive recursive, general recursive, or partial recursive, respectively.

6401 Theorem. Every general recursive function is representable in $\mathcal{Q}_0^\infty$.

Proof:

(a) The numeral 0_σ represents the number zero.

(b) The wff $S_{\sigma\sigma}$ represents the successor function, since $\vdash S\overline{n} = \overline{n+1}$ by 5200 and the definition of $\overline{n+1}$.

(c) The wff $[\lambda x^1_\sigma \ldots \lambda x^k_\sigma x^i_\sigma]$ represents the projection function $((\lambda n_1, \ldots, n_k)n_i)$, since $\vdash [\lambda x^1_\sigma \ldots \lambda x^k_\sigma x^i_\sigma]\overline{n_1}\ldots\overline{n_k} = \overline{n_i}$ by 5207.

(d) Suppose h is an m-ary function represented by $H_{\sigma\sigma\ldots\sigma}$, and $g_1, \ldots, g_m$ are k-ary functions represented by $G^1_{\sigma\sigma\ldots\sigma}, \ldots, G^m_{\sigma\sigma\ldots\sigma}$, respectively. We show that $[\lambda x^1_\sigma \ldots \lambda x^k_\sigma \centerdot H[G^1 x^1 \ldots x^k] \ldots [G^m x^1 \ldots x^k]]$ represents $((\lambda n_1, \ldots, n_k)h(g_1(n_1, \ldots, n_k), \ldots, g_m(n_1, \ldots, n_k)))$, which we shall call f.

(.1) $\vdash [\lambda x^1_\sigma \ldots x^k_\sigma \centerdot H[G^1 x^1 \ldots x^k] \ldots [G^m x^1 \ldots x^k]]\overline{n_1}\ldots\overline{n_k}$
$= H[G^1\overline{n_1}\ldots\overline{n_k}]\ldots[G^m\overline{n_1}\ldots\overline{n_k}]$ by Equality Rules and 5207

(.2^i) $\vdash G^i\overline{n_1}\ldots\overline{n_k} = \overline{g_i(n_1, \ldots, n_k)}$ for $1 \le i \le m$, since G^i represents g_i

(.3) $\vdash H\overline{g_1(n_1, \ldots, n_k)}\ldots\overline{g_m(n_1, \ldots, n_k)}$
$= \overline{h(g_1(n_1, \ldots, n_k), \ldots, g_m(n_1, \ldots, n_k))}$ since H represents h

(.4) $\vdash [\lambda x^1_\sigma \ldots \lambda x^k_\sigma \centerdot H[G^1 x^1 \ldots x^k] \ldots [G^m x^1 \ldots x^k]]\overline{n_1}\ldots\overline{n_k}$
$= \overline{f(n_1, \ldots, n_k)}$ Equality Rules: .1, .2^1 - .2^m, .3

The proof for the case where $k = 0$ is an obvious modification of the proof above.

(e) Suppose g is a k-ary function represented by $G_{\sigma\sigma\ldots\sigma}$, and h is a $(k+2)$-ary function represented by $H_{\sigma\sigma\sigma\sigma\ldots\sigma}$, and f is obtained by primitive recursion from g and h, so $f(m_1, \ldots, m_k, 0) = g(m_1, \ldots, m_k)$, and $f(m_1, \ldots, m_k, \mathrm{Sn}) = h(m_1, \ldots, m_k, n, f(m_1, \ldots, m_k, n))$. Let $F_{\sigma\sigma\sigma\ldots\sigma}$ be $[\lambda x^1_\sigma \ldots \lambda x^k_\sigma \centerdot R_{\sigma\sigma\sigma(\sigma\sigma\sigma)}[H_{\sigma\sigma\sigma\sigma\ldots\sigma}x^1_\sigma \ldots x^k_\sigma] \centerdot G_{\sigma\sigma\ldots\sigma}x^1_\sigma \ldots x^k_\sigma]$. To show that F represents f, we show by induction on n that
$\vdash F\overline{m_1}\ldots\overline{m_k}\overline{n} = \overline{f(m_1, \ldots, m_k, n)}$

(.5) $\vdash F\overline{m_1}\ldots\overline{m_k}\overline{0} = R[H\overline{m_1}\ldots\overline{m_k}][G\overline{m_1}\ldots\overline{m_k}]0$ def of $F, \overline{0}$

(.6) $\vdash R[H\overline{m_1}\ldots\overline{m_k}][G\overline{m_1}\ldots\overline{m_k}]0 = G\overline{m_1}\ldots\overline{m_k}$ 6400

(.7) $\vdash G\overline{m_1}\ldots\overline{m_k} = \overline{g(m_1, \ldots, m_k)}$ since G represents g

(.8) $\vdash F\overline{m_1}\ldots\overline{m_k}\overline{0} = \overline{f(m_1, \ldots, m_k, 0)}$ .5, .6, .7, def of f

This completes the case where $n = 0$. Next we treat the induction step.

(.9) $\vdash F\overline{m_1}\ldots\overline{m_k}\overline{n} = \overline{f(m_1, \ldots, n_k, n)}$ inductive hypothesis

(.10) $\vdash F\overline{m_1}\ldots\overline{m_k}(S\overline{n}) = R[H\overline{m_1}\ldots\overline{m_k}][G\overline{m_1}\ldots\overline{m_k}][S\overline{n}]$ def of F, $\overline{Sn}$

(.11) $\vdash R[H\overline{m_1}\ldots\overline{m_k}][G\overline{m_1}\ldots\overline{m_k}][S\overline{n}]$
$= [H\overline{m_1}\ldots\overline{m_k}]\overline{n} \centerdot R[H\overline{m_1}\ldots\overline{m_k}][G\overline{m_1}\ldots\overline{m_k}]\overline{n}$ 6400, 6110

(.12) $\vdash R[H\overline{m_1}\ldots\overline{m_k}][G\overline{m_1}\ldots\overline{m_k}]\overline{n} = \overline{f(m_1, \ldots, m_k, n)}$ .9, def of F

(.13) $\vdash H\overline{m_1}\ldots\overline{m_k}\overline{n}\overline{f(m_1,\ldots,m_k,n)} = \overline{f(m_1,\ldots,m_k,Sn)}$
H represents *h*, def of *f*

(.14) $\vdash F\overline{m_1}\ldots\overline{m_k}(\overline{Sn}) = \overline{f(m_1,\ldots,m_k,Sn)}$ .10, .11, .12, .13

(f) By 6303, a function obtained from a representable regular function by minimization is representable.

Thus the set of representable functions has properties (a) - (f), and so includes the set of general recursive functions. ■

We shall see later (Theorem 7009) that every representable numerical function is recursive.

Informally, we may regard a numerical k-ary function f as *effectively computable* iff there is a finite algorithm, or set of rules, by means of which one may calculate $f(m_1,\ldots,m_k)$ for any arguments $m_1,\ldots,m_k$, provided no limits are imposed on the amount of time one may take, or the amount of storage (memory) one may use for recording intermediate results. Of course, this notion is only as precise as the notion of an algorithm. Nevertheless, it is clear that there can be only denumerably many algorithms, since each algorithm can be described by some finite sequence of letters and symbols of a fixed language (such as English, perhaps augmented with mathematical symbols) which has a finite alphabet. Hence, there are only denumerably many effectively computable functions. Since there are uncountably many numerical functions, some (one might even say most) numerical functions are not effectively computable.

It is clear that each recursive function is effectively computable. Indeed, implicit in the proof above that every recursive function is representable is a method for effectively calculating the value of any recursive function for arbitrary arguments. The assertion that every effectively computable numerical function (which is defined everywhere on the domain of natural numbers) is recursive is known as *Church's Thesis.* This thesis was proposed by Alonzo Church in [Church, 1936b][3]. A number of quite different explications of the intuitive notion of algorithm have been proposed, and they have all turned out to define the same set of numerical functions, namely the recursive functions. On account of this and other evidence for its correctness, Church's Thesis has become generally accepted. (Additional discussion of Church's Thesis follows the proof of Theorem 7009 below.) Much of the significance of the results which we shall obtain in Chapter 7 is a consequence of Church's Thesis.

[3]See [Sieg, 1997] for an informative study of the development of Church's thesis.

Now that we have shown that in $\mathcal{Q}_0^\infty$ the natural numbers can be defined and many numerical functions can be represented, we could go on to define the integers, the rational numbers, the real numbers, and the complex numbers, and formalize much of mathematics within $\mathcal{Q}_0^\infty$. We shall not actually carry out the details of this, but we suggest that the reader examine a book such as [Gleason, 1966], and note that what is done there could be formalized in $\mathcal{Q}_0^\infty$.

EXERCISES

Prove the following theorems:

X6400. $\dot{\forall} n_\sigma \dot{\forall} m_\sigma \mathbb{N}[h_{\sigma\sigma\sigma} n_\sigma m_\sigma] \supset \dot{\forall} g_\sigma \dot{\forall} n_\sigma \centerdot \mathbb{N} \centerdot R_{\sigma\sigma\sigma(\sigma\sigma\sigma)} h_{\sigma\sigma\sigma} g_\sigma n_\sigma$.

X6401. $g_\sigma + 0 = g_\sigma$

X6402. $\dot{\forall} n_\sigma \centerdot g_\sigma + [S n_\sigma] = S[g_\sigma + n_\sigma]$

X6403. $\dot{\forall} g_\sigma \dot{\forall} n_\sigma \centerdot \mathbb{N}[g_\sigma + n_\sigma]$

X6404. $\dot{\forall} n_\sigma \centerdot 0 + n_\sigma = n_\sigma$

X6405. $\dot{\forall} n_\sigma \centerdot m_\sigma + S n_\sigma = S m_\sigma + n_\sigma$

X6406. $\dot{\forall} n_\sigma \centerdot [S m_\sigma] + n_\sigma = S[m_\sigma + n_\sigma]$

X6407. $\dot{\forall} m_\sigma \dot{\forall} n_\sigma \centerdot m_\sigma + n_\sigma = n_\sigma + m_\sigma$

X6408. $\dot{\forall} n_\sigma \dot{\forall} p_\sigma \centerdot [m_\sigma + n_\sigma] + p_\sigma = m_\sigma + [n_\sigma + p_\sigma]$

X6409. $\dot{\forall} n_\sigma \centerdot m_\sigma + n_\sigma = 0 \supset \centerdot m_\sigma = 0 \wedge n_\sigma = 0$

X6410. $\dot{\forall} m_\sigma \dot{\forall} n_\sigma \dot{\forall} p_\sigma \centerdot p_\sigma + m_\sigma = p_\sigma + n_\sigma \supset m_\sigma = n_\sigma$

X6411. $\dot{\forall} x_\sigma \dot{\forall} y_\sigma \centerdot x_\sigma \leq y_\sigma = \dot{\exists} z_\sigma \centerdot x_\sigma + z_\sigma = y_\sigma$

X6412. $\dot{\forall} x_\sigma \dot{\forall} z_\sigma \centerdot x_\sigma \leq x_\sigma + z_\sigma$

Define $\dot{+}_{(o(o\alpha))(o(o\alpha))(o(o\alpha))}$ as

$$[\lambda m_{o(o\alpha)} \lambda n_{o(o\alpha)} \lambda p_{o\alpha} \centerdot \exists q_{o\alpha} \exists r_{o\alpha} \centerdot m_{o(o\alpha)} q_{o\alpha} \wedge n_{o(o\alpha)} r_{o\alpha}$$

$$\wedge \forall x_\alpha [\sim q_{o\alpha} x_\alpha \vee \sim r_{o\alpha} x_\alpha] \wedge p_{o\alpha} = q_{o\alpha} \cup r_{o\alpha}].$$

Define $M_{o(o\alpha)} \dot{+} N_{o(o\alpha)}$ to be $\dot{+}_{(o(o\alpha))(o(o\alpha))(o(o\alpha))} M_{o(o\alpha)} N_{o(o\alpha)}$. $\dot{+}$ expresses addition of arbitrary cardinal numbers.

X6413. Prove $\dot{\forall} x_\sigma \dot{\forall} y_\sigma \,\blacksquare\, x_\sigma \dot{+} y_\sigma = x_\sigma + y_\sigma$

X6414. Let $\exp(x, y) = x^y$, where $x^0 = 1$ and $x^{y+1} = x \cdot x^y$ for all natural numbers x and y. (Thus, $0^0 = 1$, but $0^y = 0$ when $y > 0$.) Find a wff which represents exp, and prove that it does so.

X6415. The wff $R_{\sigma\sigma\sigma(\sigma\sigma\sigma)}[\lambda x_\sigma \lambda y_\sigma .\ S_{\sigma\sigma} .\ S_{\sigma\sigma}\ y_\sigma]\bar{5}$, which we shall call $F_{\sigma\sigma}$, represents a certain numerical function of one argument. Identify this function, and justify your answer with a proof.

§65. Primitive Recursive Functions and Relations

As preparation for Chapter 7, we need to show that certain numerical functions and relations are primitive recursive, and hence representable. For the sake of brevity, we shall let $\mathcal{PRF}^n$ denote the set of n-ary primitive recursive functions, $\mathcal{PRR}^n$ denote the set of n-ary primitive recursive relations, $\mathcal{PRF} = \bigcup_{n=0}^{\infty} \mathcal{PRF}^n$ (the set of primitive recursive functions), $\mathcal{PRR} = \bigcup_{n=0}^{\infty} \mathcal{PRR}^n$, and $\mathcal{PR} = \mathcal{PRF} \cup \mathcal{PRR}$. Also, we shall use $\overrightarrow{n3}$ as an abbreviation for n_1, n_2, n_3, or for (n_1, n_2, n_3) (where appropriate), $\overrightarrow{xk}$ as an abbreviation for $x_1, \ldots, x_k$, or for $(x_1, \ldots, x_k)$, etc. When $k = 0$, $\overrightarrow{xk}$ denotes the null sequence. From the definitions in §64 it is immediate that $0 \in \mathcal{PRF}^0$, $\mathrm{S} \in \mathcal{PRF}^1$, $((\lambda \overrightarrow{nk}) n_i) \in \mathcal{PRF}^k$ for $1 \leq i \leq k$, and $\mathcal{PR}$ is closed under composition and primitive recursion. Also, $r \in \mathcal{PRR}^n$ iff $\chi_r \in \mathcal{PRF}^n$.

6500 Constant Functions. For all $k \geq 0$ and $n \geq 0$, $((\lambda \overrightarrow{xk}) n) \in \mathcal{PRF}^k$. (When $k = 0$, $((\lambda \overrightarrow{xk}) n)$ means n.)

Proof by induction on k: $0 \in \mathcal{PRF}^0$, and if $n \in \mathcal{PRF}^0$ then $\mathrm{S}n \in \mathcal{PRF}^0$ by composition, so $n \in \mathcal{PRF}^0$ for all n. (Thus all numerical functions of 0 arguments, i.e., all natural numbers, are primitive recursive, so no further discussion of $\mathcal{PRF}^0$ will be necessary.)

If $((\lambda \overrightarrow{xk}) n) \in \mathcal{PRF}^k$, where $k \geq 0$, then $((\lambda \overrightarrow{x_{k+1}}) n) \in \mathcal{PRF}^{k+1}$, since $((\lambda \overrightarrow{x_{k+1}}) n)$ is obtained from $((\lambda \overrightarrow{xk}) n)$ and $((\lambda \overrightarrow{x_{k+2}}) x_{k+2})$ by primitive recursion. ■

We next wish to generalize the principle of composition to permit the following sort of inferences: if $g_1 \in \mathcal{PRF}^2$, $g_2 \in \mathcal{PRF}^3$, and $h \in \mathcal{PRF}^4$, then $((\lambda x_1, x_2, x_3)h(x_2,\ g_1(x_1, x_1),\ 5,\ g_2(x_2, 4,\ x_1))) \in \mathcal{PRF}^3$.

Let $\mathcal{S}$ be a set of numerical functions of zero or more arguments. We shall define what it means to say that a k-ary numerical function f is explicitly definable from $\mathcal{S}$. So let $\mathcal{F}^=$ be a formulation of first order logic with equality, and let $\langle \mathbb{N}, \mathcal{J}\rangle$ be an interpretation of $\mathcal{F}^=$ with the set $\mathbb{N}$ of natural numbers as domain of individuals, such that for each function g in $\mathcal{S}$ there is a constant $\bar{\mathbf{g}}$ of appropriate type of $\mathcal{F}^=$ so that $\mathcal{J}\bar{\mathbf{g}} = g$, and there is a k-ary function constant $\bar{\mathbf{f}}$ such that $\mathcal{J}\bar{\mathbf{f}} = f$. We say that f is *explicitly definable* from $\mathcal{S}$ iff there is a term $\mathbf{t}$ of $\mathcal{F}^=$ in which the only variables are the individual variables $x_1, \ldots, x_k$, and the only constants are the constants $\bar{\mathbf{g}}$ denoting functions g in $\mathcal{S}$, such that

$$\langle \mathbb{N},\ \mathcal{J}\rangle \models \forall x_1 \ldots \forall x_k \,\blacksquare\, \bar{\mathbf{f}} x_1 \ldots x_k = \mathbf{t}.$$

Note: The wff above will be called an *explicit definition* of f from $\mathcal{S}$ with respect to $\langle \mathbb{N},\ \mathcal{J}\rangle$.

6501 Explicit Definitions. If $\mathcal{S} \subseteq \mathcal{PRF}$ and f is a numerical function explicitly definable from $\mathcal{S}$, then $f \in \mathcal{PRF}$.

Proof: We hold k and $\mathcal{S}$ fixed, and prove the proposition for all k-ary functions f explicitly definable from $\mathcal{S}$. The proof is by induction on the construction of the term $\mathbf{t}$ in any explicit definition of f.

If $\mathbf{t}$ is an individual variable x_i, then the explicit definition is $\forall x_1 \ldots \forall x_k[\bar{\mathbf{f}} x_1 \ldots x_k = x_i]$, so $f = ((\lambda n_1, \ldots, n_k)n_i)$, which is in $\mathcal{PRF}$.

If $\mathbf{t}$ is an individual constant $\mathbf{c}$, then $\mathcal{J}\mathbf{c}$ must be some natural number m, so $f = ((\lambda n_1, \ldots, n_k)m)$, which is in $\mathcal{PRF}$ by 6500.

Suppose $\mathbf{t}$ has the form $\bar{\mathbf{g}}\mathbf{t}_1 \ldots \mathbf{t}_m$, where $m \geq 1$ and $\bar{\mathbf{g}}$ is an m-ary function constant, and each $\mathbf{t}_i$ is a term. For $1 \leq i \leq m$, let f_i be the k-ary numerical function with explicit definition $\forall x_1 \ldots \forall x_k[\bar{\mathbf{f}}_i x_1 \ldots x_k = \mathbf{t}_i]$, where the constants $\bar{\mathbf{f}}_1, \ldots, \bar{\mathbf{f}}_m$ are distinct from one another, $\bar{\mathbf{f}}$, and the constants in $\mathbf{t}$. By inductive hypothesis, $f_i \in \mathcal{PRF}^k$. Clearly

$$\langle \mathbb{N},\ \mathcal{J}\rangle \models \forall x_1 \ldots \forall x_k[\bar{\mathbf{f}} x_1 \ldots x_k = \bar{\mathbf{g}}(\bar{\mathbf{f}}_1 x_1 \ldots x_k) \ldots (\bar{\mathbf{f}}_m x_1 \ldots x_k)]$$

(the parentheses being inserted only for the sake of readability), so $f = ((\lambda n_1, \ldots, n_k)g(f_1(n_1, \ldots, n_k), \ldots, f_m(n_1, \ldots, n_k)))$, so $f \in \mathcal{PRF}^k$ by composition. ∎

6502 Generalized Primitive Recursion. Let $\mathcal{S} \subseteq \mathcal{PRF}$, and let g and h be k-ary and $(k+2)$-ary functions, respectively, which are explicitly definable from $\mathcal{S}$. If for all natural numbers $x_1, \ldots, x_k$, and y,

$$f(x_1, \ldots, x_i,\ 0,\ x_{i+1}, \ldots, x_k) = g(\overrightarrow{x_k}),\ \text{and}$$

$$f(x_1, \ldots, x_i, \mathrm{S}y, x_{i+1}, \ldots, x_k) = h(\overrightarrow{x_k}, f(x_1, \ldots, x_i, y, x_{i+1}, \ldots, x_k)),$$

then $f \in \mathcal{PR}$. (*Note:* As a special case, i may be 0, i.e., the sequence $x_1, \ldots, x_i$ may be void. Similarly, i may be k, so that the sequence $x_{i+1}, \ldots,$ x_k is void.)

Proof: Let t be obtained from g and h by primitive recursion. g and h are in $\mathcal{PR}$ by 6501, so $t \in \mathcal{PR}$. However,

$$f = ((\lambda x_1, \ldots, x_i, y, x_{i+1}, \ldots,\ x_k)t(\overrightarrow{x_k}, y)),$$

so f is explicitly definable from t, and so $f \in \mathcal{PR}$. ■

In applying 6502, we shall often replace g and h by their explicit definitions (so to speak). Having generalized composition and primitive recursion, we are in a position to expeditiously show that certain functions which we need are primitive recursive. We shall be somewhat informal, and combine propositions, definitions, and proofs whenever this is convenient. We shall often use 6500-6502 implicitly, without further mention of them. In general, our proofs will contain only the main ideas; other details may be supplied by the reader.

6503 $+$ (addition) $\in \mathcal{PRF}^2$.

Proof:

$$x + 0 = x$$

$$x + (\mathrm{S}y) = \mathrm{S}(x + y)$$ ■

6504. If $f \in \mathcal{PRF}^{k+1}$ (where $k \geq 0$), then $((\lambda \overrightarrow{x_k}, y) \sum_{z \leq y} f(\overrightarrow{x_k}, z))$ $\in \mathcal{PRF}^{k+1}$.

Proof:

$$\sum_{z \leq 0} f(\overrightarrow{x_k}, z) = f(\overrightarrow{x_k}, 0)$$

$$\sum_{z\leq \mathrm{S}y} f(\overrightarrow{x_k}, z) = f(\overrightarrow{x_k},\ \mathrm{S}y) + \sum_{z\leq y} f(\overrightarrow{x_k}, z) \quad \blacksquare$$

6505. $\times$ (multiplication) $\in \mathcal{PRF}^2$.

Proof:

$$x \times 0 = 0$$

$$x \times (\mathrm{S}y) = x + (x \times y) \quad \blacksquare$$

6506. If $f \in \mathcal{PRF}^{k+1}$ (where $k \geq 0$), then $((\lambda\overrightarrow{x_k}, y)\Pi_{z\leq y} f(\overrightarrow{x_k}, z)) \in \mathcal{PRF}^{k+1}$.

Proof:

$$\Pi_{z\leq 0} f(\overrightarrow{x_k}, z) = f(\overrightarrow{x_k}, 0)$$

$$\Pi_{z\leq \mathrm{S}y} f(\overrightarrow{x_k}, z) = f(\overrightarrow{x_k},\ \mathrm{S}y) \times \Pi_{z\leq y} f(\overrightarrow{x_k}, z) \quad \blacksquare$$

6507. $((\lambda x, y)x^y)$ (exponentiation) $\in \mathcal{PRF}^2$.

Proof:

$$x^0 = 1$$

$$x^{(\mathrm{S}y)} = x \times (x^y) \quad \blacksquare$$

Note: $0^0 = 1$ but $0^{(\mathrm{S}y)} = 0$.

NOTATION. We may sometimes write x^y as $x \exp y$.

6508. ! (factorial) $\in \mathcal{PRF}^1$.

Proof:

$$0! = 1$$

$$(\mathrm{S}y)! = y! \times \mathrm{S}y \quad \blacksquare$$

6509. δ (predecessor) $\in \mathcal{PRF}^1$.

Proof:

$$\delta(0) = 0$$
$$\delta(\mathrm{S}y) = y \qquad \blacksquare$$

6510. $\dot{-}$ (modified subtraction or monus) $\in \mathcal{PRF}^2$.

Proof:

$$x \dot{-} 0 = x$$
$$x \dot{-} (\mathrm{S}y) = \delta(x \dot{-} y) \qquad \blacksquare$$

Note: $x \dot{-} y = 0$ iff $x \leq y$. Also, $x \dot{-} y = x - y$ if $x \geq y$.

6511. σ (signum) $\in \mathcal{PRF}^1$.

Proof:

$$\sigma(0) = 0$$
$$\sigma(\mathrm{S}y) = 1. \qquad \blacksquare$$

6512. $\overline{\sigma}$ (antisignum) $\in \mathcal{PRF}^1$.

Proof:

$$\overline{\sigma}(0) = 1$$
$$\overline{\sigma}(\mathrm{S}y) = 0 \qquad \blacksquare$$

6513. $< \in \mathcal{PRR}^2$.

Proof: $\chi_<(x, y) = \sigma(y \dot{-} x)$ ■

6514. $\leq \in \mathcal{PRR}^2$.

Proof: $x \leq y \equiv x < \mathrm{S}y$, so $\chi_\leq(x, y) = \chi_<(x, \mathrm{S}y)$. ■

6515. $= \in \mathcal{PRR}^2$

Proof. $\chi_=(x,y) = ((\mathrm{S}x) \dot{-} y) \times ((\mathrm{S}y) \dot{-} x)$ ■

DEFINITION. Let $\mathcal{S}$ be a set of numerical relations and functions, and let r be a k-ary numerical relation. Let $\mathcal{F}$ be a formulation of first order logic, and let $\langle \mathbb{N}, \mathcal{J} \rangle$ be an interpretation of $\mathcal{F}$ with the natural numbers as individuals, such that for each function or relation g in $\mathcal{S}$ there is a constant $\overline{\mathbf{g}}$ of appropriate type of $\mathcal{F}$ so that $\mathcal{J}\overline{\mathbf{g}} = g$, and there is a k-ary predicate constant $\overline{\mathbf{R}}$ such that $\mathcal{J}\overline{\mathbf{R}} = r$. We say that r is *explicitly definable from $\mathcal{S}$ using propositional connectives* iff there is a quantifier-free wff $\mathbf{B}$ of $\mathcal{F}$ in which the only variables are the individual variables $x_1, \ldots, x_k$, and the only constants are the constants $\overline{\mathbf{g}}$ denoting relations and functions in $\mathcal{S}$, such that $\langle \mathbb{N}, \mathcal{J} \rangle \models \forall x_1 \ldots \forall x_k \centerdot \overline{\mathbf{R}} x_1 \ldots x_k \equiv \mathbf{B}$.

6516. If $\mathcal{S} \subseteq \mathcal{PR}$, and the numerical relation r is explicitly definable from $\mathcal{S}$ using propositional connectives, then $r \in \mathcal{PR}$.

EXAMPLE: Let $r(x,y) \equiv [x \neq 2y \;\wedge\; x + y < 100]$. $r \in \mathcal{PR}$ by 6500, 6503, 6505, 6513, 6515, and 6516.

Proof: We hold k and $\mathcal{S}$ fixed, and prove the proposition for all k-ary relations r explicitly definable from $\mathcal{S}$ using propositional connectives. The proof is by induction on the construction of a wff $\mathbf{B}$ explicitly defining r from $\mathcal{S}$. We may assume that $\mathbf{B}$ contains no propositional connectives other than $\sim$ and $\vee$.

If $\mathbf{B}$ is an atomic wff $\overline{\mathbf{P}}\mathbf{t}_1 \ldots \mathbf{t}_n$, and p is the relation in $\mathcal{S}$ such that $\mathcal{J}\overline{\mathbf{P}} = p$, $\langle \mathbb{N}, \mathcal{J} \rangle \models \forall x_1 \ldots \forall x_k [\overline{\mathbf{R}} x_1 \ldots x_k \equiv \overline{\mathbf{P}}\mathbf{t}_1 \ldots \mathbf{t}_n]$, so $\langle \mathbb{N}, \mathcal{J} \rangle \models \forall x_1 \ldots \forall x_k [\overline{\chi}_r x_1 \ldots x_k = \overline{\chi}_p \mathbf{t}_1 \ldots \mathbf{t}_n]$, so χ_r is explicitly definable from χ_p and functions in $\mathcal{S}$, so $\chi_r \in \mathcal{PRF}$ by 6501. Hence $r \in \mathcal{PRR}$.

Suppose $\mathbf{B}$ has the form $\sim \mathbf{C}$. Let p be the k-ary numerical relation with explicit definition $\forall x_1 \ldots \forall x_k [\overline{\mathbf{P}} x_1 \ldots x_k \equiv \mathbf{C}]$, where the predicate constant $\overline{\mathbf{P}}$ does not occur in $\mathbf{C}$. $p \in \mathcal{PRR}$ by inductive hypothesis, and for all natural numbers $\overrightarrow{n_k}, r(\overrightarrow{n_k})$ iff not $p(\overrightarrow{n_k})$. Thus $\chi_r(\overrightarrow{n_k}) = \overline{\sigma}(\chi_p(\overrightarrow{n_k}))$, so $r \in$PRR.

Suppose $\mathbf{B}$ has the form $[\mathbf{C} \;\vee\; \mathbf{D}]$. Let p and q be the k-ary numerical relations with explicit definitions $\forall x_1 \ldots \forall x_k [\overline{\mathbf{P}} x_1 \ldots x_k \equiv \mathbf{C}]$ and $\forall x_1 \ldots \forall x_k [\overline{\mathbf{Q}} x_1 \ldots x_k \equiv \mathbf{D}]$, respectively, where $\overline{\mathbf{P}}$ and $\overline{\mathbf{Q}}$ do not occur in $[\mathbf{C} \;\vee\; \mathbf{D}]$. By inductive hypothesis p and q are in $\mathcal{PRR}$, and for all natural numbers $\overrightarrow{n_k}$, $r(\overrightarrow{n_k})$ iff $p(\overrightarrow{n_k})$ or $q(\overrightarrow{n_k})$. Thus $\chi_r(\overrightarrow{n_k}) = \sigma(\chi_p(\overrightarrow{n_k}) + \chi_q(\overrightarrow{n_k}))$, so $r \in \mathcal{PRR}$. ■

We shall use the expression $(\exists z < y)B$ in our meta-language to mean $\exists z[z < y \wedge B]$, i.e., there is a number z less than y such that B. Similarly, $(\forall z < y)B$ means $\forall z[z < y \supset B]$. Analogous conventions apply when $<$ is replaced by $\leq$. Note that $(\exists z < 0)B$ is false, and $(\forall z < 0)B$ is true.

6517 Bounded Quantification. If $f \in \mathcal{PRF}^k$ (where $k \geq 0$) and $r \in \mathcal{PRR}^{k+1}$, then the following are in $\mathcal{PRR}^k$:

(a) $(\lambda \overrightarrow{x_k})(\exists z \leq f(\overrightarrow{x_k}))r(\overrightarrow{x_k}, z)$
(b) $(\lambda \overrightarrow{x_k})(\exists z < f(\overrightarrow{x_k}))r(\overrightarrow{x_k}, z)$
(c) $(\lambda \overrightarrow{x_k})(\forall z \leq f(\overrightarrow{x_k}))r(\overrightarrow{x_k}, z)$
(d) $(\lambda \overrightarrow{x_k})(\forall z < f(\overrightarrow{x_k}))r(\overrightarrow{x_k}, z)$

Proof:

(a) $\chi_{(a)} = (\lambda \overrightarrow{x_k})\sigma(\sum_{z \leq f(\overrightarrow{x_k})} \chi_r(\overrightarrow{x_k}, z))$.
(b) $(\exists z < f(\overrightarrow{x_k}, z))r(\overrightarrow{x_k}, z) \equiv (\exists z \leq f(\overrightarrow{x_k}))[z \neq f(\overrightarrow{x_k}) \wedge r(\overrightarrow{x_k}, z)]$.

By (a), 6501, 6515, and 6516, this is in $\mathcal{PRR}$.

(c) $\chi_{(c)} = (\lambda \overrightarrow{x_k})\Pi_{z \leq f(\overrightarrow{x_k})}\chi_r(\overrightarrow{x_k}, z)$.
(d) $(\forall z < f(\overrightarrow{x_k}))r(\overrightarrow{x_k}, z) \equiv (\forall z \leq f(\overrightarrow{x_k}))[z = f(\overrightarrow{x_k}) \vee r(\overrightarrow{x_k}, z)]$. ■

We shall use the expression $(\mu z \leq y)B$ in our meta-language to denote the least number $z \leq y$ such that B if such a z exists; if no such z exists, we let $(\mu z \leq y)B = \mathrm{S}y$. Also, we shall use the expression $(\mathrm{Max}\ z \leq y)B$ in our meta-language to denote the largest number $z \leq y$ such that B if such a z exists; if no such z exists , we let $(\mathrm{Max}\ z \leq y)B = 0$.

6518 Bounded Minimization and Maximization. If $f \in \mathcal{PRF}^k$ and $r \in \mathcal{PRR}^{k+1}$ (where $k \geq 0$), then the following are in $\mathcal{PRF}^k$:

(a) $(\lambda \overrightarrow{x_k})(\mu z \leq f(\overrightarrow{x_k}))r(\overrightarrow{x_k}, z)$
(b) $(\lambda \overrightarrow{x_k})(\mathrm{Max}\ z \leq f(\overrightarrow{x_k}))r(\overrightarrow{x_k}, z)$

Proof:

(a) Let $p = (\lambda \overrightarrow{x_k}, z)(\forall y \leq z) \sim r(\overrightarrow{x_k}, y)$. $p \in \mathcal{PRR}$ by 6516 and 6517, and $(\mu z \leq f(\overrightarrow{x_k}))r(\overrightarrow{x_k}, z) = \sum_{z \leq f(\overrightarrow{x_k})} \chi_p(\overrightarrow{x_k}, z)$. (See the example in Figure 6.1.)

y	0	1	2	3	4	5	6	7	...
$r(y)$	F	F	F	F	T	F	T	T	...
$py \equiv (\forall w \leq y) \sim r(w)$	T	T	T	T	F	F	F	F	...
$\chi_p(y)$	1	1	1	1	0	0	0	0	...
$\sum_{z \leq y} \chi_p(z)$	1	2	3	4	4	4	4	4	...

Figure 6.1: $(\mu z \leq y)r(z)$

(b) Let $g(\overrightarrow{x}, w) = w \dot{-} (\mu y \leq w)r(\overrightarrow{x}, w \dot{-} y)$. Clearly $g \in \mathcal{PRF}^{k+1}$. We shall show that (1) $g(\overrightarrow{x}, w) = (\text{Max } z \leq w)r(\overrightarrow{x}, z)$. From this the desired result follows.

Note that if there is no $z \leq w$ such that $r(\overrightarrow{x}, z)$, then $g(\overrightarrow{x}, w) = w \dot{-} Sw = 0$, as desired. If such a z exists, let z_0 be the largest one. Thus (2) $r(\overrightarrow{x}, z_0)$, (3) $z_0 \leq w$, and (4) $\forall z[z_0 < z \leq w \Rightarrow \sim r(\overrightarrow{x}, z)]$.

Let $y_0 = w - z_0$. We shall show that (5) $y_0 = (\mu y \leq w)r(\overrightarrow{x}, w - y)$. By (2), $r(\overrightarrow{x}, w \dot{-} y_0)$. Suppose $0 \leq y < y_0$. Then $w \geq w - y > w - y_0 = z_0$ so $\sim r(\overrightarrow{x}, w \dot{-} y)$ by (4). This establishes (5).

Now $g(\overrightarrow{x}, w) = w \dot{-} y_0$ (by (5) and the definition of g) $= z_0$ (by the definition of y_0) $= (\text{Max } z \leq w)\ r(\overrightarrow{x}, z)$ (by the definition of z_0). ∎

REMARK. It is important to note the distinction between bounded minimization, which can be used to define primitive recursive functions, and unbounded minimization, which can be used to define general recursive functions.

6519 Definition by Cases. Let $g_1, \ldots, g_n \in \mathcal{PRF}^k$ and $r_1, \ldots, r_n \in \mathcal{PRR}^k$, and suppose the relations $r_1, \ldots, r_n$ are mutually exclusive and exhaustive, i.e., for all $\overrightarrow{x_k} \in \mathbb{N}^k$, there is exactly one r_i such that $r_i(\overrightarrow{x_k})$. Let $f(\overrightarrow{x_k}) = g_i(\overrightarrow{x_k})$ iff $r_i(\overrightarrow{x_k})$ for $1 \leq i \leq n$. Then $f \in \mathcal{PRF}^k$.

Proof: $f = (\lambda \overrightarrow{x_k})((g_1(\overrightarrow{x_k}) \times \chi_{r_1}(\overrightarrow{x_k})) + \ldots + (g_n(\overrightarrow{x_k}) \times \chi_{r_n}(\overrightarrow{x_k})))$. ∎

6520. $((\lambda x, y)x \text{ Divides } y) \in \mathcal{PRR}^2$

Proof: x Divides $y \equiv (\exists z \leq y)[x \times z = y]$. ∎

6521. Let Prime(x) iff x is a prime number. Then Prime $\in \mathcal{PRR}^1$.

Proof:

$$\text{Prime}(x) \equiv \,\blacksquare\, 2 \leq x \;\wedge\; (\forall y \leq x)[y \text{ Divides } x \supset \,\blacksquare\, y = x \;\vee\; y = 1] \qquad \blacksquare$$

6522. Let Pr(n) be the nth odd prime number, where Pr(0) = 2. Then Pr $\in \mathcal{PRF}^1$.

Proof:

$$\text{Pr}(0) = 2$$

$$\text{Pr}(\text{S}n) = (\mu m \leq \text{S}(\text{Pr}(n)!))[\text{Prime}\,(m) \;\wedge\; \text{Pr}(n) < m]$$

To verify the correctness of the recursion equation, we must show that if (by inductive hypothesis) Pr(n) is prime, then there is a prime number m such that $\text{Pr}(n) < m \leq \text{S}(\text{Pr}(n)!)$. (The proof is essentially Euclid's proof that there are infinitely many primes.) Let $b = \text{Pr}(n)!$. If there is a prime p such that $\text{Pr}(n) < p \leq b$, we are done, so suppose there is no such prime. In this case we show that Sb is prime. Let p be any prime less than Sb. Then $p \leq \text{Pr}(n)$, so p divides Pr(n)!, which is b. But $p \geq 2$, so p cannot also divide Sb. Thus no prime less than Sb divides Sb, so Sb is prime.

Let $h = (\lambda x, y)(\mu m \leq \text{S}(y!))[\text{Prime}(m) \;\wedge\; y < m]$. $h \in \mathcal{PRF}^2$ by 6521, 6513, 6516, 6508, and 6518, and $\text{Pr}(\text{S}n) = h(n, \text{Pr}(n))$, so Pr $\in \mathcal{PRF}^1$ by primitive recursion. ■

Here are a few values for Pr:

n	0	1	2	3	4	5	6	7	8	9	10	11	12
Pr(n)	2	3	5	7	11	13	17	19	23	29	31	37	41

Note: $n < \text{Pr}(n)$. This is readily established by induction.

We wish to use positive integers to represent finite sequences of natural numbers, and to do this we shall make extensive use of the fundamental fact from number theory that every positive integer can be factored uniquely into a product of primes. We shall associate with each positive integer the sequence of natural numbers which constitute the exponents in its prime factorization. Of course, all but finitely many of the terms of this sequence are zero. Thus, with the number 84, which is $2^2 \times 3^1 \times 5^0 \times 7^1$, we associate the sequence $(2, 1, 0, 1, 0, 0, 0, \ldots)$, or the finite sequence $(2, 1, 0, 1)$.

6523. Let $n\ \mathcal{G}\ell\ x$ be the exponent of $\mathrm{Pr}(n)$ in the prime factorization of x, with $n\ \mathcal{G}\ell\ 0 = 0$. Then $\mathcal{G}\ell \in \mathcal{PRF}^2$.

Note: If $x = \Pi_{i=0}^{\infty}\mathrm{Pr}(i)^{a_i}$, then $n\ \mathcal{G}\ell\ x = a_n$, the nth term in the sequence $a_0, a_1, a_2, a_3, \ldots$ associated with x. The name "$\mathcal{G}\ell$" for this function was introduced in [Gödel 1931]. "Glied" means "term" in German.

Proof: $n\ \mathcal{G}\ell\ x = (\mathrm{Max}\ y \leq x)[\mathrm{Pr}(n)^y \text{ Divides } x]$. There is always some number y such that $\mathrm{Pr}(n)^y$ divides x, since $\mathrm{Pr}(n)^0 = 1$. Since $\mathrm{Pr}(n) \geq 2$, it is readily established by induction that $y < \mathrm{Pr}(n)^y$, so if $y = n\ \mathcal{G}\ell\ x$ and $x \neq 0$, then $y < \mathrm{Pr}(n)^y \leq x$, so the given bound is adequate. ■

Note: $n\ \mathcal{G}\ell\ 0 = 0 = n\ \mathcal{G}\ell\ 1$ for all n. Also, $n\ \mathcal{G}\ell\ x < x$ if $x \neq 0$.

EXAMPLE: $0\ \mathcal{G}\ell\ 84 = 2,\ 1\ \mathcal{G}\ell\ 84 = 1,\ 2\ \mathcal{G}\ell\ 84 = 0,\ 3\ \mathcal{G}\ell\ 84 = 1,\ n\ \mathcal{G}\ell\ 84 = 0$ for $n > 3$.

6524. Let $\mathrm{E}(x)$ (the extent of the sequence represented by x) be the largest number n such that $\mathrm{Pr}(n)$ divides x, with $\mathrm{E}(0) = 0 = E(1)$. Then $\mathrm{E} \in \mathcal{PRF}^1$.

Proof: $\mathrm{E}(x) = (\mathrm{Max}\ n \leq x)[\mathrm{Pr}(n) \text{ Divides } x]$. ■

Clearly if $\mathrm{Pr}(n)$ divides x, then $n < \mathrm{Pr}(n) \leq x$, so $\mathrm{E}(x) \leq x$.

Note: $x = \Pi_{i \leq \mathrm{E}(x)}\ \mathrm{Pr}(i)^{i\ \mathcal{G}\ell\ x}$ if $x \neq 0$.

Note: $\mathrm{E}(\Pi_{i \leq n}\mathrm{Pr}(i)^{a_i}) = n$ if $a_n \neq 0$.

We next define a concatenation function $*$ such that if x and y represent the sequences $(a_0, \ldots, a_n)$ and $(b_0, \ldots, b_m)$, respectively, then $x*y$ represents the sequence $(a_0, \ldots, a_n, b_0, \ldots, b_m)$. We do this in such a way that the numbers 0 and 1 can be used to represent the empty sequence.

6525. Let $x * y = x \times \Pi_{i \leq \mathrm{E}(y)}\ \mathrm{Pr}(\mathrm{E}(x) + 1 + i)^{i\ \mathcal{G}\ell\ y}$ if $1 < x$

$\qquad = y$ if $x \leq 1$.

Then $* \in \mathcal{PRF}^2$.

Proof: By 6519. ■

Note that if $x \leq 1$, then $x * y = y$, but if $x > 1$ and $y \leq 1$, then $x * y = x$ (since $\mathrm{E}(y) = 0$ and $0\ \mathcal{G}\ell\ y = 0$). Also, if $a_n \neq 0$ and $b_m \neq 0$,

$$(\mathrm{Pr}(0)^{a_0} \times \ldots \times \mathrm{Pr}(n)^{a_n}) \;*\; (\mathrm{Pr}(0)^{b_0} \times \ldots \times \mathrm{Pr}(m)^{b_m}) =$$

$$\mathrm{Pr}(0)^{a_0} \times \ldots \times \mathrm{Pr}(n)^{a_n} \times \mathrm{Pr}(n+1)^{b_0} \times \ldots \times \mathrm{Pr}(n+1+m)^{b_m}.$$

Hence, $*$ is associative, i.e., for all natural numbers x, y, and z, $x * (y * z) = (x * y) * z$.

EXAMPLE: $21 * 300 = (2^0 \times 3^1 \times 5^0 \times 7^1) * (2^2 \times 3^1 \times 5^2) = 2^0 \times 3^1 \times 5^0 \times 7^1 \times 11^2 \times 13^1 \times 17^2 = 9{,}546{,}537$.

Since the binary numerical functions $+$, $*$, and $\times$ are all associative, we shall often omit the parentheses associated with them. In contexts where two or three of these functions occur together, we shall use the convention that $+$ has the largest scope, and $\times$ the smallest. Thus, in restoring parentheses in an expression involving $+$, $*$, and $\times$, first restore parentheses associated with $\times$, giving them the smallest possible scope, and then those associated with $*$, giving them the smallest possible scope, and then (if necessary) those associated with $+$. For example, $a+b\times c*d*e+f\times g\times h+i*j$ stands for $a+(b\times c)*d*e+(f\times g\times h)+i*j$, and for $a+((b\times c)*d*e)+(f\times g\times h)+(i*j)$.

6526. Let $\mathrm{Lg} = (\lambda x)(\mu y \leq x)(\exists z \leq x)[x = z^y \;\wedge\; \mathrm{Prime}(z)]$. Then $\mathrm{Lg} \in \mathcal{PRF}^1$.

EXAMPLE: $\mathrm{Lg}(16) = 4$ since $16 = 2^4$.

Chapter 7

Incompleteness and Undecidability

Logical systems such as $\mathcal{Q}_0^\infty$ have infinitely many theorems—indeed, infinitely many essentially different theorems—but these theorems are finitely generated. There may be infinitely many axioms, but a finite description in the meta-language suffices to specify them. Wffs are finite, and proofs are finite. Such features seem to be inescapable characteristics of logical systems which can actually be used by finite, mortal men.

It is natural to ask whether such systems are inherently incapable of dealing adequately with the infinite variety and complexity of the mathematical world. One's response will obviously depend on one's notions of adequacy, but a hasty general affirmative answer may be deterred by considering the possibility of a fine balance between the powers of asking and answering questions in or about a suitable logical system, so that unanswerable questions cannot even be asked. It is natural to ask whether there is an ultimately satisfactory logical system, in which one can express the important and generally useful ideas which arise naturally in the study of the logical system itself as well as in other parts of mathematics, and in which one can prove those truths which can be expressed, but no sentences which are not true.

The results in this chapter, which involve some of the most important and profound ideas in the history of logic, should illuminate these questions. The results are formulated in terms of $\mathcal{Q}_0$ and its extensions, but the ideas behind them are of very wide applicability. With appropriate modifications, the results about $\mathcal{Q}_0^\infty$ apply to any logical system for formalizing mathematics which is not too weak, inadequately specified, inconsistent, or in some other way unsatisfactory. The significance of these results is particularly apparent in the context of Hilbert's program. In the early part of the twentieth century, the great mathematician David Hilbert set forth a set of goals for research in logic known as Hilbert's program . As applied to some particular mathematical theory, the goals of the program included:

(1) Define a formal system $\mathcal{S}$ such that the theorems of $\mathcal{S}$ are precisely the true statements of the theory.

(2) Prove the consistency of $\mathcal{S}$ by elementary methods whose correctness could not reasonably be doubted.

(3) Solve the decision problem for $\mathcal{S}$. Thus all mathematical questions in the theory would in principle be answered, and further research would simply be devoted to improving the efficiency of the decision procedure.

As we shall see, it was soon discovered that these goals are unattainable for many mathematical theories (such as the theory of natural numbers, the theory of real numbers, and extensions of them).

Without waiting for the precise definitions of the concepts involved, we now give a brief preview of the main results in this chapter. $\mathcal{Q}_0^\infty$ is essentially incomplete, i.e., it is incomplete and has no consistent complete extension which is precisely specifiable (recursively axiomatizable). Consequently, there is no recursively axiomatizable extension of $\mathcal{Q}_0$ whose theorems are precisely the wffs valid in all standard models. The consistency of $\mathcal{Q}_0^\infty$ is not provable by methods which can be formalized in $\mathcal{Q}_0^\infty$. The decision problems for $\mathcal{Q}_0$ and $\mathcal{Q}_0^\infty$ are unsolvable. On the purely semantic side, truth for $\mathcal{Q}_0^\infty$ is not definable in $\mathcal{Q}_0^\infty$, i.e., the set of sentences of $\mathcal{Q}_0^\infty$ which are true (in a given model) is not definable by any wff of $\mathcal{Q}_0^\infty$.

The crucial role of Church's Thesis in this chapter should be noted. An effective operation corresponds, via a suitable encoding, to a numerical function which, by Church's Thesis, must be recursive, and hence is representable in $\mathcal{Q}_0^\infty$. Thus, if we accept Church's Thesis and find an operation (such as deciding which wffs are theorems of $\mathcal{Q}_0$) which corresponds to a function which is not representable in $\mathcal{Q}_0^\infty$, then we must accept the fact that no algorithm can execute that operation.

§70. Gödel Numbering

We wish to be able to express certain statements about $\mathcal{Q}_0^\infty$ by wffs of $\mathcal{Q}_0^\infty$. For this purpose, we assign numbers to the type symbols, primitive symbols, formulas, and sequences of formulas of $\mathcal{Q}_0$, and use the numerals of $\mathcal{Q}_0^\infty$ denoting these numbers as names for the corresponding syntactic entities. The results we shall obtain will not really depend on our particular choice of a numbering system, since the basic ideas we shall use can be applied to obtain these results for a rather wide variety of numbering systems. Most of the numbering systems which naturally come to mind can be translated into one another by primitive recursive functions. We shall make extensive use of the results and terminology of §65.

DEFINITION. To each type symbol α we assign a natural number "α" as follows by induction on the construction of α:

(1) "o"$= 3$
(2) "$\imath$"$= 5$
(3) "$(\beta\gamma)$ "$= 3^{``\beta"} \times 5^{``\gamma"}$

Note that if α and β are distinct type symbols, then "α" $\neq$"β".

7000 Theorem. Let TypeSymbol $=((\lambda n)n$ is the number of some type symbol). Then TypeSymbol $\in \mathcal{PRR}^1$.

Proof: In order to determine whether a nonzero number Sn is the number of a type symbol, one may need to know whether certain numbers $k < Sn$ are numbers of type symbols; however, information about n may not suffice. Hence we cannot define $\chi_{\text{TypeSymbol}}$ directly by primitive recursion. Instead we define a function f such that from $f(n)$ we can determine for each number $k \leq n$ whether TypeSymbol(k).

Let $f = ((\lambda n)\Pi_{i \leq n}\text{Pr}(i)^{\chi_{\text{TypeSymbol}}(i)})$. We first show that $f \in \mathcal{PRF}^1$ by 6502.

$$f(0) = 1 \qquad (\text{since } 1 = 2^0 = 2^{\chi_{\text{TypeSymbol}}(0)})$$

$$\begin{aligned} f(\text{S}n) = {} & f(n) \times [(\text{Pr}(\text{S}n))\text{exp}(\chi_=(\text{S}n, 3) + \chi_=(\text{S}n, 5) \\ & +[\chi_=(\text{S}n, 3^{1\ \mathcal{G}\ell\ (\text{S}n)} \times 5^{2\ \mathcal{G}\ell\ (\text{S}n)}) \\ & \times (1\ \mathcal{G}\ell\ \text{S}n)\ \mathcal{G}\ell\ f(n) \times (2\ \mathcal{G}\ell\ \text{S}n)\ \mathcal{G}\ell\ f(n)])] \end{aligned}$$

since TypeSymbol(Sn) iff $Sn =$"o"$= 3$ or $Sn =$"$\imath$"$= 5$ or there exist a and b such that $Sn = 3^a \times 5^b$ (in which case $a = 1\ \mathcal{G}\ell\ Sn$ and $b = 2\ \mathcal{G}\ell\ Sn$) and TypeSymbol($a$) and TypeSymbol($b$), i.e., $a\ \mathcal{G}\ell\ f(n) = \chi_{\text{TypeSymbol}}(a) = 1$ and $b\ \mathcal{G}\ell\ f(n) = 1$. Now $\chi_{\text{TypeSymbol}} = ((\lambda n)n\ \mathcal{G}\ell\ f(n)) \in \mathcal{PRF}$. ∎

DEFINITION. To each primitive symbol **Z** of $\mathcal{Q}_0$ we assign a number '**Z**' as follows:

Z	[	]	λ	$\iota_{(\imath(o\imath))}$	$Q_{((o\alpha)\alpha)}$	f_α	g_α	h_α	...
'**Z**'	7	11	13	$17^{\text{“}(\imath(o\imath))\text{”}}$	$19^{\text{“}((o\alpha)\alpha)\text{”}}$	$23^{\text{“}\alpha\text{”}}$	$29^{\text{“}\alpha\text{”}}$	$31^{\text{“}\alpha\text{”}}$	...

x_α	y_α	nth variable $\mathbf{x}_\alpha$ of type α
$103^{\text{“}\alpha\text{”}}$	$107^{\text{“}\alpha\text{”}}$	$\text{Pr}(7+n)^{\text{“}\alpha\text{”}}$

To each formula (sequence of primitive symbols) $\mathbf{Z}_0 \ldots \mathbf{Z}_n$, we assign the number "$\mathbf{Z}_0 \ldots \mathbf{Z}_n$" $= \Pi_{i \le n}\text{Pr}(i)^{\text{‘}\mathbf{Z}_i\text{’}}$.

Note that if $\mathbf{Z}_0 \ldots \mathbf{Z}_n$ is a sequence of primitive symbols of which $\mathbf{W}_0 \ldots \mathbf{W}_m$ is a subsequence, then "$\mathbf{W}_0 \ldots \mathbf{W}_m$" $\le$ "$\mathbf{Z}_0 \ldots \mathbf{Z}_n$".

To each sequence $\mathbf{A}_0, \ldots, \mathbf{A}_n$ of formulas, we assign the number '"$\mathbf{A}_0, \ldots, \mathbf{A}_n$"' $= \Pi_{i \le n}\text{Pr}(i)^{\text{“}\mathbf{A}_i\text{”}}$.

7001 Theorem. Let $Q = (\lambda n)19 \exp((3 \exp(3^3 \times 5^n)) \times 5^n)$. Then $Q \in \mathcal{PRF}^1$, and for each type symbol α, $Q(\text{“}\alpha\text{”}) = \text{‘}Q_{o\alpha\alpha}\text{’}$.

7002 Theorem. Let Variable $= ((\lambda n)n$ is the number of a variable of $\mathcal{Q}_0)$. Then Variable $\in \mathcal{PRR}^1$.

Proof: Variable $(n) \equiv (\exists a \le n)(\exists i \le n)[8 \le i \wedge \text{TypeSymbol}(a) \wedge n = \text{Pr}(i)^a]$. ∎

7003 Theorem. Let ProperSymbol $=((\lambda n)n$ is the number of a variable or constant of $\mathcal{Q}_0)$. Then ProperSymbol $\in \mathcal{PRR}^1$.

Proof $\text{ProperSymbol}(n) \equiv [\text{Variable}(n) \vee n = \text{‘}\iota_{\imath(o\imath)}\text{’}$
$\vee (\exists a \le n)[\text{TypeSymbol}(a) \wedge n = Q(a)]$. ∎

7004 Theorem. Let Wff $= ((\lambda x)$ there is a wff $\mathbf{A}_\alpha$ of $\mathcal{Q}_0$ such that

$x = \text{"}\mathbf{A}_\alpha\text{"})$.

$$\text{Let Type}(x) = \text{"}\alpha\text{" if } x = \text{"}\mathbf{A}_\alpha\text{"}; \quad = 0 \text{ if not Wff}(x).$$

Then Wff $\in \mathcal{PRR}^1$ and Type $\in \mathcal{PRF}^1$.

Proof: $\chi_{Wff} = ((\lambda x)\sigma(\text{Type}(x)))$, so it suffices to show that Type $\in \mathcal{PRF}^1$. We can determine Type(x) from x and Type(k) for certain numbers $k < x$. Therefore we let $f(x) = \Pi_{i \leq x}\text{Pr}(i)^{\text{Type}(i)}$, and show that f can be obtained by primitive recursion from primitive recursive functions. This will establish the theorem, since Type$(x) = x\ \mathcal{G}\ell\ f(x)$.

First we make some definitions. The reader may wish to postpone studying the details of these definitions until he has surveyed the rest of the proof. The reader should also review the definition of the function $*$ (see 6525), and observe that if **A** and **B** are formulas, then "**A**"$*$ "**B**"= "**AB**".

Let A $= (\lambda y)(\exists x \leq y)[\text{ProperSymbol}(x) \wedge y = 2^x]$; $\mathbf{A}(y)$ means that there is a proper symbol $\mathbf{Z}_a$ such that $y = 2^{\text{`}\mathbf{Z}_a\text{'}} = \text{"}\mathbf{Z}_a\text{"}$. If this is the case, "$\alpha$" $= \text{Lg}(\text{`}\mathbf{Z}_a\text{'}) = \text{Lg}(\text{Lg}(y))$.

Let B $= (\lambda w, i, j, y,)[j\ \mathcal{G}\ell\ w \neq 0 \wedge i\ \mathcal{G}\ell\ w = 3^{1\ \mathcal{G}\ell\ (i\ \mathcal{G}\ell\ w)} \times 5^{j\ \mathcal{G}\ell\ w} \wedge y = 2^7 * i * j * 2^{11}]$. Suppose $w = f(n)$ and $i \leq n$ and $j \leq n$ and $\mathbf{B}(w, i, j, y)$. Then Type$(j) = j\ \mathcal{G}\ell\ w \neq 0$, so there is a wff $\mathbf{B}_\beta$ such that $j = \text{"}\mathbf{B}_\beta\text{"}$, and $j\ \mathcal{G}\ell\ w = \text{"}\beta\text{"}$. Also, there is a number a such that Type$(i) = i\ \mathcal{G}\ell\ w = 3^a \times 5^{\text{"}\beta\text{"}}$, so there is a wff $\mathbf{A}_{\alpha\beta}$ such that $i = \text{"}\mathbf{A}_{\alpha\beta}\text{"}$ and $y = \text{"[" } * \text{ "}\mathbf{A}_{\alpha\beta}\text{" } * \text{ "}\mathbf{B}_\beta\text{" } * \text{ "]" } = \text{"}[\mathbf{A}_{\alpha\beta}\mathbf{B}_\beta]\text{"}$. This wff has type α, and "α" $= 1\ \mathcal{G}\ell\ (\text{"}(\alpha\beta)\text{"}) = 1\ \mathcal{G}\ell\ (i\ \mathcal{G}\ell\ w)$.

Let C $= (\lambda w, i, j, y)[\text{Variable}(j) \wedge i\ \mathcal{G}\ell\ w \neq 0 \wedge y = (2^7 \times 3^{13} \times 5^j) * i * 2^{11}]$. Suppose $w = f(n)$ and $i \leq n$ and $j \leq n$ and $C(w, i, j, y)$. Then there is a variable $\mathbf{x}_\beta$ such that $j = \text{`}\mathbf{x}_\beta\text{'}$, so "$\beta$" $= \text{Lg}(j)$. Also, Type$(i) = i\ \mathcal{G}\ell\ w \neq 0$, so there is a wff $\mathbf{A}_\alpha$ such that $i = \text{"}\mathbf{A}_\alpha\text{"}$, and "$\alpha$" $= i\ \mathcal{G}\ell\ w$. Thus $y = \text{"}[\lambda\mathbf{x}_\beta\text{" } * \text{ "}\mathbf{A}_\alpha\text{" } * \text{ "]" } = \text{"}[\lambda\mathbf{x}_\beta\mathbf{A}_\alpha]\text{"}$. This wff has type $(\alpha\beta)$, and "$(\alpha\beta)$" $= 3^{\text{"}\alpha\text{"}} \times 5^{\text{"}\beta\text{"}} = 3^{i\ \mathcal{G}\ell\ w} \times 5^{\text{Lg}(j)}$.

$$f(0) = 2^0 = 1. \text{ Also,}$$

$$\begin{aligned} f(\text{S}n) = {} & f(n) \times [\text{Pr}(\text{S}n)\text{exp}([\text{Lg}(\text{Lg}(\text{S}n)) \times \chi_{\text{A}}(\text{S}n)] \\ & + \sum_{i \leq n} \sum_{j \leq n}([(1\ \mathcal{G}\ell\ (i\ \mathcal{G}\ell\ f(n))) \times \chi_{\text{B}}(f(n), i, j, \text{S}n)] \\ & + [3^{i\ \mathcal{G}\ell\ f(n)} \times 5^{\text{Lg}(j)} \times \chi_{\text{C}}(f(n), i, j, \text{S}n)]))]. \end{aligned}$$

To see this, note that $f(\mathrm{S}n) = f(n) \times \mathrm{Pr}(\mathrm{S}n)^{\mathrm{Type}(\mathrm{S}n)}$. $\mathrm{Type}(\mathrm{S}n) = 0$ unless $\mathrm{S}n$ is the number of a wff, in which case exactly one of the following conditions must occur:

(1) $\mathrm{S}n =$ "$\mathbf{Z}_\alpha$" for some primitive symbol $\mathbf{Z}_\alpha$, so
A("$\mathbf{Z}_\alpha$") and $\mathrm{Type}(\mathrm{S}n) =$ "α".
(2) $\mathrm{S}n =$ "$[\mathbf{A}_{\alpha\beta}\mathbf{B}_\beta]$" for some wffs $\mathbf{A}_{\alpha\beta}$ and $\mathbf{B}_\beta$, so
B($f(n)$, "$\mathbf{A}_{\alpha\beta}$", "$\mathbf{B}_\beta$", "$[\mathbf{A}_{\alpha\beta}\mathbf{B}_\beta]$"), and Type $(\mathrm{S}n) =$ "α".
(3) $\mathrm{S}n =$ "$[\lambda\mathbf{x}_\beta\mathbf{A}_\alpha]$" for some wff $\mathbf{A}_\alpha$ and variable $\mathbf{x}_\beta$, so
C($f(n)$, "$\mathbf{A}_\alpha$", '$\mathbf{x}_\beta$', "$[\lambda\mathbf{x}_\beta\mathbf{A}_\alpha]$"), and $\mathrm{Type}(\mathrm{S}n) =$ "$\alpha\beta$".

Since A, B, C $\in \mathcal{PRR}$, we see that $f \in \mathcal{PRF}$. ■

7005 Theorem. Let Axiom = $((\lambda n)$ there is an axiom $\mathbf{A}$ of $\mathcal{Q}_0^\infty$ such that $n =$ "$\mathbf{A}$"). Then Axiom $\in \mathcal{PRR}^1$.

Proof: For $i = 2, 3, 4_1, 4_2, 4_3, 4_4, 4_5$, let $\mathrm{Axiom}^i = ((\lambda x)x$ is the number of an instance of Axiom Schema i). We show that each of these is in $\mathcal{PRR}$.

Axiom 2^α is $[[\supset_{ooo} [[\mathsf{Q}_{o\alpha\alpha}x_\alpha]y_\alpha]][[\mathsf{Q}_{ooo}[h_{o\alpha}x_\alpha]][h_{o\alpha}y_\alpha]]]$.
Also, '$h_{o\alpha}$' $= 31 \exp(3^3 \times 5^{\text{“}\alpha\text{”}})$, '$x_\alpha$' $= 103^{\text{“}\alpha\text{”}}$, and '$y_\alpha$' $= 107^{\text{“}\alpha\text{”}}$. Hence

$$\begin{aligned}\mathrm{Axiom}^2(n) \equiv\ & (\exists a \leq n)(\mathrm{TypeSymbol}(a) \wedge n = \text{“}[[\supset_{ooo} [[\text{”} * 2^{\mathrm{Q}(a)} \times \\ & 3^{103 \exp a} \times 5^{11} \times 7^{107 \exp a} * \text{“}]][[\mathsf{Q}_{ooo}[\text{”} * 2^{31 \exp (3^3 \times 5^a)} \times \\ & 3^{103 \exp a} \times 5^{11} \times 7^{11} \times 11^7 \times 13^{31 \exp(3^3 \times 5^a)} \times \\ & 17^{107 \exp a} * \text{“}]]]\text{”}),\end{aligned}$$

so $\mathrm{Axiom}^2 \in \mathcal{PRR}$.

Axiom $3^{\alpha\beta}$ is

$$[[\mathsf{Q}_{ooo}[[\mathsf{Q}_{o(\alpha\beta)(\alpha\beta)}f_{\alpha\beta}]g_{\alpha\beta}]][[\mathsf{Q}_{o(o\beta)(o\beta)}[\lambda x_\beta T_o]][\lambda x_\beta[[\mathsf{Q}_{o\alpha\alpha}[f_{\alpha\beta}x_\beta]][g_{\alpha\beta}x_\beta]]]]].$$

The proof that $\mathrm{Axiom}^3 \in \mathcal{PRR}$ is similar to that for Axiom^2.

Axiom Schema 4_1 has the form $[[\mathsf{Q}_{o\beta\beta}[[\lambda\mathbf{x}_\alpha\mathbf{B}_\beta]\mathbf{A}_\alpha]]\mathbf{B}_\beta]$, where $\mathbf{B}_\beta$ is a primitive constant or variable distinct from $\mathbf{x}_\alpha$. Hence

$$\begin{aligned}\text{Axiom}^{4_1}(n) \equiv\ & (\exists x \le n)(\exists b \le n)(\exists a \le n)(\text{Variable}(x) \wedge \text{ProperSymbol}(b) \\ & \wedge\, b \ne x \wedge \text{Wff}(a) \wedge \text{Type}(a) = \text{Lg}(x) \wedge n = 2^7 \times 3^7 \\ & \times 5^{\text{Q}(\text{Lg}(b))} \times 7^7 \times 11^7 \times 13^{13} \times 17^x \times 19^b \times 23^{11} * a * \\ & 2^{11} \times 3^{11} \times 5^b \times 7^{11}),\end{aligned}$$

so $\text{Axiom}^{4_1} \in \mathcal{PRR}^1$.

Axiom Schema 4_2 has the form $[[\text{Q}_{o\alpha\alpha}[[\lambda \mathbf{x}_\alpha \mathbf{x}_\alpha]\mathbf{A}_\alpha]]\mathbf{A}_\alpha]$, so

$$\begin{aligned}\text{Axiom}^{4_2}(n) \equiv\ & (\exists x \le n)(\exists a \le n)(\text{Variable}(x) \wedge \text{Wff}(a) \wedge \text{Type}(a) = \text{Lg}(x) \\ & \wedge\, n = 2^7 \times 3^7 \times 5^{\text{Q}(\text{Lg}(x))} \times 7^7 \times 11^7 \times 13^{13} \times 17^x \times 19^x \\ & \times 23^{11} * a * 2^{11} \times 3^{11} * a * 2^{11}),\end{aligned}$$

so $\text{Axiom}^{4_2} \in \mathcal{PRR}$.

Axiom 4_3 has the form

$$[[\text{Q}_{o\beta\beta}[[\lambda \mathbf{x}_\alpha[\mathbf{B}_{\beta\gamma}\mathbf{C}_\gamma]]\mathbf{A}_\alpha]][[[\lambda \mathbf{x}_\alpha \mathbf{B}_{\beta\gamma}]\mathbf{A}_\alpha][[\lambda \mathbf{x}_\alpha \mathbf{C}_\gamma]\mathbf{A}_\alpha]]].$$

We leave further details to the reader.

Axiom Schema 4_4 has the form

$$[[\text{Q}_{o(\delta\gamma)(\delta\gamma)}[[\lambda \mathbf{x}_\alpha[\lambda \mathbf{y}_\gamma \mathbf{B}_\delta]]\mathbf{A}_\alpha]][\lambda \mathbf{y}_\gamma[[\lambda \mathbf{x}_\alpha \mathbf{B}_\delta]\mathbf{A}_\alpha]]]$$

where $\mathbf{y}_\gamma$ is distinct from $\mathbf{x}_\alpha$ and all variables in $\mathbf{A}_\alpha$.

$$\begin{aligned}\text{Axiom}^{4_4}(n) \equiv\ & (\exists x \le n)(\exists y \le n)(\exists a \le n)(\exists b \le n)(\text{Variable}(x) \wedge \\ & \text{Variable}(y) \wedge \text{Wff}(b) \wedge \text{Wff}(a) \wedge \text{Type}(a) = \text{Lg}(x) \wedge y \\ & \ne x \wedge (\forall i \le \text{E}(a))[y \ne i\ \mathcal{G}\ell\ a] \wedge n = \text{“[[”} * \\ & (2 \exp \text{Q}(3^{\text{Type}(b)} \times 5^{\text{Lg}(y)})) * \text{“[[}\lambda\text{”} * 2^x * \text{“[}\lambda\text{”} * 2^y * b * \text{“]]”} \\ & * a * \text{“]][}\lambda\text{”} * 2^y * \text{“[[}\lambda\text{”} * 2^x * b * 2^{11} * a * \text{“]]]”}).\end{aligned}$$

Thus Axiom $^{4_4} \in \mathcal{PRR}^1$.

[1] The phrase 'Wff(a)' above is actually redundant, but we leave it in as an aid to intelligibility.

Axiom Schema 4_5 has the form $[[\mathrm{Q}_{o(\delta\alpha)(\delta\alpha)}[[\lambda \mathbf{x}_\alpha[\lambda \mathbf{x}_\alpha \mathbf{B}_\delta]]\mathbf{A}_\alpha]][\lambda \mathbf{x}_\alpha \mathbf{B}_\delta]]$. We leave further details concerning this axiom to the reader.

$$\begin{aligned}\text{Axiom}(n) \equiv\ & (n = \text{“Axiom 1”} \vee \text{Axiom}^2(n) \vee \text{Axiom}^3(n) \vee \text{Axiom}^{4_1}(n)\\ & \vee \text{Axiom}^{4_2}(n) \vee \text{Axiom}^{4_3}(n) \vee \text{Axiom}^{4_4}(n) \vee \text{Axiom}^{4_5}(n)\\ & \vee n = \text{“Axiom 5”} \vee n = \text{“Axiom 6”}).\end{aligned}$$

Thus Axiom $\in \mathcal{PRR}^1$. ∎

7006 Theorem Let Rule $= (\lambda m, n, p)$ [there exist wffs $\mathbf{A}_\alpha$, $\mathbf{B}_\alpha$, $\mathbf{C}_o$, and $\mathbf{D}_o$ such that $m =$ “$[\mathbf{A}_\alpha = \mathbf{B}_\alpha]$” and $n =$ “$\mathbf{C}_o$” and $p =$ “$\mathbf{D}_o$”, and $\mathbf{D}_o$ is obtained from $[\mathbf{A}_\alpha = \mathbf{B}_\alpha]$ and $\mathbf{C}_o$ by Rule R]. Then Rule $\in \mathcal{PRR}^3$.

Proof:

$$\begin{aligned}\text{Rule}(m,n,p) \equiv\ & (\exists a \le m)(\exists b \le m)[\text{Wff}(a) \wedge \text{Wff}(b) \wedge \text{Type}(a) = \text{Type}(b)\\ & \wedge\ m = 2^7 \times 3^7 \times 5^{\mathrm{Q}(\text{Type}(a))} * a * 2^{11} * b * 2^{11} \wedge\ \text{Type}(n)\\ & = 3 \wedge\ \text{Type}(p) = 3\ \wedge\ (\exists j \le n)(\exists k \le n)[n = j * a * k\ \wedge\\ & p = j * b * k\ \wedge\ (\mathrm{E}(j))\ \mathcal{G}\ell\ j \ne 13]].\end{aligned}$$

To see this, note that if Rule(m, n, p) is true then $m =$ “$[[\mathrm{Q}_{o\alpha\alpha}\mathbf{A}_\alpha]\mathbf{B}_\alpha]$”, and there exist formulas (which need not be wffs) J and K such that $C = JA_\alpha K$ and $D = JB_\alpha K$. Of course, J and K may be empty sequences of symbols, in which case they have 1 as number. Also, the last symbol of J must not be a λ. ∎

DEFINITION. A logistic system $\mathcal{L}^2$ is a *pure extension* of a logistic system $\mathcal{L}^1$ iff the systems have the same wffs and rules of inference, and every axiom of $\mathcal{L}^1$ is a theorem of $\mathcal{L}^2$. (Hence every theorem of $\mathcal{L}^1$ is a theorem of $\mathcal{L}^2$).

DEFINITION. A logistic system $\mathcal{L}$ is [*primitive*] *recursively axiomatized* iff its set of axioms is [primitive] recursive (with respect to a suitable Gödel numbering). Note that since there are primitive recursive translations between most numbering systems, this definition does not really depend on the particular numbering system that is used.

It is a consequence of Church's Thesis that a system is recursively axiomatized iff there is an effective test for determining of an arbitrary wff whether it is an axiom. Thus we may assume that any system which will actually be used (by mortal men) to prove theorems will be recursively axiomatized.

We have shown in Theorem 7005 that $\mathcal{Q}_0^\infty$ is primitive recursively axiomatized, and hence is a (primitive) recursively axiomatized pure extension of $\mathcal{Q}_0$.

7007 Theorem. Let $\mathcal{A}$ be any [primitive] recursively axiomatized pure extension of $\mathcal{Q}_0$. Let $\text{Proof}^{\mathcal{A}} = (\lambda m, n)$ [There is a proof $\mathbf{P}^0, \ldots, \mathbf{P}^k$ in $\mathcal{A}$ such that $m =$"$\mathbf{P}^k$" and $n =$"“$\mathbf{P}^0, \ldots, \mathbf{P}^k$”"]. Then $\text{Proof}^{\mathcal{A}}$ is a [primitive] recursive relation.

Proof:

$$\text{Proof}^{\mathcal{A}}(m,n) \equiv [m = \text{E}(n)\ \mathcal{G}\ell\ n \wedge (\forall k \leq \text{E}(n))[\text{Axiom}^{\mathcal{A}}(k\ \mathcal{G}\ell\ n) \vee (\exists i < k)(\exists j < k)\text{Rule}(i\ \mathcal{G}\ell\ n,\ j\ \mathcal{G}\ell\ n,\ k\ \mathcal{G}\ell\ n)]].$$

A quick review of our results about primitive recursive functions and relations will show that they also apply to recursive functions and relations, so if $\text{Axiom}^{\mathcal{A}}$ is recursive, $\text{Proof}^{\mathcal{A}}$ is also. ■

7008 Theorem. Let $\text{Num} = ((\lambda n)\text{“}\overline{n}\text{”})$. Then $\text{Num} \in \mathcal{PRF}^1$.

Proof:

$$\text{Num}(0) = \text{“}0_\sigma\text{”}$$

$$\text{Num}(\text{S}n) = 2^7 * \text{“}S_{\sigma\sigma}\text{”} * \text{Num}(n) * 2^{11} \text{ (since } \overline{\text{S}n} = [S_{\sigma\sigma}\overline{n}]).$$ ■

7009 Theorem. If $\mathcal{A}$ is a consistent recursively axiomatized pure extension of $\mathcal{Q}_0^\infty$, then every numerical function and every numerical relation representable in $\mathcal{A}$ is recursive.

Proof: By 6113 it suffices to prove the theorem for functions. Let f be an n-ary numerical function (where we may assume $n \geq 1$), and let $\mathbf{F}_{\sigma\sigma\ldots\sigma}$ be a wff which represents f. Let us sketch the main ideas of the proof below before getting into the technical details. Given numbers $\overrightarrow{x_n}$, we can find $f(\overrightarrow{x_n})$ looking for a proof of a theorem of the form $\mathbf{F}\overline{x_1}\ldots\overline{x_n} = \overline{z}$. This involves searching simultaneously for z and the proof, but we can instead search for a number y of the form $2^z \times 3^p$, where p is the number of the proof. When we consider the least such number y for a fixed choice of $\overrightarrow{x_n}$, we have $z = 0\ \mathcal{G}\ell\ y = 0\ \mathcal{G}\ell\ ((\mu y)\mathrm{Proof}^{\mathcal{A}}(``\mathbf{F}\overline{x_1}\ldots\overline{x_n} = \overline{z}", p))$. Of course, $y$ depends on the numbers $\overrightarrow{x_n}$. We shall define a numerical relation $\mathrm{S}_n$ so that $\mathrm{S}_n(``\mathbf{F}", \overrightarrow{x_n}, 2^z \times 3^p)$ is true iff p is the number of a proof of $\mathbf{F}\overline{x_1}\ldots\overline{x_n} = \overline{z}$.

Now let us start the formal argument by defining $\mathrm{S}_n = (\lambda j, \overrightarrow{x_n}, y)[\mathrm{Proof}^{\mathcal{A}}$ $(2^7 \times 3^7 \times 5^{`\mathrm{Q}_{o\sigma\sigma}'} \times (\mathrm{Pr}(3))^7 \times \ldots \times (\mathrm{Pr}(2{+}n))^7 * j * \mathrm{Num}(x_1) * 2^{11} * \mathrm{Num}(x_2) *$ $2^{11} * \ldots * \mathrm{Num}(x_n) * 2^{11} * 2^{11} * \mathrm{Num}(0\ \mathcal{G}\ell\ y) * 2^{11}, 1\ \mathcal{G}\ell\ y)]$. We assert that $(\lambda\overrightarrow{x_n}, y)\mathrm{S}_n(``\mathbf{F}", \overrightarrow{x_n}, y)$ is regular, and that $f = (\lambda\overrightarrow{x_n})0\ \mathcal{G}\ell\ ((\mu y)\mathrm{S}_n(``\mathbf{F}", \overrightarrow{x_n}, y))$.

Given arbitrary natural numbers $\overrightarrow{x_n}$, let $z = f(\overrightarrow{x_n})$. Since $\mathbf{F}_{\sigma\sigma\ldots\sigma}$ represents f, there is a proof in $\mathcal{A}$ of the wff $\mathbf{F}_{\sigma\sigma\ldots\sigma}\overline{x_1}\ldots\overline{x_n} = \overline{z}$, i.e., $[[\mathrm{Q}_{o\sigma\sigma}[\ldots[[\mathbf{F}\overline{x_1}]\overline{x_2}]\ldots\overline{x_n}]]\overline{z}]$. Let p be the smallest possible number of such a proof, and $y = 2^z \times 3^p$. Hence $\mathrm{Proof}^{\mathcal{A}}(``\mathbf{F}\overline{x_1}\ldots\overline{x_n} = \overline{z}", p)$ is true. Note that

$$\begin{aligned}\mathrm{S}_n(``\mathbf{F}", \overrightarrow{x_n}, y) &= \mathrm{Proof}^{\mathcal{A}}(``[[Q_{o\sigma\sigma}[\ldots[[" * ``\mathbf{F}" * ``\overline{x_1}" * ``]" \\ &\qquad * ``\overline{x_2}" * ``]" * \ldots * ``\overline{x_n}" * ``]" * ``]" * ``\overline{z}" * ``]", p) \\ &= \mathrm{Proof}^{\mathcal{A}}(``[[\mathrm{Q}_{o\sigma\sigma}[\ldots[[\mathbf{F}\overline{x_1}]\overline{x_2}]\ldots\overline{x_n}]]\overline{z}]", p).\end{aligned}$$

Thus $\mathrm{S}_n(``\mathbf{F}", \overrightarrow{x_n}, y)$ is true, and the relation is regular. Moreover, since $\mathcal{A}$ is consistent, by 6112 there is no number $w \neq z$ such that $\vdash_{\mathcal{A}} \mathbf{F}\overline{x_1}\ldots\overline{x_n} = \overline{w}$, so y is the least number such that $\mathrm{S}_n(``\mathbf{F}", \overrightarrow{x_n}, y)$. Hence $f(\overrightarrow{x_n}) = z = 0\ \mathcal{G}\ell\ y = 0\ \mathcal{G}\ell\ ((\mu y)\mathrm{S}_n(``\mathbf{F}", \overrightarrow{x_n}, y))$.

Since $(\lambda\overrightarrow{x_n}, y)\mathrm{S}_n(``\mathbf{F}", \overrightarrow{x_n}, y)$ is regular and recursive, f is recursive. ∎

REMARK. Theorem 7009 also applies to extensions of $\mathcal{G}_0^\infty$ which are not pure extensions, i.e., extensions with additional primitive constants. One simply modifies the Gödel numbering (and hence the proof) to accommodate the additional constants.

7010 Theorem. If $\mathcal{A}$ is any consistent recursively axiomatized pure extension of $\mathcal{Q}_0^\infty$, then a numerical function is representable in $\mathcal{A}$ if and only if it is recursive.

Proof: By 6401 and 7009.

Thus the representable functions are precisely the recursive functions in a wide variety of extensions of $\mathcal{Q}_0^\infty$, some of which are significantly stronger than $\mathcal{Q}_0^\infty$. Actually, the situation is much more general than this. In a wide variety of logical systems in which number theory can be formalized, the representable functions are precisely the recursive functions.

Of course, if a formalization of number theory is quite weak, one may not be able to represent all recursive functions in it. However, there are finitely axiomatized formulations of number theory in first-order logic in which all recursive functions are representable. On the other hand, it is clear that every numerical function is representable in any inconsistent system.

It seems intuitively clear that if a function is computable, then an algorithm for computing its values can be expressed in some formal language. Hence, if that language is sufficiently strong, the function will actually be representable in it. On the other hand, if a function f is representable (in a consistent language where there is an effective test for deciding whether a given wff is an axiom), it must be computable. For given numbers $\overrightarrow{x_n}$, one can effectively find $f(\overrightarrow{x_n})$ simply by enumerating the theorems until one finds one of the form $\mathbf{F}\overline{x_1}\ldots\overline{x_n} = \overline{z}$ (where $\mathbf{F}$ represents f); one then knows that $f(\overrightarrow{x_n}) = z$. (Of course, this is the intuitive idea behind the proof of 7009.) Thus it seems clear that in sufficiently rich languages, the representable functions are precisely the computable functions. The fact that the representable functions are the recursive functions in a wide variety of languages, which differ among themselves in many other respects, shows that the class of recursive functions is a very natural class, and provides further evidence for Church's Thesis that it is the class of computable functions.

Before concluding this section, we note that the proof of 7009 provides certain additional information about recursive functions.

7011 Normal Form Theorem. For each natural number $n \geq 1$ there is a relation $\mathrm{S}_n \in \mathcal{PRR}^{n+2}$ such that for each n-ary recursive function f, there is a number j such that $(\lambda\overrightarrow{x_n}, y)\mathrm{S}_n(j, \overrightarrow{x_n}, y)$ is regular, and

$$f = (\lambda\overrightarrow{x_n})0\ \mathcal{G}\ell\ ((\mu y)\mathrm{S}_n(j, \overrightarrow{x_n}, y)).$$

Proof: Let $\mathcal{A}$ be $\mathcal{Q}_0^\infty$ in the proof of 7009. $\mathcal{Q}_0^\infty$ can be shown to be consistent (by semantic methods), and by 7005 and 7007 $\mathrm{Proof}^{\mathcal{Q}_0^\infty} \in \mathcal{PRR}$, so $\mathrm{S}_n \in \mathcal{PRR}^{n+2}$. ■

Note that Theorem 7011 provides an enumeration (with repetitions) of the n-ary recursive functions. Of course, some numbers do not correspond to any function in this enumeration.

7012 Theorem. Every recursive function can be defined with at most one application of minimization.

Proof: By 7011. ■

Parable. There was once a king who was quite jealous of the scholars who taught the royal school, since he did not like to be surrounded by people smarter than himself. He decided to give them a test in arithmetic, and to behead the scholars who finished it. He insisted that the scholars all work diligently on the test for four hours each day, omit no details, and take no shortcuts. The first scholar quickly did the problems using the usual algorithm for addition that is taught in grade school, left the palace, and was executed. The second scholar defined addition recursively as in 6503, carried out all the steps of the computation, and worked several days before completing the test and meeting his fate. This third scholar defined $+_{\sigma\sigma\sigma}$ as $R_{\sigma\sigma\sigma(\sigma\sigma\sigma)}[\lambda x_\sigma S_{\sigma\sigma}]$ and performed all additions by giving complete proofs of the corresponding theorems in $\mathcal{Q}_0^\infty$ without using any derived rules of inference. He lived to a ripe old age. The fourth scholar defined $+$ as $\lambda x_1 \lambda x_2\ 0\ \mathcal{G}\ell\ ((\mu y) S_2(\text{“}+_{\sigma\sigma\sigma}\text{”}, x_1, x_2, y))$. His descendants are still working on the test, and have inherited the kingdom.

EXERCISES

X7000. Show that Axiom^{4_5} $\in \mathcal{PRR}$.

X7001. Construct an example of the test the king might have given the scholars in the parable above, make reasonable assumptions about computational methods used by the scholars and their rates of work, and compute the time they would actually take to complete the test.

X7002. A set of natural numbers is said to be *recursively enumerable* (r.e.) iff it is empty or is the range of a recursive function, and a set $\mathcal{S}$ of formulas is said to be *recursively enumerable* iff $\{\text{“}\mathbf{A}\text{”} \mid \mathbf{A} \in \mathcal{S}\}$ is a recursively enumerable set of numbers. Prove that if $\mathcal{A}$ is any recursively axiomatized extension of $\mathcal{Q}_0$, then the set of theorems of $\mathcal{A}$ is recursively enumerable.

§71. Gödel's Incompleteness Theorems

DEFINITIONS. Let $\mathcal{A}$ be any extension of $\mathcal{Q}_0$.

(1) $\mathcal{A}$ is *consistent* iff not $\vdash_{\mathcal{A}} F_o$.
(2) $\mathcal{A}$ is *ω-consistent* iff there is no closed wff $\mathbf{B}_{o\sigma}$ of $\mathcal{A}$ such that $\vdash_{\mathcal{A}} \dot{\exists} y_\sigma \mathbf{B}_{o\sigma} y_\sigma$ and for each natural number k, $\vdash_{\mathcal{A}} \sim \mathbf{B}_{o\sigma}\overline{k}$.
(3) $\mathcal{A}$ is *complete* iff for each sentence $\mathbf{A}_o$ of $\mathcal{A}$, $\vdash_{\mathcal{A}} \mathbf{A}_o$ or $\vdash_{\mathcal{A}} \sim \mathbf{A}_o$.
(4) $\mathcal{A}$ is *ω-complete* iff $\vdash_{\mathcal{A}} \dot{\forall} y_\sigma \mathbf{B}_{o\sigma} y_\sigma$ for each closed wff $\mathbf{B}_{o\sigma}$ of $\mathcal{A}$ such that $\vdash_{\mathcal{A}} \mathbf{B}_{o\sigma}\overline{k}$ for each natural number k.

7100 Proposition. If $\mathcal{A}$ is an ω-consistent extension of $\mathcal{Q}_0$, then $\mathcal{A}$ is consistent.

Proof: If $\mathcal{A}$ is inconsistent, then every wff$_o$ is a theorem of $\mathcal{A}$, so $\mathcal{A}$ is ω-inconsistent. ∎

One of the most basic and ancient paradoxes, which was well known to the ancient Greeks[2], is the *liar paradox*, which is invoked by the statement "This statement is false". There are various ways in which one can construct a statement which refers to itself, as was illustrated by Example 3 near the beginning of §10. When one considers whether the statement is true or false, the paradox becomes evident. If it is false, then the assertion that the statement is false is itself false, so the statement is true. On the other hand, if it is true, then the assertion it makes, which is that the statement is false, is true, so the statement is false.

From this one may conclude that informal reasoning in natural language can lead to a contradiction, and one is led to ask what happens when one tries to imitate such an argument in a formal language. It turns out that this can lead to profound insights into the formal language.

The liar paradox, as expressed above, concerns truth, which is a semantic concept. In §73 we shall consider the consequences of the semantic form of the liar paradox for extensions of $\mathcal{Q}_0^\infty$, but first let us note that the paradox can be cast in a syntactic form if we replace the statement above by "This statement is not provable". If one has a complete axiomatization of the set of sentences true in some particular model, truth in that model should correspond to provability in the theory, so the paradox should still apply, but in a syntactic form. This idea leads to the proof of Gödel's famous Incompleteness Theorem, which we discuss next.

The theorem shows that no ω-consistent recursively axiomatized pure extension of $\mathcal{Q}_0^\infty$ can be complete. Moreover, an explicit example of a sentence which is neither provable nor refutable is provided. Clearly any satisfactory formalization of number theory must be ω-consistent. Nevertheless, in Rosser's extension (7200) of Gödel's Theorem, we shall replace the requirement of ω-consistency by the simple requirement of consistency.

By 6401, 6525, 7007, and 7008 we have established that if $\mathcal{A}$ is any recursively axiomatized pure extension of $\mathcal{Q}_0^\infty$, then the functions $*$ and Num, and the relation Proof$^{\mathcal{A}}$, are representable in $\mathcal{Q}_0^\infty$, and hence in $\mathcal{A}$. Let $*_{\sigma\sigma\sigma}, Num_{\sigma\sigma}$ and $Proof^{\mathcal{A}}_{o\sigma\sigma}$ be wffs of $\mathcal{Q}_0^\infty$ which represent them. Sometimes we will omit the type symbols from these wffs. The context will indicate

[2]See [Kleene, 1952, p. 39]

whether we are referring to the function $*$ or the wff which represents it. Note that if **A** and **B** are formulas, $\vdash \overline{\text{“}\mathbf{A}\text{”}} *_{\sigma\sigma\sigma} \overline{\text{“}\mathbf{B}\text{”}} = \overline{\text{“}\mathbf{AB}\text{”}}$.

7101 Gödel's Incompleteness Theorem [Gödel, 1931]. Let $\mathcal{A}$ be any recursively axiomatized pure extension of $\mathcal{Q}_0^\infty$. Let $\star_{o\sigma}$ be the wff

$$[\lambda x_\sigma \dot{\forall} y_\sigma \sim Proof^{\mathcal{A}}_{o\sigma\sigma}[\overline{128} * x_\sigma * [Num_{\sigma\sigma} x_\sigma] * \overline{2048}] y_\sigma].$$

Let $m \underset{\cdot}{=}$ “ $\star_{o\sigma}$ ” and let $\star\star_o$ be $[\star_{o\sigma}\overline{m}]$.

(1) If $\mathcal{A}$ is consistent, then not $\vdash_{\mathcal{A}} \star\star$.
(2) If $\mathcal{A}$ is ω-consistent, then not $\vdash_{\mathcal{A}} \sim \star\star$.

Proof: We shall write $\vdash_{\mathcal{A}}\mathbf{B}$ simply as $\vdash\mathbf{B}$.

(.1) $\vdash \overline{128} * \overline{m} * [Num_{\sigma\sigma}\overline{m}] * \overline{2048} = \overline{\text{“}\star\star\text{”}}$
since the wffs * and *Num* represent the corresponding functions, and $128 * m * Num(m) * 2048 = \text{“[”} * \text{“}\star_{o\sigma}\text{”} * \text{“}\overline{m}\text{”} * \text{“]”} = \text{“}[\star_{o\sigma}\overline{m}]\text{”} = \text{“}\star\star\text{”}$.
(.2) $\vdash \star\star = \dot{\forall} y_\sigma \sim Proof^{\mathcal{A}}\overline{\text{“}\star\star\text{”}} y_\sigma$ — λ, Def of $\star\star$; R: .1

Now suppose $\vdash \star\star$. Then

(.3) $\vdash \dot{\forall} y_\sigma \sim Proof^{\mathcal{A}}\overline{\text{“}\star\star\text{”}} y_\sigma$ — .2

Let n be the number of a proof of $\star\star$.

(.4) $\vdash Proof^{\mathcal{A}}\overline{\text{“}\star\star\text{”}}\overline{n}$
since the wff *Proof* represents the numerical relation Proof.
(.5) $\vdash \sim Proof^{\mathcal{A}}\overline{\text{“}\star\star\text{”}}\overline{n}$ — $\dot{\forall}$I: .3
(.6) $\vdash F_o$ — .4, .5

Thus if $\vdash \star\star$, then $\mathcal{A}$ is inconsistent.
Next suppose $\mathcal{A}$ is ω-consistent and $\vdash \sim \star\star$.

(.7) $\vdash \dot{\exists} y_\sigma Proof^{\mathcal{A}}\overline{\star\star} y_\sigma$ — .2, 6000

$\mathcal{A}$ is consistent by 7100, so not $\vdash \star\star$, so for each natural number k,

(.8) $\vdash \sim Proof\overline{\text{“}\star\star\text{”}}\ \overline{k}$

since k is not the number of a proof of $\star\star$, and the wff *Proof* represents the relation Proof. But this contradicts the ω-consistency of $\mathcal{A}$. Hence if $\mathcal{A}$ is ω-consistent, not $\vdash \sim \star\star$. ∎

Note from 7101.2 that with respect to our Gödel numbering, $\star\star$ does indeed express the assertion "I am not provable". The reader should study the definitions of $\star_{o\sigma}$ and $\star\star_o$ carefully in order to understand how they permit 7101.2 to be proved. $\star\star$ is a sentence about natural numbers which is true in all standard models of $\mathcal{A}$ but is not provable in $\mathcal{A}$. Of course, if one ignores the Gödel numbering, and seeks to understand $\star\star$ directly as a statement about natural numbers, one finds that it is rather complicated. For an example of a true but unprovable statement about natural numbers having a somewhat simpler form, see [Davis *et al.*, 1976, pp. 345-350].

7102 Corollary. Every consistent recursively axiomatized pure extension $\mathcal{A}$ of $\mathcal{Q}_0^\infty$ is ω-incomplete.

Proof: The wff $\star\star$ of 7101 is not provable in $\mathcal{A}$, so (as in 7101.8) for every natural number k, $\vdash \sim [Proof^{\mathcal{A}}_{o\sigma\sigma}\overline{\text{``}\star\star\text{''}}]\overline{k}$. However, the wff $\dot\forall y_\sigma \sim [Proof^{\mathcal{A}}_{o\sigma\sigma}\overline{\text{``}\star\star\text{''}}]y_\sigma$ is not provable in $\mathcal{A}$, since it is equivalent (by 7101.2) to $\star\star$. ∎

We postpone further discussion of the significance of Gödel's Incompleteness Theorem until we eliminate the condition of ω-consistency in §72.

Henceforth in this section we let $\mathcal{A}$ be any recursively axiomatized pure extension of $\mathcal{Q}_0^\infty$, and let $\vdash \mathbf{B}$ mean $\vdash_{\mathcal{A}} \mathbf{B}$.

We will next establish Gödel's Second Incompleteness Theorem by two arguments, which are presented in the proofs of Theorem 7110 and Corollary 7115 below. The theorem says that if $\mathcal{A}$ is consistent, then it cannot be proved consistent by methods which can be formalized within $\mathcal{A}$. This implies, of course, that the consistency of $\mathcal{A}$ can be proved within $\mathcal{A}$ if and only if $\mathcal{A}$ is inconsistent! The first argument essentially consists of translating from our meta-language into $\mathcal{A}$ the proof of the first part of 7101. Thus no considerations of ω-consistency arise. As is customary, we shall not carry out all the details of the formalization. The facts we use without detailed proof are expressed in Lemmas 7104 and 7106 below.

DEFINITION. Let $Theorem_{o\sigma}$ be the wff $[\lambda x_\sigma \dot\exists y_\sigma Proof^{\mathcal{A}}_{o\sigma\sigma} x_\sigma y_\sigma]$.

7103 Lemma. If $\vdash \mathbf{A}$, then $\vdash \mathit{Theorem}_{o\sigma}\overline{\text{“}\mathbf{A}\text{”}}$.

Proof: If $\vdash \mathbf{A}$, then $\mathbf{A}$ has a proof; let p be the number of such a proof. Hence the numerical relation $\text{Proof}^{\mathcal{A}}(\text{“}\mathbf{A}\text{”}, p)$ holds, so

$\vdash \mathit{Proof}^{\mathcal{A}}_{o\sigma\sigma}\overline{\text{“}\mathbf{A}\text{”}}\,\overline{p}$

$\vdash \dot{\exists} y_\sigma \mathit{Proof}^{\mathcal{A}}_{o\sigma\sigma}\overline{\text{“}\mathbf{A}\text{”}}\, y_\sigma$ 6110, $\exists$ Gen

$\vdash \mathit{Theorem}_{o\sigma}\overline{\text{“}\mathbf{A}\text{”}}$ λ

■

It should be remarked that in spite of Lemma 7103, the wff $\mathit{Theorem}_{o\sigma}$ does not represent the set of numbers of theorems of $\mathcal{A}$. In fact, we shall see (7300) that this set is not representable. Indeed, if we could prove $\vdash \sim \mathit{Theorem}\overline{\text{“}\mathbf{A}_o\text{”}}$ for some wff $\mathbf{A}_o$, we could prove within $\mathcal{A}$ the consistency of $\mathcal{A}$.

7104 Lemma. $\vdash \dot{\forall} x_\sigma \centerdot \mathit{Theorem}_{o\sigma} x_\sigma \supset \mathit{Theorem}_{o\sigma}[\overline{\text{“}[\mathit{Theorem}_{o\sigma}\text{”}} * \text{Num}\ x_\sigma * \overline{2048}]$

This theorem expresses 7103 within $\mathcal{A}$. The proof of 7103 (which depends heavily on our results about representability) is purely syntactic, and so can be formalized within $\mathcal{Q}_0^\infty$, and hence in $\mathcal{A}$. The result of this formalization is a proof of 7104. We omit the lengthy details.

In order to see that 7104 is a correct formalization of 7103, it is perhaps easiest to consider the following corollary:

7105 Corollary. For any wff$_o$ $\mathbf{A}$,

$$\vdash \mathit{Theorem}_{o\sigma}\overline{\text{“}\mathbf{A}\text{”}} \supset \mathit{Theorem}_{o\sigma}\overline{\text{“}[\mathit{Theorem}_{o\sigma}\overline{\text{“}\mathbf{A}\text{”}}]\text{”}}.$$

Proof:

(.1) $\vdash \mathit{Theorem}\overline{\text{“}\mathbf{A}\text{”}} \supset \centerdot \mathit{Theorem}[\overline{\text{“}[\mathit{Theorem}\text{”}} * \mathit{Num}\overline{\text{“}\mathbf{A}\text{”}} * \overline{2048}]$

7104

If $n = \text{“}\mathbf{A}\text{”}$, then $\text{“}[\mathit{Theorem}\text{”} * \text{Num}(n) * 2048 =$ $\text{“}[\mathit{Theorem}\text{”} * \text{“}\overline{n}\text{”} * \text{“}]\text{”} = \text{“}[\mathit{Theorem}\overline{\text{“}\mathbf{A}\text{”}}]\text{”}$, so

(.2) $\vdash \overline{\text{“}[\mathit{Theorem}\text{”}} * \mathit{Num}\overline{\text{“}\mathbf{A}\text{”}} * \overline{2048} = \overline{\text{“}[\mathit{Theorem}\overline{\text{“}\mathbf{A}\text{”}}]\text{”}}$

since $*$ and Num represent the corresponding functions.

7105 follows by Rule R. ■

7106 Lemma. $\vdash \dot{\forall} x_\sigma \dot{\forall} y_\sigma \centerdot Theorem_{o\sigma}[\overline{\text{``}[[\mathsf{Q}_{ooo}\text{''}} * x_\sigma * \overline{2048} * y_\sigma * \overline{2048}] \supset \centerdot Theorem_{o\sigma} x_\sigma \supset \centerdot Theorem_{o\sigma} y_\sigma$.

This lemma expresses within $\mathcal{A}$ the fact that if $\vdash \mathbf{A} = \mathbf{B}$ and $\vdash \mathbf{A}$, then $\vdash \mathbf{B}$, which we established in 5201.

We again omit the details of the proof. Actually the proof of this lemma, though long, is quite straightforward, since wffs representing the various numerical functions and relations we have defined can be obtained by a rather direct translation of our definitions into $\mathcal{Q}_0$. The proof would include the following steps:

(.1) $.1 \vdash \mathbb{N} x_\sigma \wedge \mathbb{N} y_\sigma \wedge Theorem[\overline{\text{``}[[\mathsf{Q}_{ooo}\text{''}} * x_\sigma * \overline{2048} * y_\sigma * \overline{2048}] \wedge Theorem\ x_\sigma$ hyp
(.2) $.1 \vdash Type\ x_\sigma = \overline{3} \wedge Type\ y_\sigma = \overline{3}$
(.3) $.1 \vdash Rule[\overline{\text{``}[[\mathsf{Q}_{ooo}\text{''}} * x_\sigma * \overline{2048} * y_\sigma * \overline{2048}] x_\sigma y_\sigma$
from the definition of Rule
(.4) $.1 \vdash Theorem\ y_\sigma$ from the definition of Proof

Actually, if one wishes to write out all the details, it may be slightly easier to prove directly the corollaries below, which are all we really need from 7106.

7107 Corollary.

$$\vdash Theorem_{o\sigma} \overline{\text{``}\mathbf{A}_o = \mathbf{B}_o\text{''}} \supset \centerdot Theorem_{o\sigma} \overline{\text{``}\mathbf{A}_o\text{''}} \supset Theorem_{o\sigma} \overline{\text{``}\mathbf{B}_o\text{''}}.$$

Proof:

(.1) $\vdash Theorem_{o\sigma}[\overline{\text{``}[[\mathsf{Q}_{ooo}\text{''}} * \overline{\text{``}\mathbf{A}\text{''}} * \overline{2048} * \overline{\text{``}\mathbf{B}\text{''}} * \overline{2048}] \supset \centerdot Theorem_{o\sigma} \overline{\text{``}\mathbf{A}\text{''}} \supset Theorem_{o\sigma} \overline{\text{``}\mathbf{B}\text{''}}$ 7106
(.2) $\vdash \overline{\text{``}[[\mathsf{Q}_{ooo}\text{''}} * \overline{\text{``}\mathbf{A}\text{''}} * \overline{2048} * \overline{\text{``}\mathbf{B}} * \overline{2048} = \overline{\text{``}[[\mathsf{Q}_{ooo}\mathbf{A}]\mathbf{B}]\text{''}}$
since the wff $*$ represents the function $*$.

Then use Rule R to complete the proof. ∎

7108 Corollary. If $\vdash \mathbf{A} = \mathbf{B}$, then $\vdash Theorem_{o\sigma} \overline{\text{``}\mathbf{A}\text{''}} \supset Theorem_{o\sigma} \overline{\text{``}\mathbf{B}\text{''}}$.

Proof:

(.1) $\vdash \mathbf{A} = \mathbf{B}$ by hypothesis
(.2) $\vdash Theorem \overline{\text{``}\mathbf{A} = \mathbf{B}\text{''}}$ 7103: .1
(.3) $\vdash Theorem \overline{\text{``}\mathbf{A}\text{''}} \supset Theorem \overline{\text{``}\mathbf{B}\text{''}}$ MP: .2, 7107

∎

DEFINITION. $Consis_o^{\mathcal{A}}$ is the wff $\sim Theorem_{o\sigma}^{\mathcal{A}}\overline{\text{“}F_o\text{”}}$.

We shall often write $Consis_o^{\mathcal{A}}$ simply as *Consis*.

7109 Corollary. For any wff **A**, $\vdash Theorem_{o\sigma}\overline{\text{“}\mathbf{A}\text{”}} \wedge Theorem_{o\sigma}\overline{\text{“}\sim \mathbf{A}\text{”}} \supset \sim Consis$.

Proof:

(.1) $\vdash \sim \mathbf{A} = \centerdot\, \mathbf{A} = F_o$ — Rule P
(.2) $\vdash Theorem\overline{\text{“}\sim \mathbf{A}\text{”}} \supset Theorem\overline{\text{“}\mathbf{A}_o = F_o\text{”}}$ — 7108: .1
(.3) $\vdash Theorem\overline{\text{“}\mathbf{A}_o = F_o\text{”}} \supset \centerdot\, Theorem\overline{\text{“}\mathbf{A}\text{”}} \supset Theorem\overline{\text{“}F_o\text{”}}$ — 7107
(.4) $\vdash Theorem\overline{\text{“}\mathbf{A}\text{”}} \wedge Theorem\overline{\text{“}\sim \mathbf{A}\text{”}} \supset \sim Consis$ — Rule P: .2, .3; def of *Consis*

■

7110 Gödel's Second Incompleteness Theorem [Gödel, 1931]. Let $\mathcal{A}$ be any recursively axiomatized pure extension of $\mathcal{Q}_0^\infty$. If $\mathcal{A}$ is consistent, then not $\vdash_{\mathcal{A}} Consis_o^{\mathcal{A}}$.

Proof: We first restate the proof of the first part of 7101 using our present terminology. Let $\star_{o\sigma}$ be $[\lambda x_\sigma \sim Theorem_{o\sigma}[\overline{128} * x_\sigma * [Num_{o\sigma} x_\sigma] * \overline{2048}]]$. (Note that this is a different wff from that defined in the proof of 7101, though the two wffs are equivalent.) Let $m =$ “$\star_{o\sigma}$” and let $\star\star_o$ be $[\star_{o\sigma}\overline{m}]$.

(a) $\vdash Theorem\overline{\text{“}\star\star\text{”}} = \sim \star\star$ by the definition of $\star\star$, and the fact that the wffs $*$ and *Num* represent the functions $*$ and Num (see 7101.1).
(b) Suppose $\vdash \star\star$. Then
(c) $\vdash Theorem\overline{\text{“}\star\star\text{”}}$ — 7103: b
(d) $\vdash \sim \star\star$ — R: a, c
(e) Hence by (b) and (d), $\mathcal{A}$ is inconsistent.
(f) Hence, if $\mathcal{A}$ is consistent, not $\vdash \star\star$.
Next we formalize the argument above:
(b1) b1 $\vdash Theorem\overline{\text{“}\star\star\text{”}}$ — hyp
(c1) b1 $\vdash Theorem\overline{\text{“}[Theorem\overline{\text{“}\star\star\text{”}}]\text{”}}$ — MP: b1, 7105
(d1) $\vdash Theorem\overline{\text{“}[Theorem\overline{\text{“}\star\star\text{”}}]\text{”}} \supset Theorem\overline{\text{“}[\sim \star\star]\text{”}}$ — 7108: a
(d2) b1 $\vdash Theorem\overline{\text{“}[\sim \star\star]\text{”}}$ — MP: c1, d1
(e1) b1 $\vdash \sim Consis$ — Rule P: 7109, b1, d2
(f1) $\vdash Theorem\overline{\text{“}\star\star\text{”}} \supset \sim Consis$ — 5240: e1
(f2) $\vdash Consis \supset \sim Theorem\overline{\text{“}\star\star\text{”}}$ — Rule P: f1
(g) $\vdash Consis \supset \star\star$ — Rule P: a, f2

Now suppose $\vdash Consis$. Then $\vdash \star\star$, so by (f), $\mathcal{A}$ is inconsistent. Thus if $\mathcal{A}$ is consistent, not $\vdash Consis$. ■

Let us consider again the proof of 7102. By a purely syntactic argument we were able to prove for every natural number k that $\vdash \sim Proof_{o\sigma\sigma}\overline{\text{“}\star\star\text{”}}\bar{k}$. Why are we unable to prove $\forall y_\sigma \sim Proof_{o\sigma\sigma}\overline{\text{“}\star\star\text{”}}y$ (which is equivalent to $\star\star$)? Simply because we cannot prove $Consis$ in the object language.

Gödel's Second (Incompleteness) Theorem shows that objective (2) of Hilbert's program, as originally conceived, is impossible to achieve for most mathematically significant theories. However, the significance of the theorem extends far beyond this. We next discuss, in an informal way, some additional consequences of the theorem. These considerations are quite general, and for purposes of this discussion we shall ignore various fine points, such as the differences between various formulations of type theory (such as the systems $\mathcal{F}^\omega$ and $\mathcal{Q}_0$ of §50, the systems $\mathcal{T}$ and $\mathcal{C}$ of X8030, and other systems in the literature) and certain technicalities regarding different orders of logic. (See [Andrews, 1974b] and [Gandy, 1956] for more details.) We shall temporarily use $\mathcal{T}^\infty$ as a somewhat ambiguous name for a system obtained by adding an axiom of infinity such as Axiom 6 to a system of type theory.

1. It is very natural to extend the system $\mathcal{G}^+$ discussed in §31 to higher-order logic, and then to try to extend Gentzen's Hauptsatz to higher-order logic by showing that the Cut rule is not actually needed. The conjecture that this could be done was known as Takeuti's Conjecture because of Takeuti's extensive work related to it. Takeuti also pointed out in [Takeuti, 1953] and [Takeuti, 1958] that this Cut-Elimination conjecture implied, by a quite elementary argument, the consistency of $\mathcal{T}^\infty$. Therefore, by Gödel's Second Theorem one could not hope to prove the Cut-Elimination Theorem for type theory by methods which could be formalized in $\mathcal{T}^\infty$. Indeed, when proofs [Takahashi, 1967] [Prawitz, 1968] for the Cut-Elimination Theorem for type theory were found, they used semantic methods whose formalization would require a richer language than (finite) type theory, such as transfinite type theory or axiomatic set theory.

2. In automated theorem proving it is useful to work with a variety of formats for proofs and translate from one to another, since the best context for seeking a proof does not necessarily correspond to the form in which one wishes to display the proof. A useful proof-format for searching for proofs and for analyzing certain theoretical issues is

an *expansion proof* [Miller, 1987]. It is desirable to prove that every proof of a theorem of type theory in natural-deduction format can be translated to an expansion proof for that theorem. However, if we had such a proof which could be formalized in $\mathcal{T}^\infty$, then we could use it to prove the consistency of $\mathcal{T}^\infty$ in $\mathcal{T}^\infty$, contradicting Gödel's Second Theorem. (See [Andrews, 1989] for more details.) Thus, the study of such translations is intrinsically difficult.

3. It is often appropriate to compare the strengths of logical systems in which mathematics can be formalized by saying that the system $\mathcal{L}_2$ is *stronger* than the system $\mathcal{L}_1$ if the consistency of $\mathcal{L}_1$ can be proved in $\mathcal{L}_2$. This relation is generally found to be transitive, and (because of Gödel's Second Theorem) irreflexive (provided the systems under consideration are actually consistent). For example, Kemeny [Kemeny, 1949] showed that Zermelo Set Theory (which is weaker than the popular Zermelo-Fraenkel Set Theory) is stronger than $\mathcal{T}^\infty$ by proving the consistency of $\mathcal{T}^\infty$ in Zermelo Set Theory. This contradicted the widely held assumption that the two systems were of essentially the same strength. For another example, let us call *nth-order arithmetic* the result of adding an Axiom of Infinity (or Peano's Postulates) to n-th order logic. One can prove the consistency of nth-order arithmetic in (n+2)th-order arithmetic, which establishes that (n+2)th-order arithmetic is stronger than nth-order arithmetic. As was mentioned in the discussion following theorem 5402, Gödel asserted in [Gödel, 1936] that "passing to the logic of the next higher order has the effect, not only of making provable certain propositions that were not provable before, but also of making it possible to shorten, by an extraordinary amount, infinitely many of the proofs already available". A complete proof of this may be found in [Buss, 1994].

4. In automated theorem proving, one tries to constrain as much as possible the space one must search while looking for a proof of a theorem. In particular, when seeking an appropriate wff of the form $[\lambda \mathbf{v}_\beta \mathbf{A}_o]$ with which to instantiate a quantifier such as $\forall \mathbf{u}_{o\beta}$, one would like to constrain the possible choices for $\mathbf{A}_o$ as much as possible. $\mathbf{A}_o$ may contain connectives and quantifiers, and it is natural to ask whether one can limit the orders of the variables in these new quantifiers to be no greater than the orders of the variables which occur in the theorem one is trying to prove. When the question is posed in this general a form, the answer is negative. For each positive integer n, let $Consis_o^n$ be a

sentence (similar to $Consis_o^{\mathcal{A}}$) which expresses the consistency of nth-order arithmetic. Note that this is simply an assertion about numbers which can be expressed in some low-order logic which is independent of n. Since $Consis_o^n$ is provable in $(n+2)$th-order arithmetic, [Axiom6 $\supset$ $Consis_o^n$] is provable in $(n+2)$th-order logic, so the wffs [Axiom6 $\supset$ $Consis_o^{10}$], [Axiom6 $\supset$ $Consis_o^{12}$], [Axiom6 $\supset$ $Consis_o^{14}$], ... are all theorems of type theory which are wffs of the same fixed low-order logic, but by Gödel's Second Theorem their proofs must be carried out in logics of increasingly high orders.

The proofs of Gödel's Incompleteness Theorems have been based upon the fact that it is possible to construct a sentence which can be interpreted as saying "I am not provable". It is also possible to construct a sentence which says "I am provable". Henkin [Henkin, 1952] raised the question whether this sentence is provable. This sentence does not seem to suggest any paradox, and at first glance it seems equally plausible that it is or is not provable. Nevertheless, Löb soon showed [Löb, 1955] that is is provable. We shall next present Löb's proof of this fact, and show how an alternative proof of Gödel's Second Theorem follows from Löb's Theorem.

We need a few more preliminary results, all of which follow from the following lemma:

7111 Lemma.

$$\vdash \dot{\forall} x_\sigma \dot{\forall} y_\sigma \,\blacksquare\, Theorem_{o\sigma}[\overline{\text{``}[[\supset_{ooo}\text{''}} * x_\sigma * \overline{2048} * y_\sigma * \overline{2048}]$$

$$\supset \,\blacksquare\, Theorem_{o\sigma} x_\sigma \supset Theorem_{o\sigma} y_\sigma.$$

This lemma expresses within $\mathcal{A}$ the fact that Modus Ponens (5224) is a derived rule of inference. The proof, which we omit, is similar in style to the proof of 7106, though longer.

7112 Corollary. For any wffs$_o$ **A** and **B**, $\vdash Theorem_{o\sigma}\overline{\text{``}[\mathbf{A} \supset \mathbf{B}]\text{''}} \supset \,\blacksquare\, Theorem_{o\sigma}\overline{\text{``}\mathbf{A}\text{''}} \supset Theorem_{o\sigma}\overline{\text{``}\mathbf{B}\text{''}}$.

Proof: By 7111. ■

7113 Corollary. If $\vdash \mathbf{A} \supset \mathbf{B}$, then $\vdash Theorem_{o\sigma}\overline{\text{``}\mathbf{A}\text{''}} \supset Theorem_{o\sigma}\overline{\text{``}\mathbf{B}\text{''}}$.

Proof: By 7103 and 7112. ■

The answer to Henkin's question follows from the next theorem. Some insight into the idea motivating the proof of this theorem may be gained by identifying the notion of provability with that of truth, and considering the following paradoxical argument in informal English: Given any statement **D**, let **E** be the statement "If I am true, so is **D**." Clearly if **E** is true, then **D** is true. Hence **E** is true. Therefore **D** is true, for *any* statement **D**.

7114 Löb's Theorem. For any sentence **D**, $\vdash \mathbf{D}$ if and only if $\vdash Theorem_{o\sigma}\ \overline{\text{“}\mathbf{D}\text{”}} \supset \mathbf{D}$.

Proof: To prove the theorem in the nontrivial direction, let $\mathbf{G}_{o\sigma}$ be the wff $[\lambda x_\sigma \,{\bullet} Theorem_{o\sigma}[\overline{128} * x_\sigma * [Num_{\sigma\sigma} x_\sigma] * \overline{2048}] \supset \mathbf{D}]$. Let $m =$ "$\mathbf{G}_{o\sigma}$" and let $\mathbf{E}_o$ be $[\mathbf{G}_{o\sigma}\overline{m}]$.

(.1) $\vdash \mathbf{E} = {\bullet}\, Theorem\overline{\text{“}\mathbf{E}\text{”}} \supset \mathbf{D}$
λ, def of **E**, representability of ∗ and Num

(.2) $\vdash Theorem\overline{\text{“}\mathbf{E}\text{”}} \supset {\bullet}\, Theorem\overline{\text{“}[Theorem\overline{\text{“}\mathbf{E}\text{”}} \supset \mathbf{D}]\text{”}}$ — 7108: .1

(.3) $\vdash Theorem\overline{\text{“}[Theorem\overline{\text{“}E\text{”}} \supset D]\text{”}} \supset {\bullet}\, Theorem\overline{\text{“}[Theorem\overline{\text{“}\mathbf{E}\text{”}}]\text{”}} \supset Theorem\overline{\text{“}\mathbf{D}\text{”}}$ — 7112

(.4) $\vdash Theorem\overline{\text{“}\mathbf{E}\text{”}} \supset Theorem\overline{\text{“}[Theorem\overline{\text{“}\mathbf{E}\text{”}}]\text{”}}$ — 7105

(.5) $\vdash Theorem\overline{\text{“}\mathbf{E}\text{”}} \supset Theorem\overline{\text{“}\mathbf{D}\text{”}}$ — Rule P: .2, .3, .4

Now suppose

(.6) $\vdash Theorem\overline{\text{“}\mathbf{D}\text{”}} \supset \mathbf{D}$

Then

(.7) $\vdash Theorem\overline{\text{“}\mathbf{E}\text{”}} \supset \mathbf{D}$ — Rule P: .5, .6

(.8) $\vdash \mathbf{E}$ — Rule P: .1, .7

(.9) $\vdash Theorem\overline{\text{“}\mathbf{E}\text{”}}$ — 7103: .8

(.10) $\vdash \mathbf{D}$ — Rule P: .7, .9

■

7115 Corollary. If **D** is any sentence such that $\vdash \mathbf{D} = Theorem\overline{\text{“}\mathbf{D}\text{”}}$, then $\vdash \mathbf{D}$.

Proof: By 7114. ■

Note that if **D** is any such sentence, then **D** asserts its own provability. Of course we can easily construct such a sentence. Simply drop the negation from the definition of ⋆⋆ in the proof of 7110.

REMARK. We now have an elegant alternative proof of Gödel's Second Theorem. Suppose $\vdash Consis$, i.e., $\vdash F_o = Theorem\overline{\text{“}F_o\text{”}}$ (by the definitions of *Consis* and ∼). Then $\vdash F_o$ by 7115, so $\mathcal{A}$ is inconsistent.

Note that the proofs of both Gödel's Second Theorem (in either version which we have given) and Löb's Theorem can be given in full detail once one carries out the details of the proofs of 7104 and 7111. If one proves 7111, one can avoid using 7106.

EXERCISES

X7100. Show that if $\mathcal{A}$ is ω-consistent and complete, then $\mathcal{A}$ is ω-complete.

X7101. Show that is $\mathcal{A}$ is consistent and ω-complete, then $\mathcal{A}$ is ω-consistent.

X7102. Use the ideas in the proofs of 7101 and 7102 to construct a theory which is consistent (if $\mathcal{Q}_0^\infty$ is consistent) but ω-inconsistent.

X7103. Let $\mathcal{A}$ be an ω-consistent recursively axiomatized pure extension of $\mathcal{Q}_0^\infty$. Let $B_{o\sigma}$ be $[\lambda x_\sigma \centerdot [x = \overline{0} \wedge \mathit{Consis}^{\mathcal{A}}] \vee \centerdot x = \overline{1} \wedge \sim \mathit{Consis}^{\mathcal{A}}]$. Prove that $\vdash_{\mathcal{A}} \dot{\exists} x_\sigma \mathbf{B}_{o\sigma} x_\sigma$ but there is no natural number n such that $\vdash_{\mathcal{A}} B_{o\sigma}\overline{n}$.

X7104. Show that if $\mathbf{D}$ is any wff$_o$, $\vdash \mathbf{D}$ iff $\vdash \mathit{Theorem}\overline{\text{“}\mathbf{D}\text{”}} = \mathbf{D}$.

X7105. Show that for any cwff $\mathbf{D}_{o\sigma}$, there is a sentence $\mathbf{E}_o$ such that $\vdash \mathbf{E}_o = \mathbf{D}_{o\sigma}\overline{\text{“}\mathbf{E}_o\text{”}}$.

X7106. Show that if $\mathcal{A}$ is a consistent recursively axiomatized pure extension of $\mathcal{Q}_0^\infty$, there is no sentence $\mathbf{B}$ such that $\vdash \sim \mathit{Theorem}\overline{\text{“}\mathbf{B}\text{”}}$.

X7107. Prove Gödel's Second Theorem by constructing a sentence $\mathbf{E}$ such that $\vdash \mathbf{E} = \mathit{Theorem}\overline{\text{“}\sim \mathbf{E}\text{”}}$. (*Hint*: prove $\vdash \mathbf{E} \supset \centerdot \mathit{Theorem}\overline{\text{“}\mathbf{E}\text{”}} \wedge \mathit{Theorem}\overline{\text{“}\sim \mathbf{E}\text{”}}$.)

X7108. Let $\mathcal{A}$ be a consistent recursively axiomatized pure extension of $\mathcal{Q}_0^\infty$. If $\vdash_{\mathcal{A}} \mathbf{E} = \sim \mathit{Theorem}^{\mathcal{A}}_{o\sigma}\overline{\text{“}\sim \mathbf{E}\text{”}}$, then $\mathbf{E}$ expresses the consistency of the system obtained by adding $\mathbf{E}$ to $\mathcal{A}$ as an additional axiom. Is $\mathbf{E}$ consistent with $\mathcal{A}$?

§72. Essential Incompleteness

We start by showing that no extension of $\mathcal{Q}_0^\infty$ can be consistent, recursively axiomatized, and complete.

7200 Gödel-Rosser Incompleteness Theorem. [Rosser, 1936] Let $\mathcal{A}$ be a consistent, recursively axiomatized pure extension of $\mathcal{Q}_0^\infty$. Let $\nabla_{o\sigma}$ be

$$[\lambda x_\sigma \dot{\forall} y_\sigma \bullet Proof^{\mathcal{A}}_{o\sigma\sigma}[\overline{128} * x_\sigma * [Num_{\sigma\sigma} x_\sigma] * \overline{2048}]y_\sigma \supset$$

$$\dot{\exists} z_\sigma \bullet z_\sigma \leq y_\sigma \wedge Proof^{\mathcal{A}}_{o\sigma\sigma}[\overline{\text{``}[\sim_{oo} [\text{''}} * x_\sigma * [Num_{\sigma\sigma} x_\sigma] * \overline{\text{``}]]\text{''}}]z_\sigma].$$

Let $m =$ "$\nabla_{o\sigma}$" and Δ_o be $[\nabla_{o\sigma}\overline{m}]$. Neither $\vdash_{\mathcal{A}} \Delta_o$ nor $\vdash_{\mathcal{A}} \sim \Delta_o$.

Proof:

(.1) $\vdash [\overline{128} * \overline{m} * [Num\ \overline{m}] * \overline{2048}] = \overline{\text{``}\Delta\text{''}}$

since $*_{\sigma\sigma\sigma}$ and $Num_{\sigma\sigma}$ represent the corresponding functions.

(.2) $\vdash \overline{\text{``}[\sim_{oo} [\text{''}} * \overline{m} * [Num\ \overline{m}] * \overline{\text{``}]]\text{''}} = \overline{\text{``}[\sim \Delta]\text{''}}$ for similar reasons.

(.3) $\vdash \Delta = \dot{\forall} y_\sigma \bullet Proof\,\overline{\text{``}\Delta\text{''}}y_\sigma \supset \dot{\exists} z_\sigma \bullet z_\sigma \leq y_\sigma \wedge Proof\,\overline{\text{``}[\sim \Delta]\text{''}}z_\sigma$

λ, def of Δ; R: .1, .2

Suppose (.4) $\vdash \Delta$. Let n be the number of a proof of Δ.

(.5) $\vdash Proof\,\overline{\text{``}\Delta\text{''}}\,\overline{n}$ since $Proof_{o\sigma\sigma}$ represents the relation Proof.

(.6) $\vdash \dot{\forall} y_\sigma \bullet Proof\,\overline{\text{``}\Delta\text{''}}y_\sigma \supset \dot{\exists} z_\sigma \bullet z_\sigma \leq y_\sigma \wedge Proof\,\overline{\text{``}[\sim \Delta]\text{''}}z_\sigma$ R: .4, .3

(.7) $\vdash Proof\,\overline{\text{``}\Delta\text{''}}\,\overline{n} \supset \dot{\exists} z_\sigma \bullet z_\sigma \leq \overline{n} \wedge Proof\,\overline{\text{``}[\sim \Delta]\text{''}}z_\sigma$ $\forall$I: .6; 6110

(.8) $\vdash \dot{\exists} z_\sigma \bullet z \leq \overline{n} \wedge Proof\,\overline{\text{``}[\sim \Delta]\text{''}}z_\sigma$ MP: .5, .7

(.9) $\vdash Proof\,\overline{\text{``}[\sim \Delta]\text{''}}0 \vee \ldots \vee Proof\,\overline{\text{``}[\sim \Delta]\text{''}}\,\overline{n}$ Sub: 6210^n; R: .8

Since $\mathcal{A}$ is consistent, by .4 we have not $\vdash \sim \Delta$. Hence for each natural number k, k is not the number of a proof of $\sim \Delta$, so for each natural number k,

($.10^k$) $\vdash \sim Proof\,\overline{\text{``}[\sim \Delta]\text{''}}\ \overline{k}$

(.11) $\vdash F_o$ Rule P: .9, $.10^0 - .10^n$

This contradicts the consistency of $\mathcal{A}$; hence Δ cannot be a theorem.

Next suppose (.12) $\vdash \sim \Delta$. Let n be the number of a proof of $\sim \Delta$.

(.13) $\vdash Proof\,\overline{\text{``}[\sim \Delta]\text{''}}\,\overline{n}$ by representability

(.14) $\vdash \dot{\exists} y_\sigma \bullet Proof\,\overline{\text{``}\Delta\text{''}}y_\sigma \wedge \dot{\forall} z_\sigma \sim \bullet z_\sigma \leq y_\sigma \wedge Proof\,\overline{\text{``}[\sim \Delta]\text{''}}z_\sigma$

Rule Q: .12, .3

(.15) $.15 \vdash \mathbb{N} y_\sigma \wedge Proof\,\overline{\text{``}\Delta\text{''}}y_\sigma \wedge \dot{\forall} z_\sigma \sim \bullet z_\sigma \leq y_\sigma \wedge Proof\,\overline{\text{``}[\sim \Delta]\text{''}}z_\sigma$

Choose y

(.16) $.15 \vdash \sim \centerdot \overline{n} \leq y_\sigma \ \wedge \ Proof\overline{``[\sim \Delta]"}\overline{n}$ Rule P, $\forall$I: .15, 6110

(.17) $.15 \vdash \overline{n} \leq y_\sigma \ \vee \ y_\sigma \leq \overline{n}$ $\forall$I: 6212; Rule P: 6110, .15

(.18) $.15 \vdash \mathbb{N}y_\sigma \ \wedge \ y_\sigma \leq \overline{n} \ \wedge \ Proof\overline{``\Delta"}y_\sigma$ Rule P: .13, .15, .16, .17

(.19) $.15 \vdash \dot{\exists} y_\sigma \centerdot y_\sigma \leq \overline{n} \ \wedge \ Proof\overline{``\Delta"}y_\sigma$ $\exists$ Gen: .18

(.20) $\vdash .19$ Rule C: .14, .19

(.21) $\vdash Proof\overline{``\Delta"}0 \ \vee \ldots \vee \ Proof\overline{``\Delta"}\overline{n}$ Sub: 6210^n; R: .20

Since $\mathcal{A}$ is consistent, by .12 we cannot have $\vdash \Delta$, so for each natural number k,

($.22^k$) $\vdash \sim Proof\overline{``\Delta"k}$

(.23) $\vdash F_o$ Rule P: .21, $.22^0 - .22^n$

This contradicts the consistency of $\mathcal{A}$, so $\sim \Delta$ cannot be a theorem. ∎

Note that with respect to our Gödel numbering, Δ says "If I am provable, so is my negation." Whether or not $\mathcal{A}$ is consistent, this is true. It may help to understand this proof if we observe that $\sim \Delta$ also asserts (or at least implies) "If I am provable, so is my negation." We can see this as follows:

(.24) $\vdash \sim \Delta \equiv \dot{\exists} y_\sigma \centerdot Proof\ \overline{``\Delta"}y_\sigma \wedge \dot{\forall} z_\sigma \centerdot \ \sim z_\sigma \leq y_\sigma \vee \ \sim \ Proof\ \overline{``[\sim \Delta]"}z_\sigma$

Rule Q: .3

(.25) $\vdash \sim \Delta \equiv \dot{\exists} y_\sigma \dot{\forall} z_\sigma \centerdot Proof\ \overline{``\Delta"}y_\sigma \wedge \centerdot \ \sim z_\sigma \leq y_\sigma \vee \ \sim \ Proof\ \overline{``[\sim \Delta]"}z_\sigma$

Rule Q (prenex): .24

(.26) $\vdash \sim \Delta \supset \dot{\forall} z_\sigma \dot{\exists} y_\sigma \centerdot Proof\ \overline{``\Delta"}y_\sigma \wedge \centerdot \ \sim z_\sigma \leq y_\sigma \vee \ \sim \ Proof\overline{``[\sim \Delta]"}z_\sigma$

Rule Q (2137): .25

(.27) $\vdash \sim \Delta \supset \dot{\forall} z_\sigma \dot{\exists} y_\sigma \centerdot \ \sim \ Proof\ \overline{``[\sim \Delta]"}z_\sigma \vee \centerdot \ \sim z_\sigma \leq y_\sigma \wedge Proof\ \overline{``\Delta"}y_\sigma$

Rule Q (Rule P, 2105): .26

(.28) $\vdash \sim \Delta \supset \dot{\forall} z_\sigma \dot{\exists} y_\sigma \centerdot Proof\ \overline{``[\sim \Delta]"}z_\sigma \supset \centerdot y_\sigma \leq z_\sigma \ \wedge \ Proof\ \overline{``\Delta"}y_\sigma$

Rule Q: .27, 6212

(.29) $\vdash \sim \Delta \supset \dot{\forall} z_\sigma \centerdot Proof\ \overline{``[\sim \Delta]"}z_\sigma \supset \dot{\exists} y_\sigma \centerdot y_\sigma \leq z_\sigma \ \wedge \ Proof\overline{``\Delta"}y_\sigma$

Rule Q (antiprenex): .28

Note how (.29) mirrors (.3). Of course, (.3) is an equivalence and (.29) is only an implication, but the implication suffices for proving that $\sim \Delta$ is not a theorem just as it was proved that Δ is not a theorem.

REMARK. As noted after the proof of theorem 7009, the requirement that $\mathcal{A}$ be a *pure* extension of $\mathcal{Q}_0^\infty$ can easily be dropped. Thus we obtain the following:

7201 Corollary.

(a) If $\mathcal{A}$ is any consistent and complete extension of $\mathcal{Q}_0^\infty$, then $\mathcal{A}$ is not recursively axiomatizable.
(b) If $\mathcal{M}$ is any model of $\mathcal{Q}_0^\infty$, the set of sentences true in $\mathcal{M}$ is not recursively axiomatizable.
(c) If $\mathcal{M}$ is any model of $\mathcal{Q}_0^\infty$, the set of sentences true in $\mathcal{M}$ is not recursive.

It is well known in informal mathematics that Peano's Postulates are categorical. In our present context, this means that all *standard* models of Peano's Postulates are isomorphic. We shall sharpen this statement slightly, and then use it with 7200 to show, as promised at the end of §55, that there is no recursively axiomatized pure extension of $\mathcal{Q}_0$ in which the theorems are precisely the wffs$_o$ which are valid in the standard sense. The sentence Inf^{III} below states Peano's Postulates and fixes the behavior of the description operator in an arbitrary but conveniently simple manner. (Recall that the wffs $\Pi_{o(o\imath)}$ and $\sum^1_{o(o\imath)}$ used below were defined in §51 and §53, respectively.)

DEFINITION. Let Inf^{III} be the sentence

$$\exists o_\imath \exists s_{\imath\imath}[\forall x_\imath[s_{\imath\imath}x_\imath \neq o_\imath] \wedge \forall x_\imath \forall y_\imath[s_{\imath\imath}x_\imath = s_{\imath\imath}y_\imath \supset x_\imath = y_\imath]$$

$$\wedge\ \forall p_{o\imath}[[p_{o\imath}o_\imath\ \wedge\ \forall x_\imath \centerdot p_{o\imath}x_\imath \supset p_{o\imath} \centerdot s_{\imath\imath}x_\imath] \supset \Pi_{o(o\imath)}p_{o\imath}]$$

$$\wedge\ \forall p_{o\imath} \centerdot \sim \textstyle\sum^1_{o(o\imath)} p_{o\imath} \supset \iota_{\imath(o\imath)}p_{o\imath} = o_\imath].$$

7202 Proposition. $\vdash_{\mathcal{Q}_0} Inf^{III} \supset$ Axiom 6.

Proof: It is easy to see that $\vdash_{\mathcal{Q}_0} Inf^{III} \supset Inf^{II}$ for the wff Inf^{II} of §60. Simply let $n_{o\imath}$ be $[\lambda x_\imath T_o]$ in Inf^{II}. Since (as discussed in §60 and §62), $\vdash_{\mathcal{Q}_0} Inf^{II} \supset$ Axiom 6, the result follows easily. (For an alternative approach, see X7200.) ∎

We next show that Inf^{III} has an essentially unique standard model. Recall that the concept of isomorphism of models was defined in §54.

7203 Proposition. There is a standard model $\mathcal{N}$ for $\{Inf^{III}\}$ such that $\mathcal{N}$ is isomorphic to every standard model for $\{Inf^{III}\}$.

Proof: Let $\mathcal{E}_\imath$ be the set $\mathbb{N}$ of natural numbers. Let $\mathcal{N} = \langle\{\mathcal{E}_\alpha\}_\alpha, \mathcal{J}\rangle$ be the standard model with $\mathcal{E}_\imath$ as domain of individuals, and with $\mathcal{J}\iota_{\imath(o\imath)}$ as that function from $\mathcal{E}_{o\imath}$ into $\mathcal{E}_\imath$ which maps each one-element set to its unique member, and every other set to the number zero.

Using X5403 it is straightforward to show, in the manner of the proof of 5402 (the Soundness Theorem), that $\mathcal{N} \models Inf^{III}$. We leave the detailed verification to the reader. Thus $\mathcal{N}$ is a model for $\{Inf^{III}\}$.

Next suppose $\mathcal{M} = \langle\{\mathcal{D}_\alpha\}_\alpha, \mathcal{I}\rangle$ is any standard model for Inf^{III}. Thus there is an assignment φ into $\mathcal{M}$ such that

$$\mathcal{M} \models_\varphi [\forall x_\imath[s_{\imath\imath}x_\imath \neq o_\imath] \land \forall x_\imath \forall y_\imath[s_{\imath\imath}x_\imath = s_{\imath\imath}y_\imath \supset x_\imath = y_\imath]$$

$$\land \forall p_{o\imath}[[p_{o\imath}o_\imath \land \forall x_\imath \centerdot p_{o\imath}x_\imath \supset p_{o\imath} \centerdot s_{\imath\imath}x_\imath] \supset \Pi_{o(o\imath)}p_{o\imath}]$$

$$\land \forall p_{o\imath} \centerdot \sim \textstyle\sum^1_{o(o\imath)} p_{o\imath} \supset \iota_{\imath(o\imath)}p_{o\imath} = o_\imath].$$

We define a map τ from $\mathcal{E}_\imath$ to $\mathcal{D}_\imath$ by recursion: $\tau 0 = \varphi o_\imath$, and $\tau(\mathrm{S}n) = (\varphi s_{\imath\imath})(\tau n)$ for each $n \in \mathbb{N}$.

We prove τ is one-one by contradiction. Suppose τ is not one-one, and k is the least natural number such that there exists $j \in \mathbb{N}$ such that $j \neq k$ and $\tau j = \tau k$. Clearly $k < j$, so there exists $m \in \mathbb{N}$ such that $j = Sm$. $\mathcal{M} \models_\varphi \forall x_\imath[s_{\imath\imath}x_\imath \neq o_\imath]$, so $\mathcal{M} \models_{(\varphi:x_\imath/\tau m)} \sim [s_{\imath\imath}x_\imath = o_\imath]$, so $\tau k = \tau j = \tau(Sm) = (\varphi s_{\imath\imath})(\tau m) \neq \varphi o_\imath = \tau 0$, so $k \neq 0$. Hence there exists $n \in \mathbb{N}$ such that $k = \mathrm{S}n$. Let $\psi = ((\varphi : x_\imath/\tau m) : y_\imath/\tau n)$. $\mathcal{V}_\psi[s_{\imath\imath}x_\imath] = (\varphi s_{\imath\imath})(\tau m) = \tau(Sm) = \tau j = \tau k = \tau(\mathrm{S}n) = (\varphi s_{\imath\imath})(\tau n) = \mathcal{V}_\psi[s_{\imath\imath}y_\imath]$, so $\mathcal{M} \models_\psi [s_{\imath\imath}x_\imath = s_{\imath\imath}y_\imath]$. However, $\mathcal{M} \models_\varphi \forall x_\imath \forall y_\imath[s_{\imath\imath}x_\imath = s_{\imath\imath}y_\imath \supset x_\imath = y_\imath]$, so $\mathcal{M} \models_\psi [s_{\imath\imath}x_\imath = s_{\imath\imath}y_\imath \supset x_\imath = y_\imath]$, so $\mathcal{M} \models_\psi [x_\imath = y_\imath]$, so $\tau m = \psi x_\imath = \psi y_\imath = \tau n$. Since $n < \mathrm{S}n = k$, this contradicts the choice of k. Thus τ must be one-one on $\mathcal{E}_\imath$.

To show that τ maps $\mathcal{E}_\imath$ onto $\mathcal{D}_\imath$, let u be that function from $\mathcal{D}_\imath$ into $\mathcal{D}_o$ such that for each $d \in \mathcal{D}_\imath$, $ud = \mathsf{T}$ iff d is in the range of τ. (Informally, we may regard u as the range of τ). Since $\mathcal{M}$ is a standard model, $u \in \mathcal{D}_{o\imath}$. Let $\psi = (\varphi : p_{o\imath}/u)$. $\mathcal{M} \models_\psi [[p_{o\imath}o_\imath \land \forall x_\imath \centerdot p_{o\imath}x_\imath \supset p_{o\imath} \centerdot s_{\imath\imath}x_\imath] \supset \Pi_{o(o\imath)}p_{o\imath}]$. $\tau 0 = \varphi o_\imath$, so $\mathcal{V}^{\mathcal{M}}_\psi[p_{o\imath}o_\imath] = u(\varphi o_\imath) = \mathsf{T}$. Also, for any $d \in \mathcal{D}_\imath$, if $ud = \mathsf{T}$, then there exists $n \in \mathcal{E}_\imath$ such that $\tau n = d$, so $\tau(\mathrm{S}n) = (\varphi s_{\imath\imath})(\tau n) = (\varphi s_{\imath\imath})d$, so $u((\varphi s_{\imath\imath})d) = \mathsf{T}$; thus $\mathcal{M} \models_\psi \forall x_\imath[p_{o\imath}x_\imath \supset p_{o\imath} \centerdot s_{\imath\imath}x_\imath]$. Therefore $\mathcal{M} \models_\psi \Pi_{o(o\imath)}p_{o\imath}$, i.e., $\mathcal{V}_\psi[\lambda x_\imath T_o] = u$. Thus u maps every element of $\mathcal{D}_\imath$ to T, so $\mathcal{D}_\imath$ is the range of τ.

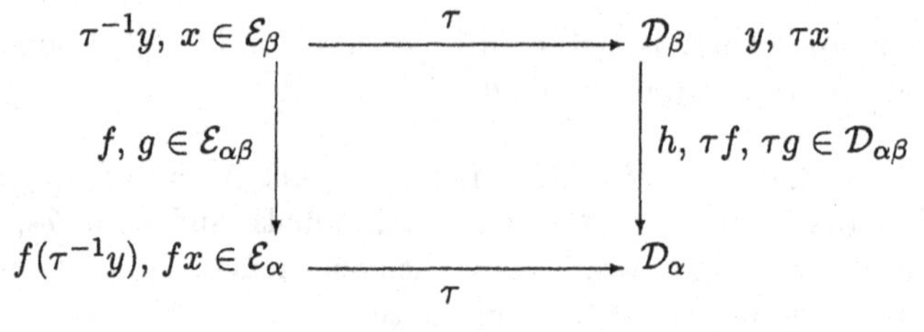

Figure 7.1: Diagram for proof of 7203

By induction on α, for each type symbol α we next extend τ to map $\mathcal{E}_\alpha$ into $\mathcal{D}_\alpha$, and show that τ is a one-one map from $\mathcal{E}_\alpha$ onto $\mathcal{D}_\alpha$.

Since $\mathcal{E}_o = \{\mathsf{T}, \mathsf{F}\} = \mathcal{D}_o$, we let τ be the identity map on $\mathcal{E}_o$.

Given that τ is a one-one map from $\mathcal{E}_\alpha$ onto $\mathcal{D}_\alpha$, and from $\mathcal{E}_\beta$ onto $\mathcal{D}_\beta$, we define τ from $\mathcal{E}_{\alpha\beta}$ into $\mathcal{D}_{\alpha\beta}$ as follows: given $f \in \mathcal{E}_{\alpha\beta}$, let τf be that function from $\mathcal{D}_\beta$ to $\mathcal{D}_\alpha$ such that for each $y \in \mathcal{D}_\beta$, $(\tau f)y = \tau(f(\tau^{-1}y))$ (See Figure 7.1). Thus for any $x \in \mathcal{E}_\beta$, $(\tau f)(\tau x) = \tau(f(\tau^{-1}\tau x)) = \tau(fx)$. Clearly $\tau f \in \mathcal{D}_{\alpha\beta}$, since $\mathcal{M}$ is a standard model. To show τ is one-one on $\mathcal{E}_{\alpha\beta}$, suppose f and g are in $\mathcal{E}_{\alpha\beta}$ and $\tau f = \tau g$. For any $x \in \mathcal{E}_\beta$, $\tau(fx) = (\tau f)(\tau x) = (\tau g)(\tau x) = \tau(gx)$, so $fx = gx$ since τ is one-one on $\mathcal{E}_\alpha$; hence $f = g$.

To show that τ maps $\mathcal{E}_{\alpha\beta}$ onto $\mathcal{D}_{\alpha\beta}$, let h be any member of $\mathcal{D}_{\alpha\beta}$. Define f from $\mathcal{E}_\beta$ to $\mathcal{E}_\alpha$ so that for each $x \in \mathcal{E}_\beta$, $fx = \tau^{-1}(h(\tau x))$. $f \in \mathcal{E}_{\alpha\beta}$ since $\mathcal{N}$ is standard. Then for any $y \in \mathcal{D}_\beta$, $(\tau f)y = \tau(f(\tau^{-1}y)) = \tau\tau^{-1}h(\tau\tau^{-1}y) = hy$, so $\tau f = h$.

It remains only to show that for each primitive constant $\mathbf{c}_\alpha$ of $\mathcal{Q}_0$, $\tau(\mathcal{J}\mathbf{c}_\alpha) = \mathcal{I}\mathbf{c}_\alpha$. Suppose $\mathbf{c}_\alpha = \mathsf{Q}_{o\beta\beta}$; then $\mathcal{J}\mathbf{c}_\alpha$ is the identity relation q on $\mathcal{E}_\beta$; we must show that τq is the identity relation $\mathcal{I}\mathsf{Q}_{o\beta\beta}$ on $\mathcal{D}_\beta$. For any z and y in $\mathcal{D}_\beta$, $((\tau q)z)y = (\tau(q(\tau^{-1}z)))y = \tau((q(\tau^{-1}z))(\tau^{-1}y)) = q(\tau^{-1}z)(\tau^{-1}y) = \mathsf{T}$ iff $\tau^{-1}z = \tau^{-1}y$ iff $z = y$, so τq is the identity relation on $\mathcal{D}_\beta$.

Finally we show that $\tau(\mathcal{J}\iota_{\imath(o\imath)}) = \mathcal{I}\iota_{\imath(o\imath)}$. Let $u \in \mathcal{D}_{o\imath}$. Since τ is the identity map on $\mathcal{E}_o$, for each $y \in \mathcal{D}_\imath$, $(\tau^{-1}u)(\tau^{-1}y) = \tau((\tau^{-1}u)(\tau^{-1}y)) = (\tau\tau^{-1}u)(\tau\tau^{-1}y) = uy$, so since τ is a one-one correspondence, $\tau^{-1}u$ is a one-element set iff u is, and if $u = \{y\}$, then $\tau^{-1}u = \{\tau^{-1}y\}$. Hence if $u = \{y\}$, $(\tau\mathcal{J}\iota_{\imath(o\imath)})u = \tau((\mathcal{J}\iota_{\imath(o\imath)})(\tau^{-1}u)) = \tau(\tau^{-1}y) = y = (\mathcal{I}\iota_{\imath(o\imath)})u$. Suppose u is not a one-element set; then $\tau^{-1}u$ is not either, so $(\tau\mathcal{J}\iota_{\imath(o\imath)})u = \tau((\mathcal{J}\iota_{\imath(o\imath)})(\tau^{-1}u)) = \tau 0 = \varphi o_\imath = (\mathcal{I}\iota_{\imath(o\imath)})u$ since $\mathcal{M} \models_\varphi \forall p_{o\imath} \centerdot \sim \sum^1_{o(o\imath)} p_{o\imath} \supset \iota_{\imath(o\imath)} p_{o\imath} = o_\imath$. Thus $\tau(\mathcal{J}\iota_{\imath(o\imath)}) = \mathcal{I}\iota_{\imath(o\imath)}$. ∎

7204 Corollary. $\mathcal{Q}_0^\infty$ is consistent.

Proof: By 7203 $\mathcal{N} \models Inf^{III}$, but by 7202 and 5402, $\mathcal{N} \models Inf^{III} \supset$ Axiom 6. Hence $\mathcal{N} \models$ Axiom 6. Therefore $\mathcal{N}$ is a model of $\mathcal{Q}_0^\infty$, and $\mathcal{Q}_0^\infty$ is consistent by 5403. ∎

Naturally, the consistency proof above is a semantic one, which can be carried out only in a language which is stronger than $\mathcal{Q}_0^\infty$, and hence more likely to be inconsistent. Therefore it cannot bring much reassurance to anyone who has serious doubts about the existence of the set of natural numbers or the possibility of reasoning in a consistent way about sets and functions. However, it may be reassuring to one who has no such fundamental doubts, but who has been concerned about the possibility of serious flaws in our formalization of these notions in $\mathcal{Q}_0^\infty$.

DEFINITION. If $\mathcal{A}$ be any extension of $\mathcal{Q}_0$, we say that $\mathcal{A}$ is *essentially incomplete* iff $\mathcal{A}$ is consistent and recursively axiomatized and every consistent recursively axiomatized pure extension of $\mathcal{A}$ is incomplete.

REMARK. We include the requirement that $\mathcal{A}$ be consistent and recursively axiomatized in the definition of essential incompleteness so that systems (such as inconsistent systems) which have no consistent recursively axiomatized pure extensions will not be regarded as essentially incomplete.

7205. Theorem. $\mathcal{Q}_0^\infty$ is essentially incomplete.

Proof: By 7204 and 7200. ∎

It can be argued that Axiom 6 is valid in every standard infinite model of $\mathcal{Q}_0$, for it seems highly plausible that if $\mathcal{M}$ is any standard model of $\mathcal{Q}_0$, then the Axiom of Choice is true in $\mathcal{M}$, so there is a well-ordering of the individuals of $\mathcal{M}$ in $\mathcal{D}_{o\imath\imath}$. If $\mathcal{M}$ is infinite and standard, from this well-ordering one can construct a relation r satisfying the requirements of Axiom 6. Thus if one is interested primarily in standard models and one wishes a system with an axiom of infinity, Axiom 6 should be derivable, so by 7205 one must have a system which is inconsistent or incomplete.

REMARK. We saw in 5404 that every finite model of $\mathcal{Q}_0$ is a standard model. Also, for any positive integer n, let C^n be $\exists x_\imath^1 \ldots \exists x_\imath^n [\forall y_\imath [\bigvee_{1 \leq i \leq n} [y_\imath = x_\imath^i]] \wedge \bigwedge_{1 \leq i < j \leq n} [x^i \neq x^j]] \wedge \forall p_{o\imath} [\sim \sum^1_{o(o\imath)} p_{o\imath} \supset \iota_{\imath(o\imath)} [\lambda x_\imath F_o] = \iota_{\imath(o\imath)} p_{o\imath}]$. Let $\mathcal{Q}_0^n$ be the system obtained from $\mathcal{Q}_0$ by adding C^n as an axiom. It can be shown that $\mathcal{Q}_0^n$ is complete and consistent, and that all models of $\mathcal{Q}_0^n$ are standard and isomorphic to one another. See Exercise X8032.

We now come to one of the most important theorems in this chapter.

7206 Theorem. There is no recursively axiomatized pure extension $\mathcal{B}$ of $\mathcal{Q}_0$ such that the theorems of $\mathcal{B}$ are precisely the wffs$_o$ which are valid in the standard sense.

Proof: Suppose there is such a system $\mathcal{B}$. Let $\mathcal{A}$ be the system obtained from $\mathcal{B}$ by adding Inf^{III} as an additional axiom. Then $\text{Axiom}^{\mathcal{A}} = (\lambda n)[\text{Axiom}^{\mathcal{B}}(n) \vee n = \text{“}Inf^{III}\text{”}]$, and $\vdash_{\mathcal{A}}$ Axiom 6 by 7202, so $\mathcal{A}$ is a recursively axiomatized pure extension of $\mathcal{Q}_0^{\infty}$. Since every standard model of $\mathcal{Q}_0$ is a model of $\mathcal{B}$, the standard model $\mathcal{N}$ of 7203 is a standard model of $\mathcal{A}$, so $\mathcal{A}$ is consistent (by 5403). Hence by 7205 $\mathcal{A}$ is incomplete.

We next note that for any sentence $\mathbf{C}$, the following are equivalent:

(a) $\vdash_{\mathcal{A}} \mathbf{C}$;
(b) $\vdash_{\mathcal{B}} [Inf^{III} \supset \mathbf{C}]$;
(c) $\mathcal{M} \models [Inf^{III} \supset \mathbf{C}]$ for every standard model $\mathcal{M}$;
(d) $\mathcal{N} \models \mathbf{C}$.

It is trivial that (a) $\Leftrightarrow$ (b) and (b) $\Leftrightarrow$ (c) and that (c) $\Rightarrow$ (d) (since $\mathcal{N} \models Inf^{III}$). To show that (d) $\Rightarrow$ (c), suppose that $\mathcal{N} \models \mathbf{C}$, and let $\mathcal{M}$ be any standard model of $\mathcal{Q}_0$. If $\mathcal{M} \models Inf^{III}$, then $\mathcal{N}$ is isomorphic to $\mathcal{M}$ by 7203, so $\mathcal{M} \models \mathbf{C}$ by 5407, so $\mathcal{M} \models Inf^{III} \supset \mathbf{C}$. One the other hand, if not $\mathcal{M} \models Inf^{III}$, then $\mathcal{M} \models Inf^{III} \supset \mathbf{C}$, so (c) follows.

Since for any sentence $\mathbf{C}$, either $\mathcal{N} \models \mathbf{C}$ or $\mathcal{N} \models \sim \mathbf{C}$, it follows that for any sentence $\mathbf{C}$, either $\vdash_{\mathcal{A}} \mathbf{C}$ or $\vdash_{\mathcal{A}} \sim \mathbf{C}$, so $\mathcal{A}$ is complete. This contradiction shows that there is no such system $\mathcal{B}$. ■

7207 Corollary. There is no decision procedure for determining of an arbitrary sentence whether it is valid in the standard sense.

Proof: Suppose there were such a procedure. We could let $\mathcal{B}$ be the extension of $\mathcal{Q}_0$ with postulates all those sentences which are valid in the standard sense. $\mathcal{B}$ would be recursively axiomatized, so it would be a counterexample to Theorem 7206.

REMARK. We can now show that the Strong Compactness conjecture mentioned after Theorem 5503 is indeed false. We shall show that there is a set

$\mathcal{S}$ of sentences such that every finite subset of $\mathcal{S}$ has a standard model, but $\mathcal{S}$ has no standard model.

Let $\mathcal{Q}_0'$ be the expansion of $\mathcal{Q}_0$ obtained by adding the constants $\overline{0}_\imath$ and $\overline{s}_{\imath\imath}$ to the constants of $\mathcal{Q}_0$, and let $\mathcal{Q}_0''$ be the expansion of $\mathcal{Q}_0'$ obtained by adding the constant $\infty_\imath$ to the constants of $\mathcal{Q}_0'$. Let $Inf^{III'}$ be the sentence

$$[\forall x_\imath[\overline{s}_{\imath\imath} x_\imath \neq \overline{0}_\imath] \wedge \forall x_\imath \forall y_\imath[\overline{s}_{\imath\imath} x_\imath = \overline{s}_{\imath\imath} y_\imath \supset x_\imath = y_\imath]$$

$$\wedge \quad \forall p_{o\imath}[[p_{o\imath}\overline{0}_\imath \wedge \forall x_\imath \,{\scriptstyle\blacksquare}\, p_{o\imath}x_\imath \supset p_{o\imath} \,{\scriptstyle\blacksquare}\, \overline{s}_{\imath\imath}x_\imath] \supset \Pi_{o(o\imath)}p_{o\imath}]$$

$$\wedge \quad \forall p_{o\imath} \,{\scriptstyle\blacksquare} \sim \textstyle\sum^1_{o(o\imath)} p_{o\imath} \supset \iota_{\imath(o\imath)}p_{o\imath} = \overline{0}_\imath].$$

$$\text{Let } \mathcal{S} \;=\; \left\{Inf^{III'}\right\} \cup \{\sim \infty_\imath = \overline{s}^n_{\imath\imath}\overline{0}_\imath \mid n \text{ is a natural number}\}$$

$$= \;\left\{Inf^{III'}\right\} \cup \{\sim \infty_\imath = \overline{0}_\imath,\ \sim \infty_\imath = \overline{s}_{\imath\imath}\overline{0},\ \sim \infty_\imath = \overline{s}_{\imath\imath}\overline{s}_{\imath\imath}\overline{0},$$

$$\sim \infty_\imath = \overline{s}_{\imath\imath}\overline{s}_{\imath\imath}\overline{s}_{\imath\imath}\overline{0},\ \ldots\}$$

Basically, an attempt to construct a standard model for $\mathcal{S}$ fails because one cannot find a denotation for $\infty_\imath$, but the details of the argument must be handled carefully.

Let $\mathcal{N}' = \langle\{\mathcal{E}_\alpha\}_\alpha, \mathcal{J}'\rangle$ be the standard model (with $\mathcal{Q}_0'$ as language) obtained by extending the model $\mathcal{N}$ in the proof of 7203 as follows: $\mathcal{J}'$ agrees with $\mathcal{J}$ on the constants of $\mathcal{Q}_0$, $\mathcal{J}'\overline{0}_\imath = 0$, and $\mathcal{J}'\overline{s}_{\imath\imath}$ is the successor function S on the natural numbers. Trivial modifications in the proof of 7203 establish that $\mathcal{N}'$ is a standard model for $\{Inf^{III'}\}$ which is isomorphic to every standard model for $\{Inf^{III'}\}$. It is easily seen by induction that $\mathcal{V}^{\mathcal{N}'}\overline{s}^n_{\imath\imath}\overline{0}_\imath = n$ for every natural number n.

To see that every finite subset of $\mathcal{S}$ has a standard model, let $\mathcal{S}_0$ be any finite subset of $\mathcal{S}$. Define $\mathcal{N}'' = \langle\{\mathcal{E}_\alpha\}_\alpha, \mathcal{J}''\rangle$, where $\mathcal{J}''$ is an extension of $\mathcal{J}'$ obtained by defining $\mathcal{J}''\infty_\imath$ to be a sufficiently large natural number that all of the (finitely many) sentences in $\mathcal{S}_0$ are true. Clearly $\mathcal{N}''$ is a standard model (with $\mathcal{Q}_0''$ as language) for $\mathcal{S}_0$.

Now suppose $\mathcal{S}$ has a standard model $\mathcal{M}'' = \langle\{\mathcal{D}_\alpha\}_\alpha, \mathcal{I}''\rangle$ (which necessarily has $\mathcal{Q}_0''$ as language). Let $\mathcal{I}'$ be the restriction of $\mathcal{I}''$ to the constants of $\mathcal{Q}_0'$ (so $\mathcal{I}'\infty_\imath$ is not defined), and let $\mathcal{M}' = \langle\{\mathcal{D}_\alpha\}_\alpha, \mathcal{I}'\rangle$. Then $\mathcal{M}'$ is a standard model (with language $\mathcal{Q}_0'$) for $\{Inf^{III'}\}$, so there is an isomorphism τ from $\mathcal{M}'$ to $\mathcal{N}'$.

$\mathcal{V}^{\mathcal{M}''}\infty_\imath \in \mathcal{D}_\imath$, so $\tau\mathcal{V}^{\mathcal{M}''}\infty_\imath$ is defined and is some natural number n in $\mathcal{E}_\imath$. $\tau\mathcal{V}^{\mathcal{M}''}\infty_\imath = n = \mathcal{V}^{\mathcal{N}'}\overline{s}^n_{\imath\imath}\overline{0}_\imath = \tau\mathcal{V}^{\mathcal{M}'}\overline{s}^n_{\imath\imath}\overline{0}_\imath$ (by 5406). Since τ is an injection,

$\mathcal{V}^{\mathcal{M}''}\infty_\iota = \mathcal{V}^{\mathcal{M}'}\bar{s}_{\iota\iota}^{n}\bar{o}_\iota = \mathcal{V}^{\mathcal{M}''}\bar{s}_{\iota\iota}^{n}\bar{o}_\iota$, so $\mathcal{M}'' \models \infty_\iota = \bar{s}_{\iota\iota}^{n}\bar{o}_\iota$ by 5401(b). This contradicts the assumption that $\mathcal{M}''$ is a model for $\mathcal{S}$. Therefore, $\mathcal{S}$ has no standard model.

EXERCISES

X7200. Give a direct proof of Proposition 7202 in the following way. Given o_ι and $s_{\iota\iota}$ as in Inf^{III}, let $R_{o\iota\iota}$ be $[\lambda x_\iota \lambda y_\iota \forall p_{o\iota} \centerdot [p_{o\iota}[s_{\iota\iota}x_\iota] \wedge \forall z_\iota \centerdot p_{o\iota}z_\iota \supset p_{o\iota} \centerdot s_{\iota\iota}z_\iota] \supset p_{o\iota}y_\iota]$. Show $R_{o\iota\iota}$ has the properties required by Axiom 6. (Show $\sim R_{o\iota\iota}x_\iota x_\iota$ by induction on x_ι.)

X7201. Show that every complete recursively axiomatized pure extension of $\mathcal{Q}_0$ is decidable (i.e., has a recursive set of theorems).

X7202. Show that there is a consistent recursively axiomatized pure extension $\mathcal{Q}_0^\infty$ which has no standard models.

X7203. Is there a consistent recursively axiomatized extension of $\mathcal{Q}_0$ whose theorems include all sentences valid in all standard models?

X7204. Suppose $\mathbf{A}_\sigma$ is a closed wff such that $\models \mathrm{N}_{o\sigma}\mathbf{A}_\sigma$. Must there exist a natural number n such that $\vdash \mathbf{A}_\sigma = \bar{n}$?

§73. Undecidability and Undefinability

A logistic system is *decidable* iff there is an effective procedure for determining of an arbitrary wff of that system whether that wff is a theorem. When an effective Gödel numbering is given, the existence of such a procedure is equivalent to the existence of an effective procedure for determining of an arbitrary natural number whether it is the number of a theorem. By Church's Thesis, we can replace the notion of an effective procedure involving natural numbers by the notion of recursiveness. Thus we are led to the following definition:

DEFINITION. A pure extension $\mathcal{A}$ of $\mathcal{Q}_0$ is (*recursively*) *decidable* iff $\{``\mathbf{B}" \mid \vdash_{\mathcal{A}} \mathbf{B}\}$ is a recursive set of numbers.

7300 Theorem. If $\mathcal{A}$ is any consistent pure extension of $\mathcal{Q}_0^\infty$, then $\mathcal{A}$ is undecidable.

Proof: Suppose $\mathcal{A}$ is decidable. Then the set $\mathcal{S}$ of numbers of theorems of $\mathcal{A}$ is recursive, and hence by 6401, there is a wff $Thm_{o\sigma}$ which represents this set. Let $G_{o\sigma}$ be $[\lambda x_\sigma \sim Thm_{o\sigma} \,\blacksquare\, \overline{128} * x_\sigma * [Num_{\sigma\sigma} x_\sigma] * \overline{2048}]$. Let $m =$ "$G_{o\sigma}$" and let E_o be $[G_{o\sigma}\overline{m}]$. We shall write $\vdash \mathbf{B}$ for $\vdash_{\mathcal{A}} \mathbf{B}$.

(.1) $\vdash E = \sim Thm\overline{\text{“}E\text{”}}$ def of E

Suppose (.2) $\vdash E$. Then

(.3) $\vdash Thm\overline{\text{“}E\text{”}}$ Since "E" $\in \mathcal{S}$
(.4) $\vdash \sim E$ .1, .3

This contradicts the consistency of $\mathcal{A}$, so not $\vdash E$. Hence

(.5) $\vdash \sim Thm\overline{\text{“}E\text{”}}$ since "E" $\notin \mathcal{S}$
(.6) $\vdash E$ .1, .5

This contradiction shows that $\mathcal{A}$ cannot be decidable. ■

Note that if $\mathcal{A}$ is inconsistent, then $\mathcal{A}$ is decidable, since every wff$_o$ is a theorem, and the set of numbers of wffs$_o$ is recursive. Thus the consistency condition in the statement of the theorem above is clearly necessary. Note that this is a theorem about $\mathcal{Q}_0^\infty$, not $\mathcal{Q}_0$. There are consistent decidable extensions of $\mathcal{Q}_0$ (see Exercise X8032). Also note that nothing was assumed about the axiomatizability of $\mathcal{A}$. Thus $\mathcal{A}$ might have as axioms (and hence as theorems) the set of all sentences true in some particular model of $\mathcal{Q}_0^\infty$.

Since the theorem above is very deep and its proof is very short, it is worth taking time to reflect on the nature of the essential ingredients of the proof, so that one will know when one can apply the same basic idea in other contexts. First of all, it was essential that every decidable set could be represented in $\mathcal{A}$. To establish this, we used Church's Thesis and the fact that recursiveness implies representability in $\mathcal{A}$. Also, it was essential that we could construct a sentence $\mathbf{E}$ such that $\vdash \mathbf{E} = \sim Thm_{o\sigma}\overline{\text{“}\mathbf{E}\text{”}}$. In the context of our Gödel numbering, this was accomplished by using the fact that * and Num are representable functions. Apart from these ingredients, the proof used only very elementary logic.

We can slightly generalize the result above. First we introduce some more terminology.

DEFINITION. A set $\mathcal{H}$ of sentences is *consistent with* a logistic system $\mathcal{S}$ iff not $\mathcal{H} \vdash_{\mathcal{S}} F_o$.

Note that $\mathcal{H}$ is consistent with $\mathcal{S}$ iff the result of adding the sentences in $\mathcal{H}$ to the postulates of $\mathcal{S}$ is a consistent system.

7301 Theorem. If $\mathcal{B}$ is an extension of $\mathcal{Q}_0$ such that {Axiom 6} is consistent with $\mathcal{B}$, then $\mathcal{B}$ is undecidable.

Proof: Let $\mathcal{A}$ be the result of adding Axiom 6 to the postulates of $\mathcal{B}$. Thus $\mathcal{A}$ is a consistent extension of $\mathcal{Q}_0^\infty$. If $\mathcal{B}$ is not a *pure* extension of $\mathcal{Q}_0$, $\mathcal{A}$ will not be a pure extension of $\mathcal{Q}_0^\infty$, but the proof of 7300 can easily be modified to show that $\mathcal{A}$ must be undecidable; one must simply introduce a suitable Gödel numbering for the wffs of $\mathcal{B}$. Note that the proof of 7300 depends primarily on the fact that every recursive function or relation is representable in $\mathcal{Q}_0^\infty$, and hence in every extension of $\mathcal{Q}_0^\infty$.

Now for each wff$_o$ $\mathbf{C}$ of $\mathcal{B}$, $\vdash_{\mathcal{A}} \mathbf{C}$ iff $\vdash_{\mathcal{B}}$ [Axiom 6] $\supset \mathbf{C}$. Thus if $\mathcal{B}$ were decidable, $\mathcal{A}$ would be also. Hence $\mathcal{B}$ must be undecidable. ∎

7302 Corollary. $\mathcal{Q}_0$ is undecidable.

Proof: By 7301 and 7204. ∎

DEFINITION. A logistic system $\mathcal{A}$ is *complete* iff for each sentence $\mathbf{B}$, either $\vdash_{\mathcal{A}} \mathbf{B}$ or $\vdash_{\mathcal{A}} \sim \mathbf{B}$.

It is worth noting that the Gödel-Rosser Theorem (7200) follows easily from 7300 and the following proposition:

7303 Proposition. Let $\mathcal{A}$ be an extension of $\mathcal{Q}_0$ (or of $\mathcal{F}$) which is recursively axiomatized and complete. Then $\mathcal{A}$ is decidable.

Proof: If $\mathcal{A}$ is inconsistent, it is clearly decidable, so we may assume that $\mathcal{A}$ is consistent. It suffices to show that there is an effective procedure for deciding whether *sentences* of $\mathcal{A}$ are theorems, since a wff is a theorem iff its universal closure is. Since for each sentence $\mathbf{B}$, either $\vdash_{\mathcal{A}} \mathbf{B}$ or $\vdash_{\mathcal{A}} \sim \mathbf{B}$, to determine whether $\mathbf{B}$ is a theorem one can simply enumerate proofs (since $\mathcal{A}$ is recursively axiomatized) until one finds a proof of $\mathbf{B}$ or of $\sim \mathbf{B}$. ∎

7200A Gödel-Rosser Incompleteness Theorem. Let $\mathcal{A}$ be a consistent, recursively axiomatized pure extension of $\mathcal{Q}_0^\infty$. Then $\mathcal{A}$ is not complete.

Proof: By 7300 and 7303. ■

REMARK. While the proof above is very elegant and does not use theorems 6210 and 6212, it does not provide an explicit example of a sentence which is neither provable nor refutable.

We next turn our attention to the semantic notion of definability.

DEFINITIONS.

(1) Let $\mathcal{M} = \langle \{\mathcal{D}_\alpha\}_\alpha, \mathcal{J}\rangle$ be a general model for $\mathcal{Q}_0$ and let $r \subseteq \mathcal{D}_{\alpha_1} \times \ldots \times \mathcal{D}_{\alpha_n}$, where $n \geq 1$. (Thus r is an n-ary relation on $\mathcal{M}$.) We say that r is a *definable* relation iff there is a cwff $\mathbf{B}_{o\alpha_n\ldots\alpha_1}$ such that for all n-tuples $\langle k_1, \ldots, k_n\rangle \in \mathcal{D}_{\alpha_1} \times \ldots \times \mathcal{D}_{\alpha_n}$, $r(k_1, \ldots, k_n)$ iff $\mathcal{M} \models \mathbf{B}_{o\alpha_n\ldots\alpha_1} x^1_{\alpha_1} \ldots x^n_{\alpha_n}$, where $\varphi x^i_{\alpha_i} = k_i$ for $1 \leq i \leq n$. In this case we can clearly identify r with $\mathcal{V}^{\mathcal{M}}\mathbf{B}_{o\alpha_n\ldots\alpha_1}$, so we can also say that r is definable iff there is a cwff $\mathbf{B}_{o\alpha_n\ldots\alpha_1}$ such that $r = \mathcal{V}^{\mathcal{M}}\mathbf{B}_{o\alpha_n\ldots\alpha_1}$.

(2) If $\mathcal{M}$ is any general model for $\mathcal{Q}_0^\infty$ and r is any n-ary numerical relation, we say that r is *definable in* $\mathcal{M}$ iff there is a cwff $\mathbf{B}_{o\sigma\ldots\sigma}$ such that for all natural numbers $k_1, \ldots, k_n$, $r(k_1, \ldots, k_n)$ iff $\mathcal{M} \models \mathbf{B}_{o\sigma\ldots\sigma}\overline{k_1}\ldots\overline{k_n}$. Note that if $\mathcal{M}$ is any general model for $\mathcal{Q}_0^\infty$, we can identify the natural numbers with the finite cardinals in $\mathcal{D}_\sigma$, and $\mathcal{V}^{\mathcal{M}}\overline{k} = k$ for each natural number k, so this definition of a definable numerical relation can be regarded as a special case of the more general definition above.

7304 Proposition. If $\mathcal{M}$ is any model for $\mathcal{Q}_0^\infty$, then every numerical relation which is representable in $\mathcal{Q}_0^\infty$ is definable in $\mathcal{M}$.

Proof: Suppose r is represented in $\mathcal{Q}_0^\infty$ by $\mathbf{B}_{o\sigma\ldots\sigma}$. If $r(k_1, \ldots, k_n)$, then $\vdash_{\mathcal{Q}_0^\infty} \mathbf{B}\overline{k_1}\ldots\overline{k_n}$, so $\mathcal{M} \models \mathbf{B}\overline{k_1}\ldots\overline{k_n}$. If not $r(k_1, \ldots, k_n)$, then $\vdash_{\mathcal{Q}_0^\infty} \sim \mathbf{B}\overline{k_1}\ldots\overline{k_n}$, so $\mathcal{M} \models \sim \mathbf{B}\overline{k_1}\ldots\overline{k_n}$, so not $\mathcal{M} \models \mathbf{B}\overline{k_1}\ldots\overline{k_n}$. ■

However, certain definable relations are not representable. For example, we showed in 7300 that the set Thm of Gödel numbers of theorems of $\mathcal{Q}_0^\infty$ is not recursive, and hence not representable. However, when we take $\mathcal{A}$ to be $\mathcal{Q}_0^\infty$, the wff $Theorem_{o\sigma}$ of §71 (namely $[\lambda x_\sigma \dot{\exists} y_\sigma Proof_{o\sigma\sigma} x_\sigma y_\sigma]$) defines this set in any standard model $\mathcal{M}$ of $\mathcal{Q}_0^\infty$. Indeed, if Thm(k), then there is a wff$_o$ $\mathbf{A}$ such that $k =$ "$\mathbf{A}$" and $\vdash \mathbf{A}$, so $\vdash Theorem_{o\sigma}\overline{k}$ by 7103, so $\mathcal{M} \models Theorem_{o\sigma}\overline{k}$. On the other hand, if not Thm(k), then for every number

m, $\vdash \sim Proof_{o\sigma\sigma}\overline{km}$, so $\mathcal{M} \models \sim Proof_{o\sigma\sigma}\overline{km}$, so $\mathcal{M} \models \dot{\forall} y_\sigma \sim Proof_{o\sigma\sigma}\overline{k}y_\sigma$ (since $\mathcal{M}$ is a *standard* model), so $\mathcal{M} \models \sim Theorem_{o\sigma}\overline{k}$.

We next show that the set of Gödel numbers of sentences *true* in $\mathcal{M}$ is not even definable in $\mathcal{M}$.

7305 Tarski's Theorem [Tarski, 1936]. For any model $\mathcal{M}$ of $\mathcal{Q}_0^\infty$, $\{$"**A**" $\mid$ **A** is a sentence and $\mathcal{M} \models \mathbf{A}\}$ is not definable in $\mathcal{M}$.

Proof: Suppose the given set $\mathcal{S}$ is definable by the wff $\mathbf{B}_{o\sigma}$. Thus for each sentence **A**, $\mathcal{M} \models \mathbf{A}$ iff "**A**" $\in \mathcal{S}$ iff $\mathcal{M} \models \mathbf{B}_{o\sigma}\overline{\text{“}\mathbf{A}\text{”}}$. Let $\mathbf{C}_{o\sigma}$ be $[\lambda x_\sigma \centerdot \sim \mathbf{B}_{o\sigma}[\overline{128} * x_\sigma * [Num_{o\sigma}x_\sigma] * \overline{2048}]]$, and let $\mathbf{A}_o$ be $[\mathbf{C}_{o\sigma}\overline{\text{“}\mathbf{C}_{o\sigma}\text{”}}]$. Then $\vdash \mathbf{A} = \sim \mathbf{B}_{o\sigma}\overline{\text{“}\mathbf{A}\text{”}}$ by the definition of **A**, so $\mathcal{M} \models \mathbf{A}$ iff not $\mathcal{M} \models \mathbf{B}_{o\sigma}\overline{\text{“}\mathbf{A}\text{”}}$. Since **A** is a sentence, this is a contradiction. ∎

REMARK. After proving 7303 we noted that 7200A (the Gödel-Rosser Theorem) follows directly from 7300 and 7303. Let us also see how to derive 7200A from Tarski's Theorem. Suppose $\mathcal{A}$ is a consistent and complete recursively axiomatized pure extension of $\mathcal{Q}_0^\infty$. By 5501 $\mathcal{A}$ has a model, which we shall call $\mathcal{M}$. Let $\mathcal{T} = \{$"**A**" $\mid$ **A** is a sentence and $\mathcal{M} \models \mathbf{A}\}$. Since $\mathcal{M}$ is a model of $\mathcal{A}$ and $\mathcal{A}$ is complete, it is clear that $\mathcal{M} \models \mathbf{A}$ iff $\vdash_{\mathcal{A}} \mathbf{A}$ for each sentence **A**, so $\mathcal{T} = \{$"**A**" $\mid$ **A** is a sentence and $\vdash_{\mathcal{A}} \mathbf{A}\}$. $\mathcal{A}$ is decidable by 7303, so $\mathcal{T}$ is recursive, and hence representable in $\mathcal{Q}_0^\infty$ (by 6401), and hence definable in $\mathcal{M}$ (by 7304). However, this contradicts Tarski's Theorem.

Let us also look at the problem of defining truth from a different perspective. When studying logic, we ordinarily try to distinguish carefully between syntactic and semantic concepts, and we only speak of a sentence as being true or false when we have a particular model in mind. However, one can also speak of truth in a more absolute sense, where one does not mention a model because one is concerned with only one model, the actual world. Let us now consider the problem of defining truth without discussing models.

Of course, questions about the nature of truth have arisen throughout the history of philosophy, so it is reasonable to ask how one is to know whether a proposed definition of truth is "adequate". Tarski has suggested [Tarski, 1936] that an adequate definition of truth for a language $\mathcal{L}^1$ which is given in a meta-language $\mathcal{L}^2$ of $\mathcal{L}^1$ should satisfy at least the following criterion: For each sentence **A** of $\mathcal{L}^1$, $\vdash_{\mathcal{L}^2} \mathbf{A} \equiv$ True "**A**", where "**A**" is the name in $\mathcal{L}^2$ for the sentence **A**. For example, if one is using a formalized version of English as a meta-language for French (the object language), and one has an adequate definition of truth for the object language, one should be able to prove that it is snowing iff the sentence "Il neige" is true. If the object language is itself English, one should be able to prove that it is snowing iff the sentence "It is snowing" is true. In [Andrews, 1965, pages 83 and 99] it is shown that a definition of truth for $\mathcal{Q}_0$ which satisfies Tarski's criterion for an adequate definition of truth can be given in the system $\mathcal{Q}$ of transfinite type theory.

To some extent a language may serve as its own meta-language. Indeed, we have shown that by using numerals representing Gödel numbers as names for formulas of $\mathcal{Q}_0$, various syntactical notions for $\mathcal{Q}_0^\infty$ such as 'wff' and 'proof' can be represented in $\mathcal{Q}_0^\infty$. It is therefore natural to ask whether the semantical notion 'truth' for $\mathcal{Q}_0^\infty$ can be adequately defined within $\mathcal{Q}_0^\infty$. Tarski showed that the answer is negative.

7306 Theorem. Let $\mathcal{A}$ be a consistent pure extension of $\mathcal{Q}_0^\infty$. There is no wff $\mathbf{B}_{o\sigma}$ of $\mathcal{A}$ such that for each sentence $\mathbf{A}$ of $\mathcal{A}$, $\vdash_{\mathcal{A}} \mathbf{A} = \mathbf{B}_{o\sigma} \overline{\text{``}\mathbf{A}\text{''}}$.

Proof: Suppose there were such a wff $\mathbf{B}_{o\sigma}$. Then as in the proof of 7305 we can construct a sentence $\mathbf{A}$ such that $\vdash_{\mathcal{A}} \mathbf{A} = \sim \mathbf{B}_{o\sigma} \overline{\text{``}\mathbf{A}\text{''}}$. (If $\mathbf{B}_{o\sigma} \overline{\text{``}\mathbf{C}\text{''}}$ says '"**C**" is true', then **A** says 'I am not true'.) Thus $\vdash_{\mathcal{A}} \mathbf{B}_{o\sigma} \overline{\text{``}\mathbf{A}\text{''}} = \sim \mathbf{B}_{o\sigma} \overline{\text{``}\mathbf{A}\text{''}}$, so $\mathcal{A}$ is inconsistent. ∎

EXERCISES

In each exercise below, discuss whether there is an algorithm for determining of an arbitrary sentence **A** whether or not it has the property in question. Justify your answers with proofs.

X7300. **A** is true in all general models.

X7301. **A** is true in all standard models.

X7302. **A** is true in all infinite general models.

X7303. **A** is true in all infinite standard models.

X7304. **A** is satisfiable.

X7305. **A** is satisfiable in some standard models.

§74. Epilogue

If one looks beyond the technical details of the theorems in this chapter and seeks to put them in philosophical perspective, one can summarize the main import of this chapter with the following conclusion:

$\mathfrak{Truth}$ $\mathfrak{is}$ $\mathfrak{Elusive}$.

The abstract concept of truth is at least partially manifested in the set of all statements (in whatever language) which are true; in particular, it is manifested in the set, which we call $\mathfrak{Truth}$, of all sentences of $\mathcal{Q}_0$ which are true in the actual existing physical and conceptual universe. By including concepts as well as physical objects as entities in our model of reality, we guarantee that it is infinite even if the physical universe and all aspects of time are finite. Thus, we may regard the actual universe as a model for $\mathcal{Q}_0^\infty$. (Of course, one might specify the domain of individuals in a variety of ways.) Since $\mathfrak{Truth}$ is closed under the rules of logical inference, we may regard $\mathfrak{Truth}$ not only as a set of sentences, but also as a consistent and complete extension of $\mathcal{Q}_0^\infty$ whose axioms are the sentences in $\mathfrak{Truth}$, and whose theorems are the wffs$_o$ whose universal closures are in $\mathfrak{Truth}$. Let us note how the results of this chapter imply that $\mathfrak{Truth}$ is elusive:

(1) By 7300, $\mathfrak{Truth}$ is not decidable. That is, there is no correct and universally applicable algorithm for determining what is true, and what is not true.
(2) By 7201a, $\mathfrak{Truth}$ is not recursively axiomatizable. By 7200, if we have correct recursively axiomatized theories about what is true, they will inevitably be incomplete.
(3) By 7305, $\mathfrak{Truth}$ is not even definable (in the language in which the truths are expressed).

Supplementary Exercises

X8000. Find disjunctive and conjunctive normal forms for the wff $[[p \wedge q] \equiv_{\bullet} \sim r \vee {}_{\bullet} q \wedge \sim p]$.

X8001. Express the following statements in symbolic form, and prove that the conclusion follows from the stated assertions.

(a) Anyone who can fly impresses everyone.
(b) Some people are not impressed by anyone.
(c) Therefore, no one can fly.

X8002. For each of the wffs below, find an equivalent wff in which the scope of each negation is an atomic wff. Here we are using a formulation of first-order logic in which $\sim$, $\wedge$, $\vee$, $\supset$, $\equiv$, $\forall$, and $\exists$ are all primitive symbols.

(a) $\sim {}_{\bullet} \sim [\forall x \exists y\, Mxy \vee \forall z \sim Qz] \supset {}_{\bullet} \forall y\, Pyy \vee \exists z\, Rz$
(b) $\sim {}_{\bullet} \forall x\, [Px \equiv Qx] \equiv \exists x\, [Rx \equiv Qy]$
(c) $\sim {}_{\bullet} \sim [\exists x\, Qx \vee \forall x \sim \exists y\, Pxy] \supset {}_{\bullet} \forall x\, Qx \wedge \exists y\, Pyy$
(d) $\sim {}_{\bullet} [\sim \forall x\, \forall y\, Pxy \supset \forall z\, Qz] \equiv \exists x\, Qx$

X8003. Let A be the wff $\forall z\, Pxz \supset \exists x\, \exists y\, [Rx \vee Qy]$. Find a wff B such that the number of quantifiers occurring in B is as small as possible and $\vdash A \equiv B$. Give a sequence $M_1, \ldots, M_n$ of wffs such that M_1 is A, M_n is B, and one can readily see that $\vdash M_i \equiv M_{i+1}$ for each $i < n$ by applying theorems from the text.

X8004. For this exercise we assume we have a formulation of first-order logic in which $\sim$, $\wedge$, $\vee$, $\supset$, $\forall$, and $\exists$ are all primitive symbols. Let D be the wff

$$\exists x\, \forall y\, \exists w\, \forall z\, {}_{\bullet} [Pxy \vee Qxy] \supset {}_{\bullet} Rwz \wedge Swz$$

Find a wff E which is equivalent to D such that each quantifier in E has a scope which contains as few occurrences of propositional connectives as possible. (The *number* of occurrences of quantifiers may be as large as you like.)

In exercises X8005–X8006, prove the given wffs. Present your proofs in the usual format, numbering each line, and indicating the hypotheses and justification for each line of a proof.

X8005. $\forall x\, \exists y_{\blacksquare}\, [Px \wedge Qy] \supset {}_{\blacksquare} Qx \vee Py$

X8006. $\forall x \exists y [Px \wedge_{\blacksquare} Qy \vee Qx] \supset \exists z_{\blacksquare} Pz \wedge Qz$

In exercises X8007–X8019, decide whether the given wff is valid or not.

X8007. $\exists x[Px \supset \forall y\, Py]$

X8008. $\exists x[Px \supset \exists y\, Py]$

X8009. $\forall x[Px \supset \forall y\, Py]$

X8010. $\forall x[Px \wedge Qx] \equiv {}_{\blacksquare} \forall x\, Px \wedge \forall x\, Qx$

X8011. $\forall x\, [Px \equiv Qx] \supset {}_{\blacksquare} \exists x\, Px \supset \exists x\, Qx$

X8012. $\exists x\, [Px \equiv Qx] \supset {}_{\blacksquare} \exists x\, Px \supset \exists x\, Qx$

X8013. $\forall y\, \exists z\, \exists w\, \forall x\, {}_{\blacksquare} [\sim Px \vee Qy] \supset {}_{\blacksquare} Pz \supset Qw$

X8014. $\forall u\, \forall v\, \forall w\, [Ruv \vee Rvw] \supset \exists x\, \forall y\, Rxy$

X8015. $\forall x\, \exists y\, [Px \wedge Qy] \supset Qx$

X8016. $\forall x[Px \vee Qx] \equiv {}_{\blacksquare} \forall x\, Px \vee \forall x\, Qx$

X8017. $\forall x[Px \supset Qx] \supset {}_{\blacksquare} \exists x[Qx \supset Rx] \supset \exists x[Px \supset Rx]$

X8018. $\forall x[Px \supset Qx] \supset {}_{\blacksquare} \exists x[Px \supset Rx] \supset \exists x[Qx \supset Rx]$

X8019. $\exists w[\forall x\, Rxx \supset \exists z\, Rzw]$

X8020. Let D be the wff $\forall x\, \exists y\, Pxy \supset \exists w\, \forall z\, Pwz$.

(a) Is D valid?
(b) Let $\mathcal{T}$ be the first-order theory obtained from the system $\mathcal{F}$ of first-order logic by adding D as an additional axiom. Is $\mathcal{T}$ consistent?

X8021. Let A be the wff $\forall x\, Rxy \supset \exists z_{\blacksquare}\, \forall y[Rxu \wedge Ryz] \supset \sim \forall u\, Ru\, z$

(a) Is x free for u in A?
(b) Is x free for y in A?
(c) Find a prenex normal form of A.

X8022. Give examples of theorems which show the relationship between syntactical concepts and semantic concepts.

X8023. Prove or refute the following conjecture: For every wff $\mathbf{C}$ of $\mathcal{F}$, $\models \mathbf{C}$ if and only if $\models [\mathbf{C} \supset \forall \mathbf{x}\, \mathbf{C}]$.

X8024. Let W be the sentence $\exists z_\imath \centerdot [\lambda x_\imath z_\imath] = [\lambda x_\imath x_\imath]$.

(a) Is W a theorem of $\mathcal{Q}_0$? Justify your answer.
(b) Is $\{W\}$ a consistent set of sentences of $\mathcal{Q}_0$? Justify your answer.

X8025. In each case below, decide whether the given wff is a theorem of $\mathcal{Q}_0$. Justify your answers with rigorous arguments.

(a) $\forall f_{\imath\imath} \forall g_{\imath\imath} \exists z_\imath [f_{\imath\imath} z_\imath = g_{\imath\imath} z_\imath] \supset \forall x_\imath \forall y_\imath [x_\imath = y_\imath]$
(b) $\exists p_{o\imath} \exists y_\imath \centerdot \sim p_{o\imath} y_\imath \wedge \forall f_{\imath\imath} \forall x_\imath \centerdot p_{o\imath} \centerdot f_{\imath\imath} x_\imath$

X8026.
(a) Explain, in an informal and intuitive way, why $\sim$Axiom 6 must be true in every finite general model.
(b) Show that there is a general model $\mathcal{M}$ for $\mathcal{Q}_0$ with an infinite domain of individuals such that $\mathcal{M} \models \sim$Axiom 6.

X8027. Give an example of a set of natural numbers which is not representable in $\mathcal{Q}_0^\infty$.

X8028. Let $\mathcal{H}$ be the set of all sentences which are true in all standard models. Does $\mathcal{H}$ have any nonstandard models?

X8029. In this exercise, the object language is $\mathcal{Q}_0^\infty$. Let $Proof_{o\sigma\sigma}$ be a wff which represents the numerical relation Proof, and $Theorem_{o\sigma}$ be $\left[\lambda x_\sigma \dot{\exists} y_\sigma \centerdot Proof_{o\sigma\sigma} x_\sigma y_\sigma\right]$. Let $M_{o\sigma\sigma}$ be the wff

$$[\lambda x_\sigma \lambda y_\sigma \centerdot Theorem_{o\sigma}[\overline{\text{“[[”}} * y_\sigma * [Num_{\sigma\sigma} x_\sigma] * \overline{\text{“]”}} * [Num_{\sigma\sigma} y_\sigma] * \overline{\text{“]”}}]]$$

and let $N_{o\sigma\sigma}$ be the wff

$$[\lambda x_\sigma \lambda y_\sigma \centerdot Theorem_{o\sigma}[\overline{\text{“[[”}} * x_\sigma * [Num_{\sigma\sigma} x_\sigma] * \overline{\text{“]”}} * [Num_{\sigma\sigma} y_\sigma] * \overline{\text{“]”}}]].$$

Let $m = \text{“}M_{o\sigma\sigma}\text{”}$ and $n = \text{“}N_{o\sigma\sigma}\text{”}$.
Let $D = [[M_{o\sigma\sigma} \overline{m}] \overline{n}]$ and let $E = [[N_{o\sigma\sigma} \overline{m}] \overline{n}]$.

(a) Prove that $\vdash D$ if and only if $\vdash E$.
(b) Is D a theorem? Justify your answer.

X8030. We start by defining the system $\mathcal{C}$, which has the same type symbols, improper symbols, variables, and formation rules as $\mathcal{Q}_0$, but has the following logical constants: $\sim_{(oo)}$ $\vee_{((oo)o)}$ $\Pi_{(o(o\alpha))}$ $\iota_{(\imath(o\imath))}$

Definitions and abbreviations for C:

1. $[\mathbf{A}_o \lor \mathbf{B}_o]$ stands for $\left[[\lor_{((oo)o)}\mathbf{A}_o]\,\mathbf{B}_o\right]$.

2. $[\mathbf{A}_o \supset \mathbf{B}_o]$ stands for $[[\sim_{oo}\,\mathbf{A}_o] \lor \mathbf{B}_o]$.

3. $[\forall \mathbf{x}_\alpha \mathbf{A}_o]$ stands for $\left[\Pi_{(o(o\alpha))}\,[\lambda \mathbf{x}_\alpha \mathbf{A}_o]\right]$.

4. Other propositional connectives, and the existential quantifier, are defined in familiar ways.

5. $\mathcal{Q}_{o\alpha\alpha}$ stands for $[\lambda x_\alpha \lambda y_\alpha \forall f_{o\alpha} \centerdot f_{o\alpha} x_\alpha \supset f_{o\alpha} y_\alpha]$.

6. $[\mathbf{A}_\alpha = \mathbf{B}_\alpha]$ stands for $\mathcal{Q}_{o\alpha\alpha}\mathbf{A}_\alpha\mathbf{B}_\alpha$.

7. $\exists_1 \mathbf{x}_\alpha \mathbf{A}_o$ stands for
$[\lambda p_{o\alpha} \centerdot \exists y_\alpha \centerdot p_{o\alpha} y_\alpha \land \forall z_\alpha \centerdot p_{o\alpha} z_\alpha \supset z_\alpha = y_\alpha]\,[\lambda \mathbf{x}_\alpha \mathbf{A}_o]$.

Rules of inference of C:

1. *Alphabetic Change of Bound Variables* (*α-conversion*). To replace any well-formed part $[\lambda \mathbf{x}_\beta \mathbf{A}_\alpha]$ of a wff by $\left[\lambda \mathbf{y}_\beta \mathsf{S}^{\mathbf{x}_\beta}_{\mathbf{y}_\beta} \mathbf{A}_\alpha\right]$, provided that $\mathbf{y}_\beta$ does not occur in $\mathbf{A}_\alpha$ and $\mathbf{x}_\beta$ is not bound in $\mathbf{A}_\alpha$.

2. *β-contraction.* To replace any well-formed part $[[\lambda \mathbf{x}_\alpha \mathbf{B}_\beta]\,\mathbf{A}_\alpha]$ of a wff by $\mathsf{S}^{\mathbf{x}_\alpha}_{\mathbf{A}_\alpha}\,\mathbf{B}_\beta$, provided that the bound variables of $\mathbf{B}_\beta$ are distinct both from $\mathbf{x}_\alpha$ and from the free variables of $\mathbf{A}_\alpha$.

3. *β-expansion.* To infer **C** from **D** if **D** can be inferred from **C** by a single application of β-contraction.

4. *Substitution.* From $\mathbf{F}_{o\alpha}\mathbf{x}_\alpha$, to infer $\mathbf{F}_{o\alpha}\mathbf{A}_\alpha$, provided that $\mathbf{x}_\alpha$ is not a free variable of $\mathbf{F}_{o\alpha}$.

5. *Modus Ponens.* From $[\mathbf{A}_o \supset \mathbf{B}_o]$ and $\mathbf{A}_o$, to infer $\mathbf{B}_o$.

6. *Generalization.* From $\mathbf{F}_{o\alpha}\mathbf{x}_\alpha$ to infer $\Pi_{o(o\alpha)}\mathbf{F}_{o\alpha}$, provided that $\mathbf{x}_\alpha$ is not a free variable of $\mathbf{F}_{o\alpha}$.

Axioms of C:

(1) $p_o \lor p_o \supset p_o$

(2) $p_o \supset p_o \lor q_o$

(3) $p_o \vee q_o \supset q_o \vee p_o$

(4) $p_o \supset q_o \supset [r_o \vee p_o \supset r_o \vee q_o]$

(5^α) $\Pi_{o(o\alpha)} f_{o\alpha} \supset f_{o\alpha} x_\alpha$

(6^α) $\forall x_\alpha [p_o \vee f_{o\alpha} x_\alpha] \supset \bullet p_o \vee \Pi_{o(o\alpha)} f_{o\alpha}$

(7) **Axioms of Extensionality:**

(7°) $[x_o \equiv y_o] \supset x_o = y_o$

($7^{\alpha\beta}$) $\forall x_\beta [f_{\alpha\beta} x_\beta = g_{\alpha\beta} x_\beta] \supset f_{\alpha\beta} = g_{\alpha\beta}$

(8) **Axiom of Descriptions:**
$\exists_1 y_\alpha p_{o\alpha} y_\alpha \supset p_{o\alpha} \left[\iota_{\alpha(o\alpha)} p_{o\alpha}\right]$

$\mathcal{C}$ is essentially the system introduced by Church in [Church, 1940], except that Church's system did not have Axiom 7° and did have axioms of infinity and choice. The system proved complete by Henkin in [Henkin, 1950] is obtained when Axiom 8 of $\mathcal{C}$ is replaced by an axiom of choice.

A notable subsystem $\mathcal{T}$ of $\mathcal{C}$ is obtained by deleting Axioms (7°), ($7^{\alpha\beta}$), and (8). $\mathcal{T}$ may also be called *elementary type theory* [Andrews, 1971][Andrews, 1974a], since $\mathcal{T}$ simply embodies the logic of propositional connectives, quantifiers, and λ-conversion in the context of type theory.

Show that $\mathcal{C}$ and $\mathcal{Q}_0$ are equivalent systems in the sense that there are type-preserving functions σ mapping wffs of $\mathcal{C}$ to wffs of $\mathcal{Q}_0$ and τ mapping wffs of $\mathcal{Q}_0$ to wffs of $\mathcal{C}$ such that

(a) For every wff $\mathbf{A}_o$ of $\mathcal{C}$, if $\vdash_{\mathcal{C}} \mathbf{A}_o$, then $\vdash_{\mathcal{Q}_0} \sigma \mathbf{A}_o$.
(b) For every wff $\mathbf{B}_o$ of $\mathcal{Q}_0$, if $\vdash_{\mathcal{Q}_0} \mathbf{B}_o$, then $\vdash_{\mathcal{C}} \tau \mathbf{B}_o$.
(c) For every wff $\mathbf{A}$ of $\mathcal{C}$, $\vdash_{\mathcal{C}} \mathbf{A} = \tau(\sigma \mathbf{A})$.
(d) For every wff $\mathbf{B}$ of $\mathcal{Q}_0$, $\vdash_{\mathcal{Q}_0} \mathbf{B} = \sigma(\tau \mathbf{B})$.

X8031. Prove the following theorem of Q_o:
$Finite_{o(o\iota)}[\lambda x_\iota \bullet x_\iota = y_\iota \vee x_\iota = z_\iota]$

X8032. Let $\mathcal{Q}_0^1$ be the system obtained by adding to $\mathcal{Q}_0$ the additional axiom $\forall x_\iota \forall y_\iota \bullet x_\iota = y_\iota$. Prove that $\mathcal{Q}_0^1$ is consistent, complete, and decidable. (*Hint:* See Exercise X5502.)

Summary of Theorems

1000 Principle of Induction on the Construction of a Wff. Let $\mathcal{R}$ be a property of formulas, and let $\mathcal{R}(\mathbf{A})$ mean that $\mathbf{A}$ has property $\mathcal{R}$. Suppose

(1) $\mathcal{R}(\mathbf{p})$ for each propositional variable $\mathbf{p}$.
(2) Whenever $\mathcal{R}(\mathbf{A})$, then $\mathcal{R}(\sim\mathbf{A})$.
(3) Whenever $\mathcal{R}(\mathbf{A})$ and $\mathcal{R}(\mathbf{B})$, then $\mathcal{R}([\mathbf{A} \vee \mathbf{B}])$.

Then every wff has property $\mathcal{R}$.

1100. If $\mathcal{H}_1 \vdash \mathbf{A}$ and $\mathcal{H}_1 \subseteq \mathcal{H}_2$, then $\mathcal{H}_2 \vdash \mathbf{A}$.

Corollary. If $\mathcal{H}_1 \vdash \mathbf{A}$ and $\mathcal{H}_2 \vdash \mathbf{A} \supset \mathbf{B}$ and $\mathcal{H}_1 \subseteq \mathcal{H}$ and $\mathcal{H}_2 \subseteq \mathcal{H}$, then $\mathcal{H} \vdash \mathbf{B}$.

1101 Rule of Substitution (Sub). If $\mathcal{H} \vdash \mathbf{A}$, and if $\mathbf{p}_1, \ldots, \mathbf{p}_n$ are distinct variables which do not occur in any wff in $\mathcal{H}$, then $\mathcal{H} \vdash \mathsf{S}^{\mathbf{p}_1 \ldots \mathbf{p}_n}_{\mathbf{B}_1 \ldots \mathbf{B}_n} \mathbf{A}$.

1102. If $\mathcal{H} \vdash \mathbf{A} \supset \mathbf{B}$ and $\mathcal{H} \vdash \mathbf{C} \vee \mathbf{A}$, then $\mathcal{H} \vdash \mathbf{B} \vee \mathbf{C}$.

1103. $\vdash p \vee \sim p$

1104. $\vdash p \supset \sim\sim p$

1105. $\vdash \sim\sim p \supset p$

1106. $\vdash p \supset p$

1107. $\vdash q \vee p \supset p \vee q$

1108. If $\mathcal{H} \vdash \mathbf{A} \vee \mathbf{B}$, then $\mathbf{B} \vee \mathbf{A}$.

1109 Transitive Law of Implication (Trans). If $\mathcal{H} \vdash \mathbf{A}_1 \supset \mathbf{A}_2$, $\mathcal{H} \vdash \mathbf{A}_2 \supset \mathbf{A}_3, \ldots,$ and $\mathcal{H} \vdash \mathbf{A}_{n-1} \supset \mathbf{A}_n$, then $\mathcal{H} \vdash \mathbf{A}_1 \supset \mathbf{A}_n$.

1110. If $\mathcal{H} \vdash \mathbf{A} \supset \mathbf{C}$ and $\mathcal{H} \vdash \mathbf{B} \supset \mathbf{C}$, then $\mathcal{H} \vdash \mathbf{A} \vee \mathbf{B} \supset \mathbf{C}$.

1111. If $\mathcal{H} \vdash \mathbf{A} \supset \mathbf{C}$ and $\mathcal{H} \vdash \sim \mathbf{A} \supset \mathbf{C}$, then $\mathcal{H} \vdash \mathbf{C}$.

1112. If $\mathcal{H} \vdash \mathbf{A} \supset \mathbf{B}$, then $\mathcal{H} \vdash \mathbf{A} \supset \centerdot \mathbf{C} \vee \mathbf{B}$ and $\mathcal{H} \vdash \mathbf{A} \supset \centerdot \mathbf{B} \vee \mathbf{C}$.

1113. $\vdash [p \lor q] \lor r \supset_{\blacksquare} p \lor_{\blacksquare} q \lor r$

1114. If $\mathcal{H} \vdash [\mathbf{A} \lor \mathbf{B}] \lor \mathbf{C}$, then $\mathcal{H} \vdash \mathbf{A} \lor {}_{\blacksquare}\mathbf{B} \lor \mathbf{C}$.

1115. If $\mathcal{H} \vdash \mathbf{A} \supset \mathbf{B}$ and $\mathcal{H} \vdash \mathbf{A} \supset {}_{\blacksquare}\mathbf{B} \supset \mathbf{C}$, then $\mathcal{H} \vdash \mathbf{A} \supset \mathbf{C}$.

1116 Deduction Theorem. If $\mathcal{H}$, $\mathbf{A} \vdash \mathbf{B}$, then $\mathcal{H} \vdash \mathbf{A} \supset \mathbf{B}$.

1117. $\vdash \sim p \supset {}_{\blacksquare} \sim q \supset \sim_{\blacksquare} p \lor q$

1118. $\vdash p \supset {}_{\blacksquare} \sim p \supset q$

1119. Principle of Induction on Proofs. Let $\mathcal{L}$ be any logistic system, and let $\mathcal{R}$ be any property of wffs of $\mathcal{L}$. If every axiom of $\mathcal{L}$ has property $\mathcal{R}$, and each rule of inference of $\mathcal{L}$ preserves property $\mathcal{R}$, then every theorem of $\mathcal{L}$ has property $\mathcal{R}$.

1200 Soundness Theorem. Every theorem of $\mathcal{P}$ is a tautology.

1201 Theorem. Let $\mathcal{S}$ be any logistic system in which $\mathbf{A} \supset {}_{\blacksquare} \sim \mathbf{A} \supset \mathbf{B}$ is a theorem schema and in which Modus Ponens is a primitive or derived rule of inference. Then $\mathcal{S}$ is absolutely consistent iff $\mathcal{A}$ is consistent with respect to negation.

1202 Consistency Theorem. $\mathcal{P}$ is absolutely consistent and consistent with respect to negation.

1203 Lemma. Let $\mathbf{A}$ be a wff and let all the variables in $\mathbf{A}$ be among $\mathbf{p}_1, \ldots, \mathbf{p}_n$. Let φ be any assignment. For any wff $\mathbf{B}$, let

$$\begin{aligned}\mathbf{B}^{\varphi} &= \mathbf{B} \text{ if } \mathcal{V}_{\varphi}\mathbf{B} = \mathsf{T}\\ &= \sim\mathbf{B} \text{ if } \mathcal{V}_{\varphi}\mathbf{B} = \mathsf{F}.\end{aligned}$$

Then $\mathbf{p}_1^{\varphi}, \ldots, \mathbf{p}_n^{\varphi} \vdash \mathbf{A}^{\varphi}$.

1204 Completeness Theorem. Every tautology is a theorem of $\mathcal{P}$.

1205 Theorem. A wff is a theorem of $\mathcal{P}$ if and only if it is a tautology.

1300 Theorem. The axiom schemata and the rule of inference of $\mathcal{P}$ are all independent.

1400 Substitution-Value Theorem. Let φ be any assignment and θ be any proper substitution. If **A** is any wff of propositional calculus, then $\mathcal{V}_\varphi \theta \mathbf{A} = \mathcal{V}_{\varphi \circ \theta} \mathbf{A}$.

1401 Corollary If $\models \mathbf{A}$, and θ is any proper substitution, then $\models \theta \mathbf{A}$.

1402 Substitutivity of Equivalence. If **A**, **B**, **C**, and **D** are wffs of propositional calculus, and **D** is obtained from **C** by replacing zero or more occurrences of **A** in **C** by occurrences of **B**, and $\models [\mathbf{A} \equiv \mathbf{B}]$, then $\models [\mathbf{C} \equiv \mathbf{D}]$.

1403 Theorem. Let h be any truth function with n arguments, where $n \geq 1$, and let $\mathbf{p}_1, \ldots, \mathbf{p}_n$ be distinct propositional variables. Then there is a wff **A** in disjunctive normal form such that $h = [\lambda \mathbf{p}_1, \ldots, \lambda \mathbf{p}_n \mathbf{A}]$.

1404 Corollary. Every wff of propositional calculus has a disjunctive normal form.

1405 Theorem. The following sets of propositional connectives are complete:

(a) $\{\lor, \sim\}$
(b) $\{\land, \sim\}$
(c) $\{\supset, \sim\}$
(d) $\{\supset, \mathsf{f}\}$
(e) $\{|\}$
(f) $\{\downarrow\}$
(g) $\{\land, \not\equiv, \mathsf{t}\}$

1406 Theorem. Let **A** be a wff in negation normal form, and let φ be an assignment. Then

(a) φ satisfies **A** iff there is a path $P \in \mathcal{VP}(\mathbf{A})$ such that φ satisfies every literal in P.
(b) φ falsifies **A** iff there is a path $P \in \mathcal{HP}(\mathbf{A})$ such that φ falsifies every literal in P.

1407 Theorem. Let **A** be a wff in negation normal form. Then

(a) $\bigvee_{P \in \mathcal{VP}(\mathbf{A})} \bigwedge P$ is a disjunctive normal form of **A**.
(b) $\bigwedge_{P \in \mathcal{HP}(\mathbf{A})} \bigvee P$ is a conjunctive normal form of **A**.

1408 Theorem. Let **A** be a wff in negation normal form. Then

(a) **A** is a contradiction iff every path in $\mathcal{VP}(\mathbf{A})$ contains complementary literals.

(b) **A** is a tautology iff every path in $\mathcal{HP}(\mathbf{A})$ contains complementary literals.

1500 Proposition. Let $\mathcal{H}$ be a finite set of wffs of $\mathcal{P}$. $\mathcal{H}$ is satisfiable iff $\mathcal{H}$ is consistent in $\mathcal{P}$.

1501 Compactness Theorem. Let $\mathcal{S}$ be an infinite set of wffs of propositional calculus such that every finite subset of $\mathcal{S}$ is satisfiable. Then $\mathcal{S}$ is satisfiable.

1502 Corollary. Let $\mathcal{S}$ be any set of wffs of $\mathcal{P}$. $\mathcal{S}$ is satisfiable iff $\mathcal{S}$ is consistent in $\mathcal{P}$.

1600 Ground Resolution Theorem. Let $\mathcal{C}$ be any set of clauses. Then $\mathcal{C}$ is contradictory iff $\mathcal{C} \vdash \Box$.

2100 Rule P. If **B** is tautologous, then $\vdash \mathbf{B}$. If $\vdash \mathbf{A}_1, \ldots, \vdash \mathbf{A}_n$, and if $[\mathbf{A}_1 \land \ldots \land \mathbf{A}_n \supset \mathbf{B}]$ is tautologous, then $\vdash \mathbf{B}$.

2101 Consistency Theorem. $\mathcal{F}$ is absolutely consistent and consistent with respect to negation.

2102. $\vdash \forall \mathbf{x}_n \ldots \forall \mathbf{x}_1 \mathbf{A} \supset \mathbf{A}$.

2103. $\supset\forall$ **Rule:** If $\vdash \mathbf{A} \supset \mathbf{B}$ and **x** is not free in **A**, then $\vdash \mathbf{A} \supset \forall \mathbf{x} \mathbf{B}$.

2104. $\vdash \forall \mathbf{x} [\mathbf{A} \supset \mathbf{B}] \supset \centerdot \forall \mathbf{x}\, \mathbf{A}$, then $\vdash \mathbf{A} \supset \forall \mathbf{x}\, \mathbf{B}$.

2105 Substitutivity of Implication. Let **M, N**, and **A** be wffs, and let $\mathbf{A}'$ be the result of replacing **M** by **N** at zero or more occurrences (henceforth called designated occurrences) of **M** in **A**. Let $\mathbf{y}_1, \ldots, \mathbf{y}_k$ be a list including all individual variables which occur free in **M** or **N**, but are bound in **A** by quantifiers whose scopes contain designated occurrences of **M**.

If the designated occurrences of **M** are all positive in **A**, then

(a) $\vdash \forall \mathbf{y}_1 \ldots \forall \mathbf{y}_k [\mathbf{M} \supset \mathbf{N}] \supset \centerdot \mathbf{A} \supset \mathbf{A}'$, and

(b) if $\vdash \mathbf{M} \supset \mathbf{N}$, then $\vdash \mathbf{A} \supset \mathbf{A}'$.

If the designated occurrences of **M** are all negative in **A**, then

(c) $\vdash \forall \mathbf{y}_1 \ldots \forall \mathbf{y}_k [\mathbf{M} \supset \mathbf{N}] \supset \centerdot \mathbf{A}' \supset \mathbf{A}$, and

(d) if $\vdash \mathbf{M} \supset \mathbf{N}$, then $\vdash \mathbf{A}' \supset \mathbf{A}$.

2106 Substitutivity of Equivalence. Let **M, N, A, A′**, and $\mathbf{y}_1, \ldots, \mathbf{y}_k$ be as in 2105.

(a) $\vdash \forall \mathbf{y}_1 \ldots \forall \mathbf{y}_k [\mathbf{M} \equiv \mathbf{N}] \supset \centerdot \mathbf{A} \equiv \mathbf{A}'$

(b) If $\vdash \mathbf{M} \equiv \mathbf{N}$ then $\vdash \mathbf{A} \equiv \mathbf{A}'$.

(c) If $\vdash \mathbf{M} \equiv \mathbf{N}$ and $\vdash \mathbf{A}$ then $\vdash \mathbf{A}'$.

2107. $\vdash \forall \mathbf{x} [\mathbf{M} \equiv \mathbf{N}] \supset \centerdot \forall \mathbf{x} \mathbf{M} \equiv \forall \mathbf{x} \mathbf{N}$

2108. $\vdash \forall \mathbf{x} [\mathbf{M} \equiv \mathbf{N}] \supset \centerdot \exists \mathbf{x} \mathbf{M} \equiv \exists \mathbf{x} \mathbf{N}$

2109. $\vdash \forall \mathbf{x} \mathbf{C} \equiv \forall \mathbf{y} \mathsf{S}^{\mathbf{x}}_{\mathbf{y}} \mathbf{C}$, provided that **y** is not free in **C** and **y** is free for **x** in **C**.

2110 Rule of Alphabetic Change of Bound Variables (α Rule). Suppose that **A′** is obtained from **A** upon replacing an occurrence of $\forall \mathbf{x} \mathbf{C}$ in **A** by an occurrence of $\forall \mathbf{y} \mathsf{S}^{\mathbf{x}}_{\mathbf{y}} \mathbf{C}$, where **y** is not free in **C** and **y** is free for **x** in **C**. If $\vdash \mathbf{A}$, then $\vdash \mathbf{A}'$.

2111 Universal Instantiation ($\forall$I) If $\mathcal{H} \vdash \forall \mathbf{x} \mathbf{A}$, then $\mathcal{H} \vdash \mathsf{S}^{\mathbf{x}}_{\mathbf{t}} \mathbf{A}$, provided that **t** is a term free for **x** in **A**.

2112 Lemma. Suppose $\mathcal{H} \vdash \mathbf{A}$, and let $\mathcal{U}$ be a finite set of individual variables which are not free in **A** or in $\mathcal{H}$. Then there is a proof of **A** from $\mathcal{H}$ in which there is no application of the rule of Generalization involving generalization on a member of $\mathcal{U}$.

2113 Theorem. If $\mathcal{H}_1 \vdash \mathbf{A}$ and $\mathcal{H}_1 \subseteq \mathcal{H}_2$, then $\mathcal{H}_2 \vdash \mathbf{A}$.

Corollary. If $\vdash \mathbf{A}$, then $\mathcal{H} \vdash \mathbf{A}$.

2114 Extended Rule P. If $\mathcal{H} \vdash \mathbf{A}_1, \ldots, \mathcal{H} \vdash \mathbf{A}_n$, and if $[\mathbf{A}_1 \wedge \ldots \wedge \mathbf{A}_n \supset \mathbf{B}]$ is tautologous, then $\mathcal{H} \vdash \mathbf{B}$. Also, if **B** is tautologous, $\mathcal{H} \vdash \mathbf{B}$.

2115 Extended $\supset \forall$ Rule. If $\mathcal{H} \vdash \mathbf{A} \supset \mathbf{B}$ and if **x** is not free in $\mathcal{H}$ or in **A**, then $\mathcal{H} \vdash \mathbf{A} \supset \forall \mathbf{x} \mathbf{B}$.

2116 Deduction Theorem (Ded). If $\mathcal{H}, \mathbf{A} \vdash \mathbf{B}$ then $\mathcal{H} \vdash \mathbf{A} \supset \mathbf{B}$.

2117 Extended Substitutivity of Implication and Equivalence. Let $\mathbf{M}$, $\mathbf{N}$, and $\mathbf{A}$ be wffs, and let $\mathbf{A}'$ be the result of replacing $\mathbf{M}$ by $\mathbf{N}$ at certain designated occurrences of $\mathbf{M}$ in $\mathbf{A}$. If $\mathcal{H} \vdash \mathbf{A}$, then $\mathcal{H} \vdash \mathbf{A}'$, provided that one of the following conditions is satisfied:

(a) $\vdash \mathbf{M} \supset \mathbf{N}$, and the designated occurrences of $\mathbf{M}$ are all positive in $\mathbf{A}$;
(b) $\vdash \mathbf{N} \supset \mathbf{M}$, and the designated occurrences of $\mathbf{M}$ are all negative in $\mathbf{A}$;
(c) $\vdash \mathbf{M} \equiv \mathbf{N}$.

2118 Rule of Substitution (Sub). Let $\mathbf{m}$ be an individual variable and $\mathbf{t}$ be a term, or let $\mathbf{m}$ be a propositional variable and let $\mathbf{t}$ be a wff. If $\mathcal{H} \vdash \mathbf{A}$, then $\mathcal{H} \vdash \mathsf{S}^{\mathbf{m}}_{\mathbf{t}}\mathbf{A}$, provided that $\mathbf{m}$ does not occur free in $\mathcal{H}$ and $\mathbf{t}$ is free for $\mathbf{m}$ in $\mathbf{A}$.

2119. $\vdash \sim \exists \mathbf{x}\mathbf{A} \equiv \forall \mathbf{x} \sim \mathbf{A}$

2120. $\vdash \sim \forall \mathbf{x}\mathbf{A} \equiv \exists \mathbf{x} \sim \mathbf{A}$

2121. $\vdash \forall \mathbf{x}\forall \mathbf{y}\mathbf{A} \equiv \forall \mathbf{y}\forall \mathbf{x}\mathbf{A}$

2122. $\vdash \exists \mathbf{x}\exists \mathbf{y}\mathbf{A} \equiv \exists \mathbf{y}\exists \mathbf{x}\mathbf{A}$

2123. $\vdash \left(\mathsf{S}^{\mathbf{x}}_{\mathbf{t}}\mathbf{A}\right) \supset \exists \mathbf{x}\mathbf{A}$ provided that $\mathbf{t}$ is a term free for $\mathbf{x}$ in $\mathbf{A}$.

2124. $\vdash \mathbf{A} \supset \exists \mathbf{x}\mathbf{A}$

2125. $\vdash \forall \mathbf{x}\mathbf{A} \supset \exists \mathbf{x}\mathbf{A}$

2126 Rule of Existential Generalization ($\exists$ Gen). If $\mathcal{H} \vdash \mathsf{S}^{\mathbf{x}}_{\mathbf{t}}\mathbf{A}$, where $\mathbf{t}$ is a term free for $\mathbf{x}$ in $\mathbf{A}$, then $\mathcal{H} \vdash \exists \mathbf{x}\ \mathbf{A}$.

2127. $\vdash \forall \mathbf{x}\mathbf{A} \equiv \mathbf{A}$ if $\mathbf{x}$ is not free in $\mathbf{A}$.

2128. $\vdash \exists \mathbf{x}\mathbf{A} \equiv \mathbf{A}$ if $\mathbf{x}$ is not free in $\mathbf{A}$.

2129. $\vdash \forall \mathbf{x}[\mathbf{A} \wedge \mathbf{B}] \equiv \centerdot \forall \mathbf{x}\mathbf{A} \wedge \forall \mathbf{x}\mathbf{B}$

2130. $\vdash \exists \mathbf{x}[\mathbf{A} \vee \mathbf{B}] \equiv \centerdot \exists \mathbf{x}\mathbf{A} \vee \exists \mathbf{x}\mathbf{B}$

2131. $\vdash \exists \mathbf{x}[\mathbf{A} \vee \mathbf{B}] \equiv \centerdot \mathbf{A} \vee \exists \mathbf{x}\mathbf{B}$ if $\mathbf{x}$ is not free in $\mathbf{A}$.
$\vdash \exists \mathbf{x}[\mathbf{B} \vee \mathbf{A}] \equiv \centerdot \exists \mathbf{x}\mathbf{B} \vee \mathbf{A}$ if $\mathbf{x}$ is not free in $\mathbf{A}$.

2132. $\vdash \forall \mathbf{x}\mathbf{A} \lor \forall \mathbf{x}\mathbf{B} \supset \forall \mathbf{x} \,\blacksquare\, \mathbf{A} \lor \mathbf{B}$.

2133. $\vdash \forall \mathbf{x}[\mathbf{A} \lor \mathbf{B}] \equiv \,\blacksquare\, \mathbf{A} \lor \forall \mathbf{x}\mathbf{B}$ if $\mathbf{x}$ is not free in $\mathbf{A}$.
$\vdash \forall \mathbf{x}[\mathbf{B} \lor \mathbf{A}] \equiv \,\blacksquare\, \forall \mathbf{x}\mathbf{B} \lor \mathbf{A}$ if $\mathbf{x}$ is not free in $\mathbf{A}$.

2134. $\vdash \forall \mathbf{x}[\mathbf{B} \supset \mathbf{A}] \equiv \,\blacksquare\, \exists \mathbf{x}\mathbf{B} \supset \mathbf{A}$ if $\mathbf{x}$ is not free in $\mathbf{A}$.

2135 Existential Rule ($\exists$ Rule). If $\mathcal{H}$, $\mathbf{B} \vdash \mathbf{A}$ and if $\mathbf{x}$ is not free in $\mathbf{A}$ or in $\mathcal{H}$, then $\mathcal{H}$, $\exists \mathbf{x}\, \mathbf{B} \vdash \mathbf{A}$.

2136 Rule C. If $\mathcal{H} \vdash \exists \mathbf{x}\, \mathbf{B}$ and $\mathcal{H}$, $\mathrm{S}^{\mathbf{x}}_{\mathbf{y}}\mathbf{B} \vdash \mathbf{A}$, where $\mathbf{y}$ is an individual variable which is free for $\mathbf{x}$ in $\mathbf{B}$ and which is not free in $\mathcal{H}$, $\exists \mathbf{x}\, \mathbf{B}$, or in $\mathbf{A}$, then $\mathcal{H} \vdash \mathbf{A}$.

2137. $\vdash \exists \mathbf{x} \forall \mathbf{z}\mathbf{A} \supset \forall \mathbf{z} \exists \mathbf{x}\mathbf{A}$.

2138. $\vdash \exists \mathbf{x}[\mathbf{A} \land \mathbf{B}] \supset \,\blacksquare\, \exists \mathbf{x}\mathbf{A} \land \exists \mathbf{x}\mathbf{B}$

2139 Rule of Cases. If $\mathcal{H} \vdash \mathbf{A} \lor \mathbf{B}$ and $\mathcal{H}$, $\mathbf{A} \vdash \mathbf{C}$ and $\mathcal{H}$, $\mathbf{B} \vdash \mathbf{C}$, then $\mathcal{H} \vdash \mathbf{C}$.

2140 Corollary. If $\mathcal{H}$, $\mathbf{A} \vdash \mathbf{C}$, and $\mathcal{H}$, $\sim \mathbf{A} \vdash \mathbf{C}$ then $\mathcal{H} \vdash \mathbf{C}$.

2141 Indirect Proof (IP). If $\mathcal{H}$, $\sim \mathbf{A} \vdash \mathbf{B}$, and $\mathcal{H}$, $\sim \mathbf{A} \vdash \sim \mathbf{B}$, then $\mathcal{H} \vdash \mathbf{A}$. Also, if $\mathcal{H}$, $\sim \mathbf{A} \vdash \mathbf{B} \land \sim \mathbf{B}$, then $\mathcal{H} \vdash \mathbf{A}$.

2200 Prenex Normal Form Theorem. Every wff has a prenex normal form. If $\mathbf{D}$ is a rectified wff and $\mathbf{G}$ is the prenex normal form of $\mathbf{D}$, then $\vdash \mathbf{D} \equiv \mathbf{G}$.

2300 Proposition. Let $\mathcal{J}$ be an interpretation, $\mathbf{Z}$ a term or wff, and φ and ψ assignments into $\mathcal{J}$ which agree on all variable which occur free in $\mathbf{Z}$. Then $\mathcal{V}^{\mathcal{J}}_{\varphi}\mathbf{Z} = \mathcal{V}^{\mathcal{J}}_{\psi}\mathbf{Z}$.

2301 Proposition. Let $\mathbf{A}$ be a wff and $\mathcal{M}$ be an interpretation.

(a) $\mathbf{A}$ is valid [in $\mathcal{M}$] iff $\sim\mathbf{A}$ is not satisfiable [in $\mathcal{M}$].
(b) $\mathbf{A}$ is satisfiable [in $\mathcal{M}$] iff $\sim\mathbf{A}$ is not valid [in $\mathcal{M}$].
(c) $\mathbf{A}$ is valid [in $\mathcal{M}$] iff $\forall \mathbf{x}\mathbf{A}$ is valid [in $\mathcal{M}$].
(d) $\mathbf{A}$ is satisfiable [in $\mathcal{M}$] iff $\exists \mathbf{x}\mathbf{A}$ is satisfiable [in $\mathcal{M}$].

2302 Substitution-Value Theorem. If $\mathbf{A}$ is a variable, constant, term, or wff, and θ is a proper substitution for free occurrences of variables such that θ is free on $\mathbf{A}$, and φ is any assignment, then $\mathcal{V}_{\varphi}\theta\mathbf{A} = \mathcal{V}_{\varphi\circ\theta}\mathbf{A}$.

2303 Soundness Theorem.

(a) Every theorem of $\mathcal{F}$ is valid.
(b) If $\mathcal{T}$ is a first-order theory and $\mathcal{M}$ is a model of $\mathcal{T}$, then every theorem of $\mathcal{T}$ is valid in $\mathcal{M}$.
(c) If $\mathcal{G}$ is a set of wffs and φ simultaneously satisfies $\mathcal{G}$ (in some interpretation) and $\mathcal{G} \vdash \mathbf{B}$, then φ satisfies $\mathbf{B}$.

2304 Theorem. Every set of sentences which has a model is consistent.

2401 Theorem. The rules of inference and axiom schemata of $\mathcal{F}$ are independent.

2500 Proposition. Every wff of $\mathcal{F}$ is a literal or is of one of the forms $[\mathbf{A} \lor \mathbf{B}]$, $\forall \mathbf{x}\mathbf{A}$, $\sim\sim \mathbf{A}$, $\sim [\mathbf{A} \lor \mathbf{B}]$, $\sim \forall \mathbf{x}\mathbf{A}$.

2501 Lemma. Let $\mathcal{G}$ be any set of wffs. $\{\mathbf{A}_1, \ldots, \mathbf{A}_n\}$ is inconsistent with $\mathcal{G}$ iff $\mathcal{G} \vdash \sim \mathbf{A}_1 \lor \ldots \lor \sim \mathbf{A}_n$. The wff $\mathbf{A}$ is inconsistent with $\mathcal{G}$ iff $\mathcal{G} \vdash \sim \mathbf{A}$.

2502 Proposition. Let $\mathcal{F}^1$ and $\mathcal{F}^2$ be first order theories, with $\mathcal{F}^2$ a conservative extension of $\mathcal{F}^1$. Then $\mathcal{F}^2$ is consistent iff $\mathcal{F}^1$ is consistent.

2503 Proposition. Let $\mathcal{F}^1$ and $\mathcal{F}^2$ be formulations of $\mathcal{F}$, so that $\mathcal{F}^2$ is an expansion of $\mathcal{F}^1$ obtained by adding additional individual constants to the set of constants of $\mathcal{F}^1$. Let $\mathcal{G}$ be any set of sentences of $\mathcal{F}^1$. Then

(1) $\mathcal{F}^2 \cup \mathcal{G}$ is a conservative extension of $\mathcal{F}^1 \cup \mathcal{G}$.
(2) $\mathcal{G}$ is consistent in $\mathcal{F}^1$ iff it is consistent in $\mathcal{F}^2$.

2504 Lemma. If Γ is a class of sets which is of finite character, then Γ is closed under subsets.

2505 Proposition. $\{ \mathcal{S} \mid \mathcal{S}$ is a consistent set of sentences of $\mathcal{F} \}$ is an abstract consistency class.

2506 Proposition. Let Γ be an abstract consistency class. Let $\Delta = \{\mathcal{S} \mid$ every finite subset of $\mathcal{S}$ is in $\Gamma\}$. Then $\Gamma \subseteq \Delta$ and Δ is an abstract consistency class of finite character.

2507 Smullyan's Unifying Principle. If Γ is an abstract consistency class and $\mathcal{S}$ is a sufficiently pure set of sentences and $\mathcal{S} \in \Gamma$, then $\mathcal{S}$ has a frugal model.

2508 Theorem. Every consistent set of sentences has a frugal model.

2509 Generalized Completeness Theorem. If $\mathcal{S}$ is a set of sentences and $\mathbf{A}$ is valid in all frugal models of $\mathcal{S}$, then $\mathcal{S} \vdash \mathbf{A}$.

2510 Gödel's Completeness Theorem. Every valid wff of $\mathcal{F}$ is a theorem.

2511 Löwenheim's Theorem. If $\mathbf{A}$ is valid in every countable interpretation, then $\mathbf{A}$ is valid.

2512 Theorem. A set of sentences is consistent iff it has a model.

2513 Löwenheim-Skolem Theorem. Every set of sentences which has a model has a frugal model.

2514 Compactness Theorem. A set $\mathcal{S}$ of sentences has a model iff every finite subset of $\mathcal{S}$ has a model.

2515 Proposition. Let $\mathcal{F}^1$ and $\mathcal{F}^2$ be formulations of $\mathcal{F}$, so that $\mathcal{F}^2$ is an expansion of $\mathcal{F}^1$ obtained by adding additional constants (of any type) to the set of constants of $\mathcal{F}^1$. Let $\mathcal{G}$ be any set of sentences of $\mathcal{F}^1$. Then

(1) $\mathcal{F}^2 \cup \mathcal{G}$ is a conservative extension of $\mathcal{F}^1 \cup \mathcal{G}$.
(2) $\mathcal{G}$ is consistent in $\mathcal{F}^1$ iff it is consistent in $\mathcal{F}^2$.

2600. $\vdash x = y \supset y = x$.

2601. $\vdash x = y \supset \centerdot\, y = z \supset x = z$

2602. $\vdash \mathbf{x} = \mathbf{y} \supset \centerdot \left(\mathsf{S}^{\mathbf{z}}_{\cdot\mathbf{x}}\mathbf{A}\right) \equiv \mathsf{S}^{\mathbf{z}}_{\cdot\mathbf{y}}\mathbf{A}$, where $\mathbf{x}$ and $\mathbf{y}$ are free for $\mathbf{z}$ in $\mathbf{A}$.

2603. $\vdash \mathbf{x} = \mathbf{y} \supset \left(\mathsf{S}^{\mathbf{z}}_{\mathbf{x}}\mathbf{t}\right) = \left(\mathsf{S}^{\mathbf{z}}_{\mathbf{y}}\mathbf{t}\right)$ for any term $\mathbf{t}$.

2604. $\vdash x_1 = y_1 \wedge \ldots \wedge x_n = y_n \supset \mathbf{f}x_1 \ldots x_n = \mathbf{f}y_1 \ldots y_n$ for any n-ary function symbol $\mathbf{f}$.

2605. $\vdash x_1 = y_1 \wedge \ldots \wedge x_n = y_n \supset \centerdot\, \mathbf{P}x_1 \ldots x_n \equiv \mathbf{P}y_1 \ldots y_n$ for any n-ary predicate symbol $\mathbf{P}$.

2606 Proposition. Let $\mathcal{T}$ be a first-order theory with a binary predicate constant $=$ such that

(a) $\vdash_{\mathcal{T}} x = x$

and for each n-ary function symbol $\mathbf{f}^n$

(b) $\vdash_{\mathcal{T}} x_1 = y_1 \ \wedge \ldots \wedge \ x_n = y_n \supset \mathbf{f}^n x_1 \ldots x_n = \mathbf{f}^n y_1 \ldots y_n$

and for each n-ary predicate symbol $\mathbf{P}^n$

(c) $\vdash_{\mathcal{T}} x_1 = y_1 \ \wedge \ldots \wedge \ x_n = y_n \ \supset \ \centerdot \ \mathbf{P}_1^n \ldots x_n \ \supset \ \mathbf{P}^n y_1 \ldots y_n.$

Then $\mathcal{T}$ is a first-order theory with equality.

2607. $\vdash \exists_1 x P x \equiv \exists x \centerdot P x \ \wedge \ \forall y \centerdot P y \supset x = y$

2608. $\vdash \exists_1 x P x \equiv \exists y \forall x \centerdot P x \equiv x = y$

2609 Soundness Theorem for Logic with Equality. If $\mathcal{M}$ is an equality-model for a set $\mathcal{S}$ of sentences and $\mathcal{S} \vdash_{\mathcal{F}^=} \mathbf{A}$, then $\mathcal{M} \models \mathbf{A}$.

2610 Proposition. Let $\mathcal{M}$ be any model for Axioms 6 and 7. Then there is an equality-model $\mathcal{N}$ for Axioms 6 and 7 such that the cardinality of $\mathcal{N}$ is $\leq$ the cardinality of $\mathcal{M}$, and $\mathcal{N}$ is elementarily equivalent to $\mathcal{M}$.

2611 Theorem. Let $\mathcal{G}$ be a set of sentences consistent in $\mathcal{F}^=$. Then $\mathcal{G}$ has a frugal equality-model.

2612 Extended Completeness and Soundness Theorem. Let $\mathcal{S}$ be a set of sentences and $\mathbf{A}$ be a wff of $\mathcal{F}^=$. Then $\mathcal{S} \vdash_{\mathcal{F}^=} \mathbf{A}$ iff $\mathbf{A}$ is valid in all frugal equality-models of $\mathcal{S}$.

2613 Löwenheim-Skolem-Tarski Theorem. If $\mathcal{S}$ is a set of sentences of $\mathcal{F}^=$ which has an infinite equality-model, and K is any cardinal number such that $K \geq \text{card}(\mathcal{L}(\mathcal{F}^=))$, then $\mathcal{S}$ has an equality-model of cardinality K.

2614 Proposition. Suppose $\mathcal{F}$ is a formulation of first-order logic such that $\mathcal{L}(\mathcal{F}) = \mathcal{L}(\mathcal{F}^=)$. For any wff $\mathbf{D}$ of $\mathcal{F}$, let $\mathbf{B}$ be the conjunction of all wffs of the form

$$\forall x_1 \ldots \forall x_n \forall y_1 \ldots \forall y_n \left[x_1 = y_1 \ \wedge \ldots \wedge \ x_n = y_n \ \supset \ \mathbf{f} x_1 \ldots x_n = \mathbf{f} y_1 \ldots y_n\right],$$

where $\mathbf{f}$ is a function symbol which occurs in $\mathbf{D}$, and let $\mathbf{C}$ be the conjunction of all wffs of the form

$$\forall x_1 \ldots \forall x_n \forall y_1 \ldots \forall y_n \left[x_1 = y_1 \ \wedge \ldots \wedge \ x_n = y_n \ \supset \ \centerdot \ \mathbf{P} x_1 \ldots x_n \supset \mathbf{P} y_1 \ldots y_n\right],$$

where $\mathbf{P}$ is $=$ or a predicate symbol which occurs in $\mathbf{D}$. Then $\vdash_{\mathcal{F}^=} \mathbf{D}$ iff $\vdash_{\mathcal{F}} \forall x\,[x = x] \wedge \mathbf{B} \wedge \mathbf{C} \supset \mathbf{D}$.

3100 Lemma. If $\vdash_{\mathcal{G}} \mathbf{M} \vee \sim\sim \mathbf{E}$ then $\vdash_{\mathcal{G}} \mathbf{M} \vee \mathbf{E}$.

3101 Theorem. For every wff $\mathbf{A}$ of $\mathcal{F}$, $\vdash_{\mathcal{F}} \mathbf{A}$ iff $\vdash_{\mathcal{G}} \mathbf{A}$.

3102 Corollary. If $\vdash_{\mathcal{G}} \mathbf{M} \vee \mathbf{A}$ and $\vdash_{\mathcal{G}} \sim \mathbf{A} \vee \mathbf{N}$, then $\vdash_{\mathcal{G}} \mathbf{M} \vee \mathbf{N}$.

3200 Theorem. Let $\mathcal{S}$ be a set of sentences of $\mathcal{F}$. $\mathcal{S}$ has no model iff $\mathcal{S}$ has a closed semantic tableau.

3300 Lemma. For any sentence $\mathbf{B}$ of $\mathcal{F}$, $\vdash_{\mathcal{F}^\star} (\bigstar\mathbf{B}) \supset \mathbf{B}$.

3301 Theorem. Let $\mathcal{G}$ be a set of sentences of $\mathcal{F}$ and let $\mathcal{G}^\star = \{\bigstar\mathbf{B} \mid \mathbf{B} \in \mathcal{G}\}$. Then $\mathcal{G}^\star$ has a model iff $\mathcal{G}$ has a model, and $\mathcal{F}^\star \cup \mathcal{G}^\star$ is a conservative extension $\mathcal{F} \cup \mathcal{G}$.

3302 Corollary. A sentence $\mathbf{B}$ of $\mathcal{F}$ is satisfiable iff $\bigstar\mathbf{B}$ is satisfiable.

3303 Corollary. Let $\mathbf{A}$ be any wff, and let $\overline{\mathbf{A}}$ be its universal closure. $\mathbf{A}$ is valid iff $\bigstar \sim \overline{\mathbf{A}}$ is unsatisfiable.

3400 Theorem. Let $\mathcal{S}$ be any set of universal sentences. $\mathcal{S}$ is $\mathcal{R}$-refutable iff $\mathcal{S}$ is unsatisfiable.

3401 Theorem. Let $\mathbf{A}$ be any wff, let $\overline{\mathbf{A}}$ be its universal closure, and let $\mathbf{M}$ be the quantifier-free wff obtained by deleting all quantifiers from $\bigstar \sim \overline{\mathbf{A}}$. Let $\mathbf{N}_1 \wedge \ldots \wedge \mathbf{N}_k$ be a conjunctive normal form of $\mathbf{M}$ (so each $\mathbf{N}_i$ is a disjunction of literals).

The following conditions are equivalent:

(a) $\models \mathbf{A}$.
(b) $\{\mathbf{N}_1, \ldots, \mathbf{N}_k\}$ has no model.
(c) $\Box$ can be derived from $\{\mathbf{N}_1, \ldots, \mathbf{N}_k\}$ by using rules $\mathcal{R}1$, $\mathcal{R}2$, $\mathcal{R}3$, and $\mathcal{R}4$.

3500 Lemma. Let $\mathbf{D}$ be universal nnf, and let $\mathbf{H}$ be a c-instance of $\mathbf{D}$. Then there is a conjunction $\mathbf{M}$ of simple instances of $\mathbf{D}$ such that $\mathbf{M} \equiv \mathbf{H}$ is tautologous; also, $\mathbf{M}$ and $\mathbf{H}$ contain exactly the same atomic wffs.

3501 Lemma. Let $\mathbf{G}$ and $\mathbf{H}$ be c-instances of a universal nnf $\mathbf{D}$. Then there is a c-instance $\mathbf{K}$ of $\mathbf{D}$ such that $\models \mathbf{K} \supset [\mathbf{G} \wedge \mathbf{H}]$.

3502 Lemma. Let $\mathbf{D}$ be a universal nnf. If no c-instance of $\mathbf{D}$ is a *t-f* contradiction, there is a truth assignment which verifies every c-instance of $\mathbf{D}$.

3503 Herbrand's Theorem. Let $\mathbf{A}$ be a wff, let $\overline{\mathbf{A}}$ be its universal closure, and let $\mathbf{D}$ be a negation normal form of the Skolemization $\bigstar \sim \overline{\mathbf{A}}$ of $\sim \overline{\mathbf{A}}$. Then $\mathbf{A}$ is valid if and only if $\mathbf{D}$ has a compound Herbrand instance which is a truth-functional contradiction.

3504 Corollary. Let $\mathbf{A}$ be a wff, let $\overline{\mathbf{A}}$ be its universal closure, and let $\mathbf{D}$ be a negation normal form of the Skolemization $\bigstar \sim \overline{\mathbf{A}}$ of $\sim \overline{\mathbf{A}}$. Then $\mathbf{A}$ is valid if and only if some conjunction of simple Herbrand instances of $\mathbf{D}$ is a truth-functional contradiction.

3505 Corollary. Let $\mathbf{B}$ be a sentence, and let $\mathbf{D}$ be a negation normal form of the Skolemization $\bigstar\mathbf{B}$ of $\mathbf{B}$. Then $\mathbf{B}$ is satisfiable if and only if the set of simple Herbrand instances of $\mathbf{D}$ is t-f satisfiable.

3600 Proposition. If $\mathcal{W}$ is a set of expressions and θ and τ are most general unifiers of $\mathcal{W}$, then $\theta \approx \tau$.

3601 Unification Theorem. There is an algorithm which can be applied to an arbitrary nonempty finite set $\mathcal{W}$ of expressions to determine whether $\mathcal{W}$ is unifiable. The algorithm always terminates, and produces a most general unifier of $\mathcal{W}$ if $\mathcal{W}$ is unifiable.

4000 Proposition. If $\mathbf{B}$ and $\mathbf{D}$ are duals of $\mathbf{A}$, then $\vdash \mathbf{B} \equiv \mathbf{D}$.

4001 Corollary. For any wff $\mathbf{A}$ of $\mathcal{F}$, $\vdash \mathbf{A}^{dd} \equiv \mathbf{A}$.

4002 Theorem. Let $\mathcal{M} = \langle \mathcal{D}, \mathcal{J} \rangle$ be any interpretation. Define $\mathcal{J}'$ to agree with $\mathcal{J}$ on all individual and function constants, and for all n-ary predicate constants $\mathbf{P}$, let $(\mathcal{J}'\mathbf{P})d_1 \dots d_n = \sim (\mathcal{J}\mathbf{P})d_1 \dots d_n$ for all $d_1, \dots, d_n \in \mathcal{D}$. Let $\mathcal{M}' = \langle \mathcal{D}, \mathcal{J}' \rangle$. Then for any wff $\mathbf{A}$ and assignment φ, $\mathcal{V}_\varphi^{\mathcal{M}}\mathbf{A} = \sim \mathcal{V}_\varphi^{\mathcal{M}'}\mathbf{A}$.

4003 Corollary. If $\vdash \mathbf{A}$, and $\mathbf{B}$ is any dual of $\mathbf{A}$, then $\vdash \sim \mathbf{B}$.

4004 Corollary. If $\vdash \mathbf{A} \supset \mathbf{B}$ then $\vdash \mathbf{B}^d \supset \mathbf{A}^d$.

4005 Corollary. If $\vdash \mathbf{A} \equiv \mathbf{B}$, then $\vdash \mathbf{A}^d \equiv \mathbf{B}^d$.

4100 Craig-Lyndon Interpolation Theorem. If $\mathbf{H} \supset \mathbf{K}$ is a valid sentence of $\mathcal{F}_f$, then there is an interpolation sentence for $\mathbf{H} \supset \mathbf{K}$.

4200 Beth's Definability Theorem. Let $\mathbf{H}$ be a sentence containing a k-ary predicate constant $\mathbf{P}$. Then $\mathbf{H}$ defines $\mathbf{P}$ implicitly iff $\mathbf{P}$ is explicitly definable from $\mathbf{H}$.

5100 Principle of Induction on the Construction of a Wff. Let $\mathcal{R}$ be a property of formulas, and let $\mathcal{R}(\mathbf{A})$ mean that $\mathbf{A}$ has property $\mathcal{R}$. Suppose

(1) Every formula consisting of a single variable or constant standing alone has property $\mathcal{R}$.
(2) Whenever $\mathbf{A}_{\alpha\beta}$ is a wff$_{(\alpha)}$ and $\mathbf{B}_\beta$ is a wff$_\beta$ and $\mathcal{R}(\mathbf{A}_{\alpha\beta})$ and $\mathcal{R}(\mathbf{B}_\beta)$, then $\mathcal{R}([\mathbf{A}_{\alpha\beta}\mathbf{B}_\beta])$.
(3) Whenever $\mathbf{x}_\beta$ is a variable and $\mathcal{R}(\mathbf{A}_\alpha)$, then $\mathcal{R}([\lambda \mathbf{x}_\beta \mathbf{A}_\alpha])$.

Then every wff of $\mathcal{Q}_0$ has property $\mathcal{R}$.

5200. $\vdash \mathbf{A}_\alpha = \mathbf{A}_\alpha$

5201 Equality Rules (= Rules).
If $\mathcal{H} \vdash \mathbf{A}$ and $\mathcal{H} \vdash \mathbf{A} \equiv \mathbf{B}$ then $\mathcal{H} \vdash \mathbf{B}$.
If $\mathcal{H} \vdash \mathbf{A}_\alpha = \mathbf{B}_\alpha$ then $\mathcal{H} \vdash \mathbf{B}_\alpha = \mathbf{A}_\alpha$.
If $\mathcal{H} \vdash \mathbf{A}_\alpha = \mathbf{B}_\alpha$ and $\mathcal{H} \vdash \mathbf{B}_\alpha = \mathbf{C}_\alpha$ then $\mathcal{H} \vdash \mathbf{A}_\alpha = \mathbf{C}_\alpha$.
If $\mathcal{H} \vdash \mathbf{A}_{\alpha\beta} = \mathbf{B}_{\alpha\beta}$ and $\mathcal{H} \vdash \mathbf{C}_\beta = \mathbf{D}_\beta$ then $\mathcal{H} \vdash \mathbf{A}_{\alpha\beta}\mathbf{C}_\beta = \mathbf{B}_{\alpha\beta}\mathbf{D}_\beta$.
If $\mathcal{H} \vdash \mathbf{A}_{\alpha\beta} = \mathbf{B}_{\alpha\beta}$ then $\mathcal{H} \vdash \mathbf{A}_{\alpha\beta}\mathbf{C}_\beta = \mathbf{B}_{\alpha\beta}\mathbf{C}_\beta$.
If $\mathcal{H} \vdash \mathbf{C}_\beta = \mathbf{C}_\beta = \mathbf{D}_\beta$ then $\mathcal{H} \vdash \mathbf{A}_{\alpha\beta}\mathbf{C}_\beta = \mathbf{A}_{\alpha\beta}\mathbf{D}_\beta$.

5202 Rule RR. If $\vdash \mathbf{A}_\alpha = \mathbf{B}_\alpha$ or $\vdash \mathbf{B}_\alpha = \mathbf{A}_\alpha$, and if $\mathcal{H} \vdash \mathbf{C}$, then $\mathcal{H} \vdash \mathbf{D}$, where $\mathbf{D}$ is obtained from $\mathbf{C}$ by replacing one occurrence of $\mathbf{A}_\alpha$ in $\mathbf{C}$ by an occurrence of $\mathbf{B}_\alpha$, and the occurrence of $\mathbf{A}_\alpha$ is not an occurrence of a variable immediately preceded by λ.

5203. $\vdash [\lambda \mathbf{x}_\alpha \mathbf{B}_\beta]\, \mathbf{A}_\alpha = \underset{\centerdot}{\mathrm{S}}^{\mathbf{x}_\alpha}_{\mathbf{A}_\alpha} \mathbf{B}_\beta$ provided no variable in $\mathbf{A}_\alpha$ is bound in $\mathbf{B}_\beta$.

5204. If $\vdash \mathbf{B}_\beta = \mathbf{C}_\beta$, then $\vdash \underset{\centerdot}{\mathrm{S}}^{\mathbf{x}_\alpha}_{\mathbf{A}_\alpha} [\mathbf{B}_\beta = \mathbf{C}_\beta]$, provided no variable in $\mathbf{A}_\alpha$ is bound in $\mathbf{B}_\beta$ or $\mathbf{C}_\beta$.

5205. $\vdash f_{\alpha\beta} = [\lambda \mathbf{y}_\beta \centerdot f_{\alpha\beta}\mathbf{y}_\beta]$

5206 $\vdash [\lambda \mathbf{x}_\beta \mathbf{A}_\alpha] = [\lambda \mathbf{z}_\beta \underset{\centerdot}{\mathrm{S}}^{\mathbf{x}_\beta}_{\mathbf{z}_\beta} \mathbf{A}_\alpha]$, provided that $\mathbf{z}_\beta$ does not occur free in $\mathbf{A}_\alpha$ and $\mathbf{z}_\beta$ is free for $\mathbf{x}_\beta$ in $\mathbf{A}_\alpha$.

5207 $\vdash [\lambda \mathbf{x}_\alpha \mathbf{B}_\beta]\mathbf{A}_\alpha = \mathsf{S}^{\mathbf{x}_\alpha}_{\mathbf{A}_\alpha} \mathbf{B}_\beta$, provided that $\mathbf{A}_\alpha$ is free for $\mathbf{x}_\alpha$ in $\mathbf{B}_\beta$.

5208 $\vdash \left[\lambda \mathbf{x}^1_{\alpha 1} \ldots \lambda \mathbf{x}^n_{\alpha_n} \mathbf{B}_\beta\right] \mathbf{x}^1_{\alpha_1} \ldots \mathbf{x}^n_{\alpha_n} = \mathbf{B}_\beta$ where $n \geq 1$.

5209. If $\vdash \mathbf{B}_\beta = \mathbf{C}_\beta$, then $\vdash \mathsf{S}^{\mathbf{x}_\alpha}_{\mathbf{A}_\alpha} [\mathbf{B}_\beta = \mathbf{C}_\beta]$, provided $\mathbf{A}_\alpha$ is free for $\mathbf{x}_\alpha$ in $[\mathbf{B}_\beta = \mathbf{C}_\beta]$.

5210 $\vdash T = [\mathbf{B}_\beta = \mathbf{B}_\beta]$

5211. $\vdash [T \wedge T] = T$

5212. $\vdash T \wedge T$

5213. If $\vdash \mathbf{A}_\alpha = \mathbf{B}_\alpha$ and $\vdash \mathbf{C}_\beta = \mathbf{D}_\beta$ then $\vdash [\mathbf{A}_\alpha = \mathbf{B}_\alpha] \wedge [\mathbf{C}_\beta = \mathbf{D}_\beta]$.

5214. $\vdash T \wedge F = F$

5215 Universal Instantiation ($\forall$I). If $\mathcal{H} \vdash \forall \mathbf{x}_\alpha \mathbf{B}$, then $\mathcal{H} \vdash \mathsf{S}^{\mathbf{x}_\alpha}_{\mathbf{A}_\alpha} \mathbf{B}$, provided that $\mathbf{A}_\alpha$ is free for $\mathbf{x}_\alpha$ in $\mathbf{B}$.

5216. $\vdash [T \wedge \mathbf{A}] = \mathbf{A}$

5217. $\vdash [T = F] = F$

5218 $\vdash [T = \mathbf{A}] = \mathbf{A}$

5219 Rule T. $\mathcal{H} \vdash \mathbf{A}$ if and only if $\mathcal{H} \vdash T = \mathbf{A}$; $\mathcal{H} \vdash \mathbf{A}$ if and only if $\mathcal{H} \vdash \mathbf{A} = T$.

5220 Rule of Universal Generalization (Gen). If $\mathcal{H} \vdash \mathbf{A}$ then $\mathcal{H} \vdash \forall \mathbf{x}_\alpha \mathbf{A}$, provided that $\mathbf{x}_\alpha$ is not free in any wff in $\mathcal{H}$.

5221n Rule of Substitution (Sub). If $\mathcal{H} \vdash \mathbf{B}$, then $\mathcal{H} \vdash \mathsf{S}^{\mathbf{x}^1_{\alpha_1} \ldots \mathbf{x}^n_{\alpha_n}}_{\mathbf{A}^1_{\alpha_1} \ldots \mathbf{A}^n_{\alpha_n}} \mathbf{B}$, provided that the variables $\mathbf{x}^1, \ldots, \mathbf{x}^n$ are distinct from one another and from all free variables of wffs in $\mathcal{H}$, and that $\mathbf{A}^i_{\alpha_i}$ is free for $\mathbf{x}^i_{\alpha_i}$ in $\mathbf{B}$ for all i $(1 \leq i \leq n)$.

5222 Rule of Cases. If $\mathcal{H} \vdash \mathsf{S}^{\mathbf{x}_o}_{\mathbf{T}} \mathbf{A}$ and $\mathcal{H} \vdash \mathsf{S}^{\mathbf{x}_o}_{\mathbf{F}} \mathbf{A}$, then $\mathcal{H} \vdash \mathbf{A}$.

5223. $\vdash [T \supset y_o] = y_o$

5224 Modus Ponens (MP). If $\mathcal{H} \vdash \mathbf{A}$ and $\mathcal{H} \vdash \mathbf{A} \supset \mathbf{B}$, then $\mathcal{H} \vdash \mathbf{B}$.

5225. $\vdash \Pi_{o(o\alpha)} f_{o\alpha} \supset f_{o\alpha} x_\alpha$

5226. $\vdash \forall \mathbf{x}_\alpha \mathbf{B} \supset \mathsf{S}^{\mathbf{x}_\alpha}_{\mathbf{A}_\alpha} \mathbf{B}$ provided that $\mathbf{A}_\alpha$ is free for $\mathbf{x}_\alpha$ in $\mathbf{B}$.

5227. $\vdash F \supset x_o$

5228. $\vdash [T \supset T] = T$; $\vdash [T \supset F] = F$; $\vdash [F \supset T] = T$; $\vdash [F \supset F] = T$

5229. $\vdash [T \wedge T] = T$; $\vdash [T \wedge F] = F$; $\vdash [F \wedge T] = F$; $\vdash [F \wedge F] = F$

5230. $\vdash [T = T] = T$; $\vdash [T = F] = F$; $\vdash [F = T] = F$; $\vdash [F = F] = T$

5231. $\vdash \sim T = F$; $\vdash \sim F = T$

5232. $\vdash T \vee T = T$; $\vdash T \vee F = T$; $\vdash F \vee T = T$; $\vdash F \vee F = F$

5233. If $\mathbf{A}$ is a tautology, then $\vdash \mathbf{A}$.

5234 Rule P. If $\mathcal{H} \vdash \mathbf{A}^1, \ldots, \mathcal{H} \vdash \mathbf{A}^n$, and if $\left[\mathbf{A}^1 \wedge \ldots \wedge \mathbf{A}^n\right] \supset \mathbf{B}$ is tautologous, then $\mathcal{H} \vdash \mathbf{B}$. Also, if $\mathbf{B}$ is tautologous, then $\mathcal{H} \vdash \mathbf{B}$.

5235. $\vdash \forall \mathbf{x}_\alpha [\mathbf{A} \vee \mathbf{B}] \supset \centerdot\, \mathbf{A} \vee \forall \mathbf{x}_\alpha \mathbf{B}$ provided $\mathbf{x}_\alpha$ is not free in $\mathbf{A}$.

5236 "Rule Q". If $\mathcal{H} \vdash \mathbf{A}^1, \ldots,$ and $\mathcal{H} \vdash \mathbf{A}^n$ (where $n \geq 0$), and if $\bar{\mathbf{A}}^i$ can be obtained from $\mathbf{A}^i$ by generalizing on zero or more free variables of $\mathbf{A}^i$ which do not occur free in $\mathcal{H}$, and if $[\bar{\mathbf{A}}^1 \wedge \ldots \wedge \bar{\mathbf{A}}^n \supset \mathbf{B}]$ is a substitution instance of a theorem of quantification theory, then $\mathcal{H} \vdash \mathbf{B}$.

5237 $\supset \forall$ Rule. If $\mathcal{H} \vdash \mathbf{A} \supset \mathbf{B}$, then $\mathcal{H} \vdash \mathbf{A} \supset \forall \mathbf{x} \mathbf{B}$, provided that $\mathbf{x}_\alpha$ is not free in $\mathbf{A}$ or in $\mathcal{H}$.

5238. $\vdash [\lambda \mathbf{x}^1_{\beta_1} \ldots \lambda \mathbf{x}^n_{\beta_n} \mathbf{A}_\alpha] = [\lambda \mathbf{x}^1_{\beta_1} \ldots \lambda \mathbf{x}^n_{\beta_n} \mathbf{B}_\alpha] = \forall \mathbf{x}^1_{\beta_1} \ldots \forall \mathbf{x}^n_{\beta_n} [\mathbf{A}_\alpha = \mathbf{B}_\alpha]$

5239. $\vdash \forall \mathbf{x}^1_{\beta_1} \ldots \forall \mathbf{x}^n_{\beta_n} [\mathbf{A}_\alpha = \mathbf{B}_\alpha] \supset \centerdot\, \mathbf{C} = \mathbf{D}$, where $n \geq 0$ and

(a) $\mathbf{D}$ is obtained from $\mathbf{C}$ and $[\mathbf{A}_\alpha = \mathbf{B}_\alpha]$ as in Rule R;

(b) $\mathbf{x}^1_{\beta_1}, \ldots, \mathbf{x}^n_{\beta_n}$ is a complete list of those variables $\mathbf{x}_\beta$ such that $\mathbf{x}_\beta$ occurs free in $[\mathbf{A}_\alpha = \mathbf{B}_\alpha]$, and the occurrence of $\mathbf{A}_\alpha$ in $\mathbf{C}$ (which is replaced by $\mathbf{B}_\alpha$ in $\mathbf{D}$) is in a wf part of $\mathbf{C}$ of the form $[\lambda \mathbf{x}_\beta \mathbf{E}_\gamma]$.

5240 Deduction Theorem. If $\mathcal{H}$, $\mathbf{H} \vdash \mathbf{P}$ then $\mathcal{H} \vdash \mathbf{H} \supset \mathbf{P}$.

5241. If $\mathcal{H} \vdash \mathbf{A}$ and $\mathcal{H} \subseteq \mathcal{G}$ then $\mathcal{G} \vdash \mathbf{A}$.

5242 Rule of Existential Generalization (**$\exists$ Gen**). If $\mathcal{H} \vdash \mathsf{S}^{\mathbf{x}_\alpha}_{\mathbf{A}_\alpha}\mathbf{B}$, and $\mathbf{A}_\alpha$ is free for $\mathbf{x}_\alpha$ in $\mathbf{B}$, then $\mathcal{H} \vdash \exists \mathbf{x}_\alpha \mathbf{B}$.

5243 Comprehension Theorem.
$\vdash \exists \mathbf{u}_{\beta\alpha_n\ldots\alpha_1} \forall \mathbf{x}^1_{\alpha_1} \ldots \forall \mathbf{x}^n_{\alpha_n} \,\blacksquare\, \mathbf{u}_{\beta\alpha_n\ldots\alpha_1} \mathbf{x}^1_{\alpha_1} \ldots \mathbf{x}^n_{\alpha_n} = \mathbf{B}_\beta$
when $n \geq 0$ and $\mathbf{u}_{\beta\alpha_n\ldots\alpha_1}$ is not free in $\mathbf{B}_\beta$.

5244 Existential Rule (**$\exists$ Rule**). If $\mathcal{H}, \mathbf{B} \vdash \mathbf{A}$, and $\mathbf{x}_\alpha$ is not free in $\mathcal{H}$ or in $\mathbf{A}$, then $\mathcal{H}$, $\exists \mathbf{x}_\alpha \mathbf{B} \vdash \mathbf{A}$.

5245 Rule C. If $\mathcal{H} \vdash \exists \mathbf{x}_\alpha \mathbf{B}$ and $\mathcal{H}$, $\mathsf{S}^{\mathbf{x}_\alpha}_{\mathbf{y}_\alpha}\mathbf{B} \vdash \mathbf{A}$, where $\mathbf{y}_\alpha$ is free for $\mathbf{x}_\alpha$ in $\mathbf{B}$ and $\mathbf{y}_\alpha$ is not free in $\mathcal{H}$, $\exists \mathbf{x}_\alpha \mathbf{B}$, or $\mathbf{A}$, then $\mathcal{H} \vdash \mathbf{A}$.

5300. $\vdash [x_\alpha = y_\alpha] \supset \,\blacksquare\, h_{\beta\alpha} x_\alpha = h_{\beta\alpha} y_\alpha$

5301. $\vdash [x_\alpha = y_\alpha] \wedge [f_{\beta\alpha} = g_{\beta\alpha}] \supset \,\blacksquare\, f_{\beta\alpha} x_\alpha = g_{\beta\alpha} y_\alpha$

5302. $\vdash [x_\alpha = y_\alpha] = [y_\alpha = x_\alpha]$

5303. $\vdash [x_\alpha = y_\alpha] \supset \,\blacksquare\, [x_\alpha = z_\alpha] = [y_\alpha = z_\alpha]$

5304. $\vdash \exists_1 y_\alpha p_{o\alpha} y_\alpha = \exists y_\alpha \,\blacksquare\, p_{o\alpha} = \mathsf{Q}_{o\alpha\alpha} y_\alpha$

5305. $\vdash \exists_1 y_\alpha p_{o\alpha} y_\alpha = \exists y_\alpha \forall z_\alpha \,\blacksquare\, p_{o\alpha} z_\alpha = \,\blacksquare\, y_\alpha = z_\alpha$

5306. $\vdash \exists_1 y_\alpha p_{o\alpha} y_\alpha = \exists y_\alpha \,\blacksquare\, p_{o\alpha} y_\alpha \wedge \forall z_\alpha \,\blacksquare\, p_{o\alpha} z_\alpha \supset y_\alpha = z$

5307. $\vdash \exists_1 y_\alpha p_{o\alpha} y_\alpha = \,\blacksquare\, \exists y_\alpha p_{o\alpha} y_\alpha \wedge \forall y_\alpha \forall z_\alpha \,\blacksquare\, p_{o\alpha} y_\alpha \;\wedge\; p_{o\alpha} z_\alpha \supset y_\alpha = z_\alpha$

5308. $\vdash \iota_{o(oo)} [\mathsf{Q}_{ooo} y_o] = y_o$

5309$^\gamma$. $\vdash \iota_{\gamma(o\gamma)} [\mathsf{Q}_{o\gamma\gamma} y_\gamma] = y_\gamma$ for each type symbol γ.

5310. $\vdash \forall z_\alpha [p_{o\alpha} z_\alpha = \,\blacksquare\, y_\alpha = z_\alpha] \supset \,\blacksquare\, \iota_{\alpha(o\alpha)} p_{o\alpha} = y_\alpha$

5311. $\vdash \exists_1 y_\alpha p_{o\alpha} y_\alpha \supset p_{o\alpha} \left[\iota_{\alpha(o\alpha)} p_{o\alpha}\right]$

5312. $\vdash \exists_1 y_\alpha p_{o\alpha} y_\alpha \supset \forall z_\alpha \,\blacksquare\, p_{o\alpha} z_\alpha = \,\blacksquare\, \iota_{\alpha(o\alpha)} p_{o\alpha} = z_\alpha$

5313. $\vdash [C_{\gamma o\gamma\gamma} x_\gamma y_\gamma T_o = x_\gamma] \wedge \,\blacksquare\, C_{\gamma o\gamma\gamma} x_\gamma y_\gamma F_o = y_\gamma$

5400 Proposition. Let $\mathcal{M}$ be a general model, $\mathbf{A}_\alpha$ a wff, and φ and ψ assignments which agree on all free variables of $\mathbf{A}_\alpha$. Then $\mathcal{V}_\varphi^{\mathcal{M}}\mathbf{A}_\alpha = \mathcal{V}_\psi^{\mathcal{M}}\mathbf{A}_\alpha$.

5401 Lemma. Let $\mathcal{M}$ be a general model, and φ be an assignment into $\mathcal{M}$.

(a) $\mathcal{V}_\varphi^{\mathcal{M}}[[\lambda \mathbf{x}_\alpha \mathbf{B}_\beta]\mathbf{A}_\alpha] = \mathcal{V}_{(\varphi:\mathbf{x}_\alpha/\mathcal{V}_\varphi^{\mathcal{M}}\mathbf{A}_\alpha)}^{\mathcal{M}}\mathbf{B}_\alpha$.
(b) $\mathcal{V}_\varphi^{\mathcal{M}}[\mathbf{A}_\alpha = \mathbf{B}_\alpha] = \mathsf{T}$ iff $\mathcal{V}_\varphi^{\mathcal{M}}\mathbf{A}_\alpha = \mathcal{V}_\varphi^{\mathcal{M}}\mathbf{B}_\alpha$.
(c) $\mathcal{V}^{\mathcal{M}}T_o = \mathsf{T}$.
(d) $\mathcal{V}^{\mathcal{M}}F_o = \mathsf{F}$.
(e) If $x, y \in \mathcal{D}_o$, then $(\mathcal{V}^{\mathcal{M}}\wedge_{ooo})xy = \mathsf{T}$ if $x = \mathsf{T}$ and $y = \mathsf{T}$;
$= \mathsf{F}$ otherwise.
(f) If $x,\ y \in \mathcal{D}_o$, then $(\mathcal{V}^{\mathcal{M}} \supset_{ooo})xy = \mathsf{T}$ if $x = \mathsf{F}$ or $y = \mathsf{T}$;
$= \mathsf{F}$ if $x = \mathsf{T}$ and $y = \mathsf{F}$.
(g) $\mathcal{M} \models_\varphi \forall \mathbf{x}\mathbf{A}$ iff $\mathcal{M} \models_\psi \mathbf{A}$ for all assignments ψ which agree with φ off $\mathbf{x}_\alpha$.

5402 Soundness Theorem.

(a) Every theorem of $\mathcal{Q}_0$ is valid in the general sense, and hence in the standard sense.
(b) If $\mathcal{G}$ is a set of wffs$_o$ and $\mathcal{M}$ is a model for $\mathcal{G}$ and $\mathcal{G} \vdash \mathbf{A}$, then $\mathcal{M} \models \mathbf{A}$.

5403 Consistency Theorem $\mathcal{Q}_0$ is consistent, and every set of wffs$_o$ which has a model is consistent.

5404 Theorem. Every finite model for $\mathcal{Q}_0$ is standard.

5405 Proposition. An isomorphism of general models is a one-to-one mapping.

5406 Proposition. Let $\mathcal{N}$ and $\mathcal{M}$ be general models for $\mathcal{Q}_0$, and let τ be an isomorphism from $\mathcal{N}$ to $\mathcal{M}$. For each assignment φ into $\mathcal{N}$, let $\tau \circ \varphi$ be the assignment into $\mathcal{M}$ such that for each variable $\mathbf{x}_\gamma$, $(\tau \circ \varphi)\mathbf{x}_\gamma = \tau(\varphi \mathbf{x}_\gamma)$. Then for every wff $\mathbf{C}_\gamma$ of $\mathcal{Q}_0$, $\tau(\mathcal{V}_\varphi^{\mathcal{N}}\mathbf{C}_\gamma) = \mathcal{V}_{\tau\circ\varphi}^{\mathcal{M}}\mathbf{C}_\gamma$.

5407 Corollary. If $\mathcal{N}$ and $\mathcal{M}$ are isomorphic general models of $\mathcal{Q}_0$, then $\mathcal{M} \models \mathbf{C}$ iff $\mathcal{N} \models \mathbf{C}$.

5500 Extension Lemma. Let $\mathcal{G}$ be any consistent set of sentences of $\mathcal{Q}_0$. Then there is an expansion $\mathcal{Q}_0^+$ of $\mathcal{Q}_0$ and a set of $\mathcal{H}$ of sentences of $\mathcal{Q}_0^+$ such that

(1) $\mathcal{G} \subseteq \mathcal{H}$.
(2) $\mathcal{H}$ is consistent.
(3) $\mathcal{H}$ is complete in $\mathcal{Q}_0^+$.
(4) $\mathcal{H}$ is extensionally complete in $\mathcal{Q}_0^+$.
(5) $\text{card}(\mathcal{L}(\mathcal{Q}_0^+)) = \text{card}(\mathcal{L}(\mathcal{Q}))$.

5501 Henkin's Theorem. Every consistent set of sentences of $\mathcal{Q}_0$ has a frugal general model.

5502 Henkin's Completeness and Soundness Theorem. Let $\mathbf{A}$ be a wff$_o$ and let $\mathcal{G}$ be a set of sentences.

(a) $\vdash \mathbf{A}$ iff $\models \mathbf{A}$.
(b) $\mathcal{G} \vdash \mathbf{A}$ iff $\mathbf{A}$ is valid in every general model for $\mathcal{G}$.
(c) $\mathcal{G} \vdash \mathbf{A}$ iff $\mathbf{A}$ is valid in every frugal general model for $\mathcal{G}$.

5503 Weak Compactness Theorem. If every finite subset of the set $\mathcal{S}$ of sentences has a general model, then $\mathcal{S}$ has a general model.

5504 Lemma. Let $\mathcal{M} = \langle\{\mathcal{D}_\alpha\}_\alpha, \mathcal{J}\rangle$ be a frugal model for $\mathcal{Q}_0$ such that $\text{card}(\mathcal{D}_\iota) = \text{card}(\mathcal{L}(\mathcal{Q}_0))$. Then $\mathcal{M}$ is nonstandard.

5505 Corollary. Every infinite frugal model for a countable formulation of $\mathcal{Q}_0$ is nonstandard.

5506 Theorem. Let $\mathcal{H}$ be any set of sentences of $\mathcal{Q}_0$ which has infinite models. Then $\mathcal{H}$ has nonstandard models.

6000. $\vdash \sim (\forall \mathbf{x}_\alpha \in \mathbf{M}_{o\alpha})\, \mathbf{A} = (\exists \mathbf{x}_\alpha \in \mathbf{M}_{o\alpha}) \sim \mathbf{A}$ provided $\mathbf{x}_\alpha$ is not free in $\mathbf{M}_{o\alpha}$.

6001. $\vdash \sim (\exists \mathbf{x}_\alpha \in \mathbf{M}_{o\alpha})\, \mathbf{A} = (\forall \mathbf{x}_\alpha \in \mathbf{M}_{o\alpha}) \sim \mathbf{A}$ provided $\mathbf{x}_\alpha$ is not free in $\mathbf{M}_{o\alpha}$.

6100. $\vdash \mathbb{N}_{o\sigma} 0_\sigma$ by the def of $\mathbb{N}$

6101. $\vdash \forall x_\sigma \centerdot \mathbb{N}_{o\sigma} x_\sigma \supset \mathbb{N}_{o\sigma} \centerdot S_{\sigma\sigma} x_\sigma$

6102. The Induction Theorem.

$$\vdash \forall p_{o\sigma} \centerdot \left[p_{o\sigma} 0_\sigma \wedge \dot{\forall} x_\sigma \centerdot p_{o\sigma} x_\sigma \supset p_{o\sigma} \centerdot S_{\sigma\sigma} x_\sigma\right] \supset \dot{\forall} x_\sigma p_{o\sigma} x_\sigma$$

6103. $\vdash p_{o\sigma} 0_\sigma \wedge \dot{\forall} x_\sigma p_{o\sigma} S_{\sigma\sigma} x_\sigma \supset \dot{\forall} n_\sigma p_{o\sigma} n_\sigma$

6104. $\vdash \dot{\forall} n_\sigma \centerdot n_\sigma = 0 \ \lor \ \dot{\exists} m_\sigma \centerdot n_\sigma = S_{\sigma\sigma} m_\sigma$

6105. $\vdash \forall n_\sigma \centerdot S_{\sigma\sigma} n_\sigma \neq 0_\sigma$

6106. $\vdash \dot{\forall} n_\sigma \centerdot S_{\sigma\sigma} n_\sigma \neq 0_\sigma$

6107. $\vdash \dot{\forall} n_\sigma \forall p_{o\iota} \centerdot S n_\sigma p_{o\iota} \ \land \ p_{o\iota} w_\iota \supset n_\sigma \left[\lambda t_\iota \centerdot t_\iota \neq w_\iota \ \land \ p_{o\iota} t_\iota\right]$

6108. $\vdash \dot{\forall} n_\sigma \centerdot n_\sigma p_{o\iota} \supset \exists w_\iota \sim p_{o\iota} w_\iota$

6109. $\vdash \dot{\forall} n_\sigma \dot{\forall} m_\sigma \centerdot S_{\sigma\sigma} n_\sigma = S_{\sigma\sigma} m_\sigma \supset n_\sigma = m_\sigma$

6110 $\vdash \mathbb{N}\bar{n}$ where n is any natural number.

6111. If k is a natural number greater than 0, $\vdash \dot{\forall} x_\sigma \centerdot S^k x_\sigma \neq x_\sigma$.

6112 Theorem. $\mathsf{Q}_{o\sigma\sigma}$ represents the equality relation. Thus, if m and n are distinct natural numbers, $\vdash \bar{m} \neq \bar{n}$.

6113 Theorem. Let r be any relation between natural numbers. Then r is representable iff χ_r is representable.

6200. $\vdash \dot{\forall} y_\sigma \centerdot 0 \leq y_\sigma$

6201. $\vdash x_\sigma \leq x_\sigma$

6202. $\vdash x_\sigma = y_\sigma \supset x_\sigma \leq y_\sigma$

6203. $\vdash x_\sigma \leq S x_\sigma$

6204. $\vdash x_\sigma \leq y_\sigma \ \land \ y_\sigma \leq z_\sigma \supset x_\sigma \leq z_\sigma$

6205. $\vdash \dot{\forall} x_\sigma \centerdot x_\sigma \leq y_\sigma \supset \ \centerdot x_\sigma = y_\sigma \ \lor \ \dot{\exists} z_\sigma \centerdot x_\sigma \leq z_\sigma \ \land \ S z = y$

6206. $\vdash \dot{\forall} x_\sigma \dot{\forall} y_\sigma \centerdot x_\sigma \leq S y_\sigma \equiv \ \centerdot x_\sigma \leq y_\sigma \ \lor \ x_\sigma = S y_\sigma$

6207^0. $\vdash \dot{\forall} x_\sigma \centerdot x \leq 0 \equiv \ \centerdot x = 0$

6207^1. $\vdash \dot{\forall} x_\sigma \centerdot x \leq \bar{1} \equiv \ \centerdot x = 0 \ \lor \ x = \bar{1}$

6207^k. $\vdash \dot{\forall} x_\sigma \centerdot x \leq \bar{k} \equiv \ \centerdot x = 0 \ \lor \ x = \bar{1} \ \lor \ldots \lor \ x = \bar{k}$

6208 Theorem. $\leq_{o\sigma\sigma}$ represents the relation $\leq$ between natural numbers.

6209^k. $\vdash \dot{\forall} y_\sigma \left[y_\sigma \leq \overline{k} \supset p_{o\sigma} y_\sigma\right] = \; \centerdot \, p_{o\sigma} 0 \; \wedge \ldots \wedge \; p_{o\sigma} \overline{k}$

6210^k. $\vdash \dot{\exists} y_\sigma \left[y_\sigma \leq \overline{k} \; \wedge \; p_{o\sigma} y_\sigma\right] = \; \centerdot \, p_{o\sigma} 0 \; \vee \ldots \vee \; p_{o\sigma} \overline{k}$

6211. $\vdash \dot{\forall} y_\sigma \dot{\forall} x_\sigma \leq y_\sigma \supset S x_\sigma \leq S y_\sigma$

6212. $\vdash \dot{\forall} x_\sigma \dot{\forall} y_\sigma \, \centerdot \, x \leq y \; \vee \; y \leq x$

6213. $\vdash \dot{\forall} y_\sigma \dot{\forall} x_\sigma \, \centerdot \, x \leq y \; \wedge \; y \leq x \supset x = y$

6214. $\vdash \dot{\forall} x_\sigma \dot{\forall} y_\sigma \, \centerdot \; \sim x \leq y \supset S y \leq x$

6215. $\vdash$*Infin*$^\sigma$

6300. $\vdash \mathbb{N}_{o\sigma} t_\sigma \wedge p_{o\sigma} t_\sigma \supset \; \centerdot \, \mathbb{N}_{o\sigma} [\mu_{\sigma(o\sigma)} p_{o\sigma}] \; \wedge \; p_{o\sigma} [\mu_{\sigma(o\sigma)} p_{o\sigma}] \; \wedge \; \mu_{\sigma(o\sigma)} p_{o\sigma} \leq t_\sigma$

6301. $\vdash \mathbb{N}_{o\sigma} t_\sigma \wedge p_{o\sigma} t_\sigma \wedge \dot{\forall} y_\sigma [y_\sigma \leq t_\sigma \supset \; \centerdot \, p_{o\sigma} y_\sigma \supset y_\sigma = t_\sigma] \supset \mu_{\sigma(o\sigma)} p_{o\sigma} = t_\sigma$

6302 Theorem. If r is a representable $(k+1)$-ary regular numerical relation, then $((\lambda n_1, \ldots, n_k)(\mu m) r(n_1, \ldots, n_k, m))$ is a representable function.

6303 Theorem. If g is a representable $(k+1)$-ary regular numerical function, and h is obtained from g by minimization, then h is representable.

6400. $\vdash [R_{\sigma\sigma\sigma(\sigma\sigma\sigma)} h_{\sigma\sigma\sigma} g_\sigma] 0_\sigma = g_\sigma \; \wedge \; \dot{\forall} n_\sigma \, \centerdot \, [R h_{\sigma\sigma\sigma} g_\sigma][S_{\sigma\sigma} n] = h_{\sigma\sigma\sigma} n_\sigma \, \centerdot \, [R h_{\sigma\sigma\sigma} g_\sigma] n_\sigma$

6401 Theorem. Every general recursive function is representable in $\mathcal{Q}_0^\infty$.

6500 Constant Functions. For all $k \geq 0$ and $n \geq 0$ $((\lambda \vec{x_k})\, n) \in \mathcal{PRF}^k$.

6501 Explicit Definitions. If $\mathcal{S} \subseteq \mathcal{PRF}$ and f is a numerical function explicitly definable from $\mathcal{S}$, then $f \in \mathcal{PRF}$.

6502 Generalized Primitive Recursion. Let $\mathcal{S} \subseteq \mathcal{PRF}$, and let g and h be k-ary and $(k+2)$-ary functions, respectively, which are explicitly definable from $\mathcal{S}$. If for all natural numbers $x_1, \ldots, x_k$, and y,

$$f(x_1, \ldots, x_i, \; 0, \; x_{i+1}, \ldots, x_k) = g(\overrightarrow{x_k}), \text{ and}$$

$$f(x_1, \ldots, x_i, \mathrm{S}y, x_{i+1}, \ldots, x_k) = h(\overrightarrow{x_k}, f(x_1, \ldots, x_i, y, x_{i+1}, \ldots, x_k)),$$

then $f \in \mathcal{PR}$.

6503 $+$ (addition) $\in \mathcal{PRF}^2$.

6504. If $f \in \mathcal{PRF}^{k+1}$ (where $k \geq 0$), then $\left((\lambda \vec{x_k}, y) \sum_{z \leq y} f(\vec{x_k}, z)\right) \in \mathcal{PRF}^{k+1}$.

6505. $\times$ (multiplication) $\in \mathcal{PRF}^2$.

6506. If $f \in \mathcal{PRF}^{k+1}$ (where $k \geq 0$), then $((\lambda \vec{x_k}, y)\, \Pi_{z \leq y} f(\vec{x_k}, z)) \in \mathcal{PRF}^{k+1}$.

6507. $((\lambda x, y)\, x^y)$ (exponentiation) $\in \mathcal{PRF}^2$.

6508. ! (factorial) $\in \mathcal{PRF}^1$.

6509. δ (predecessor) $\in \mathcal{PRF}^1$.

6510. $\dot{-}$ (modified subtraction or monus) $\in \mathcal{PRF}^2$.

6511. σ (signum) $\in \mathcal{PRF}^1$.

6512. $\overline{\sigma}$ (antisignum) $\in \mathcal{PRF}^1$.

6513. $< \in \mathcal{PRR}^2$.

6514. $\leq \in \mathcal{PRR}^2$.

6515. $= \in \mathcal{PRR}^2$

6516. If $\mathcal{S} \subseteq \mathcal{PR}$, and the numerical relation r is explicitly definable from $\mathcal{S}$ using propositional connectives, then $r \in \mathcal{PR}$.

6517 Bounded Quantification. If $f \in \mathcal{PRF}^k$ (where $k \geq 0$) and $r \in \mathcal{PRR}^{k+1}$, then the following are in $\mathcal{PRR}^k$:

(a) $(\lambda \vec{x_k})(\exists z \leq f(\vec{x_k}))r(\vec{x_k}, z)$
(b) $(\lambda \vec{x_k})(\exists z < f(\vec{x_k}))r(\vec{x_k}, z)$
(c) $(\lambda \vec{x_k})(\forall z \leq f(\vec{x_k}))r(\vec{x_k}, z)$
(d) $(\lambda \vec{x_k})(\forall z < f(\vec{x_k}))r(\vec{x_k}, z)$

6518 Bounded Minimization and Maximization. If $f \in \mathcal{PRF}^k$ and $r \in \mathcal{PRR}^{k+1}$ (where $k \geq 0$), then the following are in $\mathcal{PRF}^k$:

(a) $(\lambda \vec{x_k})(\mu z \leq f(\vec{x_k}))r(\vec{x_k}, z)$

(b) $(\lambda \overrightarrow{x_k})(\text{Max } z \leq f(\overrightarrow{x_k}))r(\overrightarrow{x_k}, z)$

6519 Definition by Cases. Let $g_1, \ldots, g_n \in \mathcal{PRF}^k$ and $r_1, \ldots, r_n \in \mathcal{PRR}^k$, and suppose the relations $r_1, \ldots, r_n$ are mutually exclusive and exhaustive, i.e., for all $\overrightarrow{x_k} \in \mathbb{N}^k$, there is exactly one r_i such that $r_i(\overrightarrow{x_k})$. Let $f(\overrightarrow{x_k}) = g_i(\overrightarrow{x_k})$ iff $r_i(\overrightarrow{x_k})$ for $1 \leq i \leq n$. Then $f \in \mathcal{PRF}^k$.

6520. $((\lambda x, y)x \text{ Divides } y) \in \mathcal{PRR}^2$

6521. Let Prime(x) iff x is a prime number. Then Prime $\in \mathcal{PRR}^1$.

6522. Let Pr(n) be the nth odd prime number, where Pr(0) = 2. Then Pr $\in \mathcal{PRF}^1$.

6523. Let $n \, \mathcal{G}\ell \, x$ be the exponent of Pr(n) in the prime factorization of x, with $n \, \mathcal{G}\ell \, 0 = 0$. Then $\mathcal{G}\ell \in \mathcal{PRF}^2$.

6524. Let E(x) (the extent of the sequence represented by x) be the largest number n such that Pr(n) divides x, with E(0) $= 0 = E(1)$. Then E $\in \mathcal{PRF}^1$.

6525. Let $x * y = x \times \Pi_{i \leq \mathrm{E}(y)} \Pr(\mathrm{E}(x) + 1 + i)^{i \mathcal{G}\ell y}$ if $1 < x$
$= y$ if $x \leq 1$.
Then $* \in \mathcal{PRF}^2$.

6526. Let $Lg = (\lambda x)\,(\mu y \leq x)\,(\exists z \leq x)\,[x = z^y \;\wedge\; \text{Prime}(z)]$. Then $Lg \in \mathcal{PRF}^1$.

7000 Theorem. Let TypeSymbol =$((\lambda n)\, n$ is the number of some type symbol). Then TypeSymbol $\in \mathcal{PRR}^1$.

7001 Theorem. Let Q = $(\lambda n)19 \exp((3 \exp(3^3 \times 5^n)) \times 5^n)$. Then Q $\in \mathcal{PRF}^1$, and for each type symbol α, Q("α") = '$\mathrm{Q}_{o\alpha\alpha}$'.

7002 Theorem. Let Variable = $((\lambda n)n$ is the number of a variable of $\mathcal{Q}_0$). Then Variable $\in \mathcal{PRR}^1$.

7003 Theorem. Let ProperSymbol= $((\lambda n)n$ is the number of a variable or constant of $\mathcal{Q}_0$). Then ProperSymbol $\in \mathcal{PRR}^1$.

7004 Theorem. Let Wff $= ((\lambda x)$ there is a wff $\mathbf{A}_\alpha$ of $\mathcal{Q}_0$ such that $x =$ "$\mathbf{A}_\alpha$").

$$\text{Let Type}(x) = \text{“}\alpha\text{” if } x = \text{“}\mathbf{A}_\alpha\text{”}; \\ = 0 \text{ if not Wff}(x).$$

Then Wff $\in \mathcal{PRR}^1$ and Type $\in \mathcal{PRF}^1$.

7005 Theorem. Let Axiom $= ((\lambda n)$ there is an axiom $\mathbf{A}$ of $\mathcal{Q}_0^\infty$ such that $n =$ "$\mathbf{A}$"). Then Axiom $\in \mathcal{PRR}^1$.

7006 Theorem Let Rule $= (\lambda m, n, p)$ [there exist wffs $\mathbf{A}_\alpha$, $\mathbf{B}_\alpha$, $\mathbf{C}_o$, and $\mathbf{D}_o$ such that $m =$ "$[\mathbf{A}_\alpha = \mathbf{B}_\alpha]$" and $n =$ "$\mathbf{C}_o$" and $p =$ "$\mathbf{D}_o$", and $\mathbf{D}_o$ is obtained from $[\mathbf{A}_\alpha = \mathbf{B}_\alpha]$ and $\mathbf{C}_o$ by Rule R]. Then Rule $\in \mathcal{PRR}^3$.

7007 Theorem. Let $\mathcal{A}$ be any [primitive] recursively axiomatized pure extension of $\mathcal{Q}_0$. Let Proof$^\mathcal{A}$ $= (\lambda m, n)$ [There is a proof $\mathbf{P}^0, \ldots, \mathbf{P}^k$ in $\mathcal{A}$ such that $m =$"$\mathbf{P}^k$" and $n =$"'$\mathbf{P}^0, \ldots, \mathbf{P}^k$'"]. Then Proof$^\mathcal{A}$ is a [primitive] recursive relation.

7008 Theorem. Let Num $= ((\lambda n)$"$\overline{n}$"). Then Num $\in \mathcal{PRF}^1$.

7009 Theorem. If $\mathcal{A}$ is a consistent recursively axiomatized pure extension of $\mathcal{Q}_0^\infty$, then every numerical function and every numerical relation representable in $\mathcal{A}$ is recursive.

7010 Theorem. If $\mathcal{A}$ is any consistent recursively axiomatized pure extension of $\mathcal{Q}_0^\infty$, then a numerical function is representable in $\mathcal{A}$ if and only if it is recursive.

7011 Normal Form Theorem. For each natural number $n \geq 1$ there is a relation $\mathrm{S}_n \in \mathcal{PRR}^{n+2}$ such that for each n-ary recursive function f, there is a number j such that $(\lambda \overrightarrow{x_n}, y)\mathrm{S}_n(j, \overrightarrow{x_n}, y)$ is regular, and

$$f = (\lambda \overrightarrow{x_n})0 \; \mathcal{G}\ell \; ((\mu y)\mathrm{S}_n(j, \overrightarrow{x_n}, y)).$$

7012 Theorem. Every recursive function can be defined with at most one application of minimization.

7100 Proposition. If $\mathcal{A}$ is an ω-consistent extension of $\mathcal{Q}_0$, then $\mathcal{A}$ is consistent.

7101 Gödel's Incompleteness Theorem [Gödel, 1931]. Let $\mathcal{A}$ be any recursively axiomatized pure extension of $\mathcal{Q}_0^\infty$. Let $\star_{o\sigma}$ be the wff

$$[\lambda x_\sigma \dot\forall y_\sigma \sim Proof^{\mathcal{A}}_{o\sigma\sigma}[\overline{128} * x_\sigma * [Num_{\sigma\sigma} x_\sigma] * \overline{2048}] y_\sigma].$$

Let $m =$ "$\star_{o\sigma}$" and let $\star\star_o$ be $[\star_{o\sigma}\overline{m}]$.

(1) If $\mathcal{A}$ is consistent, then not $\vdash_{\mathcal{A}} \star\star$.
(2) If $\mathcal{A}$ is ω-consistent, then not $\vdash_{\mathcal{A}} \sim \star\star$.

7102 Corollary. Every consistent recursively axiomatized pure extension $\mathcal{A}$ of $\mathcal{Q}_0^\infty$ is ω-incomplete.

7103 Lemma. If $\vdash \mathbf{A}$, then $\vdash Theorem_{o\sigma}\overline{\text{“}\mathbf{A}\text{”}}$.

7104 Lemma. $\vdash \dot\forall x_\sigma \,.\, Theorem_{o\sigma} x_\sigma \supset Theorem_{o\sigma}[\overline{\text{“}[Theorem_{o\sigma}\text{”}} * \text{Num } x_\sigma * \overline{2048}]$

7105 Corollary. For any wff$_o$ $\mathbf{A}$,

$$\vdash Theorem_{o\sigma}\overline{\text{“}\mathbf{A}\text{”}} \supset Theorem_{o\sigma}\overline{\text{“}[Theorem_{o\sigma}\overline{\text{“}\mathbf{A}\text{”}}]\text{”}}.$$

7106 Lemma. $\vdash \dot\forall x_\sigma \dot\forall y_\sigma \,.\, Theorem_{o\sigma}[\overline{\text{“}[[Q_{ooo}\text{”}} * x_\sigma * \overline{2048} * y_\sigma * \overline{2048}] \supset \,.\, Theorem_{o\sigma} x_\sigma \supset \,.\, Theorem_{o\sigma} y_\sigma$.

7107 Corollary.

$$\vdash Theorem_{o\sigma}\overline{\text{“}\mathbf{A}_o = \mathbf{B}_o\text{”}} \supset \,.\, Theorem_{o\sigma}\overline{\text{“}\mathbf{A}_o\text{”}} \supset Theorem_{o\sigma}\overline{\text{“}\mathbf{B}_o\text{”}}.$$

7108 Corollary. If $\vdash \mathbf{A} = \mathbf{B}$, then $\vdash Theorem_{o\sigma}\overline{\text{“}\mathbf{A}\text{”}} \supset Theorem_{o\sigma}\overline{\text{“}\mathbf{B}\text{”}}$.

7109 Corollary. For any wff $\mathbf{A}$, $\vdash Theorem_{o\sigma}\overline{\text{“}\mathbf{A}\text{”}} \wedge Theorem_{o\sigma}\overline{\text{“}\sim \mathbf{A}\text{”}} \supset \sim Consis$.

7110 Gödel's Second Incompleteness Theorem. Let $\mathcal{A}$ be any recursively axiomatized pure extension of $\mathcal{Q}_0^\infty$. If $\mathcal{A}$ is consistent, then not $\vdash_{\mathcal{A}} Consis$.

7111 Lemma.

$$\vdash \dot\forall x_\sigma \dot\forall y_\sigma \,.\, Theorem_{o\sigma}[\overline{\text{“}[[\supset_{ooo}\text{”}} * x_\sigma * \overline{2048} * y_\sigma * \overline{2048}]$$

$$\supset \,.\, Theorem_{o\sigma} x_\sigma \supset Theorem_{o\sigma} y_\sigma.$$

7112 Corollary. For any wffs$_o$ $\mathbf{A}$ and $\mathbf{B}$, $\vdash Theorem_{o\sigma}\overline{\text{“}[\mathbf{A} \supset \mathbf{B}]\text{”}} \supset$ $\centerdot\, Theorem_{o\sigma}\overline{\text{“}\mathbf{A}\text{”}} \supset Theorem_{o\sigma}\overline{\text{“}\mathbf{B}\text{”}}$.

7113 Corollary. If $\vdash \mathbf{A} \supset \mathbf{B}$, then $\vdash Theorem_{o\sigma}\overline{\text{“}\mathbf{A}\text{”}} \supset Theorem_{o\sigma}\overline{\text{“}\mathbf{B}\text{”}}$.

7114 Löb's Theorem. For any sentence $\mathbf{D}$, $\vdash \mathbf{D}$ if and only if $\vdash Theorem_{o\sigma}$ $\overline{\text{“}\mathbf{D}\text{”}} \supset \mathbf{D}$.

7115 Corollary. If $\mathbf{D}$ is any sentence such that $\vdash \mathbf{D} = Theorem\overline{\text{“}\mathbf{D}\text{”}}$, then $\vdash \mathbf{D}$.

7200 Gödel-Rosser Incompleteness Theorem. [Rosser, 1936] Let $\mathcal{A}$ be a consistent, recursively axiomatized pure extension of $\mathcal{Q}_0^\infty$. Let $\nabla_{o\sigma}$ be

$$[\lambda x_\sigma \dot{\forall} y_\sigma \centerdot Proof^{\mathcal{A}}_{o\sigma\sigma}[\overline{128} * x_\sigma * [Num_{\sigma\sigma}x_\sigma] * \overline{2048}]y_\sigma \supset$$

$$\dot{\exists} z_\sigma \centerdot z_\sigma \leq y_\sigma \wedge Proof^{\mathcal{A}}_{o\sigma\sigma}[\overline{\text{“}[\sim_{oo}[\text{”}} * x_\sigma * [Num_{\sigma\sigma}x_\sigma] * \overline{\text{“}]]\text{”}}]z_\sigma].$$

Let $m =$ “$\nabla_{o\sigma}$” and Δ_o be $[\nabla_{o\sigma}\overline{m}]$. Neither $\vdash_{\mathcal{A}} \Delta_o$ nor $\vdash_{\mathcal{A}} \sim \Delta_o$.

7201 Corollary.

(a) If $\mathcal{A}$ is any consistent and complete extension of $\mathcal{Q}_0^\infty$, then $\mathcal{A}$ is not recursively axiomatizable.
(b) If $\mathcal{M}$ is any model of $\mathcal{Q}_0^\infty$, the set of sentences true in $\mathcal{M}$ is not recursively axiomatizable.
(c) If $\mathcal{M}$ is any model of $\mathcal{Q}_0^\infty$, the set of sentences true in $\mathcal{M}$ is not recursive.

7202 Proposition. $\vdash_{\mathcal{Q}_0} Inf^{III} \supset$ Axiom 6.

7203 Proposition. There is a standard model $\mathcal{N}$ for $\{Inf^{III}\}$ such that $\mathcal{N}$ is isomorphic to every standard model for $\{Inf^{III}\}$.

7204 Corollary. $\mathcal{Q}_0^\infty$ is consistent.

7205. Theorem. $\mathcal{Q}_0^\infty$ is essentially incomplete.

7206 Theorem. There is no recursively axiomatized pure extension $\mathcal{B}$ of $\mathcal{Q}_0$ such that the theorems of $\mathcal{B}$ are precisely the wffs$_o$ which are valid in the standard sense.

7207 Corollary. There is no decision procedure for determining of an arbitrary whether it is valid in the standard sense.

7300 Theorem. If $\mathcal{A}$ is any consistent pure extension of $\mathcal{Q}_0^\infty$, then $\mathcal{A}$ is undecidable.

7301 Theorem. If $\mathcal{B}$ is an extension of $\mathcal{Q}_0$ such that {Axiom 6} is consistent with $\mathcal{B}$, then $\mathcal{B}$ is undecidable.

7302 Corollary. $\mathcal{Q}_0$ is undecidable.

7303 Proposition. Let $\mathcal{A}$ be an extension of $\mathcal{Q}_0$ (or of $\mathcal{F}$) which is recursively axiomatized and complete. Then $\mathcal{A}$ is decidable.

7304 Proposition. If $\mathcal{M}$ is any model for $\mathcal{Q}_0^\infty$, then every numerical relation which is representable in $\mathcal{Q}_0^\infty$ is definable in $\mathcal{M}$.

7305 Tarski's Theorem. For any model $\mathcal{M}$ of $\mathcal{Q}_0^\infty$,
$\{``\mathbf{A}" \mid \mathbf{A}$ is a sentence and $\mathcal{M} \models \mathbf{A}\}$ is not definable in $\mathcal{M}$.

7306 Theorem. Let $\mathcal{A}$ be a consistent pure extension of $\mathcal{Q}_0^\infty$. There is no wff $\mathbf{B}_{o\sigma}$ of $\mathcal{A}$ such that for each sentence $\mathbf{A}$ of $\mathcal{A}$, $\vdash_{\mathcal{A}} \mathbf{A} = \mathbf{B}_{o\sigma}\overline{``\mathbf{A}"}$.

Bibliography

[Anderson and Bledsoe, 1970] Robert Anderson and W. W. Bledsoe. A Linear Format for Resolution with Merging and a New Technique for Establishing Completeness. *Journal of the ACM*, 17:525–534, 1970.

[Andrews and Bishop, 1996] Peter B. Andrews and Matthew Bishop. On Sets, Types, Fixed Points, and Checkerboards. In Pierangelo Miglioli, Ugo Moscato, Daniele Mundici, and Mario Ornaghi, editors, *Theorem Proving with Analytic Tableaux and Related Methods. 5th International Workshop. (TABLEAUX '96)*, volume 1071 of *Lecture Notes in Artificial Intelligence*, pages 1–15, Terrasini, Italy, May 1996. Springer-Verlag.

[Andrews *et al.*, 1984] Peter B. Andrews, Dale A. Miller, Eve Longini Cohen, and Frank Pfenning. Automating Higher-Order Logic. In W. W. Bledsoe and D. W. Loveland, editors, *Automated Theorem Proving: After 25 Years*, Contemporary Mathematics series, vol. 29, pages 169–192. American Mathematical Society, 1984.

[Andrews *et al.*, 1996] Peter B. Andrews, Matthew Bishop, Sunil Issar, Dan Nesmith, Frank Pfenning, and Hongwei Xi. TPS: A Theorem Proving System for Classical Type Theory. *Journal of Automated Reasoning*, 16:321–353, 1996.

[Andrews, 1963] Peter B. Andrews. A Reduction of the Axioms for the Theory of Propositional Types. *Fundamenta Mathematicae*, 52:345–350, 1963.

[Andrews, 1965] Peter B. Andrews. *A Transfinite Type Theory with Type Variables*. Studies in Logic and the Foundations of Mathematics. North-Holland, 1965.

[Andrews, 1971] Peter B. Andrews. Resolution in Type Theory. *Journal of Symbolic Logic*, 36:414–432, 1971.

[Andrews, 1972a] Peter B. Andrews. General Models and Extensionality. *Journal of Symbolic Logic*, 37:395–397, 1972.

[Andrews, 1972b] Peter B. Andrews. General Models, Descriptions, and Choice in Type Theory. *Journal of Symbolic Logic*, 37:385–394, 1972.

[Andrews, 1974a] Peter B. Andrews. Provability in Elementary Type Theory. *Zeitschrift fur Mathematische Logic und Grundlagen der Mathematik*, 20:411–418, 1974.

[Andrews, 1974b] Peter B. Andrews. Resolution and the Consistency of Analysis. *Notre Dame Journal of Formal Logic*, 15(1):73–84, 1974.

[Andrews, 1981] Peter B. Andrews. Theorem Proving via General Matings. *Journal of the ACM*, 28:193–214, 1981.

[Andrews, 1989] Peter B. Andrews. On Connections and Higher-Order Logic. *Journal of Automated Reasoning*, 5:257–291, 1989.

[Andrews, 2001] Peter B. Andrews. Classical Type Theory. In Alan Robinson and Andrei Voronkov, editors, *Handbook of Automated Reasoning*, volume 2, chapter 15, pages 965–1007. Elsevier Science, 2001.

[Appel and Haken, 1976] K. Appel and W. Haken. Every Planar Map is Four Colorable. *Bulletin of the American Mathematical Society*, 82:711–712, 1976.

[Baader and Siekmann, 1994] Franz Baader and Jörg Siekmann. Unification theory. In Dov M. Gabbay, C.J. Hogger, and J.A. Robinson, editors, *Handbook of Logic in Artificial Intelligence and Logic Programming*, pages 41–125. Oxford University Press, 1994.

[Baader and Snyder, 2000] Franz Baader and Wayne Snyder. Unification theory. In Alan Robinson and Andrei Voronkov, editors, *Handbook of Automated Reasoning*. Elsevier Science, 2000.

[Bernays and Schönfinkel, 1928] P. Bernays and M. Schönfinkel. Zum Entscheidungsproblem der mathematischen Logik. *Mathematische Annalen*, 99:342–372, 1928.

[Beth, 1953] Evert W. Beth. On Padoa's Method in the Theory of Definition. *Indag. Math.*, 15:330–339, 1953.

[Beth, 1959] E. W. Beth. *The Foundations of Mathematics*. North-Holland Publishing Co., 1959.

[Bibel, 1987] Wolfgang Bibel. *Automated Theorem Proving.* Vieweg, Braunschweig, second edition, 1987.

[Boolos, 1998] George Boolos. *Logic, logic, and logic.* Harvard University Press, 1998.

[Buss, 1994] Samuel R. Buss. On Gödel's Theorem on Lengths of Proofs I: Number of Lines and Speedup for Arithmetic. *Journal of Symbolic Logic,* 59:737–756, 1994.

[Church, 1936a] Alonzo Church. A Note on the Entscheidungsproblem. *Journal of Symbolic Logic,* 1:40–41, 1936. Correction ibid., 101–102.

[Church, 1936b] Alonzo Church. An Unsolvable Problem of Elementary Number Theory. *Amer. J. Math.,* 58:345–363, 1936. Reprinted in [Davis, 1965].

[Church, 1940] Alonzo Church. A Formulation of the Simple Theory of Types. *Journal of Symbolic Logic,* 5:56–68, 1940.

[Church, 1956] Alonzo Church. *Introduction to Mathematical Logic.* Princeton University Press, Princeton, N.J., 1956.

[Craig, 1957] W. Craig. Linear Reasoning. A New Form of the Herbrand-Gentzen Theorem. *Journal of Symbolic Logic,* 22:250–268, 1957.

[Davis *et al.*, 1976] M. Davis, Y. Matijasevic, and J. Robinson. Hilbert's Tenth Problem Diophantine Equations: Positive Aspects of a Negative Solution. In *Proc. Sympos. Pure Math,* volume 28, pages 323–378, 1976.

[Davis, 1965] M. Davis. *The Undecidable.* Raven Press, Hewlett, N. Y., 1965.

[Degtyarev and Voronkov, 1998] Anatoli Degtyarev and Andrei Voronkov. What You Always Wanted to Know About Rigid E-Unification. *Journal of Automated Reasoning,* 20:47–80, 1998.

[Dreben and Denton, 1966] Burton Dreben and John Denton. A Supplement to Herbrand. *Journal of Symbolic Logic,* 31:393–398, 1966.

[Dreben and Goldfarb, 1979] Burton Dreben and Warren D. Goldfarb. *The Decision Problem: Solvable Classes of Quantificational Formulas.* Addison-Wesley Publishing Company, Reading, Massachusetts, 1979.

[Dreben *et al.*, 1963] Burton Dreben, Peter Andrews, and Stål Aanderaa. False Lemmas in Herbrand. *Bulletin of the American Mathematical Society*, 69:699–706, 1963.

[Feferman, 1964] Solomon Feferman. Systems of Predicative Analysis. *Journal of Symbolic Logic*, 29:1–30, 1964.

[Frege, 1879] Gottlob Frege. *Begriffsschrift, eine der arithmetischen nachgebildete Formelsprache des reinen Denkens.* Halle, 1879. Translated in [van Heijenoort, 1967].

[Gallier and Snyder, 1989] Jean H. Gallier and Wayne Snyder. Complete Sets of Transformations for General E-Unification. *Theoretical Computer Science*, 67:203–260, 1989.

[Gallier *et al.*, 1992] Jean H. Gallier, Paliath Narendran, Stan Raatz, and Wayne Snyder. Theorem Proving Using Equational Matings and Rigid E-Unification. *Journal of the ACM*, 39:377–429, 1992.

[Gandy, 1956] R. O. Gandy. On the Axiom of Extensionality - Part I. *Journal of Symbolic Logic*, 21:36–48, 1956.

[Gentzen, 1935] G. Gentzen. Untersuchungen über das Logische Schließen I und II. *Mathematische Zeitschrift*, 39:176–210,405–431, 1935. Translated in [Gentzen, 1969].

[Gentzen, 1969] G. Gentzen. Investigations into Logical Deductions. In M. E. Szabo, editor, *The Collected Papers of Gerhard Gentzen*, pages 68–131. North-Holland Publishing Co., Amsterdam, 1969.

[Gleason, 1966] Andrew M. Gleason. *Fundamentals of Abstract Analysis.* Addison Wesley, 1966.

[Gödel, 1930] Kurt Gödel. Die Vollstandigkeit der Axiome des logischen Funktionenkalküls. *Monatsh. Math. Phys.*, 37:349–360, 1930.

[Gödel, 1931] Kurt Gödel. Über formal unentscheidbare Sätze der Principia Mathematica und verwandter Systeme I. *Monatsh. Math. Phys.*, 38:173–198, 1931.

[Gödel, 1936] Kurt Gödel. Über die Länge von Beweisen. *Ergebnisse eines Mathematischen Kolloquiums*, 7:23–24, 1936. Translated in [Gödel, 1986], pp. 396–399.

[Gödel, 1986] Kurt Gödel. *Collected Works, Volume I.* Oxford University Press, 1986.

[Goldson and Reeves, 1993] Doug Goldson and Steve Reeves. Using Programs to Teach Logic to Computer Scientists. *Notices of the American Mathematical Society*, 40:143–148, 1993.

[Goldson *et al.*, 1993] Douglas Goldson, Steve Reeves, and Richard Bornat. A Review of Several Programs for the Teaching of Logic. *The Computer Journal*, 36:373–386, 1993.

[Gordon, 1986] Mike Gordon. Why higher-order logic is a good formalism for specifying and verifying hardware. In G. J. Milne and P. A. Subrahmanyam, editors, *Formal Aspects of VLSI Design*, pages 153–177. North-Holland, 1986.

[Hanna and Daeche, 1985] F. K. Hanna and N. Daeche. Specification and Verification using Higher-Order Logic. In Koomen and Moto-oka, editors, *Computer Hardware Description Languages and their Applications*, pages 418–433. North Holland, 1985.

[Hanna and Daeche, 1986] F. K. Hanna and N. Daeche. Specification and Verification Using Higher-Order Logic: A Case Study. In G. J. Milne and P. A. Subrahmanyam, editors, *Formal Aspects of VLSI Design*, pages 179–213. North-Holland, 1986.

[Hatcher, 1968] William S. Hatcher. *Foundations of Mathematics.* W. B. Saunders Co., 1968.

[Hazen, 1983] A.P. Hazen. Ramified Type Theories. In D.M. Gabbay and F. Guenthner, editors, *Handbook of Philosophical Logic*, volume I. Reidel, Dordrecht, 1983.

[Henkin, 1949] Leon Henkin. The Completeness of the First-Order Functional Calculus. *Journal of Symbolic Logic*, 14:159–166, 1949.

[Henkin, 1950] Leon Henkin. Completeness in the Theory of Types. *Journal of Symbolic Logic*, 15:81–91, 1950.

[Henkin, 1952] Leon Henkin. A Problem Concerning Provability. *Journal of Symbolic Logic*, 17:160, 1952.

[Henkin, 1960] Leon Henkin. On Mathematical Induction. *American Mathematical Monthly*, 67:323–338, 1960.

[Henkin, 1963a] Leon Henkin. An Extension of the Craig-Lyndon Interpolation Theorem. *Journal of Symbolic Logic*, 28:201–216, 1963.

[Henkin, 1963b] Leon Henkin. A Theory of Propositional Types. *Fundamenta Mathematicae*, 52:323–344, 1963.

[Herbrand, 1930] Jacques Herbrand. Recherches sur la théorie de la démonstration. *Travaux de la Société des Sciences et des Lettres de Varsovie, Classe III Sciences Mathematiques et Physiques*, 33, 1930. Translated in [Herbrand, 1971].

[Herbrand, 1971] Jacques Herbrand. *Logical Writings*. Harvard University Press, 1971. Edited by Warren D. Goldfarb.

[Hindley, 1997] J. Roger Hindley. *Basic Simple Type Theory*. Cambridge University Press, 1997.

[Hintikka, 1955] K. J. J. Hintikka. Notes on Quantification Theory. *Soc. Sci. Fenn. Comment. Phys. Math*, 17, 1955..

[Huet and Oppen, 1980] Gérard Huet and D. C. Oppen. Equations and Rewrite Rules: a Survey. In Ronald V. Book, editor, *Formal Language Theory: Perspectives and Open Problems*, pages 349–405. Academic Press, 1980.

[Huet, 1975] Gérard P. Huet. A Unification Algorithm for Typed λ-Calculus. *Theoretical Computer Science*, 1:27–57, 1975.

[Jacobs, 1999] Bart Jacobs. *Categorical Logic and Type Theory*. Elsevier, 1999.

[Jensen and Pietrzykowski, 1976] D.C. Jensen and T. Pietrzykowski. Mechanizing ω-Order Type Theory Through Unification. *Theoretical Computer Science*, 3:123–171, 1976.

[Jouannaud and Kirchner, 1991] J.-P. Jouannaud and C. Kirchner. Solving Equations in Abstract Algebras: A Rule-Based Survey of Unification. In J.-L. Lassez and G. Plotkin, editors, *Computational logic : essays in honor of Alan Robinson*. MIT Press, Cambridge, MA, 1991.

[Kemeny, 1949] John G. Kemeny. *Type Theory vs. Set Theory*. PhD thesis, Princeton University, 1949. (abstract in [Kemeny, 1950]).

[Kemeny, 1950] John G. Kemeny. Type theory vs. set theory. *Journal of Symbolic Logic*, 15:78, 1950. (abstract).

[Kleene, 1952] S. C. Kleene. *Introduction to Metamathematics.* Van Nostrand, 1952.

[Leivant, 1994] Daniel Leivant. Higher Order Logic. In Dov M. Gabbay, C.J. Hogger, and J.A. Robinson, editors, *Handbook of Logic in Artificial Intelligence and Logic Programming*, volume 2, pages 229–321. Oxford University Press, 1994.

[Löb, 1955] M.H. Löb. Solution to a Problem of Leon Henkin. *Journal of Symbolic Logic*, 20:115–118, 1955.

[Löwenheim, 1915] Leopold Löwenheim. Über Möglichkeiten im Relativkalkül. *Mathematische Annalen*, 76:xque447–470, 1915.

[Lyndon, 1959] Roger C. Lyndon. An Interpolation Theorem in the Predicate Calculus. *Pacific J. Math*, 9:129–142, 1959.

[Marshall and Chuaqui, 1991] M. Victoria Marshall and Rolando Chuaqui. Sentences of type theory: the only sentences preserved under isomorphisms. *Journal of Symbolic Logic*, 56:932–948, 1991.

[Mendelson, 1964] E. Mendelson. *Introduction to Mathematical Logic.* D. Van Nostrand Company Ltd., Princeton, N.J., 1964.

[Mendelson, 1987] Elliott Mendelson. *Introduction to Mathematical Logic.* Wadsworth & Brooks/Cole Advanced Books & Software, third edition, 1987.

[Meredith, 1953] Carew A. Meredith. Single Axioms for the Systems $(C,\ N), (C,\ O)$, and $(A,\ N)$ of the Two-Valued Propositional Calculus. *J. Comput. Systems*, 1:155–164, 1953.

[Miller, 1987] Dale A. Miller. A Compact Representation of Proofs. *Studia Logica*, 46(4):347–370, 1987.

[Moore, 1982] Gregory H. Moore. *Zermelo's Axiom of Choice : Its Origins, Development, and Influence.* Springer-Verlag, 1982.

[Paterson and Wegman, 1978] M.S. Paterson and M.N. Wegman. Linear Unification. *Journal of Computer and System Sciences*, 16:158–167, 1978.

[Prawitz, 1965] Dag Prawitz. *Natural Deduction.* Almqvist & Wiksell, 1965.

[Prawitz, 1968] Dag Prawitz. Hauptsatz for Higher Order Logic. *Journal of Symbolic Logic*, 33:452–457, 1968.

[Quine, 1950] W. V. Quine. On Natural Deduction. *Journal of Symbolic Logic*, 15:93–102, 1950.

[Quine, 1963] Willard Van Orman Quine. *Set Theory and Its Logic*. Belknap Press of Harvard University Press, 1963.

[Rasiowa, 1949] H. Rasiowa. Sur un certain systèm d'axiomes du calcul des propositions. *Norsk matematisk Tidsskrift*, 31:1–3, 1949.

[Robinson, 1965] J. A. Robinson. A Machine-Oriented Logic Based on the Resolution Principle. *Journal of the ACM*, 12:23–41, 1965.

[Rosser, 1936] J. B. Rosser. Extensions of some Theorems of Gödel and Church. *Journal of Symbolic Logic*, 1:87–91, 1936.

[Rubin and Rubin, 1985] Herman Rubin and Jean E. Rubin. *Equivalents of the Axiom of Choice, II.* North-Holland, 1985.

[Russell, 1908] Bertrand Russell. Mathematical Logic as Based on the Theory of Types. *American Journal of Mathematics*, 30:222–262, 1908. Reprinted in [van Heijenoort, 1967], pp. 150–182.

[Schönfinkel, 1924] M. Schönfinkel. Über die Bausteine der mathematischen Logik. *Mathematische Annalen*, 92:305–316, 1924.

[Schütte, 1950] K. Schütte. Schlussweisen Kalküle der Prädikatenlogik. *Mathematische Annalen*, 122:47–65, 1950.

[Schütte, 1956] Kurt Schütte. Ein System des Verknüpfenden Schliessens. *Archiv für Mathematische Logik und Grundlagenforschung*, 2, 1956.

[Shapiro, 1991] Stewart Shapiro. *Foundations without Foundationalism : A Case for Second-order Logic*. Clarendon Press, Oxford, 1991.

[Sieg, 1997] Wilfried Sieg. Step by recursive step: Church's analysis of effective calculabilty. *Bulletin of Symbolic Logic*, 3:154–180, 1997.

[Siekmann, 1984] Jörg Siekmann. Universal Unification. In R. E. Shostak, editor, *Proceedings of the 7th International Conference on Automated Deduction*, volume 170 of *Lecture Notes in Computer Science*, pages 1–42, Napa, California, USA, 1984. Springer-Verlag.

[Siekmann, 1989] Jörg Siekmann. Unification Theory. *Journal of Symbolic Computation*, 7:207–274, 1989.

[Skolem, 1920] Thoralf Skolem. Logisch-kombinatorische Untersuchungen über die erfüllbarkeit oder beweisbarkeit mathematischer Sätze nebst einem Theoreme über dichte Mengen. *Skr. av Videnkapsselskapet i Kristiana, I. Matem.-natur klasse*, 4, 1920.

[Skolem, 1928] Thoralf Skolem. Über die Mathematische Logik. *Norse matematisk tidsskrift*, 10:125–142, 1928. Translated in [van Heijenoort, 1967, pp. 508–524].

[Smullyan, 1963] Raymond M. Smullyan. A Unifying Principle in Quantification Theory. *Proceedings of the National Academy of Sciences, U.S.A.*, 49:828–832, 1963.

[Smullyan, 1968] R. M. Smullyan. *First-Order Logic.* Springer-Verlag, Berlin, 1968.

[Snyder, 1991] Wayne Snyder. *A Proof Theory for General Unification.* Birkhäuser Boston, 1991.

[Statman, 1978] Richard Statman. Bounds for Proof Search and Speed-up in the Predicate Calculus. *Annals of Mathematical Logic*, 15:225–287, 1978.

[Takahashi, 1967] Moto-o Takahashi. A Proof of Cut-Elimination Theorem in Simple Type Theory. *Journal of the Mathematical Society of Japan*, 19:399–410, 1967.

[Takeuti, 1953] Gaisi Takeuti. On a Generalized Logic Calculus. *Japanese Journal of Mathematics*, 23:39–96, 1953. Errata: ibid, vol. 24 (1954), 149–156.

[Takeuti, 1958] Gaisi Takeuti. Remark on the fundamental conjecture of GLC. *Journal of the Mathematical Society of Japan*, 10:44–45, 1958.

[Tarski, 1936] Alfred Tarski. Der Wahrheitsbegriff in den formalisierten Sprachen. *Studia Philosophica*, 1:261–405, 1936. translated in [Tarski, 1956].

[Tarski, 1956] A. Tarski. The Concept of Truth in Formalized Languages. In J. H. Woodger, editor, *Logic, Semantics, Metamathematics*, pages 152–278. Oxford University Press, Oxford, 1956.

[Turing, 1948] A. M. Turing. Practical Forms of Type Theory. *Journal of Symbolic Logic*, 13:80–94, 1948.

[van Benthem and Doets, 1983] J.F.A.K. van Benthem and K. Doets. Higher-Order Logic. In D. M. Gabbay and F. Günthner, editors, *Handbook of Philosophical Logic*, volume I, pages 275–329. Reidel, Dordrecht, 1983.

[van Heijenoort, 1967] Jean van Heijenoort. *From Frege to Gödel. A Source Book in Mathematical Logic 1879–1931.* Harvard University Press, Cambridge, Massachusetts, 1967.

[van Vaalen, 1975] Jophien van Vaalen. An Extension of Unification to Substitutions with an Application to Automatic Theorem Proving. In *Proceedings of the Fourth International Joint Conference on Artificial Intelligence*, pages 77–82, Tbilisi, Georgia, 1975.

[Whitehead and Russell, 1913] Alfred North Whitehead and Bertrand Russell. *Principia Mathematica.* Cambridge University Press, Cambridge, England, 1913. 3 volumes; first edition 1913, second edition 1927.

[Wolfram, 1993] D. A. Wolfram. *The Clausal Theory of Types*, volume 21 of *Cambridge Tracts in Theoretical Computer Science.* Cambridge University Press, 1993.

List of Figures

Index